MOLECULAR DIAGNOSTICS IN PATHOLOGY

UNITED STATES AND CANADIAN ACADEMY OF PATHOLOGY, INC.
TECHNIQUES IN DIAGNOSTIC PATHOLOGY

SERIES EDITOR, **Nathan Kaufman, M.D.**
Secretary–Treasurer

No. 1. Molecular Diagnostics in Pathology
C. M. FENOGLIO-PREISER, C. L. WILLMAN, *and* N. KAUFMAN, *Editors*

No. 2. Diagnostic Flow Cytometry
J. S. COON, R. S. WEINSTEIN, *and* N. KAUFMAN, *Editors*

MOLECULAR DIAGNOSTICS IN PATHOLOGY

EDITED BY

Cecilia M. Fenoglio-Preiser, M.D.

MacKenzie Professor and Director
Department of Pathology and Laboratory Medicine
University of Cincinnati College of Medicine
Cincinnati, Ohio

Cheryl L. Willman, M.D.

Director, Center for Molecular and Cellular Diagnostics
Associate Professor, Departments of Pathology and Cell Biology
University of New Mexico School of Medicine
Albuquerque, New Mexico

WILLIAMS & WILKINS

Baltimore • Hong Kong • London • Sydney

Editor: Timothy S. Satterfield
Associate Editor: Marjorie Kidd Keating
Copy Editor: Ann Schwartz
Designer: JoAnne Jonowiak
Illustration Planner: Wayne Hubbel
Production Coordinator: Raymond E. Reter

Copyright © 1991
United States and Canadian
Academy of Pathology, Inc.

Printed in the United States of America

Library of Congress Cataloging-in-Publication Data

Molecular diagnostics in pathology / edited by Cecilia M. Fenoglio
-Preiser, Cheryl L. Willman.
    p. cm. — (Techniques in diagnostic pathology ; no. 1)
    Includes bibliographical references.
    Includes index.
    ISBN 0-683-03152-X
    1. Molecular probes—Diagnostic use. 2. Cancer—Molecular
aspects. 3. Cancer—Diagnosis. 4. Oncogenes. 5. Pathology,
Molecular. I. Fenoglio-Preiser, Cecilia M., 1943- II. Willman,
Cheryl L. III. Series.
    [DNLM: 1. Genetics, Biochemical. 2. Neoplasms—diagnosis.
3. Oncogenes—physiology. QZ 241 M7176]
RB43.7.M634 1991
616.99'20756—dc20
DNLM/DLC
for Library of Congress   90-12400
                                                  CIP

90 91 92 93 94
1 2 3 4 5 6 7 8 9 10

FOREWORD

The series on *Techniques in Diagnostic Pathology* is published in recognition of the rapid advances being made in technology transfer from the research and development stages and research applications to applications in the practice of diagnostic pathology.

This particular volume, *Molecular Diagnostics in Pathology*, focuses on the application of the new molecular biology to diagnostic pathology. It recognizes the complex and central role that the pathologist must be prepared to play, and indeed plays, in applying the new and developing concepts and techniques of molecular biology to the diagnostic laboratory. The subjects covered are extensive. The course material on which this volume is based has been enhanced and expanded. The appendices, a list of abbreviations and a glossary, should be helpful to those readers who are relatively new to this field.

The Academy wishes to express its appreciation to Drs. Cecilia M. Fenoglio-Preiser and Cheryl L. Willman and to the other distinguished contributors for making this publication a reality. It also wishes to thank the publisher, Williams & Wilkins, for support and cooperation during the publishing process.

Nathan Kaufman, M.D.
Series Editor

PREFACE

The monograph series *Techniques in Diagnostic Pathology* addresses the applications of new technologies to diagnostic pathology and laboratory medicine. This volume, *Molecular Diagnostics in Pathology*, focuses on one of the most exciting new areas of pathology: molecular pathology. Over the past few years, scientific investigations have provided exciting new insights into the molecular basis of many human neoplastic and nonneoplastic diseases. These investigations have ranged from the study of the role of oncogenes and tumor suppressor genes in human neoplasia, to the cloning of genes involved in consistent chromosomal abnormalities in human tumors such as the BCR and ABL genes in the t(9;22) in chronic myelogenous leukemia, the identification of the genes involved in inherited diseases such as cystic fibrosis and Duchenne's muscular dystrophy, the molecular characterization of infectious agents, and finally the discovery of human DNA polymorphisms that are unique for each individual and that have been exploited in the discovery of disease-associated genes and the development of DNA "fingerprinting" techniques and molecular epidemiology. The development of molecular probes from these scientific studies and the development of new molecular-based technologies such as *in situ* hybridization and the polymerase chain reaction have provided the pathologist with powerful new tools to study human disease.

The utilization of molecular-based technologies in the clinical laboratory setting is already having a significant impact on disease diagnosis, prediction of risk for a specific disease, prediction of prognosis, therapeutic monitoring, and identity determination. It is essential that the pathologist play a central role in the rapid technology transfer from the molecular research laboratory to the bedside. Functioning as a scientist in this pivotal role, the pathologist is already involved in the detailed molecular analysis of human tissues and human disease. Functioning as a physician in the clinical laboratory, the pathologist will perform detailed studies comparing molecular diagnostic approaches with more established technologies in the expanding diagnostic armamentarium so that molecular-based tests assume their appropriate place in the diagnostic arena. Functioning as a clinician, the pathologist will play an essential daily role in diagnosis and monitoring of patients.

Because of the central role of the pathologist in this new era of modern molecular pathology, the United States and Canadian Academy of Pathology decided that it was essential to establish a Special Course in Diagnostic Molecular Pathology. This course was first presented in 1989 at the Academy Meeting in San Francisco and again at its meeting in 1990 in Boston. This volume in part reflects the content of these courses, but it is more than just a collection of lectures. Every attempt has been made to provide those interested in molecular pathology with a timely, comprehensive, and well-referenced review of the subject.

Additional authors have kindly provided chapters in specific areas to round out this volume. Following a general introduction to molecular biology and the technologies currently used on diagnostic specimens are chapters featuring the polymerase chain reaction, cytogenetics and molecular cytogenetics, and a discussion of disease-associated polymorphisms and genetic linkage analysis. Several chapters highlight and extensively review the currently known role of oncogene expression and mutation in various human tumors and the genetic regulation of invasion and metastasis. Five practical chapters discuss the role of molecular diagnostics in lymphoid neoplasms, myeloid neoplasms, analysis of gene expression in endocrine cells and tumors, diagnosis of infectious diseases, and identity determination in forensic medicine. Where possible, illustrative case material has been included.

It is not recommended that the book be read cover to cover. Rather, we suggest that you skim each chapter to gain an idea of the information that it contains and then refer to specific chapters in greater detail, depending upon your area of interest. Because we recognize that not all readers will have the same background, we have provided an extensive glossary to assist those less knowledgeable in molecular concepts.

It is hoped that this book will serve as a resource for those interested in molecular pathology and will open new vistas that can be incorporated into your daily practice. Although this volume is very timely, new advances in molecular biology occur on a daily basis in these exciting times. It is time for all of us to participate.

Cecilia M. Fenoglio-Preiser, M.D.
Cheryl L. Willman, M.D.

CONTRIBUTORS

Ronald A. DeLellis, M.D.
Department of Pathology
Tufts University School of Medicine
New England Medical Center Hospital
Boston, Massachusetts

Cecilia M. Fenoglio-Preiser, M.D.
MacKenzie Professor and Director
Department of Pathology and
  Laboratory Medicine
University of Cincinnati College of
  Medicine
Cincinnati, Ohio

Anthony A. Killeen, M.D.
Department of Laboratory Medicine
  and Pathology
Institute of Human Genetics
University of Minnesota
Minneapolis, Minnesota

Elise C. Kohn, M.D.
Laboratory of Pathology
  and Medicine Branch
National Cancer Center
National Institutes of Health
Bethesda, Maryland

Lance A. Liotta, M.D., Ph.D.
Laboratory of Pathology
National Cancer Center
National Institutes of Health
Bethesda, Maryland

Margaret B. Listrom, M.D.
Chief, Cytopathology
Veterans Administration Medical
  Center
Assistant Professor
Department of Pathology
University of New Mexico School of
  Medicine
Albuquerque, New Mexico

Teri A. Longacre, M.D.
Research Fellow
Veterans Administration Medical
  Center
Department of Pathology
University of New Mexico School of
  Medicine
Albuquerque, New Mexico

W. John Martin, M.D., Ph.D.
Professor of Pathology
Chief of Immunology/Molecular
  Pathology
Section of Laboratories and Pathology
Los Angeles County and University of
  Southern California Medical Center
Los Angeles, California

James K. McDougall, Ph.D.
Fred Hutchinson Cancer Research
  Center
Department of Pathology
University of Washington
Seattle, Washington

ix

Harry T. Orr, Ph.D.
Department of Laboratory Medicine
   and Pathology
Institute of Human Genetics
University of Minnesota
Minneapolis, Minnesota

Michael H. Whittaker, M.D.
Hematopathology Fellow
Department of Pathology
Center for Molecular and Cellular
   Diagnostics
University of New Mexico School of
   Medicine
Albuquerque, New Mexico

Cheryl L. Willman, M.D.
Associate Professor of Pathology and
   Cell Biology
Director, Center for Molecular and
   Cellular Diagnostics
University of New Mexico School of
   Medicine
Albuquerque, New Mexico

Hubert J. Wolfe, M.D.
Department of Pathology
Tufts University School of Medicine
New England Medical Center Hospital
Boston, Massachusetts

Sandra R. Wolman, M.D.
Director, Levy-Stone Program of
   Cancer Genetics
Associate Medical Director
Michigan Cancer Foundation
Detroit, Michigan

Ross E. Zumwalt, M.D.
Assistant Chief Medical Investigator
State of New Mexico
Associate Professor of Pathology
University of New Mexico School of
   Medicine
Albuquerque, New Mexico

CONTENTS

Molecular Diagnostics: Scientific Fundamentals and Technical Approaches

Cheryl L. Willman

Introduction

Recent scientific and technological advances in molecular biology hold promise to reveal the molecular basis of many human diseases. The tools of recombinant DNA technology are now being utilized in virtually every subspecialty of diagnostic medicine and pathology, including:

1. Identification of genetic sequences involved in human neoplastic and nonneoplastic diseases, *e.g.*, the cystic fibrosis and Duchenne's muscular dystrophy genes[1-3,6,10];

2. Prenatal and antenatal diagnosis of inherited diseases[6,10];

3. Determination of genetic susceptibility and predisposition to diseases such as atherosclerosis and diabetes[2];

4. Accurate diagnosis and classification of neoplastic disease, as well as prediction of prognosis and therapeutic monitoring of the cancer patient[4,5,23,24];

5. Diagnosis of acquired infectious diseases[29,30];

6. Assessment of drug sensitivity and drug resistance in neoplastic and infectious disease[4];

7. Determination of relatedness and identity in transplantation, paternity testing, and forensic medicine.[25-28]

Molecular diagnostics are likely to play a role in every aspect of patient management, from initial diagnosis to monitoring the therapeutic response. Most molecular techniques will soon be performed routinely in the clinical laboratory setting. Recent and continued progress in the automation of molecular technologies (such as automated nucleic acid extraction, amplification, electrophoresis, synthesis, and sequencing)[7] will facilitate the technology transfer from the research to the clinical diagnostic laboratory. Wide application of these techniques will result from their sensitivity, specificity, speed, and relatively inexpensive cost. Although the ethics and economics of some molecular tests will spark intensive discussion, recombinant DNA technologies are likely to play an ever-increasing role in disease diagnosis and will be an essential tool for the pathologist to assimilate into the clinical laboratory.

Molecular Biology Fundamentals

DNA STRUCTURE AND THE GENETIC CODE

Transcription and Translation

Genetic information in human cells is encoded in the nucleus in deoxyribonucleic acid (DNA). DNA is a high molecular weight polymer consisting of individual nucleotide bases that are chemically linked by a backbone of deoxyribose sugar residues (Fig. 1.1).[8,11] These deoxyribose residues are joined by regular phosphodiester bonds that always link the 3' carbon of one deoxyribose residue with the 5' carbon of the next successive nucleotide in the chain (Fig. 1.1). In the nucleus, DNA exists as a double-stranded molecule composed of two complementary strands of nucleotide base pairs. Each adenine (A) nucleotide on one strand is base paired through hydrogen bonds to a thymidine (T) nucleotide on the complementary strand, while cytosine (C) nucleotides base pair with complementary guanine (G) nucleotides (Fig. 1.1). For hydrogen bonds to form between any two complementary nucleic acid strands (DNA-DNA, DNA-RNA, or RNA-RNA), the regular phosphodiester bonds of each deoxyribose-phosphate backbone must be oriented in opposite ("antiparallel") directions, as shown in Fig. 1.1.

The high specificity of base pairing between complementary nucleotides (A:T; G:C) is the basis for the transfer of genetic information from DNA to messenger

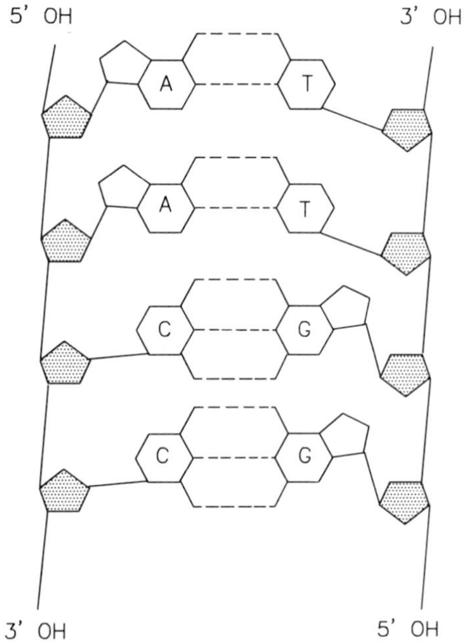

Figure 1.1. Structure of double-stranded DNA. The deoxyribonucleotide residues (*shaded pentagons*) of each single strand of DNA are chemically linked by phosphodiester bonds between the 3' carbon of the first deoxyribose residue in the chain and the 5' carbon of the next succeeding residue. Two single-stranded DNA molecules associate through the formation of hydrogen bonds (*dashed lines*) between complementary nucleotides with base pairs being formed between purine (adenine; A and guanine; G) and pyrimidine residues (cytosine; C and thymidine; T). A pairs with T through the formation of two hydrogen bonds, while G pairs with C through three hydrogen bonds. For stable hydrogen bonds to form between complementary nucleotides, the phosphodiester bonds of the two DNA chains must be oriented in opposite or "antiparallel" directions, as shown in the figure.

RNA (mRNA) during transcription and for the synthesis of a new complementary DNA strand from a single DNA strand (or template) during DNA replication.

The genetic information that specifies the sequence of cellular proteins is imprinted in the DNA in the form of a triplet code. The flow of information from DNA to protein (Fig. 1.2) involves the synthesis of a complementary single-stranded mRNA polymer from one of the DNA strands in a double-stranded stretch of DNA (a gene), in a genetic process called *transcription* (Fig. 1.2).[8,11] The mRNA molecule is then transported from the nucleus to the cytoplasm, where its sequence is *translated* on the ribosomes from a series of nucleotides to a linear sequence of amino acids comprising the protein (Fig. 1.2).[8,11] Each of three successive bases in the mRNA (the triplet code) specifies a particular amino acid. Thus the linear, directional sequence of nucleotide bases in the DNA coding sequence of a gene specifies the mRNA sequence that ultimately determines the linear order of amino acids in the protein.

Chromatin Structure

DNA in the nucleus exists in a compact arrangement called chromatin, which results from the packaging of DNA with a large array of nuclear proteins. The basic unit of chromatin is the nucleosome particle, which is a core structure composed of

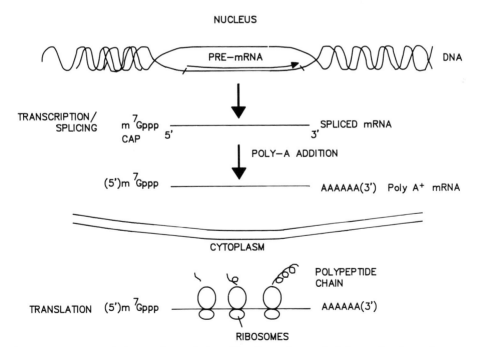

Figure 1.2. Transcription and translation. The flow of genetic information from the nucleus to the cytoplasm occurs through transcription of an mRNA molecule from the DNA template (the gene) in the nucleus, transport of the mRNA to the cytoplasm, and translation of the mRNA into protein on the ribosomes of the endoplasmic reticulum.

two copies of each of the four basic histone proteins (H-2A, H-2B, H-3, and H-4) (Fig. 1.3).[8] Approximately 140 base pairs of DNA are wrapped around each nucleosome core (a "mononucleosome") (Fig. 1.3).

Another 60–100 base pairs of DNA separate each nucleosome core, and this intervening DNA is associated with the basic histone protein H-1. Chromatin packaging structurally compresses DNA into the confined nuclear space. However, dynamic changes in chromatin structure in and around human genes undoubtedly influence the regulation of gene expression.

Chromosome Structure

Human genomic DNA is composed of approximately 3×10^9 nucleotide base pairs and is divided among 24 different chromosomes in normal somatic cells. Each human chromosome consists of a linear double-stranded DNA molecule and its associated proteins. Each normal somatic cell contains 2 copies of 22 different autosomes (1 copy of each autosome is inherited from each parent) and 2 sex chromosomes (X and Y; one contributed by the maternal (XX) and one by the paternal (XY) parent), for a total of 46 chromosomes.

Individual human chromosomes are easily identifiable in metaphase spreads of dividing cells, at a time when the chromatin is condensed in preparation for cell division. Metaphase chromosomes may be stained with DNA-binding dyes that resolve different banding patterns (due to DNA sequence and packaging variations) along the length of each chromosome. Banding patterns obtained on different human chromosomes have been standardized into a conventional cytogenetic nomenclature.

Advances in understanding the molecular pathophysiology of many human diseases have resulted from the mapping of human genes to chromosomal bands (loci) that are consistently involved in inherited and neoplastic disorders by the hybridization of probes for human genes to metaphase chromosomes. However, even at the limits of cytogenetic resolution, when extended chromosomes can be produced allowing the detection of approximately 700 dispersed bands in the haploid genome, each band on average still contains over 250 genes. Thus, even though a gene has been mapped to a particular chromosomal locus at the cytogenetic level of resolution, molecular studies are essential for more sensitive mapping and the precise determination of whether a gene is involved in the chromosomal abnormality at the molecular level.

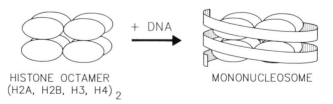

HISTONE OCTAMER
(H2A, H2B, H3, H4)$_2$ + DNA   MONONUCLEOSOME

Figure 1.3. Chromatin structure. See text for full discussion.

MOLECULAR HYBRIDIZATION

Major advances in the cloning of human genes and in molecular diagnostics have been facilitated by exploiting the specificity of the complementary base pairing of nucleotide bases. The basis of all molecular hybridization assays is complementary base pairing between two nucleic acid strands (Table 1.1). Under conditions of appropriate stringency (defined by the conditions of the hybridization reaction: temperature, salt concentration, time, DNA sequence content), two single-stranded DNA molecules that are complementary in base sequence will reassociate ("anneal") through the formation of hydrogen bonds between the complementary base pairs to form a stable duplex molecule. Because of this high degree of specificity, a radioactively labeled single-stranded DNA molecule may be utilized as a probe for its complementary target sequence, even in a complex mixture in which the target sequence is present in only 1 in 10^6 DNA molecules.

Since most DNA exists as double-stranded DNA, it must be denatured into single strands by heating or with base (NaOH) before being hybridized to single-stranded probes. RNA that exists naturally as a single-stranded molecule will form stable duplexes with single-stranded DNA or RNA under appropriate denaturing and annealing conditions.

If one has a full or partial amino acid sequence for a protein of interest, then a specific nucleic acid sequence may be designed using the genetic code. This sequence could then be utilized as a probe to isolate or hybridize to the gene encoding the protein. The ability to chemically synthesize short DNA fragments (called oligonucleotides) has dramatically facilitated the design of suitable probes for hybridization assays and molecular cloning. Due to the high specificity of base pair-

Table 1.1.
Molecular Hybridization: Definition of Terms

Molecular basis for hybridization assays	Complementary base pairing between two single-stranded nucleic acid strands.
Denaturation	Separation of two base-paired strands by heat or extremes of pH; "melting."
Renaturation	Reannealing of two nucleic acid strands by hydrogen bonding between complementary nucleotide base pairs; "hybridization."
Stringency	Conditions of the hybridization reaction that determine the stability of a duplex formed between two complementary nucleic acid strands, including: Temperature, Salt concentration, DNA sequence composition, Presence of organics; 75% stringency implies that 25% of the nucleotide base pairs in a duplex may be mismatched, but a stable duplex will form.

ing, oligonucleotides of only 15–20 base pairs can hybridize to a specific comple-mentary sequence in genomic DNA under conditions of high stringency.[7]

RESTRICTION ENDONUCLEASES

One of the fundamental developments that made recombinant DNA technol-ogy possible was the discovery of restriction endonucleases. Restriction enzymes are DNA-cleaving enzymes that cut double-stranded DNA at specific base se-quences. Thus, fragmentation of the human genome at specific sequences was made possible, allowing the development of molecular cloning technologies and the structural analysis of human genes.

Restriction endonucleases are isolated from bacteria and are named according to the bacterial strain from which they are derived (for example, Eco RI was derived from the *E. coli* strain RI; Fig. 1.4). Restriction endonucleases cleave specific 4, 6, or 8 base pair DNA recognition sequences wherever these sequences occur in the human genome (Fig. 1.4). Restriction enzyme recognition sequences are palin-dromic, *i.e.*, there is an axis of symmetry in the middle of the recognition sequence (Fig. 1.4). The restriction enzyme may cut in the middle of this sequence to create "blunt" (smooth) ends (as with Hpa I, Fig. 1.4), or it may cut unevenly to create stag-gered or "sticky" ends (as with Eco RI, Fig. 1.4).

Utilizing restriction enzymes, genomic DNA may be cut into millions of frag-ments, making the structural analysis of human genes technically feasible. Restric-tion maps may be created for genes whose genomic sequence is unknown by exam-ining the fragments generated by cleaving DNA with different restriction enzymes. If the restriction map of a gene is known, then digestion of that gene with a particu-lar restriction enzyme will generate DNA fragments of predictable lengths. These fragments of DNA may then be examined in hybridization assays (such as South-

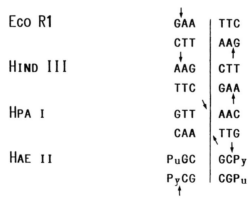

Figure 1.4. Recognition sequences for restriction endonucleases. Restriction enzymes recognize specific 4, 6, or 8 base-pair recognition sequences in double-stranded DNA. These sequences, such as the EcoRI recognition sequence GAATTC, always have an axis of symmetry in the middle of the sequence (*solid line*). Restriction enzymes may cut double-stranded DNA in a staggered fashion (*arrows*) to create overlapping ends, like EcoRI, HindIII, or HaeII; or, enzymes may cut in the middle of a double-stranded DNA sequence to create smooth or "blunt" ends, like Hpal.

ern analysis, described below) to determine if the structure of the gene is normal or abnormal (deleted, rearranged, amplified, etc.). (See references 4, 23, and 24 for a detailed discussion.)

MOLECULAR CLONING

The discovery of restriction endonucleases and the purification of other enzymes involved in molecular reactions such as DNA-joining enzymes (ligases), DNA-synthesizing enzymes (polymerases and reverse transcriptases—which synthesize a complementary DNA copy from an RNA template), and RNA-synthesizing enzymes (RNA polymerases) have dramatically facilitated the cloning of human genes (Table 1.2).

Isolation of a *genomic clone* for a human gene may be initiated by recombining restriction fragments of human DNA with a "cloning vector" (see below) cleaved with the same restriction endonuclease (Fig. 1.5). The human fragments and vector are mixed, and vector sequences may reassociate ("reanneal") with each individual human DNA fragment because the vector ends will be complementary to the ends of each human DNA fragment after restriction endonuclease digestion. Vectors utilized in cloning strategies include bacteriophages, bacterial plasmids, or vectors that combine features of both plasmids and phages (cosmids), each with particular advantages. As discussed extensively in references 13–15, different types of vectors may be chosen for cloning sequences of interest, and the choice of vector depends upon the frequency of the sequence in the starting material, the length of the DNA fragment to be cloned, and the method to be used for the isolation of the specific sequence.

The array of human DNA sequences incorporated into vector sequences during a cloning procedure is referred to as a library. In order to clone a specific sequence, the population of vectors in the library, each containing a single DNA insert derived from the original restriction digest, is introduced into bacterial cells so that

Table 1.2.
Molecular Cloning: Definition of Terms

Genomic clones	Fragments of human DNA containing exons and introns, derived from restriction digestion or mechanical shearing of genomic DNA, cloned into vector sequences.
cDNA clones	DNA clones containing only exons, derived from cellular mRNA sequences that were copied into cDNA with reverse transcriptase and cloned into vector sequences.

Vector sequences	DNA molecules that replicate autonomously in host cells and which can accept a foreign DNA sequence, including:
	    Bacterial plasmids,
	    Bacteriophages (such as phage λ),
	    Cosmids.

each cell contains a single recombinant DNA molecule (Fig. 1.5). After replication of the entire bacterial mixture, the bacterial colony or bacteriophage plaque containing the desired gene insert may be identified by molecular hybridization studies utilizing a sequence-specific radiolabeled nucleic acid probe, as described extensively in references 13–15.

Human genetic sequences may also be isolated from *cDNA libraries*, utilizing mRNA as the starting material in the cloning procedure. The population of mRNA molecules isolated from cells expressing the gene of interest is copied into DNA (referred to as *cDNA* for "copy DNA") utilizing reverse transcriptase. After synthesis of the complementary second DNA strand, the double-stranded cDNAs may be cloned into suitable vectors (as above). To isolate a genetic sequence of interest, genomic or cDNA libraries may be screened with a radiolabeled nucleic acid probe complementary to a portion of the gene.

For certain genes of interest, there may be no related nucleic acid probe, partial gene sequence, or protein sequence information that could be used to develop a nucleic acid probe to isolate the gene from a library. In this case, a library could be constructed using a vector capable of directing the mRNA synthesis of the inserted

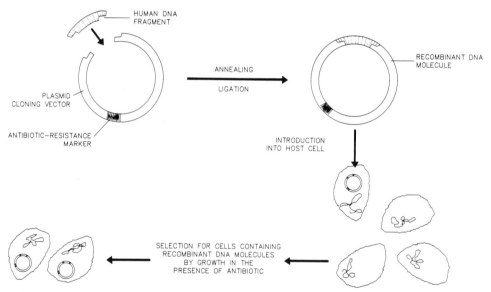

Figure 1.5.   Molecular cloning. Human DNA fragments, isolated from human DNA digested with a particular restriction enzyme, may be cloned into a vector sequence (in this case a circular bacterial plasmid that also contains an antibiotic-resistance gene) that has been linearized with the same restriction enzyme. Human DNA fragments and linearized vector sequences are mixed and allowed to reanneal through base pairing between their complementary overlapping ends. After ligation, the circularized vectors (each containing a human DNA insert fragment) are introduced into bacterial host cells. Each bacterial cell takes up only one plasmid. The population of bacterial cells are then "selected" for those that took up recombinant plasmids, by growing the cells in the presence of antibiotic. The remaining bacterial cells each contain a recombinant DNA molecule containing a specific DNA insert fragment. The population of such sequences is referred to as a library.

human DNA fragment. Because of the universal genetic code, this human mRNA will be translated into protein in the bacterial host cell. The libraries constructed from these so-called "expression" vectors may then be screened with antibodies specific for the protein product encoded by the gene of interest. Individual bacterial colonies containing the protein (and thus the gene sequence of interest) may then be isolated.

DNA STRUCTURE IN HIGHER ORGANISMS

Noncoding DNA

The human genome contains only 100,000 distinct genes. These genes, encoding all of the structural and functional proteins expressed by all types of human cells, are referred to as "single-copy" genes because only one copy of each gene exists in a single specific chromosome location in the haploid human genome. Each human cell thus has two copies (alleles) of each distinct genetic sequence, one contributed by the maternal and one by the paternal chromosome. Strikingly, these single-copy genes comprise only 3–5% of the 3.2×10^9 nucleotide base pairs that make up the haploid human genome. Thus, the vast majority of the human genome (>95%) is composed of "noncoding" DNA, the function of which has not been completely elucidated.

Present within noncoding DNA are families of human repetitive (or "reiterated") sequences. Human repetitive elements can be divided into two main categories: (*a*) interspersed DNA sequences and (*b*) tandemly repeated DNA sequences.[25,26] Interspersed DNA sequences are either short interspersed segments (SINES; approximately 500 base pairs (bp) in length) or long interspersed segments (LINES; averaging 5–6 kilobases (kb) (1 kb = 1000 bp) in length). The SINE and LINE sequences are dispersed throughout the genome as individual units. The most ubiquitous human SINE element is the Alu repetitive sequence, a unit of 300 bp present in 300,000–900,000 dispersed locations in the haploid genome.[25]

In contrast to interspersed DNA sequences that are often repeated throughout the genome, tandemly repeated DNA loci are sites in which a block or "core" sequence of nucleotides is directly repeated over and over in tandem. Tandemly repeated DNA sequences are found in "satellites" (large blocks of tandemly repeated DNA sequences located primarily in the centromeres of human chromosomes) and "minisatellites" (smaller blocks of tandemly repeated core sequences scattered throughout the genome).

Most variations in the amount and sequence of DNA between individuals (DNA or genetic polymorphisms) occur in noncoding and repetitive DNA. Out of the 3 billion base pairs of DNA inherited from each parent, it is estimated that 3 million bases will differ between any two individuals.[25] These naturally occurring variations in DNA sequence (polymorphisms) have been exploited in the identification of disease-associated genes and the diagnosis of genetic disorders (as discussed extensively in Chapter 4 and in references 1–3, 6, 9–10) as well as in the development of DNA "fingerprinting" techniques used in forensic medicine and paternity testing (as discussed extensively in Chapter 16 and references 25–28).

Polymorphisms in human DNA occur predominantly in two forms, (*a*) restriction fragment length polymorphisms (RFLPs) and (*b*) variable number tandem repeats (VNTRs).

Restriction fragment length polymorphisms (RFLPs) are variations in DNA sequence between individuals (usually point mutations and base additions in noncoding DNA) which abolish or create new restriction-endonuclease cleavage sites. Using a DNA probe that is complementary to a polymorphic region of DNA, variation in a restriction-endonuclease cleavage site in this region in different individuals may be detected by Southern blot analysis (as diagrammed in Fig. 1.6 RFLP (*top*); see references 1–3, 6, and 10 for a review).

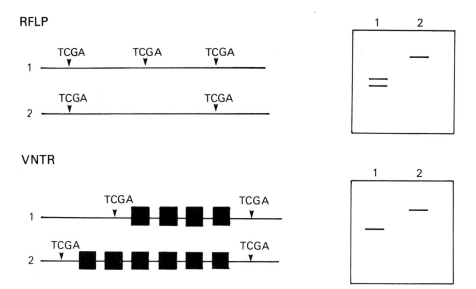

Figure 1.6. Detection of genetic polymorphisms. *RFLP*: restriction fragment length polymorphisms. Two individuals in the population, each of whom is homozygous for a particular allele at a polymorphic locus (alleles denoted 1 and 2) may be distinguished by these genetic polymorphisms in Southern analysis. This polymorphic locus in individual 1 contains three restriction enzyme recognition sites (TCGA). Cleavage of the DNA of individual 1 with the enzyme recognizing this site and subsequent hybridization with the probe complementary to this region will yield two bands (denoted on lane 1 of the schematic Southern blot). In contrast, individual 2 has a point mutation in the middle TCGA restriction site such that this sequence is now CCGA, which will no longer be recognized by the restriction enzyme. Thus this individual effectively has only two restriction sites at this locus. Cleavage of the DNA of individual 2 with the enzyme recognizing TCGA and subsequent hybridization with the probe will yield only one band of larger size (lane 2 in the schematic Southern blot), thus distinguishing the two alleles and the two individuals at this locus. *VNTR*: Variable number of tandem repeats. Two individuals are each homozygous for a particular allele at this VNTR locus (alleles denoted 1 and 2 respectively), *i.e.*, each allele is distinguished by a variable number of tandem repeats of a core sequence of nucleotides (*dark boxes*). Individual 1 has 4 repeats, while individual 2 has 6 repeats at this site. Digestion of the DNA of each individual with a restriction enzyme that cuts outside of the repeat region (TCGA in this example) and subsequent hybridization of the digested DNA with a probe complementary to the tandem repeat will yield bands of different lengths in Southern analysis (as shown in lanes 1 and 2, corresponding to the DNA from individuals 1 and 2 respectively). Thus, the two individuals or alleles may be distinguished at this VNTR locus.

VNTRs (including some of the "minisatellite" loci) exhibit a high degree of polymorphism. The number of tandemly repeated core sequences at a particular chromosome location may vary greatly in different individuals in the population. Using a DNA probe complementary to the core sequence, one can detect variations in the *length* of DNA at this site in different individuals due to variations in the number of tandem repeats, using Southern blot analysis (as diagrammed in Fig. 1.6 VNTR (*bottom*); see references 9, 26, and 27 for a review).

Polymorphisms in DNA sequence surrounding single-copy genes have no effect on gene function. However, if these polymorphisms are adjacent to ("tightly linked" with) a gene of interest (such as the gene for cystic fibrosis, Huntington's disease, or polycystic kidney disease), they are said to be "informative" and may be used for disease diagnosis.[6,10] Polymorphisms are inherited in a Mendelian fashion just like the disease-associated locus that they lie near. In the study of a family with a genetic disease, it may be found that a polymorphism (an RFLP or a polymorphic VNTR) always cosegregates with the clinical manifestations of the disorder. This would imply that the particular polymorphism is lying near the gene containing the disease mutation. This polymorphism can then be used as a marker of the disease gene even though the precise location, structure, sequence, and function of the disease gene is unknown, as discussed in detail in Chapter 4.

Single-Copy Genes

Several years ago, it was determined that the single-copy genes encoding proteins in higher eucaryotic organisms are discontinuous; in other words, a sequence of nucleotide bases in DNA specifying a particular protein sequence was disrupted by intervening, noncoding DNA.[8,11] The actual coding sequences are referred to as *exons* (Fig. 1.7, *black boxes*), while the intervening noncoding sequences are referred to as *introns*. The number and length of individual introns and exons in every unique gene is highly variable. In general, different exons tend to encode distinct functional domains in a protein.

It has been theorized that human genetic sequences were initially colinear and existed as primordial RNA molecules, evolving later to DNA molecules for greater

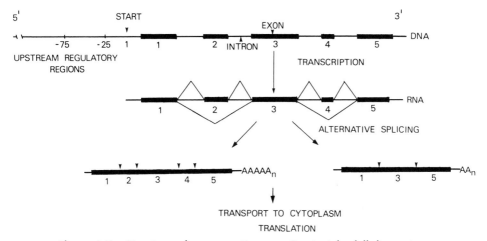

Figure 1.7. Structure of a eucaryotic gene. See text for full discussion.

stability. These colinear sequences were then disrupted by noncoding intervening sequences ("mobile introns," or other mobile genetic elements) that could splice themselves in and out of coding sequences.[8] The disruption of human genes into discrete exons presumably allowed faster evolution of proteins, as new proteins could evolve by exchange and duplication of different exons ("exon-shuffling").[8]

By convention, the first nucleotide that is transcribed from DNA to mRNA during the process of transcription is referred to as position 1 in a gene, or the "start" site of transcription (Fig. 1.7). Transcription of an mRNA molecule always proceeds in the 5' to 3' direction, i.e., the first nucleotide in the mRNA has a free 5' OH group and is linked to the next nucleotide by a phosphodiester bond between the 3' carbon atom of the ribose sugar of the first ribonucleotide and the 5' carbon atom of the second ribonucleotide in the mRNA chain. The mRNA is synthesized from a single, opened strand of the DNA duplex, using the 3' to 5' strand as a template. (Remember that any two paired nucleotide strands must be complementary and "antiparallel" for stable duplex formation; refer to Fig. 1.1 and accompanying discussion). Successive nucleotides "downstream," or 3' from the start site of transcription, are numbered from 1 to +X. Similarly, nucleotides upstream, or 5' from the start site of transcription, are given negative numbers, from −1 to −X. Upstream, or 5' to the transcriptional start site of every gene, are the regulatory regions of the gene (referred to as the *promoter*), which control the binding of RNA polymerase, an enzyme that transcribes RNA from the DNA template strand.[8] Other regulatory elements that control transcription include *enhancer* elements, which are DNA sequences that bind regulatory proteins that may affect transcription from the promoter site.[8] Unlike promoters, enhancers may function over long distances and may be located upstream or downstream of the actual coding sequence.[8]

The first mRNA transcribed in the nucleus (*heterogeneous nuclear RNA* or *pre-mRNA*) contains all of the exon and intron sequences in linear order (Fig. 1.7). Intron sequences are removed from the pre-mRNA in the nucleus by a process called "splicing." All eucaryotic introns have identical conserved consensus sequences at both the 5' and the 3' ends of the intron, at the intron-exon "splice junctions." The invariant nucleotides GT at the 5' end of the intron (which are always the first two nucleotides of any intron) are surrounded by the following consensus sequence of nucleotides at the exon-intron junction (the invariant GT residues are in italics): AG*GT*(A or G)AGT. Similarly, the last two nucleotides at the 3' end of every intron are always AG, and these nucleotides are surrounded by the consensus sequence: Pyrimidines (C or T: > 11 residues)N(which stands for "any nucleotide")(C or T)*AG*G, in which the invariant AG residues at the 3' intron-exon splice junction are in italics.

The "spliceosome" is the structure responsible for the removal of introns from nuclear pre-mRNA. The spliceosome is formed when the ribonucleoprotein complexes U1, U2, U5, and U4/6 (formed by small nuclear RNAs and proteins; commonly referred to as SNRPs for "small nuclear ribonucleoproteins") associate with the conserved consensus sequences at the intron-exon splice junctions (Fig. 1.8). The small RNAs in the SNRPs base pair with the conserved consensus sequences at the splice junctions (Fig. 1.8). The association of each of the SNRPs with the conserved intron sequences and with each other gives the spliceosome complex an enzymatic function, leading to cleavage of the introns from the pre-mRNA molecule

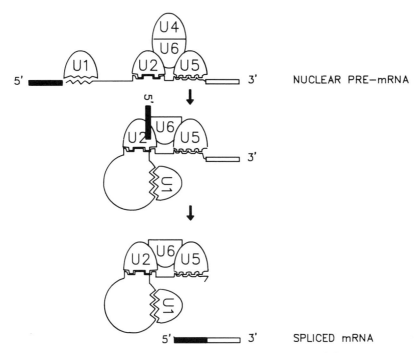

Figure 1.8. Mechanism of the splicing of introns from mRNA and the spliceosome complex. Small nuclear ribonucleoproteins ("SNRPs"; U1–U6) bind to the conserved consensus sequences present at each intron-exon splice junction. The binding and association of the SNRPs produces an enzymatic function that leads to intron cleavage and the fusion of the proximal (*black box*) and distal (*white box*) exons. See text for full discussion.

precisely at the intron-exon borders (Fig. 1.8). The mechanism of intron excision by this process is an area of intensive investigation.

Abnormalities in intron splicing cause human disease. Point mutations in the conserved intron-exon splice junction sequences in the β-globin gene lead to defective splicing of β-globin mRNA, producing an mRNA that cannot be appropriately translated into protein, resulting in the disease β thalassemia.[2] Interestingly, the autoantibodies Ro, La, and Sm, diagnostic of certain connective tissue diseases, are directed against the U1–U6 ribonucleoprotein complexes that are part of the spliceosome complex.

Nuclear pre-mRNAs can be spliced by alternative mechanisms, referred to as "alternative splicing." All coding exons of a gene may be spliced in linear order to encode a protein specified by the entire coding sequence (Fig. 1.7). Alternatively, certain exons may be spliced out of the nuclear RNA (at any position) to create an mRNA that would encode a related, but different protein (Fig. 1.7). Alternative splicing from individual genes allows multiple proteins to be encoded by one DNA sequence. Thus, the old dogma "one gene–one protein" is no longer true. In fact, most human genes encode multiple, related proteins, each with a different function, by alternative exon splicing. Exactly how alternative splicing is regulated in a cell is unclear. Certain cells of the same lineage but at different stages of differentiation may splice a gene in an alternative fashion. Similarly, cells of distinct lineages may splice a gene in different ways, each lineage thus producing a different, but related protein from the same DNA coding sequence.

As each nuclear RNA is spliced to form a functional mRNA, polyadenylation of the mRNA transcript (the addition of adenine (A) ribonucleotide residues) at the 3′ end is also completed. Polyadenylation may be essential for transport of the spliced mRNA from the nucleus to the cytoplasm and for mRNA stability in the cytoplasm. All eucaryotic mRNAs are also "capped" (i.e., a GTP residue is added to the 5′ end of the mRNA through a 5′–5′ triphosphate linkage). Capping increases the efficiency of translation of an mRNA molecule to protein on the cytoplasmic ribosomes associated with the endoplasmic reticulum (Fig. 1.2).

Molecular Techniques for the Analysis of Gene Structure

DETECTION OF IMMOBILIZED TARGET SEQUENCES

Southern Blots

When high molecular weight DNA (DNA fragments greater than 25–50 kb in length) is isolated from human cells and tissues (utilizing fresh, quick-frozen, or dried samples, *not* fixed or paraffin-embedded) and is digested with a particular restriction endonuclease, the enzyme will cleave at every site where its base recognition sequence occurs in the double-stranded DNA molecule. The millions of DNA fragments generated, ranging in size from a few hundred to several thousand base pairs, can be separated according to size by electrophoresis in a conventional agarose gel. In order to identify the fragment(s) containing the sequence of interest, the gel is first treated with base (NaOH) to denature the double-stranded DNA fragments in the gel to single-stranded DNA. The DNA fragments in the neutralized gel are then transferred to and immobilized on a solid support such as nitrocellulose or nylon (the *Southern blot* technique; see references 4, 13–15, 21, and 24). The membrane is then incubated under annealing conditions with a single-stranded radioactively labeled or nonisotopic probe[16] that forms a stable duplex with its target sequence on the membrane. After washing the membrane to eliminate nonspecific binding, a piece of X-ray film is laid on the membrane and an autoradiograph is produced to detect an isotopically labeled probe (or a colorimetric reaction is carried out to detect the binding of a nonisotopic probe).

Southern blot analysis can yield information on (a) *the structure of a gene; i.e.,* has the gene been deleted or rearranged? (as in immunoglobulin and T cell receptor gene rearrangements in lymphoid cells or by chromosomal alterations in neoplastic cells; discussed in Chapters 6 and 7 and in references 23 and 24); (b) *the number of copies of a gene; i.e.,* has the gene sequence been amplified from the normal number of two copies per cell? (as in the amplification of proto-oncogenes in human tumor cells; discussed in Chapters 5 and 8 and reference 4); and (c) *the presence of polymorphisms in the DNA sequence; i.e.,* utilizing a particular restriction enzyme, are all the restriction sites present in the stretch of DNA that hybridizes to the probe? (as in RFLP analysis in human genetic diseases; discussed in Chapter 4, or, utilizing a restriction enzyme that cuts outside a tandem repeat, is the length of the repeat sequence the same or different in two individuals? (as in the use of VNTR probes in DNA fingerprinting or identity determination; discussed in Chapter 16 and references 9, 26, and 27).

Pulsed-Field Gel Electrophoresis

Conventional agarose gel electrophoresis separates DNA from a few hundred base pairs in size to a maximal size of approximately 20–25 kb (*i.e.,* 20,000 bp). Recent studies have shown that DNA rearrangements may occur over very large distances (>100 kb) from a gene of interest and still alter the expression of the gene. To examine long-range DNA structure, the technique of pulsed-field gel electrophoresis has been developed.[17] Pulsed-field gels differ from conventional gels with a horizontal current flow (which allows the DNA to move in only one direction) by the presence of an alternating electrical current during electrophoresis. The alternating current allows very large DNA fragments (ranging from 500 bp to 10 million bp) to "snake" around the agarose matrix and be resolved during electrophoresis (Fig. 1.9).

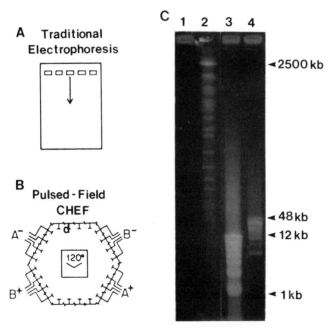

Figure 1.9. Pulsed-field gel electrophoresis. Traditional electrophoresis is characterized by current flow in a single direction (**A**), while in pulsed-field gel electrophoresis (**B**), (a schematic of the CHEF system from BioRad; see reference 17), electrodes are placed around the gel matrix and current is alternated between the various electrodes producing an alternating electrical field which allows for the separation of large DNA fragments as discussed in the text. Using the CHEF system, one can separate DNA fragments of a large size range (**C**) such as the separation of whole yeast chromosomes ranging in size from 2500–245 kilobases (kb) (*lane 2*); DNA fragments ranging from 12–1 kb (*lane 3*); and DNA fragments ranging from 48–8 kb (*lane 4*).

When DNA is isolated by traditional methods,[4,24] it is subjected to shearing forces during isolation, which lead to random breakage. The highest molecular weight DNA achieved under these conditions is still only 25–50 kb in size. To examine DNA structure over long distances (100–2500 kb) in the pulsed-field gel technique, the DNA fragments must be very long. DNA isolated under the best traditional methods is still too fragmented.[4,24] For pulsed-field gels, DNA is isolated from 10^5–10^6 *viable* cells (either fresh or cryopreserved) by embedding the cells in an agarose matrix ("an agarose plug") to prevent fragmentation of DNA during cell lysis and DNA isolation. Cell lysis and restriction endonuclease digestion are then carried out in the agarose plug. DNA is digested in the agarose plug with "rare-base cutters," restriction endonucleases that recognize only rarely occurring sites in human DNA (such as enzymes like NotI or SfiI, which have 8-base recognition sequences), to produce very long DNA fragments. This agarose plug is then embedded in the regular gel matrix and electrophoresed with alternating current. After pulsed-field gel electrophoresis, the gel is handled exactly like a traditional gel for Southern blot analysis.

DETECTION OF POINT MUTATIONS

RFLPs

If a point mutation in DNA abolishes or creates a restriction endonuclease cleavage site, then the point mutation may be detected by digestion of DNA with the appropriate restriction endonuclease, followed by Southern hybridization (as demonstrated in Fig. 1.6).[2,4–6,10]

Allele-Specific Oligonucleotide Probes (ASO)

Immobilized DNA sequences may be probed with ASO probes, which are synthetic oligonucleotides approximately 20 base pairs in length. These probes are of sufficient length to specifically hybridize to unique DNA sequences in the genome, but are short enough to be destabilized by a single internal mismatch in the duplex between the probe and the target sequence under conditions of high hybridization stringency.[2] Thus, if an ASO probe is exactly complementary to a sequence of interest and that sequence has undergone a point mutation in the diseased patient, then the duplex will be less stable than a perfectly matched duplex, and it can be "washed off" the membrane under annealing conditions that only allow perfect duplexes to form. Mismatched duplexes can also be resolved from perfectly matched duplexes during denaturing gradient gel electrophoresis (Fig. 1.10B). A mismatched duplex is more easily denatured than a stable duplex. During electrophoresis in a denaturing (formamide) gradient, mismatched duplexes migrate more slowly than perfect duplexes, and thus, mismatched duplexes containing point mutations may be resolved.

ASO probes have been utilized to diagnose point mutations in the globin genes in sickle cell anemia and thalassemia, and to detect point mutations in α-1-antitrypsin deficiency, phenylketonuria, and the *ras* family of proto-oncogenes in malignant cells.[2]

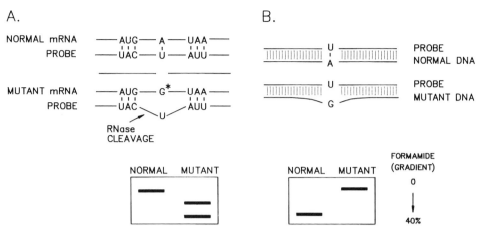

Figure 1.10. **A**, RNase cleavage technique for detecting point mutations in mRNA. A probe is synthesized exactly complementary to a target mRNA sequence. If a point mutation occurs in the mRNA (denoted by an A --> G mutation in the mRNA), then the complementary probe will have a single base-pair mismatch. Since only the mismatched duplex is susceptible to cleavage with RNase H, the duplex with the point mutation will be cleaved while the normal, nonmutated mRNA remains intact. These normal and mutant mRNAs may then be distinguished by their cleavage patterns in RNA gels. **B**, Denaturing gels for the detection of point mutations in DNA. A probe exactly complementary to a target sequence is synthesized. If a point mutation occurs in the DNA in the target sequence (such as an A ---> G point mutation in the example), then the duplex formed between the probe and the mutant DNA may be more easily denatured. After electrophoresis in a denaturing formamide gel, the duplex formed between the mutant and the probe will migrate more slowly and may thus be distinguished from the duplex formed between the probe and the normal DNA.

Ribonuclease A (RNase A) Cleavage Method

This technique can detect single base mismatches between a radioactively labeled RNA probe and the patient's genomic or expressed mRNA sequences.[2] A mismatched hybrid will be cleaved at the mismatched nucleotides by RNase A. Thus, the mismatched duplex will be cleaved into multiple fragments wherever a mismatch due to a mutation has occurred. These fragments may be resolved on denaturing gels and compared to the full-length normal fragment (Fig. 1.10A).[2] However, specific information regarding the mutation can only be determined by sequencing. This technique has been utilized to diagnose mutations in the HPRT gene in Lesch-Nyhan syndrome and point mutations in the retinoblastoma gene.[2]

GENE AMPLIFICATION TECHNIQUES

Polymerase Chain Amplification Techniques (PCR)

Typically, 5–10 micrograms (10^{-6} grams) of genomic DNA is required for the detection of single-copy genes in Southern analysis using standard or ASO probes. This amount of DNA is present in about 10^6 nucleated cells. The most sensitive

Southern blots can reveal an abnormal sequence (such as a genomic rearrangement) only when it is present in 2–5% of the cells in a heterogeneous mixture.[24] As discussed in Chapter 2, the PCR technique circumvents the problem of detecting DNA abnormalities that occur at low frequency in the target population or the problem of too little DNA for analysis, by enzymatically amplifying target sequences prior to Southern hybridization. In this technique, a denatured target DNA is incubated with two oligonucleotide primers complementary to the ends of the target sequence. These primers direct the DNA polymerase-dependent synthesis of a complementary copy of each of the two strands of the target DNA. With approximately 25 repeated rounds of heat denaturation, primer annealing, cooling, and DNA synthesis, a target sequence can easily be exponentially amplified 10^6-fold.[18,19] Use of the thermostable TaqI DNA-polymerase enzyme obviates the need to add new enzyme during each 15-minute round as the temperature is varied. PCR may also be performed with RNA as the starting material. The RNA is copied into a single cDNA strand with reverse transcriptase, and the cDNA copy is then amplified.

PCR amplification has been utilized (a) to detect rare cells (1 in 10^6) with a molecular abnormality such as a chromosomal translocation or mutation and will be useful for the detection of minimal residual disease[18,19]; (b) to identify infectious organisms[29,30]; (c) to amplify target DNA sequences from single human hairs and blood stains for forensic identification[25]; and (d) to amplify target sequences for ASO probes[2], thereby avoiding problems with signal-to-noise ratios during hybridization of ASO probes to genomic DNA during Southern analysis.

To perform PCR, the sequence of the target must be partially known so that appropriate primers may be synthesized for the amplification procedure. Because the sites of primer annealing to the target DNA are known, the size of the fragment of DNA amplified between the primers is predictable. Often this amplified DNA fragment may be run on a gel and visualized with ethidium bromide staining to make the diagnosis without subsequent hybridization with a gene-specific probe. If a fragment of appropriate size is visualized, then the target sequence must have been present in the starting material. DNA obtained from PCR amplification may be isolated from the gel and directly sequenced, allowing highly specific characterization of molecular abnormalities.

Molecular Techniques for the Analysis of Gene Expression

NORTHERN ANALYSIS

Total RNA may be isolated from fresh or quick-frozen cells and tissues. This RNA is then fractionated according to size and separated in a denaturing gel during electrophoresis. In a technique similar to the Southern blot analysis of DNA, RNA is then transferred after electrophoresis to a nylon or nitrocellulose membrane, followed by hybridization with a specific radiolabeled probe (the Northern blot; see references 4, 13–15, and 22 for a more thorough discussion).

In normal cells, total cellular RNA consists primarily of ribosomal RNA with some transfer RNA. mRNA, the transcripts of actual single-copy genes, only comprises 1–2% of the total cellular RNA. However, with Northern analysis, electrophoresis of 10 micrograms of total RNA is sufficient to detect hybridization between the probe and the mRNA of interest in the total RNA population. mRNA may be isolated from total RNA by virtue of its poly A tail, using oligo-dT columns that specifically bind the poly A tails of mRNA; however, this is usually not necessary to detect cellular transcripts.

Northern analysis provides information about *the size of an mRNA transcript* and *the level of expression* of a particular transcript.[4]

IN SITU HYBRIDIZATION

RNA molecules (or DNA molecules) in a tissue section may be identified by hybridization of the tissue affixed to a glass slide, with a radiolabeled probe. This technology gives information regarding the level of expression but no qualitative information regarding the size or structure of the mRNA. An extensive discussion of the technique and applications of *in situ* hybridization follows in Chapters 14 and 15.

REFERENCES AND SUGGESTED READINGS

GENERAL MOLECULAR BIOLOGY REVIEWS AND TEXTS

1. Botstein, D., White, R. L., Skolnick, M., and Davis, R. W. Construction of a genetic linkage map in man using restriction fragment length polymorphisms. *Am. J. Hum. Genet.* 32:314–331, 1980.
2. Caskey, C. T. Disease diagnosis by recombinant DNA methods. *Science (Wash DC)* 236:1223–1228, 1987.
3. Donis-Keller, H., Green, J., Helms, C., *et al.* A genetic linkage map of the human genome. *Cell* 51:319–337, 1987.
4. Fenoglio-Preiser, C., and Willman, C.L. Molecular biology and the pathologist. *Arch. Pathol. Lab. Med.* 111:601–619, 1987.
5. Grody, W. W., Gatti, R. A., and Naiem, F. Diagnostic molecular pathology. *Mod. Pathol.* 2:553–568, 1989.
6. Killeen, A. A., and Orr, H. T. Molecular basis of human genetic disease: Clinical applications of genetic linkage analysis. In *Molecular Diagnostics in Pathology,* Fenoglio-Presier, C. M., Willman, C. L. and Kaufman, N., eds. Baltimore, Williams & Wilkins, 1990.
7. Landegren, U., Kaiser, R., Caskey, C. T., and Hood, L. DNA diagnostics, molecular techniques and automation. *Science (Wash DC)* 242:229–237, 1988.
8. Lewin, B. *Genes IV,* 4th ed. Cambridge, Mass., Cell Press, and New York, Oxford University Press, 1990.
9. Nakamura, Y. M., Leppert, P., O'Connell, *et al.* Variable number of tandem repeat (VNTR) markers for human gene mapping. *Science (Wash DC)* 235:1616, 1987.
10. Sommer, S. S., and Sobell, J. S. Application of DNA-based diagnosis to patient care: The example of hemophilia A. *Mayo Clin. Proc.* 62:387–404, 1987.
11. Watson, J. D., Tooze, J., and Kurtz, D. T. *Recombinant DNA—A Short Course.* New York, W. H. Freeman and Co., 1983.
12. White, R., and Lalouel, J. M. Chromosome mapping with DNA markers. *Sci. Am. 258*: 40–48, 1988.

TECHNICAL REFERENCES

General Laboratory References

13. Ausubel, F. M., *et al.,* eds. *Current Protocols in Molecular Biology.* New York, Greene Publishing Associates and Wiley-Interscience, 1989.

14. Rickwood, D., and Hames, B. D., eds. *The Practical Approach Series.* Washington, D.C., IRL Press.
Including the following titles:
   1. Gel Electrophoresis of Nucleic Acids
   2. Human Genetic Diseases
   3. DNA Cloning, vols. 1 and 2
   4. Transcription and Translation
   5. Nucleic Acid and Protein Sequence Analysis
15. Sambrook, J., Fritsch, E. F., and Maniatis, T. *Molecular Cloning—A Laboratory Manual,* vols. 1–3, 2nd ed. U.S.A., Cold Spring Harbor Press, 1989.

Specific Technical References

16. Gilliam, I. C. Non-radioactive probes for specific DNA sequences. *Trends Biotechnol.* 5:332, 1987.
17. Lai, E., Birren, B. W., Clark, S. M., Simon, S. I., and Hood, L. Pulse field electrophoresis. *Biotechniques* 7:1–9, 1989.
18. Lum, J. B. Visualization of mRNA transcription of specific genes in human cells and tissues using *in situ* hybridization. *Biotechniques* 4:32, 1986.
19. Marx, J. Multiplying genes by leaps and bounds. *Science (Wash DC)* 240:1408–1410, (PCR), 1988.
20. Saiki, R. Primer-directed enzymatic amplification of DNA with a thermostable DNA polymerase. *Science (Wash DC)* 239:487–491, 1988.
21. Southern, E. M. Detection of specific sequences among DNA fragments separated by gel electrophoresis. *J. Mol. Biol.* 98:503, 1975.
22. Thomas, P. Hybridization of denatured RNA and small DNA fragments transferred to nitrocellulose. *Proc. Nat. Acad. Sci. U.S.A.* 77:5201, 1980.

SPECIAL REVIEWS AND APPLICATIONS

Lymphoid Lesions and Immunoglobulin Genes

23. Cossman, J., Uppenkamp, M., Sundeen, J., *et al.* Molecular genetics and the diagnosis of lymphoma. *Arch. Pathol. Lab. Med.* 112:117–127, 1988.
24. Willman, C. L., Griffith, B. B., and Whittaker, M. Molecular genetic approaches for the diagnosis of clonality in lymphoid neoplasms. *Clin. Lab. Med.* 10:119–149, 1990.

Forensic Applications and Identity Determination

25. Ballantyne, J., Sensabaugh, G., and Witkowski, J., eds. *Banbury Report 32: DNA Technology and Forensic Sciences.* U.S.A., Cold Spring Harbor Press, 1989.
26. Jeffreys, A. J., Wilson, V., and Thein, S. L. Hypervariable "minisatellite" regions in human DNA. *Nature (Lond)* 314:67–73, 1985.
27. Jeffreys, A. J., Wilson, V., and Thein, S. L. Individual-specific "fingerprints" of human DNA. *Nature (Lond)* 316:76–79, 1985.
28. Zumwalt, R. E. Applications of molecular biology to forensic pathology. *Hum. Pathol.* 20:303, 1989.

Molecular Microbiologic Diagnosis

29. Koblet, H. Contributions of molecular biology to diagnosis, pathogenesis, and epidemiology of infectious diseases. *Experientia* 43:1185–1192, 1987.
30. Tenover, F. C. Diagnostic deoxyribonucleic acid probes for infectious disease. *Clin. Microbiol. Rev.* 1:82–101, 1988.

2

Polymerase Chain Reaction: A Tool for the Modern Pathologist

W. John Martin

Introduction

Molecular biology is having a profound influence on the basic understanding of disease processes. No longer in the realm of esoteric science, the newer concepts of disease have far-reaching practical applications for the diagnosis of disease at the level of clearly defined genetic alterations. The polymerase chain reaction (PCR) has provided a very powerful tool for the widespread application of molecular technology to the practice of pathology. PCR refers to an *in vitro* enzymatic amplification of a defined DNA sequence by repeated rounds of heat denaturation, primer annealing, and DNA polymerase-mediated primer extension. This chapter reviews the principles of the PCR and provides practical details on the applications and the potential pitfalls encountered in using this technology in the clinical laboratory.

Principle of the PCR

DNA Structure. Genetic information in DNA is coded by a linear sequence of nucleotides. Nucleotides consist of either a purine (adenine or guanine) or a pyrimidine (thymine or cytosine) base bound to a phosphorylated deoxyribose (sugar) molecule. The deoxyribose molecules are linked by a phosphate group, which is attached to the fifth (5') carbon atom of one deoxyribose molecule and to the third (3') carbon atom of the adjoining deoxyribose molecule by phosphodiester bonds. The 5'–3' linkage gives a single-stranded nucleic acid polymer. Normally, DNA is double-stranded, with two polymers associating with each other to form a helical structure. The two strands run antiparallel to each other with respect to the direction of the 5'–3' linkage. The helix is established by hydrogen bonding between complementary base pairs. Two hydrogen bonds form between adenosine and thymidine (A:T bonding), while three hydrogen bonds form between cytosine and guanosine (G:C bonding). Hydrogen bonding is disrupted by heat or elevated pH, leading to separation (denaturation) of the two strands. Neither heat nor alkali disrupts the phosphodiester bonds linking the nucleotides within single-stranded DNA. On cooling or lowering the pH, the separated single strands of DNA will reanneal (hybridize) by A:T and G:C base pairing to reform the double-stranded, helical structure. Any mismatching

21

of individual bases between the two strands will impair reformation of the helix, necessitating a further lowering of the temperature for effective hybridization between the matched bases (1, see Chapter 1.)

DNA Replication. During DNA replication, double-stranded DNA becomes progressively denatured into single-stranded molecules. The exposed single strands act as templates for the synthesis of complementary DNA strands. This reaction is mediated by enzymes termed DNA polymerases. For most DNA polymerases, a short region of double-stranded nucleic acid is needed to initiate (or prime for) DNA synthesis. The primer is extended in the 5′ to 3′ direction by the incorporation of the appropriately matched deoxynucleotide triphosphates (dNTP) with the release of a diphosphate residue.[47]

Repetitive DNA Synthesis. *In vitro* primer-initiated, unidirectional DNA synthesis has been used extensively in DNA-sequencing reactions[71] (Fig. 2.1). What makes the PCR distinctive is the use of a pair of primers for repetitive, bidirectional synthesis of a segment of double-stranded DNA[55] (Fig. 2.2). The primers are chosen such that the extension product of each primer will form template molecules in all subsequent DNA synthesis cycles. For this to occur, the primers must be complementary to the flanking regions, on opposing DNA strands, of the particular segment of double-stranded DNA to be replicated. The 3′ ends of the primers are directed to-

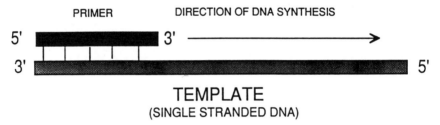

Figure 2.1. Unidirectional DNA synthesis. Schematic representation of primer-initiated DNA synthesis. The region of double-stranded DNA, created by the hybridization of an oligonucleotide (primer) to the template DNA, is recognized by the DNA polymerase enzyme. The primer is extended in a 5′ to 3′ direction by the sequential addition of nucleotide bases complementary to the corresponding bases on the template DNA.

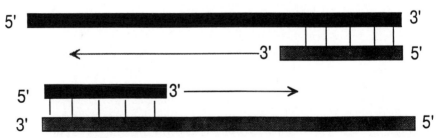

Figure 2.2. Bidirectional DNA synthesis. Schematic representation of the effect of a bidirectional DNA synthesis reaction involving two primers, reactive on opposing strands of double-stranded DNA, with flanking regions of the DNA. Note that the extended primers become templates for the alternate primers.

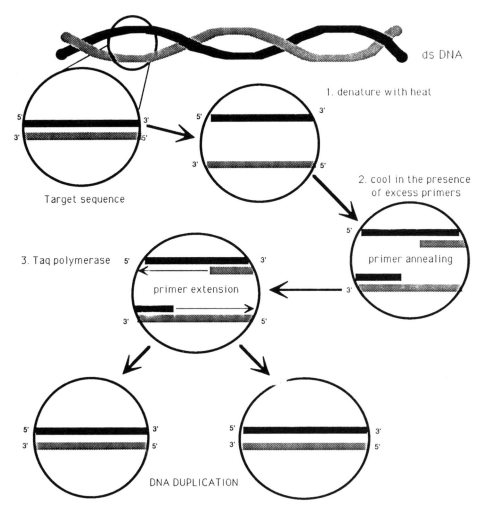

Figure 2.3. Polymerase chain reaction. Schematic representation of the first cycle of the PCR.

ward the center of the DNA segment. PCR proceeds by heat denaturing the double-stranded DNA molecule and cooling in the presence of the oligonucleotide primers (Fig. 2.3). Because of their high concentration and their size-determined greater mobility in solution, the primers bind more rapidly to the target DNA than the slower reannealing of the larger complementary DNA strands. The primer:DNA complex provides a substrate for DNA polymerase. In the presence of dNTP, the polymerase will extend the primers in a DNA synthesis reaction. Each newly synthesized strand will be complementary to the template DNA and will acquire, at its 3′ end, the sequence complementary to the other primer used in the PCR. On reheating, the newly formed hybrids will denature, thereby providing two additional template molecules during the next primer-annealing step (Fig. 2.4). Each successive cycle of heating, primer annealing, and primer extension gives an exponential (2-fold) increase in the

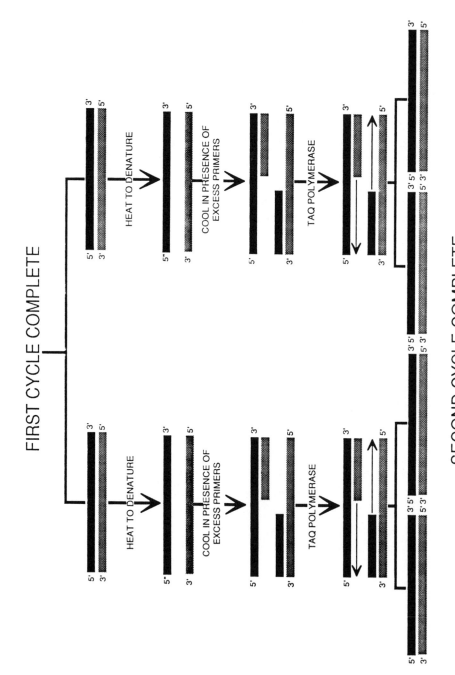

Figure 2.4. Schematic representation of the second cycle of the PCR.

targeted segment of DNA (Fig. 2.5). Eventually, the reaction will become rate limiting, due mainly to competition between primer binding and reannealing of the greatly amplified single DNA molecules synthesized during the PCR. In a typical reaction, however, amplification of the order of 10^6-fold can be readily achieved ($2^{20} = 10^6$). The product will be of uniform size, corresponding to the distance separating the 5′ ends of the two primer-binding sites on the opposing strands of the target segment of DNA.

Performance of the PCR

An in-depth understanding of the role of the various components used in the PCR will enhance the efficiency of running the PCR assays in a clinical laboratory and help avoid some of the potential pitfalls. The most important components are the DNA polymerase, oligonucleotide primers, dNTP, buffer, and target DNA. The reactions require thermal cycling, which can be achieved using an automated machine. Several alternative methods are available to analyze the PCR product. Finally in this section, various modifications of the basic PCR format are discussed which increase the versatility and clinical usefulness of this technology.

DNA POLYMERASE

Taq polymerase from the thermophilic bacterium *Thermus aquaticus* has essentially replaced the use of the Klenow fragment of *E. coli* DNA polymerase I in the PCR.[68] *Taq* withstands the 94–96°C required to denature DNA and needs to be added only at the start of the PCR. The optimal temperature for *Taq* is 72°C, although significant activity can be observed at temperatures as low as 55°C. The enzymatic activity of *Taq* requires Mg^{++} ions and is potentiated by low concentrations of KCl. Unfortunately, *Taq* is readily inhibited by heme compounds present in lysed blood and by cellular enzymes present in tissue extracts. Recombinant DNA-derived *Taq* (Amplitaq) is available with a specific activity of 200,000 units/mg. One unit of *Taq*, corresponding to approximately 5×10^{-8}M, is required in the typical PCR.[26]

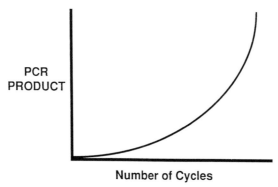

Number of Cycles

Figure 2.5. Exponential increase in PCR product. Theoretical exponential relationship between the production of PCR product and the number of cycles. In many types of PCR at least 80% efficiency can be achieved over a range of 10–20 cycles.

PRIMERS

The relative efficiency of PCR applied to clinical medicine depends to a large extent on the selection of a set of primers appropriate for the clinical problem being addressed. Following several important principles underlying the design of PCR primer pairs will help alleviate potential problems of sensitivity and specificity.

Specificity.   The most important quality for a primer is its specificity for the desired target sequence. Specificity is related to primer length. Various combinations of nucleotides will occur by chance throughout the human genome at the frequency of 4^N, where N is the length of the sequence. If a primer is too short it will be complementary to multiple genes and will specifically bind to inappropriate regions of DNA.[85] Statistically, an oligonucleotide should be at least 17 bases long to avoid "cross-priming" within the human genome (3×10^9 base pairs). For added stability and because certain sequences may occur more often than by chance, primers for PCR are generally about 20 nucleotides in length. Increasing the length of the primer beyond 25 nucleotides is counterproductive, since it impairs mobility and extends the time required for annealing to the target DNA. Longer primers may, however, be required if there is possible mismatching between primer and target DNA. It is best to avoid stretches of the same nucleotide, since this can promote misalignment of the primer to target DNA. Sequences coding for regions of a protein molecule crucial for biological activity are likely to be more conserved and, therefore, more reliable primer targets than either introns or sequences coding for biologically less important regions of the molecule. The selected sequences can be checked for relatedness to other known genetic sequences using computer programs available at various national centers such as the Genetic Computer Group (GCG) at the University of Wisconsin and the Human Gene Mapping Library at Yale University.

Composition.   Strong binding of primer to target DNA is important for hybrid stability, especially at the higher temperatures used for primer extension with *Taq* polymerase. Stability is directly related to both primer length and its G:C versus A:T nucleotide content. A useful formula for approximating the influence of length and G:C content on the temperature of hybrid dissociation is as follows: Td (expressed as °C) = $4(G + C) + 2(A + T)$ for oligomers from 11–20 nucleotides in length.[85] The Td is also influenced by salt concentration and by various organic agents. Since the optimal temperature for the *Taq* enzyme is 72°C, it is desirable to select matched primers with a relatively high G:C content. G:C residues comprise only about 40% of the human genome, and, in practice, it is generally difficult to go much beyond 50% G:C content. Occasionally, primers may be needed that can tolerate sequence variability in specific positions within the template DNA. Single base mismatches can reduce the Td by as much as 5°C. A useful solution to this problem is to use the nucleotide base inosine. This base acts as a neutral insert that can hydrogen bond with each of the 4 possible deoxynucleotides in DNA.[58] Inosine is read as a guanosine base during DNA replication. Inosine does not contribute to hybrid stability and should not be included when calculating the Td of the primer.

3′ End.   It is important that the 3′ termini of the two primers are non-complementary to each other, because if they self-anneal they will act as a substrate for the DNA polymerase and yield the predominant product in the PCR. This complication, termed "primer-dimer" formation, can sometimes occur even when the 3′ termini are not complementary, which presumably reflects an enzyme dysfunction.[67]

It is most important that the 3' base of the primer be complementary to the target DNA to allow efficient interaction with *Taq* polymerase, which lacks the 3' to 5' exonuclease-editing activity present in most other DNA polymerases.[26] The degenerate nature of the genetic code is most often reflected in the third base of the codon sequence.[47] It is, therefore, preferable to have the terminal base of the primer complementary to the first or second base of a codon. If possible, preference might be given to codons specifying amino acids for which there are no coding alternatives (tryptophan and methionine), while avoiding amino acids for which there are multiple alternative codes (*e.g.*, arginine, which has six codons).

5' End. Primers can tolerate major additions to their 5' ends without significant loss of activity in the PCR. This has allowed the synthesis of PCR products with functional end regions to facilitate detection, cloning, and sequencing.[89]

Spacing. The size of the PCR product is determined by the distance separating the 5' termini of the two primer-binding sites on the double-stranded DNA. Providing that the template DNA is not markedly degraded, very large stretches (*e.g.*, 6–10 kb) of DNA can be amplified.[38] For most uses, however, stretches of 100–200 bases are much more efficiently replicated. In addition, long stretches of targeted DNA can yield partially extended primers that tend to erroneously cross-prime in subsequent cycles (primer-jumping).

Synthesis and Purification. Primers can be readily synthesized using automated equipment. Alternatively, they can be purchased from commercial sources. For optimal results, primers should be homogeneous in size. This can be achieved by high performance liquid chromatography (HPLC) and/or polyacrylamide gel purification of full length primers. Primers are used at a concentration of 0.2–1.0 μM.[67]

DEOXYNUCLEOTIDE TRIPHOSPHATES

The recommended PCR mixture contains 200 μM of each of the four dNTP.[67] This is sufficient material to synthesize over 10 μg of DNA.[26] A typical PCR generally yields less than 1 μg of product, and lower concentrations of dNTP can be employed. The pH of the nucleotide solution should be adjusted to 7.0. *Taq* will incorporate into the PCR product various nucleotide analogues including biotin- and digoxigenin-labeled deoxyuridine triphosphates with approximately the same efficiency as thymidine (unpublished data).

BUFFER

The remaining ingredients in the PCR are 10mM Tris-Cl to maintain the pH at 8.3, 2.5 mM $MgCl_2$, gelatin (0.01 mg/ml) to help stabilize *Taq*, 50 mM KCl (which enhances *Taq* activity), and a combination of 0.45% Nonidet P-40 and 0.45% Tween-20 to help release target DNA.[67] Although these amounts are optimal for many primer: target DNA combinations, slight modifications can significantly increase the quantity and/or specificity of the PCR products obtained with different primers. Useful data can be obtained by simple titrations of pH, Mg^{++}, dNTP, and primer concentration. Since Mg^{++} is bound in an equimolar relationship by dNTP, the effect of reducing the dNTP concentration is to effectively increase the Mg^{++} concentration. It is sometimes advantageous to include additional components in the PCR, especially when multiple primer sets are used. These components, including low concentrations of dimethyl sulfoxide, dithiothreitol, mercaptoethanol, and ammonium sulfate, appear to provide a more uniform response by different primers when used in a single reaction. A thin layer of mineral oil or paraffin may be overlaid to prevent evaporation. The total volume of the reaction should be kept small, however, to facilitate heat exchange between the tubes and the heating block. All of the reactants required for the PCR, with the exception of mineral oil and target DNA, can be premixed and stored at −20°C for at least 2 weeks. Repeated freezing and thawing, however, will result in loss of enzyme activity and hydrolysis of the dNTP. The reaction mix, including *Taq*, can be lyophilized and stored at room temperature.

TARGET DNA

Extensive purification of DNA is not required for it to act as a target in the PCR. The ability to use single slices of paraffin-embedded, formalin-fixed tissue in the PCR is of significance to pathologists.[78] Tissues fixed in paraformaldehyde and glutaraldehyde have also been used. A 6–10 micron tissue slice, containing an approximately 1 cm square tissue section, is placed directly into a 1.5 ml Eppendorf tube. The tissue is deparaffinized with a single wash in xylene. This is followed by two washes in absolute ethanol. The specimen is desiccated to dryness for 1–2 hours. The resulting small white pellet is resuspended in 50 µl PCR buffer and placed in boiling water for 10 minutes. A larger volume of buffer is used if the original section exceeds 1 cm square. Also, for liver, lymph nodes, and other cellular tissues, an increase in sensitivity of the PCR can be obtained by adding 50–100 µg (in 5–10 µl) of Proteinase K and incubating for 1–4 hours at 60°C. This step is performed to inactivate tissue proteases and release tissue DNA. Proteinase K is inactivated by placing the tubes in boiling water for 10 min. All tubes are centrifuged and 5–25 µl of the sample is added to the PCR mix. One should avoid using too much sample in the PCR, for this may nonspecifically inhibit the *Taq* enzyme and lead to false negative results.

For whole blood, either EDTA or heparin can be used as an anticoagulant. For analysis of cellular DNA, the erythrocytes in 200 µl of blood are lysed by the addition of 1 ml of a solution consisting of 1% Triton X-100, 0.32 M sucrose, 5 mM $MgCl_2$, and 10 mM Tris-HCl, pH 7.5.[33] The white blood cell nuclear pellet is washed twice in saline and resuspended in 40 µl PCR buffer. Proteinase K (100 µg in 10 µl) is added for 1 hour at 60°C. After heat inactivation and centrifugation, 10 µl aliquots are used for PCR analysis. Where indicated, mononuclear cells can be separated from whole blood using Ficoll-Hypaque. The cells are washed, resuspended in PCR buffer, and treated with Proteinase K.

Cells derived from cervical swabs and from fine needle aspirates can be similarly processed. Other material suitable for PCR includes serum, CSF, urine, saliva, chorionic villus samplings, and swabs from various body sites. As little as a single cell can be used in the PCR.[48] For most applications, an estimated 10^3 cells is satisfactory. Because of the potential concern with enzyme inhibitors, it is advisable to test each sample for amplification of a normal cellular DNA sequence (*e.g.*, β-globin gene).

THERMAL CYCLING

The PCR can be conveniently performed in an automated thermal cycling machine such as the one manufactured by Perkin Elmer–Cetus. A 48-well block is heated by electrical resistance and cooled by a circulating refrigerant. Other thermal cycling machines use Peltier electronic heat exchange (MJ Research, Cambridge, MA), water (Techne, Princeton, NJ) or air cooling (Bio Therm, Arlington, VA). The typical temperature cycling for the PCR involves 3 steps.

DNA denaturation is at 94–96°C. The higher temperature is necessary when longer hybrids are being formed, especially if they are rich in G:C sequences. Thirty seconds is adequate for denaturation of most samples. Longer periods or temperatures in excess of 96°C can result in significant loss of enzyme activity.

The annealing temperature will vary somewhat with the length, G:C content, and fidelity of the primers. The higher the annealing temperature, the less cross-priming is likely to occur. For many primers, a range of 50–55°C is satisfactory. For some primers, annealing can occur even at 70–72°C, allowing a two-step PCR.

Primer extension is best achieved at 72°C. Although 1 min is usually recommended for annealing and for primer extension, shorter periods often prove to be equally satisfactory. The number of cycles can vary from 20 to 40, depending upon the sensitivity required. As the number of cycles increases, production of the specific product can begin to plateau, while products resulting from cross-priming continue to increase. There is little or no benefit, therefore, in exceeding 40 cycles. To facilitate complete extension of the primers during the latter cycles of a PCR, the extension times can be progressively increased from 30 to 90 seconds, using one of the programs available with the automated DNA Thermal Cycler. A 3-minute extension time is sometimes recommended in the final cycle to help ensure that all PCR products will be double-stranded DNA.[67]

DETECTION METHODS

The synthesized products from the PCR can be quantitated by a wide variety of methods. Agarose electrophoresis of the PCR products should yield a clearly definable band that can be stained with ethidium bromide and visualized using a UV light source. Figure 2.6A shows well-defined single bands obtained with a primer set spe-

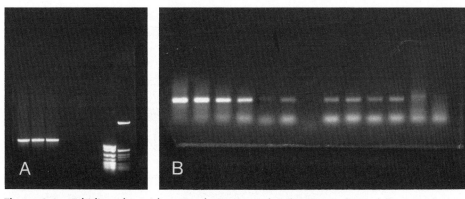

Figure 2.6. Ethidium bromide-stained agarose gels of PCR products following electrophoresis. The *left* photograph (**A**) shows identically migrating bands in wells 1–3, resulting from an optimized PCR. No extraneous PCR products are synthesized, and no primer-dimers are formed. Size markers are shown on the *right-hand side* of the photograph. The *right* photograph (**B**) shows a less than optimal PCR with extraneous products being synthesized and extensive primer-dimer formation.

cific for the human globin gene. The two right-hand lanes contain DNA size markers run concurrently with the PCR products shown in the three left-hand lanes. A PCR for human herpesvirus-6, which shows more extensive primer-dimer formation but still with identifiable specific products, is shown in Figure 2.6B. (This primer set has since been replaced.)

Quantitative incorporation assays, using either labeled primers, labeled dNTP, dNTP analogues, or a combination of labeled reagents, have been used successfully (Fig. 2.7). Primers can be synthesized with an amino group at the 5' end for coupling to fluorescent or biotin residues.[90] *Taq* enzyme can utilize several nucleic acid analogues with little loss of efficiency. Incorporation techniques are not suitable if there is significant cross-priming or primer-dimer formation (*e.g.*, as shown in Fig. 2.6B) or if information is needed on the sequence specificity of the amplified product. Even minor sequence variations between PCR-generated hybrids can result in detectable changes in mobility when the products are electrophoresed through a polyacrylamide gel containing a gradient of a denaturing agent such as formamide. This

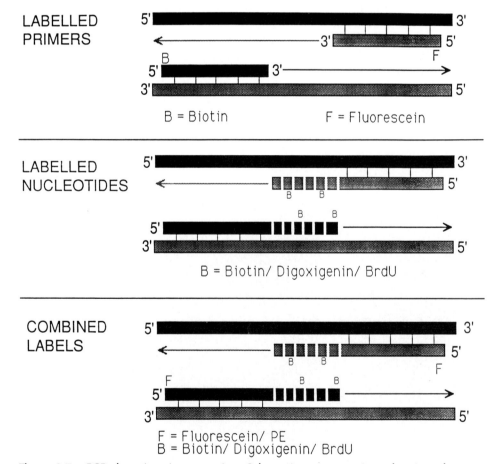

Figure 2.7. PCR detection: incorporation. Schematic representation of various formats used to quantitate the synthesis of PCR products, based on incorporation of either labeled nucleotides, labeled primers or a combination of both types of labeled reactants.

technique is termed denaturating-gradient gel electrophoresis (DGGE). The resolving power of this technique can be greatly enhanced if a 5′ GC-rich extension is added to one of the primers. The 5′ region becomes incorporated into the PCR product as a GC "clamp". The clamp allows the PCR products to migrate farther into the denaturating gel and helps to maximize the influence of even a single nucleotide difference between PCR products generated from normal and mutant alleles.

Sequence information on the PCR products is most conveniently obtained by a secondary hybridization reaction using labeled probes specific for sequences contained within one of the strands of the amplified segment (Fig. 2.8). Hybridization

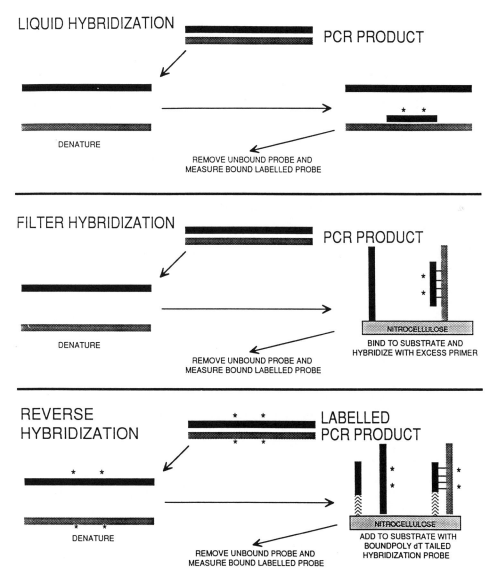

Figure 2.8.   PCR detection: hybridization. Schematic representation of various formats used for hybridizing PCR products.

can be performed in the liquid phase and the hybridized probe distinguished from nonhybridized probe by agarose electrophoresis. A recently introduced format uses the chemiluminescent label acridinium bound to single-stranded detecting probe. The acridinium becomes protected from chemical hydrolysis when hybridized. Chemiluminescence can be subsequently detected to determine the extent of hybridization (Gen Probe, San Diego, CA). Rather than liquid hybridization, one of the reactants can be immobilized on nitrocellulose or nylon filters. Usually, it is the PCR products that are denatured in alkali and bound onto the substrate using heat and/or UV irradiation. A labeled single-stranded probe is added. Nonhybridized probe is washed away, and the amount of retained probe determined. Alternatively, in reverse hybridization, the detecting probe is bound to the substrate via a 3' poly dT tail and labeled PCR products are added.[69] Many of the newer techniques will likely be incorporated into automated machines to facilitate detection and accurate quantitation. The hybridization reactions can be run under varying degrees of stringency. Highly stringent conditions, using short, allele-specific oligonucleotide probes can be used to differentiate PCR products differing at a single nucleotide. For maximum information, the PCR products can be sequenced either directly or after cloning in a suitable vector.[89]

VARIATIONS IN THE PCR

There are an increasing number of innovative variations of the basic PCR procedure. Probably the most useful is the analysis of RNA.[2] The RNA is converted to DNA using the reverse transcriptase enzyme. Because mRNA molecules have a 3' polyadenylation sequence, a poly dT oligomer can be used for priming the initial round of DNA single-strand synthesis (Fig. 2.9). The single-stranded DNA molecules can be extended at their 3' ends with poly dC using the enzyme terminal transferase. Reverse synthesis of the second DNA strand can then be achieved using poly dG primer with *Taq*. Selected amplification of specific mRNA molecules can be subsequently achieved using primers reactive with defined nucleotide sequences present in the mRNA molecules of interest. The addition of a poly dG primer binding site at the end of first-round synthesis has also been employed in a technique to derive information on sequences external to those amplified in the conventional PCR. This has been termed "anchored PCR."[50] In an alternative approach, termed "inverse PCR," genomic DNA fragments are circularized by restriction enzyme cutting and ligation. Primers reactive with the flanking regions of a DNA segment of interest are chosen so that they are extended around the circularized DNA rather than internally along the known DNA segment.[57] The amplified PCR products using both techniques can be subsequently sequenced and additional primers generated.

Multiple primers can be used in a single "multiplex" reaction.[13] The multiple primers can be used to coamplify independent genetic regions or to provide a two-step amplification of a single region. In the latter procedure, one set of primers is "nested" within a set of outside primers. The outside primers are chosen so as to preamplify the genetic region of interest (Fig. 2.10A). Ideally, they should anneal at a higher temperature than the other primer set. When the annealing temperature is lowered, the internal primer set will function to yield larger amounts of PCR products than if only the internal set of primers were used. Alternatively, the products of the

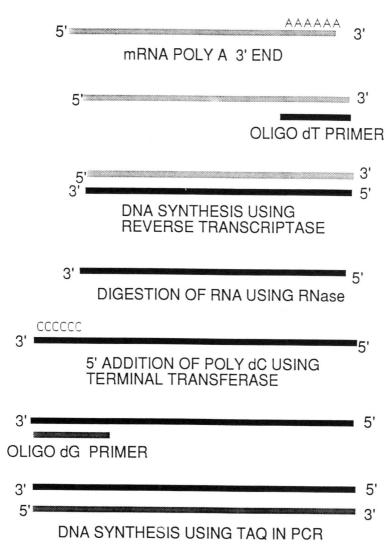

Figure 2.9. Variations of the PCR: mRNA analysis. In the example shown, the entire pool of mRNA molecules is amplified by using a generic oligo dT primer for the initial conversion of mRNA to cDNA. Alternatively, a specific primer reactive with a sequence contained within the mRNA coding a particular protein can be used to selectively amplify a single species of mRNA or viral RNA lacking a poly A tail. The creation of a second generic primer-binding site using the enzyme terminal transferase can also be applied to direct analysis of DNA.

first PCR can be partially purified and subjected to re-PCR using the internal primer set. Differentially labeled multiple primers can also provide sequence information when used as competing primers to define alleleic variants. Primers that differ at their 3′ ends (3′ mismatched primers) can be used to probe for point mutations in the template DNA[20] (Fig. 2.10B). Reducing the relative concentration of one primer will lead to the predominant synthesis of single-stranded product. This variation has

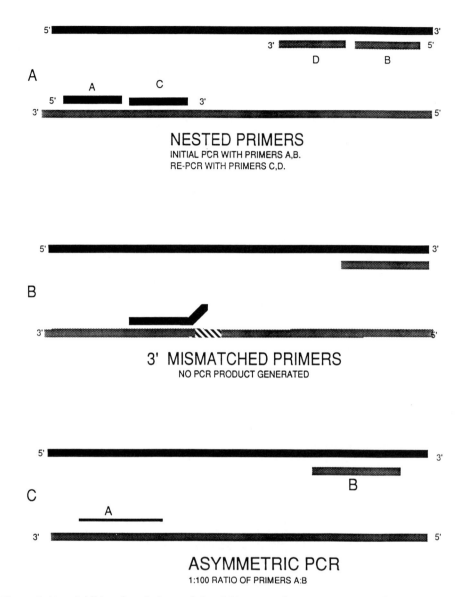

Figure 2.10. Additional variations of the PCR. Use of a two-step procedure employing nested primers can add significantly to both the sensitivity and the specificity of the assay. Specificity can also be achieved by the use of a 3' mismatched primer. In this situation, it is necessary to reduce the nucleotide concentration to 2 μM. Preferential production of single-stranded DNA can be achieved using the asymmetric PCR. An alternative approach is to biotinylate one of the primers and to capture single-stranded DNA using avidin-coated beads.

been termed the asymmetric PCR[29] (Fig. 2.10C). The single-stranded products are easier to sequence than is double-stranded DNA. Alternatively, one primer can be labeled with biotin and the single strands formed from this primer can be extracted from the final reaction products by affinity coupling to avidin.[82]

POTENTIAL PITFALLS WHEN PERFORMING PCR

Because of its exquisite sensitivity, the issue of DNA contamination or "carry over" is of extreme importance. Meticulous care is necessary to avoid contact between PCR products and solutions used to set up the PCR. Quality-control testing of reagents includes testing for possible inclusion of human DNA and products of commonly run PCR assays. Reagents are aliquoted into single assay portions and are clearly labeled, "Not for reuse once opened." An adequate initial training program is provided for all individuals participating in the laboratory, with periodic follow-up.

In performing the PCR, it is advisable to use separate rooms for initial processing of specimens, setting up the assays, running the amplifications, and analyzing the PCR products. Each room should have its own equipment including refrigerators, centrifuges, and pipettes. When processing paraffin-embedded blocks, it is important to cleanse the microtome knife between blocks. Blood processing and reagent aliquoting should be performed under a laminar-flow hood. The use of prealiquoted reaction mix helps to simplify assay setup. Negative controls consisting of all reactants with the exception of the patient's sample should be run in each assay. A positive control consisting of a known quantity of the template DNA should be included to test the primers, the *Taq* enzyme, and the detection method. Appropriately diluted products from a previously positive PCR can be used as the positive control. This approach has the advantage that if contamination does occur, one can switch to a set of external primers that would not amplify the contaminant.

The various items of equipment used in the PCR need to be subjected to quality-assurance testing. The actual temperatures reached by individual tubes placed at differing locations within the heating block can be periodically checked using a thermal wire probe, so as to determine uniformity of temperature between wells and consistency with machine readout. Several of the thermal cycling machines have a built-in program to evaluate the performance of the heating, cooling, and temperature-recording functions. Such programs should be run at least weekly.

Even with these precautions, two major problems may arise in developing a clinically useful PCR. The first problem is a lack of sensitivity, even with known positive DNA. This may reflect an instability of the primer:template binding at the high temperature required for optimal activity of the *Taq* enzyme. To address this problem, it may be advisable to lower the annealing temperature to 37°C and prolong the duration of the temperature rise to 72°. This modification will allow partial primer extension to occur, with a corresponding increase in the strength of template binding. A primer extension temperature of less than 72°C (*e.g.*, 65–68°C) can also be tried. A lack of sensitivity due to improper sample preparation can be readily tested by spiking the sample with a small amount of known positive DNA. When using tissue slices from paraffin-embedded blocks, it is useful to include a second set of primers reactive with a normal cellular DNA sequence. Inhibitory activity in the samples may be avoided by proteinase K predigestion or by simply using diluted samples. The lack of sensitivity in the PCR may also be due to incomplete denaturation of the template molecules. This problem can be addressed by increasing the denaturation temperature from 94 to 96°C. The pH of the reaction mixture should be checked. The pH can be affected by the buffer used to dissolve the dNTP. Improper storage conditions can lead to hydrolysis of the dNTP, loss of enzyme activity, and precipitation of primers. The $MgCl_2$ concentration as well as pH may need to be titrated to obtain optimal amplification. If sensitivity is still a problem, a two-step PCR can be performed. In the first reaction, a set of outside primers is chosen that will amplify a large segment of DNA. Internal or nested primers, specific for the segment of DNA of interest, are subsequently added.

A second major problem that can be encountered in performing PCR is that of extensive cross-priming. This will lead to the production of nonspecific products including primer-dimers. Multiple bands will be seen on gel electrophoresis and unacceptably high backgrounds may occur with hybridization analyses of PCR products. Low-level false-positive responses may occur due to single-primer reactions on multiple sites of the genome. It can be difficult to distinguish this type of response from the presence of a low level of the specific target sequence. One approach to these problems is to increase the temperature of primer annealing. The size homogeneity of the primers should be checked, and if necessary, an additional round of HPLC or gel purification instituted. It is sometimes helpful to reduce the concentrations of template, primers, enzyme, and dNTP and to use fewer cycles of shorter duration. If hybridization is being used to detect the PCR products, it should be run under highly stringent conditions. Southern blotting, rather than dot blotting, will reveal if the hybridizing PCR product has the expected size. For reasons that are not well understood, some regions of DNA appear to be easier to amplify than others. Increased specificity can often be achieved by the use of an alternative set of primers, directed against a slightly different region of the genome.

Once suitable parameters are established for a particular PCR application, it is important that they be adhered to in all subsequent assays. It is appropriate to maintain detailed records showing compliance in each clinical assay and to retain aliquots of samples both prior to and following amplification for subsequent review.

Diagnostic Applications of the PCR

Pathology will likely be the first clinical discipline to benefit from the explosive growth in molecular biology. There has been an enormous increase in information relating abnormal gene function to specific disease entities. PCR technology will foster the partnership between basic molecular biologists and pathologists. A minimal contribution by pathologists could be simply to provide samples of abnormal tissues, microbial isolates, etc., to the basic scientist. The ease with which PCR technology can be applied within the pathologist's laboratory should, however, encourage the pathologist to play a much more active role in this partnership.

PCR technology has already opened up many exciting opportunities for routine molecular diagnostic pathology. The greatest impact to date has been in the sensitive identification of organisms that are refractory to simple *in vitro* cultivation. Molecular microbiology will undoubtedly grow once virulence and chemosensitivity characteristics are related to defined genetic sequences within the genome of microbial pathogens. PCR also holds great promise for surgical pathology, especially for the identification of mutated and/or rearranged oncogenes within tumor cells. PCR will also greatly enhance the scope of molecular genetics for the identification of disease susceptibilities and for forensic applications. Recent progress in each of these areas of molecular pathology is briefly reviewed below.

MOLECULAR MICROBIOLOGY

PCR has been applied to the detection of an ever-increasing number of human pathogens. Examples include human papillomavirus,[74,76,87] human immunodeficiency virus-1[36] and -2,[62] human T lymphotrophic virus types I and II,[21] cytomegalovirus,[59,77] Epstein-Barr virus,[66] Herpes simplex virus,[6] human herpesvirus-6,[10] hepatitis B virus,[40,81] B16 parvovirus,[70] JC and BK viruses,[3] rubella virus,[12] mycobacteria,[30] *Toxoplasma gondii*,[29] *Trypanosoma cruzi*,[54] and malaria.[64] PCR can be applied to the detection of virtually any pathogen for which even limited DNA (or RNA) sequence information is known and for which a specimen of infected tissue can be readily obtained.

Using well-standardized procedures, PCR can be considered to be at least a semiquantitative assay. With the excess enzyme and primer concentrations commonly used and a reduced number of cycles, the rate-limiting component in the PCR is the initial availability of template DNA. This quantitative relationship can be demonstrated by running a graded series of control samples within an assay. The intensity of the signal obtained with a patient's sample can be compared with that shown by the controls. This type of semiquantitative assay is particularly suited for following infections with viruses for which an asymptomatic "latent" state has been identified (*e.g.*, human herpesviruses) and for which a low-level signal can sometimes be encountered in healthy individuals.

The decision of whether to develop and apply PCR for routine diagnosis will depend on the cost, speed, and reliability of more conventional culture and/or serological methods. PCR also needs to be compared with alternative molecular probe and monoclonal antibody-based assays. Although lacking the sensitivity of PCR, *in situ* hybridization and immunohistochemistry can provide useful information concerning the tissue localization of certain pathogens. To be detectable by these techniques, there must be an abundance of target nucleic acid or protein products within individual infected cells. Rapid assays, based on the detection of species-specific ribosomal RNA sequences, are also becoming available for several bacteria and parasites. These assays are quite sensitive because each organism contains up to 10^3–10^5 ribosomes. Overall, none of the alternative assays has the sensitivity or versatility of PCR to become established as a central core technique within most clinical microbiology departments.

Serological responses, especially IgM-based assays or antigen detection assays are appropriate when blood testing is used to test for an infection present elsewhere in the body. PCR can provide a useful backup where questions are raised concerning the reliability of conventional serological tests in certain individual patients. For example, although not common, cases of HIV infection, detectable by both PCR and culture, have been described in the absence of seroconversion.[22] Similarly, HTLV-I and HBV infections are not always accompanied by serological manifestations, yet can be detected by PCR.[21,86] PCR has been used to document neonatal transmission of infection with HIV[45] and HBV.

PCR can be much more discriminative than conventional serology. Most clinical serological reactions are directed against a mixture of antigenic epitopes that can be variously shared by closely related pathogens. Thus, for example, it is difficult to distinguish HIV-1 from HIV-2 or HTLV-I from HTLV-II by serology, yet such distinctions can be readily made on the basis of PCR amplifications of type-specific genetic sequences.[21,62] Recently, PCR analysis has led to the detection of different strains of EBV[79] and of *Plasmodium vivax* malaria.[64]

PCR holds its greatest promise for extending diagnosis beyond simple detection. Important genetic characteristics such as virulence and responsiveness to chemotherapy are amenable to direct analysis by PCR. This area of endeavor is proceeding rapidly, and it is likely that many types of organisms will be categorized pathologically according to subtle genetic changes. For example, genetic alterations in the gene coding the e antigen of hepatitis B virus have recently been described in a patient with severe chronic liver disease.[11] Chloroquine resistance in *Plasmodium falciparum* shows several characteristics in common with the phenomenon of multidrug resistance (*mdr*) in certain human tumor cells. PCR was used to screen malarial parasites for a gene related to the *mdr* gene in human cells. The corresponding gene was identified and shown to be involved in at least some cases of chloroquine resistance.[23,88] Pyrimethamine resistance in *P. falciparum* is associated with a point mutation in the dihydrofolate reductase-thymidine synthetase gene detectable by PCR.[61] There is currently a great deal of interest in seeking possible changes in the polymerase gene of human immunodeficiency virus as a marker for Zidovudine resistance as well as markers for DHPG resistance in cytomegaloviral isolates.

Genetic heterogeneity among different isolates of an organism can potentially present a problem for PCR analysis. Thus, if the pathogen carries a minor deletion or even a point mutation at a critical primer-binding site, a false-negative PCR could result. In addition, one needs to be alert to the possibility of potential genetic homologies between distinct microorganisms, giving rise to common reactivity with a given set of primers. To help avoid these possibilities, several sets of primers may be used to probe for common pathogens. As experience is gained, however, the diagnostic value of any particular primer set, as well as the optimal conditions for use, will become firmly established.

Conventional culture techniques have been established to more or less provide a broad screen for a range of microorganisms. In principle, PCR can also be used as an initial screen for major categories of pathogens. The primer sets required would need to be reactive with highly conserved regions of bacterial, viral, or fungal genomes. This same type of approach can lead to the detection of new types of pathogens.[52] Although somewhat preliminary, there have been several recent reports that tissues from individuals with various autoimmune diseases, including multiple sclerosis, rheumatoid arthritis, and autoimmune thyroiditis[14,28,63] may contain exogenous DNA sequences related to those of HIV and HTLV-1. These results are by no means conclusive[5] and even the normal human genome can be shown to contain retroviral-like sequences using low-stringency hybridization reactions.[60] While caution is necessary in interpreting these provocative findings, they do underscore another major advantage of PCR over conventional culture techniques. Thus, PCR allows the identification of portions of an exogenously derived genome in the absence of a complete pathogen. Residual genes from the pathogen could account for some of the chronic inflammatory and/or degenerative diseases in which there has been no evidence for overt infection using techniques less sensitive than the PCR.

MOLECULAR ONCOLOGY

Understanding the nature of cancer has been the central theme of much of the molecular biological research performed during the last two decades. This work has led to the identification of a series of oncogenes and more recently antioncogenes, and to the description of potent growth factors and growth-factor receptors. Using laboratory models, it is now possible to identify the genetic elements responsible for the abnormal growth characteristics of malignant cells. Extrapolating these findings to the everyday practice of diagnostic surgical pathology represents an important challenge for pathologists. The ability to perform PCR on formalin-fixed, paraffin-embedded tissues has greatly facilitated this task. As with many other special procedures, it is possible to review histological findings on routinely processed samples before deciding if additional sections need to be taken for more detailed study. Some of the suggested applications of PCR to diagnostic surgical pathology such as oncogene amplification[24] and detection of oncogenic viruses[43,49,51,75] can probably be addressed equally well using the rapidly evolving alternative strategies of *in situ* hybridization and immunochemistry. These techniques provide greater precision than PCR for studies on transcription and expression of specific growth factors, amplification or deletion of whole genes, and testing for the presence of oncogenic viruses.

The major unique applications of PCR in medical oncology are (*a*) the ability to detect minor structural changes and rearrangements in single copy genes and (*b*) the capacity to detect minimal residual disease following therapy. PCR has been used effectively to test tumors for single nucleotide mutations within the H-, K-, and N-*ras* cellular genes. Approximately 50% of colon cancers and over 90% of carcinomas of the exocrine pancreas will show evidence of a specific K-*ras* mutation at codons 12, 13, or 61.[1,8] The presence of the mutation is not diagnostic of malignancy, since a similar change can occur in benign colonic polyps. Nevertheless, finding a mutated *ras* oncogene provides information concerning one of the abnormal growth-controlling mechanisms operating within the tumor cells. At present there is considerable interest in analyzing the gene coding a nuclear protein termed p53. This protein is thought to possess antioncogenic activity similar to that postulated for the retinoblastoma gene product. By PCR analysis, the p53 coding gene is mutated in the majority of lung and colon cancers tested.[4,83] Detailed studies on other growth-regulatory genes in human tumors are underway in many laboratories. It is predictable that these studies will allow morphologically indistinguishable tumors to be classifiable into biologically distinct categories on the basis of their differing genetic abnormalities. Once this type of correlation becomes established, the use of PCR will become a near-essential component in the diagnostic workup of malignant lesions.

Of immediate diagnostic value is the ability to use PCR to test for rearrangement between single genetic loci. For example, t 14:18 translocation is a common feature of follicular lymphomas and can be readily detected using PCR.[16] One primer is reactive with a consensus sequence contained in the J_H gene from chromosome 14, while the other primer is reactive with a sequence 3' to the major breakpoint sequence of the *bcl*-2 gene on chromosome 18. A PCR product is formed only if a t 14:18 translocation is present. Furthermore, the size of the product is a distinguishing characteristic of each individual tumor, allowing a direct comparison between a primary and a suspected recurrent tumor in the same patient.[80] It has recently been shown that rearrangement within DNA coding for the δ polypeptide of the T cell receptor (TCR) is not particularly specific for cells of the T lineage. This type of rearrangement has been seen in the majority of B cell and even in some myeloid leukemias.[32] It should be possible to tailor the PCR to identify clonogenic rearrangements of this locus and to prepare tumor-specific probes for monitoring disease progression in acute leukemias. For translocations with more variability in the junctional regions, it is preferable to perform RNA analysis by PCR. For example, PCR has been used to detect the hybrid mRNA reflecting the breakpoint cluster region (*bcr*) to c-ALB, t 9:22, translocation in chronic myelogenous leukemias.[41]

The exquisite sensitivity of PCR has provided an unparalleled opportunity for the molecular detection of rare residual or recurrent tumor cells. Tumor-specific primers, reactive with either a specifically mutated or a rearranged genetic locus can be designed for an increasing array of tumor cell types. Detection sensitivities of a single tumor cell in the presence of 10^7 normal cells have been recorded using tumor-specific primers. Clinical studies will be required to assess the prognostic and therapeutic significance of finding abnormal cellular DNA sequences in an otherwise healthy individual, since it cannot be assumed that all of the cells have neoplastic potential.

MEDICAL GENETICS

PCR can detect virtually all of the common genetically inherited diseases in which the defective genetic locus has been identified. Examples include sickle cell anemia, thalassemia, Duchenne muscular dystrophy, Lesch-Nyhan syndrome, hemophilia A and B, α_1-antitrypsin inhibitor deficiency, phenylketonuria, Tay-Sachs disease, and adenine deaminase deficiency.[7,17–19,27,42,44,56] Because of the need for limited cellular material, PCR is ideally suited for direct analysis on the cells obtained by amniocentesis or chorionic villus samplings.[72] PCR can also trace the inheritance of disease in which the defective locus has only been defined in terms of linkage to other cellular genes. Using PCR, allelic forms of many cellular genes can be identified by sizing selected introns between the coding exons.[84] This approach should allow much more detailed linkage studies than are presently possible using restriction fragment length polymorphism (RFLP) analysis. Moreover, because PCR-amplified DNA is not methylated, sequences allowing differential restriction by certain enzymes can be increased in comparison with native DNA. PCR-defined polymorphisms in loci linked to the cystic fibrosis gene have been defined,[37,65] and many more informative loci linked to other disease-associated genes can be expected in the very near future.

Sequencing of PCR-amplified genes has led to the discovery of disease-associated single-nucleotide differences in the genes encoding histocompatibility antigens[39,73] and various hormone and growth-factor receptor molecules. PCR-based analysis will undoubtedly lead to the identification of a vast array of genetically based diseases as well as provide valuable insights into disease pathogenesis. Even for those diseases in which the nature of the genetic defect is highly variable among different affected individuals, specific PCR-based assays can be devised to test family members for inheritance of the abnormal genotype characteristic of that particular family.[53,72] The power of PCR is most convincingly demonstrated by the ability to genotype a single cell obtained from an 8–16 cell blastomer derived by *in vitro* fertilization[31,35] or even the polar body from a mature unfertilized ovum[15] (personal communication). These approaches hold great promise for couples at risk for transmitting an inherited disease and provide yet another example of the revolution in medical science consequent to the application of molecular technology.

FORENSIC APPLICATIONS

The capacity to perform PCR on minute amounts of starting material, including a single hair,[34] sperm cells,[48] throat washings,[24] and exfoliated epithelial cells in normal urine,[46] offers enormous advantages to forensic pathologists. In addition, PCR applied to formalin-fixed, paraffin-embedded tissues allows retrospective analysis on samples from deceased individuals on whom autopsies have been performed. Even tissue from human remains exhumed many years after death has yielded PCR-amplifiable material (unpublished observations). At present, the major targets that have been used in DNA-typing by PCR have been mitochondrial DNA[89] and DNA coding the HLA-DQ locus. A difficulty occasionally experienced is the failure of one of the two loci in a heterozygous individual to amplify in the PCR. This can cause an erroneous conclusion of homozygosity. This type of problem is likely to be corrected by better-prepared DNA samples and the selection of targets in which heterozygosity is to be expected. Sequence information is becoming available on several of the more highly polymorphic loci within the human genome. In addition, the frequency distribution of the various alleles of multiple independent loci is being documented. With these data, unequivocable genetic identification of an individual, based on PCR of minimal amounts of tissue samples, should become readily achievable by forensic molecular pathologists.

Conclusion

PCR represents an important breakthrough of molecular technology from the research to the clinical laboratory. PCR is an extremely useful adjunct to other molecular diagnostic techniques. The simplicity, sensitivity, and widespread applicability of PCR should enable the interested pathologist to actively participate in the molecular biological revolution that is sweeping through the biomedical research community. The discipline of molecular pathology will flourish if residents, technologists, and continuing educational training programs include a strong emphasis on molecular technology and if centralized, well-equipped, core facilities for DNA-related studies become available within medical centers. Pathologists have an opportunity, and perhaps even an obligation, to keep abreast of this field, so as to assume an important leadership role in identifying clinically relevant problems. By doing so they will also help invigorate the practice of pathology.

ACKNOWLEDGMENTS

The quality of the clinical laboratory reflects the skill and dedication of the technologists involved in its day-to-day operation. I am, therefore, especially pleased to acknowledge the outstanding contributions of Anton Mayr, M.T., Deogracias Delfin, M.T., and Peyman Javaherbin, M.T., to the development of molecular pathology at the LAC-USC Medical Center and toward establishing the highest standards possible for performing clinical testing using PCR technology.

Supported by a grant from Diatech (Cooperative Agreement DPE-5935-A00-5065-00 between the Program for Appropriate Technology in Health and U.S. Agency for International Development).

REFERENCES

1. Almoguera, C., Shibata, D., Forrester, K., et al. Most human carcinomas of the exocrine pancreas contain mutant c-K-ras genes. Cell 53:549–554, 1988.
2. Arrigo, S. I., Weitsman, S., Rosenblatt, J. D., et al. Analysis of rev gene function on human immunodeficiency virus type 1 replicating in lymphoid cells by using a quantitative polymerase chain reaction. J. Virol. 63:4875–4881, 1989.
3. Arthur, R. R., Dogostin, S., and Shah, K. V. Detection of BK virus and JC virus in urine and brain tissue by the polymerase chain reaction. J. Clin. Microbiol. 27:1174–1179, 1989.
4. Baker, S. J., Feron, E. R., Nigro, J. M., et al. Chromosome 17 deletions and p53 gene mutations in colorectal carcinomas. Science (Wash DC) 44:217–221, 1989.
5. Bangham, C. R. M., Nightingale, S., Cruickshank, J. K., et al. PCR analysis of DNA from multiple sclerosis patients for the presence of HTLV-I. Science (Wash DC) 246:821–822, 1989.
6. Boerman, R. H., Arnoldus, F. R., Raap, A. K., et al. Polymerase chain reaction and viral culture techniques to detect HSV in small volumes of cerebrospinal fluid, an experimental mouse encephalitis study. J. Viol. Methods 25:189–197, 1989.
7. Bottema, C. D., Koeberl, D. D., and Sommer, C. S. Direct carrier testing in 14 families with hemophilia B. Lancet ii:526–529, 1989.
8. Bos, J. L., Fearon, E. R., Hamilton, S. R., et al. Prevalence of ras gene mutations in human colorectal cancers. Nature (Lond) 327:293–297, 1987.
9. Burg, J. L., Grover, C. M., Pouletty, P., et al. Direct and sensitive detection of a pathogenic protozoan, Toxoplasma gondii, by polymerase chain reaction. J. Clin. Microbiol. 27:1787–1792, 1989.
10. Bushbinder, A., Josephs, S. F., Ablashi, D., et al. Polymerase chain reaction amplification and in situ hybridization for the detection of human B-lymphotropic virus. J. Virol. Methods 21:191–197, 1988.
11. Carman, W. F., Jacyna, M. R., Hadziyannis, S., et al. Mutation preventing formation of hepatitis B e antigen in patients with chronic hepatitis B infection. Lancet ii:588–590, 1989.
12. Carman, W. F., Williamson, C., Cunliffe, B. A., et al. Reverse transcription and subsequent DNA amplification of rubella virus RNA. J. Virol. Methods 25:21–29, 1989.
13. Chamberlain, J. S., Gibbs, R. A., Ranier, J. E., et al. Deletion screening of the Duchenne muscular dystrophy locus via multiplex DNA amplification. Nucleic Acids Res. 16:11141–11156, 1988.
14. Ciampolillo, A., Marini, V., Mirakian, R., et al. Retrovirus-like sequences in Graves' disease: Implications for human autoimmunity. Lancet i:1096–1100, 1989.
15. Coutelle, C., Williams, C., Handyside, A., et al. Genetic analysis of DNA from single human oocytes: A model for preimplantation diagnosis of cystic fibrosis. Br. Med. J. 299:22–24, 1989.
16. Crescenzi, M., Seto, M., Herzig, G. P., et al. Thermostable DNA polymerase chain amplification of the (14:18) chromosome breakpoints and detection of minimal residual disease. Proc. Natl. Acad. Sci. U.S.A. 85:4869–4873, 1988.

17. Crisan, D., Diven, W. F., Hartla, K., *et al.* Prenatal diagnosis of hemoglobinopathies by using the polymerase chain reaction and allele specific probes. *Clin. Chem. 35*:1854, 1989.

18. Dermer, S. J., and Johnson, E. M. Rapid DNA analysis of α-1-antitrypsin deficiency. Application of an improved method for amplifying mutated gene sequences. *Lab. Invest. 59*: 403–408, 1988.

19. DiLelia, A. D., Huang, W.-H., and Woo, S. L. C. Screening for phenylketonuria mutations by DNA amplification with the polymerase chain reaction. *Lancet i*:497–499, 1988.

20. Ehlen, T., and DuBeau, L. Detection of ras gene mutations by polymerase chain reaction using mutation specific, inosine containing oligonucleotide primers. *Biochem. Biophys. Res. Commun. 160*:441–447, 1989.

21. Ehrlich, G. D., Glaser, J. B., LaVigne, K., *et al.* Prevalence of human T-cell leukemia/lymphoma virus (HTLV) Type II infection among high-risk individuals: Type-specific identification of HTLVs by polymerase chain reaction. *Blood 74*:1658–1664, 1989.

22. Farzadegan, H., Polis, M., Wolinsky, S. M., *et al.* Loss of human immunodeficiency type I (HIV-1) antibodies with evidence of viral infection in asymptomatic homosexual men. *Ann. Intern. Med. 108*:785–790, 1988.

23. Foote, S. J., Thompson, J. K., Cowman, A. F., *et al.* Amplification of the multidrug resistance gene in some chloroquine-resistant isolates of P. falciparum. *Cell 57*:921–930, 1989.

24. Frye, R. A., Benz, C. C., and Liu, E. Detection of amplified oncogenes by differential polymerase chain reaction. *Oncogene 4*:1153–1157, 1989.

25. Gasparini, P., Savoia, A., Pignatti, P. F., *et al.* Amplification of DNA from epithelial cells in urine. *N. Engl. J. Med. 320*:809, 1989.

26. Gelfand, D., Taq DNA polymerase. In: *PCR Technology. Principles and Applications for DNA Amplification*, edited by H. A. Erlich, New York, Stockton, 1989, pp. 17–22.

27. Gibbs, R. A., Chamberlain, J. S., and Caskey, C. T. Diagnosis of new mutation disease using the PCR. In: *PCR Technology. Principles and Applications for DNA Amplification*, edited by H. A. Erlich, New York, Stockton, 1989, pp. 171–192.

28. Greenberg, S. J., Ehrlich, G. D., Abbott, M. A., *et al.* Detection of sequences homologous to human retroviral DNA in multiple sclerosis by gene amplification. *Proc. Natl. Acad. Sci. U.S.A. 86*:2878–2882, 1989.

29. Gyllenstein, U. B., and Erlich, H. A. Generation of single-stranded DNA by the polymerase chain reaction and its application to direct sequencing of the HLA-DQA locus. *Proc. Natl. Acad. Sci. U.S.A. 85*:7652–7656, 1988.

30. Hance, A. J., Grandchamp, B., Levy-Frebault, V., *et al.* Detection and identification of mycobacteria by amplification of mycobacterial DNA. *Mol. Microbiol. 3*:843–849, 1989.

31. Handyside, A. H., Patinson, J. K., Penketh, R. J. A., *et al.* Biopsy of human preimplantation embryos and sexing by DNA amplification. *Lancet i*:347–349, 1989.

32. Hansen-Hagge, T. E., Yokota, S., and Bartram, C. R. Detection of minimal residual disease in acute lymphoblastic leukemia by in vitro amplification of rearranged T-cell receptor delta chain sequences. *Blood 74*:1762–1767, 1989.

33. Higuchi, R. Simple and rapid preparation of samples for PCR. In: *PCR Technology. Principles and Applications for DNA Amplification*, edited by H. A. Erlich, New York, Stockton, 1989, pp. 31–38.

34. Higuchi, R., von Beroldingen, C. H., Sensabaugh, G. H., *et al.* DNA typing from single hairs. *Nature (Lond) 332*:543–546, 1988.

35. Holding, C., and Monk, M. Diagnosis of beta-thalassaemia by DNA amplification in single blastomers from mouse preimplantation embryos. *Lancet ii*:532–535, 1989.

36. Hufert, F. T., Laer, Dv., Schramm, C., *et al.* Detection of HIV-1 DNA in different subsets of human peripheral blood mononuclear cells using the polymerase chain reaction. *Arch. Virol. 106*:341–345, 1989.

37. Huth, A., Estivill, X., Grade, K., *et al.* Polymerase chain reaction for detection of the pMP6d-9/MspI RFLP, a marker closely linked to the cystic fibrosis mutation. *Nucleic Acids Res. 17*:7118, 1989.

38. Jeffreys, A. J., Wilson, V., Neumann, R., *et al.* Amplification of human minisatellites by the polymerase chain reaction: Towards DNA fingerprinting single cells. *Nucleic Acids Res.* 16:10953–10972, 1988.
39. Kagnoff, M. F., Harwood, J. I., Bugawan, T. L., *et al.* Structural analysis of the HLA-DR, -DQ and -DP alleles on the celiac disease associated HLA-DR3 (DRw17) haplotype. *Proc. Natl. Acad. Sci. U.S.A.* 86:6274–6278, 1989.
40. Kaneko, S., Feinstone, S. M., and Miller, R. H. Rapid and sensitive method for the detection of serum hepatitis B virus DNA using the polymerase chain technique. *J. Clin. Microbiol.* 27:1930–1933, 1989.
41. Kawasaki, E. S., Clark, S. S., Coyne, M. Y., *et al.* Diagnosis of chronic and acute lymphocytic leukemias by detection of leukemia specific mRNA sequences amplified in vitro. *Proc. Natl. Acad. Sci. U.S.A.* 85:5698–5702, 1988.
42. Kazizian, H. H. Use of PCR in the diagnosis of monogenic disease. In: *PCR Technology. Principles and Applications for DNA Amplification,* edited by H. A. Erlich, New York, Stockton, 1989, pp. 153–170.
43. Kiyabu, M. T., Shibata, D., Arnheim, N., *et al.* Detection of human papillomavirus in formalin-fixed invasive squamous cell carcinomas using the polymerase chain reaction. *Am. J. Surg. Pathol.* 13:221–224, 1989.
44. Kogan, S. C., Doherty, M., and Gitschier, J. An improved method for prenatal diagnosis of genetic diseases by analysis of amplified DNA sequences. Prenatal diagnosis of hemophilia A. *N. Engl. J. Med.* 317:985–990, 1987.
45. Laure, F., Courgnaud, V., Rouzioux, C., *et al.* Detection of HIV1 DNA in infants and children by means of the polymerase chain reaction. *Lancet ii:*538–541, 1988.
46. Lench, N., Stanier, P., and Williamson, R. Simple non-invasive method to obtain DNA for gene analysis. *Lancet i:*1356–1358, 1988.
47. Levin, B. *Genes II.* New York, John Wiley & Sons, 1988.
48. Li, H., Gyllensten, U. B., Cui, X., *et al.* Amplification and analysis of DNA sequences in single human sperm and diploid cells. *Nature (Lond)* 335:414–417, 1988.
49. Lo, Y. M., Mehal, W. Z., Fleming, K. A. In vitro amplification of hepatitis B virus sequences from liver tumor DNA and from paraffin embedded tissues using the polymerase chain reaction. *J. Clin. Pathol.* 42:840–846, 1989.
50. Loh, E. Y., Elliott, J. F., Cwirla, S., *et al.* Polymerase chain reaction with single-sided specificity: Analysis of T cell receptor delta chain. *Science (Wash DC)* 243:217–220, 1989.
51. McDonnell, J. M., Mayr, A. J., and Martin, W. J. DNA of human papilloma-virus type 16 in dysplastic and malignant lesions of the conjunctiva and cornea. *N. Engl. J. Med.* 320:1442–1446, 1989.
52. Mach, D. H., and Sninsky, J. J. A sensitive method for the identification of uncharacterized viruses related to known virus groups: Hepadnavirus model system. *Proc. Natl. Acad. Sci. U.S.A.* 85:6977–6981, 1988.
53. Miyano, M., Nanjo, K., and Chan, S. J. Use of in vitro gene amplification to screen family members for an insulin gene mutation. *Diabetes* 37:862–866, 1988.
54. Moser, D. R., Kirchhoff, L. V., and Donelson, J. E. Detection of *Trypanosoma cruzi* by DNA amplification using the polymerase reaction. *J. Clin. Microbiol.* 27:1477–1482, 1989.
55. Mullis, K. B., and Faloona, F. A. Specific synthesis of DNA in vitro via a polymerase catalyzed chain reaction. *Methods Enzymol.* 255:335–350, 1987.
56. Navon, R., and Proia, L. The mutations in Ashkenazi Jews with adult G_{M2} gangliosidosis, the adult form of Tay-Sachs disease. *Science (Wash DC)* 243:1471–1474, 1989.
57. Ochman, H., Ajioka, J. W., Garza, D., *et al.* Inverse polymerase chain reaction. In: *PCR Technology. Principles and Applications for DNA Amplification,* edited by H. A. Erlich, New York, Stockton, 1989, pp. 105–112.
58. Ohtsuka, E., Matsuki, S., Ikehara, M., *et al.* An alternative approach to deoxyoligonucleotides as hybridization probes by insertion of deoxyinosine at ambiguous codon positions. *J. Biol. Chem.* 260:2605–2608, 1985.

59. Olive, D. M., Simsek, M., and Al-Mufti, S. Polymerase chain reaction assay for detection of human cytomegalovirus. *J. Clin. Microbiol.* 27:1238–1242, 1989.

60. Perl, A., Rosenblatt, J. D., Chen, I. Y. S., *et al.* Detection and cloning of new HTLV-related endogenous sequences in man. *Nucleic Acids Res.* 17:6841–6853, 1989.

61. Peterson, D. S., Walker, D., and Wellens, T. E. Evidence that a point mutation in dihydrofolate reductase-thymidylate synthase confers resistance to pyrimethamine in falciparum malaria. *Proc. Natl. Acad. Sci. U.S.A.* 85:9144–9188, 1988.

62. Rayfield, M., Cock, K. D., Heyward, W., *et al.* Mixed human immunodeficiency virus (HIV) infection in an individual: Demonstration of both HIV type 1 and type 2 proviral sequences by using polymerase chain reaction. *J. Infect. Dis.* 158:1170–1176, 1988.

63. Reddy, E. P., Sandberg-Wollheim, M., Mettus, R. V., *et al.* Amplification and molecular cloning of HTLV-I sequences from DNA of multiple sclerosis patients. *Science (Wash DC)* 243:529–533, 1989.

64. Rosenberg, R., Wirtz, R. A., Lanar, D. E., *et al.* Circumsporozoite protein heterogeneity in the human malaria parasite *Plasmodium vivax. Science (Wash DC)* 245:973–976, 1989.

65. Rosenbloom, C. L., Kerem, B. S., Rommens, J. M., *et al.* DNA amplification for detection of the XV-2c polymorphism linked to cystic fibrosis. *Nucleic Acids Res.* 17:7117, 1989.

66. Saito, I., Servenius, B., Compton, T., *et al.* Detection of Epstein-Barr virus by polymerase chain reaction in blood and tissue biopsies from patients with Sjogren's syndrome. *J. Exp. Med.* 169:2191–2197, 1989.

67. Saiki, R. K. The design and optimization of the PCR. In: *PCR Technology. Principles and Applications for DNA Amplification,* edited by H. A. Erlich, New York, Stockton, 1989, pp. 7–16.

68. Saiki, R. K., Gelfand, D. H., Stoffel, S., *et al.* Primer directed enzymatic amplification of DNA with a thermostable DNA polymerase. *Science (Wash DC)* 239:487–491, 1987.

69. Saiki, R. K., Walsh, P. S., Levenson, C. H., *et al.* Genetic analysis of amplified DNA with immobilized sequence-specific oligonucleotide probes. *Proc. Natl. Acad. Sci. U.S.A.* 86: 6230–6234, 1989.

70. Salimans, M. M., van de Ryke, F. M., Raap, A. K., *et al.* Detection of parvovirus B19 DNA in fetal tissue by in situ hybridization and polymerase chain reaction. *J. Clin. Pathol.* 42:525–529, 1989.

71. Sanger, F., Nicklen, S., and Coulson, A. R. DNA sequencing with chain terminating inhibitors. *Proc. Natl. Acad. Sci. U.S.A.* 74:5463–5467, 1987.

72. Sarkar, C., Evans, M. T., Kogans, S., *et al.* Accurate prenatal diagnosis with novel polymerase chain reaction primers in a family with sporadic hemophilia A. *Obstet. Gynecol.* 74:414–417, 1989.

73. Scharf, S. J., Friedmann, A., Steinman, L., *et al.* Specific HLA-DQB and HLA-DRB1 alleles confer sensitivity to pemphigus vulgaris. *Proc. Natl. Acad. Sci. U.S.A.* 86:6215–6219, 1989.

74. Shibata, D., Arnheim, N., and Martin, W. J. Detection of human papillomavirus in paraffin embedded tissue using the polymerase chain reaction. *J. Exp. Med.* 158:225–230, 1988.

75. Shibata, D., Cosgrove, M., Arnheim, N., *et al.* Detection of human papillomavirus DNA in fine needle aspirations of metastatic squamous-cell carcinoma of the uterine cervix using the polymerase chain reaction. *Diagn. Cytopathol.* 5:40–43, 1989.

76. Shibata, D., Fu, Y. S., Gupta, J. W., *et al.* Detection of human papillomavirus in normal and dysplastic tissue by the polymerase chain reaction. *Lab. Invest.* 59:555–559, 1988.

77. Shibata, D., Martin, W. J., Appleman, M. D., *et al.* Detection of cytomegaloviral DNA in peripheral blood of patients infected with human immunodeficiency virus. *J. Infect. Dis.* 158:1185–1192, 1988.

78. Shibata, D., Martin, W. J., and Arnheim, N. Analysis of DNA sequences in forty-year-old paraffin-embedded thin-tissue sections. A bridge between molecular biology and conventional histology. *Cancer Res.* 48:4564–4566, 1988.

79. Sixbey, J. W., Shirley, P., Chesney, P. J., *et al.* Detection of a second widespread strain of Epstein-Barr virus. *Lancet ii*:761–765, 1989.

80. Stetler-Stevenson, M., Raffeld, M., Cohen, P., *et al.* Detection of occult follicular lymphoma by specific DNA amplification. *Blood* 72:1822–1825, 1988.
81. Sumazaki, R., Motz, M., Wolf, H., *et al.* Detection of hepatitis B virus in serum using amplification of viral DNA by means of the polymerase chain reaction. *J. Med. Virol.* 27:304–308, 1989.
82. Syvanen, A. C., Laaksonen, M., and Soderlund, H. Fast quantification of nucleic acid hybrids of affinity-based hybrid collection. *Nucleic Acids Res.* 14:5037–5048, 1986.
83. Takahashi, T., Nau, M. M., Chiba, I., *et al.* p53: A frequent target for genetic abnormalities in lung cancer. *Science (Wash DC)* 246:491–493, 1989.
84. Tautz, D. Hypervariability of simple sequences as a general source for polymorphic DNA markers. *Nucleic Acids Res.* 17:6463–6471, 1989.
85. Thein, S. L., and Wallace, R. B. The use of synthetic oligonucleotides as specific hybridization probes in the diagnosis of genetic disorders. In: *Human Genetic Diseases: A Practical Approach,* edited by K. E. Davis, New York, IRL Press, 1986, pp. 33–50.
86. Thiers, V., Nakajima, E., Kremsdorf, D., *et al.* Transmission of hepatitis B from hepatitis-B-seronegative subjects. *Lancet ii:*1273–1276, 1988.
87. Tidy, J. A., Vousden, K. H., and Farrell, P. J. Relation between infection with a subtype of HPV16 and cervical neoplasia. *Lancet ii:*1225–1227, 1989.
88. Wilson, C. M., Serrano, A. E., Wasley, A., *et al.* Amplification of a gene related to mammalian mdr genes in drug-resistant *Plasmodium falciparum. Science (Wash DC)* 244:1184–1186, 1989.
89. Wrischnik, L. A., Higuchi, R. G., Stoneking, M., *et al.* Length mutations in human mitochondrial DNA: Direct sequencing of enzymatically amplified DNA. *Nucleic Acids Res.* 15:529–542, 1987.
90. Zimran, A., Glass, C., Thorp, V. S., *et al.* Analysis of "color PCR" by automatic DNA sequencer. *Nucleic Acids Res.* 17:7538, 1989.

3

Cytogenetics and Molecular Cytogenetics in the Diagnosis of Cancer

Sandra R. Wolman

The current and future practice of pathology must be prepared to take advantage of cytogenetics as a diagnostic tool and to interpret cytogenetic findings in relation to molecular techniques. The recent explosion in areas involving the interplay of cytogenetics and molecular biology has yielded new diagnostic markers that will lead to major improvements in the definition and classification of cancers, and eventually to an understanding of the basic biology and phylogeny of tumors. Continuing advances at this interface are proceeding simultaneously with an expansion of interest and activity in traditional cytogenetics. DNA content determination, the subject of numerous other symposia, offers important perspectives on cytogenetic studies, including the ability to refer back to the original tissue, even after paraffin embedding, and the monitoring of populations as they evolve in cell culture. Both flow cytometry and image analysis provide background data on larger (and nondividing) populations of cells against which the highly selective cytogenetic information should be viewed.

This review illustrates the types of analyses and results achievable with cytogenetics alone, describes some of the advances made possible by a combination of cytogenetic and molecular approaches, and projects and speculates on future coordination of these disciplines in tumor diagnosis and prognostication.

Cytogenetics

Conventional cytogenetic studies, because of their dependence on viable, disaggregated, dividing cells, are limited to direct examination of cells and tissues freshly removed from the living person or to cells that have been grown in tissue culture for various time periods. For practical purposes, the most readily available cells for study are those of the blood and bone marrow, and this is reflected in the large amount of cytogenetic data on the leukemias and the frequency with which such studies are performed. The 1988 edition of the *Catalog of Chromosome Aberrations in Cancer*[45] lists 9069 neoplasms of which 5211 are nonlymphocytic leukemias. Of the remainder, another 2460 cases are other hematologic disorders or lymphomas. Of the 1398 nonhematopoietic solid tumors, 354 are neural tumors (including 130 benign tumors) and 140 are soft-tissue sarcomas.

Cytogenetic information is scanty for the most common malignancies, lung, breast, colon, and prostate, whereas for many types of tumors essentially no data are yet available. Direct analyses of most solid tumors have been hampered by low spontaneous rates of mitosis and difficulties of cellular disaggregation, even when adequate amounts of fresh tumor tissue are available. In contrast, because of their relative ease of adaptation to growth in culture, nervous system tumors and the rare soft-tissue sarcomas continue to be comparatively well-studied. It is important to note that the catalog excludes most tumors analyzed after prolonged culture, all cases with normal karyotypes, and all cases of chronic myelogenous leukemia with the typical 9;22 translocation as the sole abnormality.[a] Therefore, the relative underrepresentation of solid tumors is far greater than is immediately apparent.

The well-known tumor-chromosome associations (Table 3.1) have substantial diagnostic value and, increasingly, are part of the routine workup and differential diagnosis for cancer patients. Their impact is such that many investigators now assume that all tumors are marked by chromosome aberrations and that identification of specific tumor-chromosome associations is critical to understanding tumor etiology. It is widely assumed that the marked genes are relevant to carcinogenesis or to lineage-specific development and that the chromosome aberrations constitute a common pathogenetic mechanism. However, as noted by Teyssier,[73,74] the number of solid tumors that have been reported to show consistent changes without major discrepancy by two or more laboratories are few, and there are well-documented exceptions even for the tumors listed in Table 3.1. The information and references included in this review are not intended to be a comprehensive view of the relevant chromosomal and molecular data bases, but only to provide appropriate examples that illustrate the points under discussion.

DIAGNOSIS, PROGNOSIS, AND CLASSIFICATION

Specificity. The tumor-specific chromosome associations that have been described are useful in identification of tumor cells, in diagnosis of clinical remission or relapse, in correlation with prognosis, and in better classification of tumor subtype. Among the myeloid leukemias, the first described and best known t(9;22) of chronic myelogenous leukemia (CML) has been joined by the t(8;21) of acute nonlymphocytic leukemia (ANLL) M2, the t(15;17) of acute promyelocytic leukemia (APL), and many less frequently seen translocations involving chromosomes 6, 9, 11, and 14 in the other categories of leukemia. Correlations with prognosis have been documented in many studies (*e.g.*, t(9;11) in monocytic M5 proliferations and poor prognosis[44] and inv(16) in eosinophilic M4 leukemia with its relatively high rate of complete remission[63]). The important lesson is that many of the specific chromosome aberrations among the leukemias correlate with specific clinical pictures and convey useful prognostic information.

The identification of stage- and diagnosis-specific chromosome markers of lymphomas is more controversial, largely because of the greater difficulty in obtaining adequate cytogenetic representation of the tumor population. Some better-known examples are included in Table 3.1. Clinical correlations have proved elu-

---

[a] Cytogenetic nomenclature is described in *An International System for Human Cytogenetic Nomenclature (1985)*, ISCN, 1985, published by S. Karger AG, Basel and New York.

Table 3.1.
Chromosome Aberrations of Malignant Tumors[a]

Myeloid tumors	
Chronic myelogenous leukemia	t(9;22)
Acute nonlymphocytic leukemia;M2	t(8;21)
Acute promyelocytic leukemia	t(15;17)
Lymphoid tumors	
Burkitt's lymphoma	t(8;14) [t(2;8),t(8;22)]
Follicular lymphoma	t(14;18)
Diffuse lymphoma (also CLL, myeloma)	t(11;14)
Solid tumors	
Alveolar(?) rhabdomyosarcoma	t(2;13)
Synovial sarcoma	t(X;18)
Ewing's sarcoma	t(11;22)
Myxoid liposarcoma	t(12;16)
Transitional cell carcinoma of bladder	? 11p-
Renal carcinoma	del (3p)
Small cell carcinoma of lung	del (3p)
Ovarian adenocarcinoma	? t(6;14)
Testicular tumor (seminoma)	i(12p)
Retinoblastoma	del (13) (q14)
Wilms' tumor	del (11) (p13)
Neuroblastoma	1p-

[a] For more comprehensive listings, see Hecht[28] and Teyssier.[74]

sive. For example, three large series of patients with chronic lymphocytic leukemia were examined for the prognostic associations of trisomy 12, a relatively frequent concomitant of the disease; two studies reported that trisomy 12 influenced survival (negatively), whereas the third did not find any correlation.[24,27,51] In Hodgkin's disease (HD), where the cell of origin is still poorly understood, cytogenetic data have suggested origin from a B lymphocyte. This conclusion was based on the observation that 35% of HD cases involved chromosome 14q and that the most common breakpoint occurred at 14q32, which is the locus of the IGH chain gene. Furthermore, most of the commonly observed chromosome aberrations in cases of HD have been those observed in other B cell lymphoid disorders.[11]

Because of convincing evidence of specific chromosomal associations with a number of soft-tissue tumors, karyotypic aberrations have been used in confirmation or reconsideration of diagnoses that were questionable histologically, and for identification of morphologic and/or behavioral subcategories within a tumor type. Evidence of chromosomal aberration is often interpreted as indicating that the lesion in question is neoplastic.[17] This type of reasoning was shown in a case (Fig. 3.1) originally described as a desmoid tumor but reclassified as a low-grade fibrosarcoma, in part because of the existence of extensive clonal chromosomal aberrations. A clinically similar desmoid or low-grade fibrosarcoma was reported recently in which dissimilar but equally aberrant cytogenetic findings influenced the classification of the tumor as malignant.[35] In desmoid tumors, which are known to recur

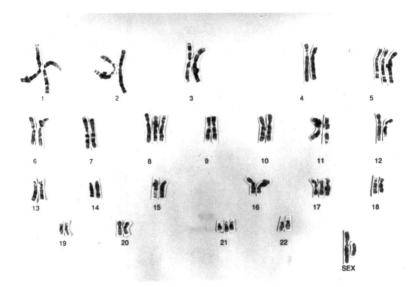

Figure 3.1. Karyotype of a desmoid tumor of the retroperitoneum from a 42-year-old male. The patient had a history of previous desmoid tumor resection from the anterior chest wall 5 years earlier and of osteosarcoma of the left femur 3 years earlier. The tumor was placed in culture, and chromosome harvests were obtained from the second subculture after 14 days. The pattern shown, typical of the tumor stemline, is 50,XY,+5,del(6)(q13q16.1),+8,inv(13)(p11.2q31),+17,+21,t(7;14)(p22;q31). Neither the t(7;14) nor the inv(13) were known to be constitutional rearrangements. (Overlaps partially obscure the banding pattern of several chromosomes.)

locally, the distinction from fibrosarcoma may be difficult histologically, the prognosis is uncertain, and cytogenetic aberrations can be influential in reclassification of the lesion as malignant.

When the tumor is uncommon and a chromosome aberration is found routinely or as the sole abnormality in otherwise diploid cells, the diagnostic association is powerful. When cultured tumor cells from the marrow proved to have the typical translocation, t(2;13), of alveolar rhabdomyosarcoma the diagnosis was considered to be established without examination or extirpation of the primary tumor.[23] Further, cytogenetic assay may help to estimate the extent of tumor progression. The evolution from a chronic to a blastic phase of myeloid leukemia is routinely accompanied by acquisition of additional chromosome aberrations superimposed on the t(9;22). Among melanotic lesions, congenital nevi are diploid, as are most dysplastic nevi, whereas melanomas uniformly show cytogenetic abnormality.[14]

Direct Analysis versus Culture Systems. Both the advantages and disadvantages of direct analyses and of short or prolonged periods of culture prior to cytogenetic analysis should be considered. Up to the present, however, only the hematologic tumors have been studied in substantial numbers by direct harvest, and regardless of the disadvantages, the bulk of our cytogenetic data on human solid tumors has been derived from culture studies. With the identification and characterization of new growth factors, our ability to culture and study specialized tumor cell types should expand.

Different problems of interpretation arise when fresh tumors are analyzed as compared with tumors after brief or prolonged culture. Such problems are greatly reduced with material from blood or bone marrow because fairly direct morphologic correlations can be made and putative dividing populations are generally identifiable. Moreover, in short-term cultures, known hemopoietic or lymphopoietic mitogens can be added to assist in stimulating the cell type of particular interest. In contrast, considerable karyotypic diversity often characterizes direct material from solid tumors. Since stromal and inflammatory cells are often present in the tumor, the presence of diploid mitoses is usually interpreted as representing those normal elements rather than the tumor cell population. Evidence that some of the diploid cells do indeed represent tumor cells is discussed below. Some of the chromosomally aberrant cells may be either irrelevant or nonrepresentative of the tumor, either because the genetic alteration is lethal to future cell divisions or because it interferes with mitotic separation and is, therefore, overrepresented in a metaphase population.

In addition, the number of spontaneously dividing cells in fresh solid tumors varies greatly. Limitations in numbers of mitotic cells and poor quality of the metaphases available for study are among the reasons why direct analysis of many tumors has thus far yielded unsatisfactory data. Interpretation is further diluted by the often extensive heterogeneity of genetic pattern within an individual tumor and the great heterogeneity among tumors of the same morphologic classification.

Culture systems also have inherent disadvantages. Selection in culture can result from tumor-processing techniques, from pretreatment of cells prior to culture, and from the methods employed in initial culturing and for subculturing. Further selection is effected by ambient culture conditions, including those that are preferential and intended to enhance tumor cell concentrations. Perhaps the greatest problem is contamination by overgrowth of fibroblasts derived from the normal supporting stromal tissues. In addition to selection biases, there is often evidence of spontaneous evolution in culture of cytogenetically marked subpopulations that may have been absent, or present only in limited numbers, in the original tumor cell population. Thus, the greater the time in culture prior to chromosome analysis, the less likely are the cells to represent those of the original tumor. The various factors that may affect selection for growth in culture have been reviewed recently.[81]

A substantial limitation on interpretation of many cytogenetic analyses is the paucity of available markers of tissue or tumor specificity which can be applied to cells in metaphase and which are not altered or lost in cells growing in culture. The method employed by Knuutila and co-workers is unique in that it permits some phenotyping of the cell in which metaphase chromosomes are identified.[72] This method, by labeling mitotic cells with monoclonal antibodies against B or T cell antigens, was employed to investigate the karyotype of Reed-Sternberg cells in Hodgkins lesions.[71] The method is not, however, optimal for karyotypic preparations, and even though it permits the identification of an abnormality, the quality of the metaphases obtained limits analysis and interpretation of that abnormality.

Absence of Specificity.   Despite the examples described above and a rapidly expanding literature, ample evidence exists that not all tumors of a given histotype show similar cytogenetic aberrations, that within individual tumors the findings may not be internally consistent, and moreover, that some tumor cells appear to show no aberrations.

Difficulties in histologic diagnosis and the paucity of differentiation markers have hampered accurate distinction among the soft-tissue sarcomas. The unusual translocations often present in pseudodiploid tumors suggest an important source of new diagnostic discriminants. However, although many cases of association between t(X;18) and synovial sarcoma have been reported, the same aberration has been seen in other types of tumors and is occasionally absent from synovial sarcoma.[8,39] Does this mean that a tumor lacking that aberration should be reclassified?[46] Or does it mean that a subvisible deletion or even a point mutation is present that alters the same function? Or does it indicate phylogenetic relationships among cells of origin of the tumors so marked?

The importance of chromosome abnormalities for malignant tumor diagnosis is vitiated by increasingly frequent reports of aberrations, sometimes specific, in benign proliferations and premalignant tissues or conditions (Table 3.2). Many benign but cytogenetically aberrant tumors are presumed to be at increased risk of evolution to frank malignancy. The pleomorphic adenomas of salivary glands often show chromosome rearrangement, but these tumors recur at high frequency and may eventually emerge as malignant. Similarly, among the lipomas, morphologic variants with differing probabilities of recurrence or sarcomatous degeneration may correlate with specific karyotypes. However, the suspicion that 12q rearrangements in lipomas could evolve to the t(12;16) observed in liposarcomas is at variance with the infrequent evolution of lipomas to frank malignancy. Uterine leiomyomata, which rarely evolve to malignant tumors, have demonstrated a remarkably high frequency (and specificity) of cytogenetic aberration.[40] More than half show abnormal stemlines, and translocations involving 12q13–15 are common. As in the case of malignant tumors, and despite the apparent specificity of chromosome 12 involvement, the frequency with which any clonal aberration is reported has varied considerably in different laboratories. Some tumors show preferential involvement

Table 3.2.
Chromosome Aberrations in Benign Proliferations

Tumor-specific aberrations	
Meningioma	−22
Lipoma	12q13-14
Uterine leiomyomas	rearr 12q
Parotid adenoma	rearr 8q12
Colonic adenoma	+8, 12q-
Occasional reports	
Nasal papilloma	46,XY,t(1;3)/46,XY,t(11;?)
Follicular thyroid adenoma	46,XY,t(10;19)
Angiomyolipoma	+7, −13, del(6),del(21)
Dysplastic nevi	−9(9p-)
Carcinoid of lung	+7

of chromosome 14 rather than chromosome 12,[48] and in some cases deletion 7 has also been observed as the sole clonal change.[7]

Angioleiomyomata, often considered as hamartomas rather than true neoplasms, have also been reported to show clonal aberrations in culture.[19,30,50] Even normal tissues adjacent to tumor have in a few instances been reported to show clonal aberrations in culture. Trisomy 7, which appears relatively commonly in solid tumors,[73] was also noted in lung and kidney tissues uninvolved by tumor.[36,37] Other clonal aberrations have been found in fibroblast cultures from individuals without any evidence of tumor.[4] And finally, when tissues of nontumor origin adapt to long-term growth in culture, they are usually characterized by chromosome aberrations[2] (Fig. 3.2). These observations lead to substantial concern about the interpretation of similar findings in tumors, about selection for growth *in vitro* (since trisomy 7 is common, and chromosome 7 is the locus for epidermal growth factor receptor), and about whether the observed genetic instability relates to the cancer potential of the normal tissue.

Examples of diploidy in tumors are variously interpreted. In breast cancer there

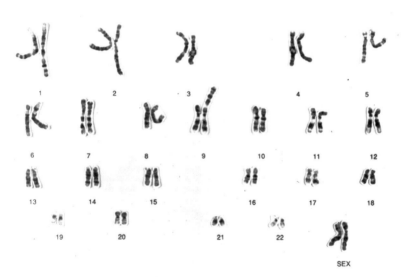

Figure 3.2.   Karyotype of 46,XX,3p−,6p+,9p+ from a line of breast epithelial cells originated from surgically excised fibrocystic disease of the breast. The cells were normal diploid for well over a year in culture. With further time in culture, there was gradual acquisition of several cytogenetic changes, although the cells remained pseudodiploid. The pattern shown was obtained from cells harvested after 1229 days in culture. It had evolved as follows: By the 846th day, a balanced translocation t(3;9)(p13p22) was present. The der(3) has persisted but the der(9) was further modified before the 1009th day. The additional changes were interpreted as duplication and addition of 9q(q21−q33) and 5q(q31−qter) to the original 9p22 breakpoint. The other aberration, the 6p+, was first observed after 1009 days in culture and was interpreted as addition and translocation of a segment representing 19q(q12−qter). Overlapping chromosomes distort the banding patterns of the 3p− marker (q arm) and of 4A (q arm).

are so few spontaneously dividing cells in primary tumors that, until recently, fewer than 20 tumors had been analyzed directly with banding methods.[80] One of the largest and best-analyzed series dismissed diploid cells as contaminants.[58] A more recent report, based on analysis of 12 tumors (out of 50 attempts), found diploid metaphases in 3 cases.[25] Moreover, in one patient from whom both primary and metastatic tumor cells were evaluated, the abnormal stemlines from the two lesions showed nonoverlapping aberrations, suggesting that the original tumor stemline may have been diploid. Short-term cultures of breast cancers in different laboratories have been predominantly normal diploid, even though the cultured cells showed mammary epithelial markers and invasive, tumor-like behavior.[68,82,84] Similar results may emerge from studies of prostate cancers.[10]

Cytogenetic analysis of six leiomyosarcomas analyzed after short-term culture revealed four that were exclusively diploid, one tumor with a pseudodiploid clonal aberration, and a tumor that was nonclonal (and also had diploid cells); flow cytometry indicated that the DNA content in all cases was diploid.[49] A study of pediatric osteosarcomas reported marked structural and numerical aberrations in six of eight tumors; however, all of the cultured tumors had some normal diploid metaphases, and the two cases that were exclusively diploid appeared to have relatively benign clinical courses.[5] Similarly, a study of Wilms' tumors[69] revealed that many were entirely composed of diploid cells or showed nonclonal aberrations yet had cytologic markers indicative of tumor origin. More significantly, several cultured Wilms' tumors have demonstrated the typical 11p13 molecular lesion.[16] At least one human cell line originating from a mixed mesodermal malignant tumor was diploid by karyotype but highly tumorigenic in nude mice,[13] providing another line of evidence that some tumor cells are cytogenetically diploid.

Cytogenetic specificity has contributed to the perception that tumors are clonal proliferations and that the cytogenetic alteration marking the tumor also confers the biologic modification that alters behavior and assures continuance of the tumor. This assumption is challenged by evidence of polyclonality in tumors. A recent review has emphasized that cytogenetic polyclonality is relatively frequent in some forms of myeloid and lymphoid tumors (5–7%),[32] and it appears far more commonly in some solid tumors.[34,55] A typical illustration of polyclonality in tumors was the study of an in situ squamous cell carcinoma of the skin, in which a number of cytogenetically unrelated clones were identified.[29] Polyclonality has also been reported in several other benign and malignant neoplasms, particularly those of squamous origin.[31] Similar observations in highly malignant glial tumors were reported much earlier.[65]

Many confusing aspects of these problems are well illustrated by data from a single patient. This individual had a benign chondromyxoid fibroma with normal diploid cells and two abnormal clones, one marked by a t(2;13) and the other by trisomy 5. A second tumor, a malignant extraskeletal myxoid chondrosarcoma, showed cytogenetically normal cells and an abnormal clone marked by a t(2;13) with breakpoints totally unrelated to those observed in the benign lesion.[9]

In summary, the much-vaunted specificity of certain chromosome aberrations is considerably less apparent when one examines a sufficient number of cases. It makes the meaningful associations that have led to tremendous advances in molecular work even more impressive. For example, the chromosome 13 deletion that is characteristic of retinoblastoma and that led eventually to the isolation and characterization of the retinoblastoma gene was actually observed in only approximately 20% of the tumors that have been investigated.[53] Similarly, the known deletion of 11p13 in Wilms' tumor was observed in only 2 examples of a recent series of 20 tumors, all of which showed other chromosome abnormalities.[69]

The potential for definition of chromosome-specific tumor associations is further diluted by the very small proportion of tumors in which any detailed karyotyping can be done. For example, a recent investigation of lung cancer permitted detailed examination of only five tumors from among the 90 tumors entered into the study. All five tumors were aneuploid, with increased chromosome modal numbers, and each had from 4 to 20 marker chromosomes.[3] Thus, many tumors are not amenable to the definition of cytogenetic specificity by virtue of the extent and complexity of the cytogenetic changes.

The central role of chromosome aberration in tumorigenesis is challenged by the existence of diploid tumor cells, the absence of clonality in some tumors, and the chromosome aberrations that appear common to many tumors. Recent observations linking specific chromosome abnormalities with benign tumors and even with nontumor tissues indicate that we should pursue other more powerful means to define and clarify the critical genetic events in tumor formation.

Molecular Methods and Applications

The advent of molecular analysis has greatly expanded and modified interpretations of the roles of chromosome aberrations in tumors. The finding and molecular analysis of a consistent chromosomal breakpoint led ultimately to the first identification of a biologically altered protein in tumor cells that was thought to be responsible for aspects of unregulated growth. Analyses of interphase cells from fresh tumors and of archival tissues are now possible. Some techniques permit the recognition of very small numbers of tumor cells within much larger populations. Molecular approaches have increased the power and extended the range of cytogenetic diagnosis in ways not imagined a decade ago. (See Chapter 1).

Techniques.   High molecular weight DNA can be isolated from fresh or fixed cells and then digested into small fragments by bacterial enzymes that recognize and cleave specific nucleotide sequences. Some enzymes cleave at frequent intervals, yielding fragments detectable in a range from 500 to 28,000 base pairs in length; others that result in longer nucleotide chains must be separated somewhat differently. Small digested fragments can be separated by size in an electrophoretic field in an agarose gel. The DNA is then transferred to a membrane or filter so that fragments of differing size are immobilized. The membrane is then reacted with a suitable probe or marker to identify the region or gene of interest. The reaction with a probe is called hybridization, because it is based on the preferential binding of like or highly similar linear nucleotide sequences in DNA or RNA. This method of analysis is known as Southern blotting if DNA is localized on the filter and Northern blotting if RNA is used.

Probes for DNA may be generated by radioactive or nonradioactive labeling. The radioactive probes usually involve using tritium, ^{32}P, or ^{35}S to label bases that are incorporated into DNA. Nonradioactive probes may depend upon chemical modification of nucleic acids; for example, with bromodeoxyuridine or enzymic incorporation of biotinylated bases or, alternatively, by conjugation with a fluorochrome. Yet another nonradioactive probe can be developed by labeling DNA fragments with dinitrophenyl residues and detecting hybridization using anti-DNP antibodies.[47] In any event, the labeled nucleotide chain is hybridized to single-stranded DNA, and the bound probe is detected by its label. If a gene is isolated and sufficiently reactive copies can be generated, it can serve as a probe to locate the native gene within metaphase chromosome preparations, a method known as hybridization *in situ*.

In combination with the digestion of DNA by bacterial enzymes, a probe can be used to detect known gene loci or cloned DNA fragments that have been localized to chromosomal regions, whether or not the gene function is understood. When whole-cell DNA is digested and analyzed by Southern blotting, the probe may be used to define different alleles at the same genetic locus. The differing alleles when cut by some enzymes will generate DNA fragments of differing sizes that hybridize to the same probe—*i.e.*, have a high degree of internal sequence in common (homology)—and are known as restriction fragment length polymorphisms (RFLPs). In this way, individual differences in the primary DNA sequence of normal genes can serve as markers when they alter the cleavage pattern. Using the same technology, gene deletions (losses) or amplifications (increases in copy number) in tumors may be detected by comparison with the degree of hybridization to the same gene in normal tissues from the same individual.

Cloning is the mechanism by which one attempts to isolate a unique DNA fragment that contains a functional gene. Many strategies have been devised for this purpose, but ultimately cloning is intended as a tool for gene isolation as a preface to determination of its function. Chromosome sorting by size, using flow cytometry, can be the first step by providing material enriched for a single chromosome; a library of probes from that chromosome can then be isolated and reproduced. When a tumor-specific chromosome breakpoint has been recognized, probes that localize near the breakpoint can be identified, and overlapping pieces of DNA can be used to map the gene that is actually interrupted and therefore dysfunctional. The gene alteration could result in loss of product, increased or decreased amount of product, or a related but different product.

A different type of DNA probe has been used to detect centromeric DNA in tumors and tissues. Antibodies developed against centromeric proteins can be used to demonstrate that micronuclei contain centromere-reacting material and are likely to have resulted from whole chromosome aneuploidy. If micronuclei lack centromere-positive material, they are interpreted as having resulted from clastogenic events.[18]

The new technique of polymerase chain reaction (PCR) depends upon small probes to generate hybrids that can be copied rapidly.[60] Both strands of the hybrid serve as templates for the reaction. Amplification within DNA and the specificity of the reaction are only limited by the oligonucleotide chosen. The replication of small nucleotide chains is speeded by repeat cycles of heat denaturation that produce more than a million copies of the desired sequence within a few hours.[52] The excitement generated by PCR is clearly justified; with an appropriate and well-labeled probe the target sequence is detectable in one cell in a population of ten thousand.

A probe can represent a chromosome aberration; for example, the *bcl-2* gene rearrangement characteristic of some B cell lymphomas originally was cloned from the chromosome breakpoint of lymphoma cells carrying a t(14;18). With this probe, the PCR reaction can be used to demonstrate whether gene rearrangement is present in the original tumor, and can be used to screen the patient for the effectiveness of therapy in terms of tumor regrowth. In fact, cells carrying the t(14;18) were detected by PCR at a dilution of 1:100,000 in a patient with follicular lymphoma in remission, when morphology and Southern blot studies failed to disclose the presence of residual tumor.[38] As additional tumor-specific gene rearrangements are defined and suitable probes are developed, the PCR method can use them for diagnosis, tumor staging, monitoring of treatment, and long-term patient follow-up.

CONSIDERATIONS FOR THE PATHOLOGIST

A major advantage of molecular techniques is that they, like DNA content determinations, are not dependent on the availability of cells in mitosis. Thus, molecular studies are critical to the analysis of genetic abnormality in certain tumors such as breast where long tumor-doubling times limit mitotic harvest from fresh tissues. However, they should only be applied to fresh solid-tumor tissues in concert with pathologic assessment. In assaying any tumor, it is essential to be aware of morphologic heterogeneity within the tumor, but also to note areas of necrosis or hemorrhage within the mass. Even more important, with techniques that depend on DNA extraction, is the evaluation of the dilution of tumor by normal cells of the host. Inflammatory cells and supporting stroma are present in varying amounts in all tumors, and the numbers of such cells are often grossly underestimated.[41] The amount of stromal material (*e.g.*, collagen) is not as important as is the number of nontumor nuclei, such as those of lymphocytes.

Another major advance in the ability to define and investigate genetic lesions in tumors was the demonstration that DNA could be extracted from formalin-fixed, paraffin-embedded tissues.[26] The quality of the extracted DNA varies with many factors including the viability of the tissue prior to surgery, the time to fixation, characteristics of the fixative, and possibly the heat of tissue embedding. The total time after embedding also appears to have some effect on the quality of extractable DNA. In good material the DNA is double-stranded, cleavable with restriction enzymes, and clonable after cleavage, and also can provide identifiable substrates for PCR, *in situ* hybridization, and other techniques. There is usually greater variability in fragment size than would be observed from comparable fresh tissues. The application of genetic analyses to paraffin-embedded tumors is now feasible even for rare tumors that are not available in sufficient numbers from any single institution. Of equal importance, retrospective tumor studies can be combined with long-term patient follow-up for evaluation of clinical correlations.

MOLECULAR MODIFICATIONS OF CYTOGENETIC RESULTS

The "specific associations" almost invariably are absent from some tumors where they are expected, and occasionally are present in tumors of an unexpected histotype. Approximately 90% or more cases of CML show the classic t(9;22) or variant translocations involving other chromosomes in addition to 9 and 22. Cloning of the breakpoint region on chromosome 22, known as the *bcr*, has provided a number of probes and, more importantly, has led to identification of the altered product and some understanding of how that product alters behavior in the leukemic cells. Even Philadelphia chromosome (Ph)-negative cases often show *bcr* rearrangement, confirming the diagnosis of CML, and undermining earlier beliefs that the Ph-negative cases differed in clinical presentation and prognosis from those that were Ph-positive.[6,79] (Refer to Chapter 6.)

The t(9;22) generates a unique hybrid transcript of mRNA containing information from both *bcr* and the *abl* oncogene that was translocated from chromosome 9. The encoded protein has enhanced tyrosine kinase activity in comparison to that of the nontranslocated *abl* kinase. Translocations involving three or more chromosomes, found in about 5% of CML cases, display the same *bcr-abl* rearrangement. Not only the existence but the subdivision of breaks in the *bcr* region may have prognostic significance in CML patients. The issue of whether the chronic phase of disease is prolonged when the breakpoint is in the early 5' portion of the *bcr* region is the subject of considerable debate and investigation.[22,43,54,67] (See Chapter 6).

The Ph chromosome seen in acute lymphocytic leukemia (ALL), although cytogenetically indistinguishable, usually demonstrates a breakpoint on chromosome 22 that differs by a small genetic distance from that seen in CML. When the rearrangement in ALL involves the *bcr* region, some authors have suggested those cases may, in fact, be variants of CML arising in a multipurpose stem cell that differentiated in a lymphoblastic direction, whereas the *bcr*-negative cases arose in lymphoid-committed progenitor cells.[62]

The example that historically preceded that of CML in demonstrating the importance of breakpoint analysis in the identification and localization of genes important in cancer was that of Burkitt's lymphoma, where the typical t(8;14) results in juxtaposition of the c-*myc* oncogene next to the actively transcribing immunoglobulin region. Other B cell tumors with consistent translocation sites are the subject of active investigation to identify new oncogenes.[66] Support for the role of breakpoints is also derived from gene localization studies; for example, the native site of a growth factor is 7q11, a region frequently altered in certain tumors.[70] It should be noted that molecular markers of lymphoid rearrangement that occur in normal lymphoid maturation are also valuable in tumor diagnosis, but in those instances they serve as markers of clonal expansion of a cell population rather than as markers of abnormal gene function.

Analysis of a matrix of cell markers such as chromosome aberrations, oncogene amplification, and specific enzyme products can clarify morphogenetic relationships and the assessment of tumor-cell origins. A t(11;22), first observed as a constitutional chromosome aberration, was later identified in several cases of Ewing's sarcoma,[1,76] and the identical translocation was observed in neuroectodermal tumors.[78] Although the tumors are histologically distinct, similarities of chromosome aberration and oncogene expression suggested a common neural origin.[42] The converse is also demonstrable; histologically similar tumors can be distinguished by differing patterns of oncogene expression, karyotype, and neurotransmitter enzyme activities.[75] Thus, genetic markers can contribute to better definition of etiology and development among similar and closely related lesions.

Molecular studies may reinforce and extend the specificity of tumor-chromosome associations. Meningioma, the first solid tumor for which a specific aneuploidy was demonstrated (approximately 50% of tumors show loss of chromosome 22), can also be investigated by molecular means. In one study, loss of heterozygosity (LOH) for loci on chromosome 22 was found in 43% of informative patients, showing good correlation with cytogenetic results.[64] Approximately 21% of retinoblastoma tumors show a chromosomal deletion in the region including 13q14,[53] but when molecular probes are used, the frequency of allelic loss (LOH) is increased to over 70% by a variety of mechanisms.[12] This association has been so strong that only recently was the first case reported in which the 13q14 deletion in two generations did not result in either retinal abnormality or tumor formation.[15] Wilms' tumors, even when unmarked by any detectable cytogenetic aberration, show loss of heterozygosity for the 11p13 locus reported to be the site of chromosome-specific damage.[16]

The demonstration of LOH has been extended to many other tumor types,[83] but certain losses, notably 3p, 11p, 13q, and 17p, are common to several tumors, and it appears possible that LOH may be widely dispersed and apparently nonspecific.[21] Cytogenetic evaluation of carcinomas of the colon has been remarkable for nonconformity of results from different laboratories. Suggested specificities have included structural abnormalities of chromosomes 7 and 12, trisomies of chromosomes 7 and 8, duplication of 1q, and losses of 18 and 17p. Even though widespread losses may be "random," genomic studies in colon tumors indicate certain critical loci are lost at frequencies far above background and suggest that the involvement of these loci need not occur in a predetermined sequence.[77]

In the face of this complexity, definition of potential or probable sites of cytogenetic loss, breakage, recombination, or other involvement is essential as the first step, prior to genomic approaches. It is only after localization to one or a limited number of sites within the enormous, complex, and as yet ill-defined human genome that one may begin to utilize the powerful and precise tools of molecular analysis. The function of cytogenetics is to provide the signposts or flags that will lead ultimately to definition and understanding of the genes responsible for the patterns of cellular behavior we recognize as cancer.

Molecular studies in cancer genetics have altered our understanding of tumor biology and classification and have expanded our tumor diagnostic capabilities. Demonstration by molecular means that cancers lacking a "specific" chromosome aberration have submicroscopic lesions at the same locus has tended to unify results and explain anomalous cases. Differences at the molecular level have clarified the contradiction inherent in apparently similar chromosomal markers for morphologically and clinically dissimilar disorders. Molecular tools permit the exploration of sequence in cancer development and the construction of tissue lineage relationships. Ultimately, as with "the Philadelphia story," precise dissection of genomic lesions in cancer will lead to definition of their structural and functional consequences.

New Directions

The dissection of individual lesions is not sufficient; cancer develops in many steps, and interactions of the component cells with each other and with the host determine the outcome. We need tools to connect data on individual genes with the entire genomic map. Methods that currently bridge the gap between molecular and cytogenetic studies include a combination of (*a*) bacterial enzymes that cut DNA infrequently and (*b*) gel systems devised for physical separation of the resultant larger pieces of DNA.[61] The next step should be aimed at a larger picture—definition and expression of individual genetic alterations in the contexts of whole nuclear architecture and whole cell and tissue organization.

The tools of chromosome labeling are the same as those widely used for other immunohistochemical procedures; *i.e.*, immunofluorescence, enzyme-labeled antibodies, etc. Fluorescent-labeled antibodies to nucleolar, centromeric, and mitotic spindle components are already marketed commercially. The problems relate to the relatively poor immunogenicity of chromosomal components and some uncertainty about the nature, localization, and relative concentrations of chromosomal proteins. A valuable set of alternative methods depends on the development of probes that hybridize to regions of homology within DNA strands. Two important categories of such probes have already been developed, those that identify chromosome-specific repetitive sequences and those that are composites or collections of unique sequences all of which identify regions from the same chromosome. Repetitive DNA sequences have been used to probe interphase nuclei by *in situ* hybridization applied to cell lines and fresh tissues. The targets of these probes in human chromosomes are the centromeric regions of chromosomes, which are rich in repetitive DNA. The technical approaches that permitted construction of the probes involved mass culture and collection of mitotic human cells, differential centrifugation and flow sorting of the chromosomes, and separation of the satellite DNA from individual-chromosome enriched peaks.

In fixed and stained cell preparations, the number of target sites per nucleus that react with the probe reflects the copy number of the particular chromosome. This method, aptly described as "interphase cytogenetics,"[20] is a powerful adjunct to current techniques, permitting assessment of aneuploidy in poorly mitotic tumors and of cytogenetic heterogeneity within tumor cell populations. Probes for the centromeric regions of chromosome 18 and the X were applied to the study of breast cancer. It was first demonstrated that the number of spots corresponded to the number of X chromosomes per nucleus in cells from individuals with differing chromosome constitutions. The spots per nucleus (S/N) show a modal peak that corresponds to the expected, but with fewer S/N in one-third or one-fourth of the cells and a small fraction of the population showing a larger than expected S/N. Artificial, defined admixtures of aneuploid cells can be detected at levels of 25% or higher. The probe also reacted with structurally altered chromosomes in tumor cells and potentially could be used to determine the component chromosomes of origin of recombinant markers. At present, the method is still limited by interobserver variability.

Another approach to definition of cellular chromosomal aberration that can be applied to cells in interphase has been developed recently.[52] When a substantial number of genes have been mapped to a particular chromosome, then all the known genes for that chromosome can be considered as probes. If they are combined, together with a suitable marker such as a linked fluorescent dye, they may be used to "paint" the particular chromosome. Whereas no single gene can bind in sufficient quantity to be visualized, the combined attachment of 100 or 200 probes to a single chromosome permits visual identification. When many chromosome-specific probes were linked to fluorescent dyes and a "cocktail" of probes was applied to the examination of cells, the first and most surprising observation was that the compact spots of dye indicated that some chromosomes are actually quite condensed even during interphase. The number of spots corresponds to the number of chromosomes that have regions complementary to the probe.

These techniques can be used to demonstrate trisomy (three spots instead of two) and translocation (one large and two smaller spots within a nucleus), and have potential for demonstrating chromosomal heterogeneity within tumors.[33] Several probes linked to different fluorochromes can be used simultaneously to investigate more than a single chromosome in the same cell; for example, probes for chromosomes 9 and 22 seen adjacent could detect the t(9;22). Other projected uses include sex-chromosome probes for rapid evaluation of recurrence in patients who have received crossed-sex bone marrow transplants. In future the selection of regional probes within chromosomes could increase the resolving power for detection of translocations. Difficulties in the determination of histologic type, grade, and extent of heterogeneity within the tumor all contribute to problems in defining patient prognosis. The potential of these methods for detecting aneuploidy without the necessity for cell culture and for demonstration of heterogeneity within tumors, may have widespread application to the development of individualized tumor therapy.

Automated image analysis of cells in tissue sections is an alternative route to the measurement of nuclear DNA content. The analysis may be based on measuring the nuclear diameter or area, with a suitable model for projection of spherical dimension. The semi-quantitative reaction of DNA with Feulgen stain permits estimation of DNA content and demonstration of cells with increased ploidy.[56] Moreover, this method can be applied to the evaluation and demonstration of abnormalities in precancerous lesions.[57]

Finally, although the genetic bases for some tumors are clearly established, major questions remain unresolved. We do not know in most common tumors whether chromosome changes are necessary or sufficient for tumor initiation or progression. Relationships between benign and malignant proliferations of the same tissues are uncertain. We do not understand the cause(s) of genetic heterogeneity in tumors. What has been gained from the studies described here is widespread acceptance that both the type and the extent of genetic aberration are informative of tumor diagnosis and progression and offer tantalizing clues to etiology.

ACKNOWLEDGMENTS

My thanks to Mary-Ann Lane for helpful discussions and to Kathy Dauphinais for library research.

REFERENCES

1. Aurias, A., Rimbaut, C., Buffe, D., Zucker, J., and Mazabraud, A. Translocation involving chromosome 22 in Ewing's sarcoma: A cytogenetic study of four fresh tumors. *Cancer Genet. Cytogenet.* 12:21–25, 1984.

2. Baden, H. P., Kubilus, J., Kvedar, J. C., Steinberg, M. L., and Wolman, S. R. Isolation and characterization of a spontaneously arising long-lived line of human keratinocytes (NMI). *In Vitro* 23:205–213, 1987.

3. Benfield, J. R., Wain, J. C., Derrick, M., *et al.* Biochemical and cytogenetic studies of human lung cancers. *J. Thorac. Cardiovasc. Surg.* 96:840–848, 1988.

4. Benn, P. A. Specific chromosome aberrations in senescent fibroblast cell lines derived from human embryos. *Am. J. Human Genet.* 28:465–473, 1976.

5. Biegel, J. A., Womer, R. B., and Emanuel, B. S. Complex karyotypes in a series of pediatric osteosarcomas. *Cancer Genet. Cytogenet.* 38:89–100, 1989.

6. Blennerhassett, G. T., Furth, M. E., Anderson, A., *et al.* Clinical evaluation of a DNA probe assay for the Philadelphia (PH1) translocation in chronic myelogenous leukemia. *Leukemia* 2:648–657, 1988.

7. Boghosian, L., Dal Cin, P., and Sandberg, A. A. An interstitial deletion of chromosome 7 may characterize subgroups of uterine leiomyoma. *Cancer Genet. Cytogenet.* 34:207–208, 1988.

8. Bridge, J. A., Bridge, R. S., Borek, D. A., Shaffer, B., and Norris, C. W. Translocation t(X;18) in orofacial synovial sarcoma. *Cancer* 62:935–937, 1988.

9. Bridge, J. A., Sanger, W. G., and Neff, J. R. Translocation involving chromosomes 2 and 13 in benign and malignant cartilaginous neoplasms. *Cancer Genet. Cytogenet.* 38:83–88, 1989.

10. Brothman, A. R., Lesho, L. J., Somers, K. D., Schellhammer, P. F., Ladaga, L. E., and Merchant, D. J. Cytogenetic analysis of four primary prostatic cultures. *Cancer Genet. Cytogenet.* 37:241–248, 1989.

11. Cabanillas, F. A review and interpretation of cytogenetic abnormalities identified in Hodgkin's disease. *Hematol. Oncol.* 6:271–274, 1988.

12. Cavanee, W. K., Dryja, T. P., Phillips, R. A., *et al.* Expression of recessive alleles by chromosomal mechanisms in retinoblastoma. *Nature (Lond)* 305:779–784, 1983.

13. Chen, R. T. SK-UT-1B, a human tumorigenic diploid cell line. *Cancer Genet. Cytogenet.* 33:77–81, 1988.

14. Cowan, J. M., Halaban, R., and Francke, U. Cytogenetic analysis of melanocytes from premalignant nevi and melanomas. *J. Natl. Cancer Inst.* 80:1159–1164, 1988.

15. Cowell, J. K., Rutland, P., Hungerford, J., and Jay, M. Deletion of chromosome region 13q14 is transmissible and does not always predispose to retinoblastoma. *Hum. Genet.* 80:43–45, 1988.

16. Dao, D. D., Schroeder, W. T., Chao, L., *et al.* Genetic mechanisms of tumor-specific loss of 11p DNA sequences in Wilms tumor. *Am. J. Hum. Genet.* 41:202–217, 1987.

17. da Silva, M. A. P., Heerema, N., Schwenk, G. R., and Hoffman, R. Evidence for the clonal nature of hypereosinophilic syndrome. *Cancer Genet. Cytogenet.* 32:109–115, 1988.

18. Degrassi, F., and Tanzarella, C. Immunofluorescent staining of kinetochores in micronuclei: A new assay for the detection of aneuploidy. *Mutat. Res.* 203:339–345, 1988.

19. de Jong, B., Castedo, S. M., Oosterhuis, J. W., and Dam, A. Trisomy 7 in a case of angiomyolipoma. *Cancer Genet. Cytogenet.* 34:219–222, 1988.

20. Devilee, P., Thierry, R. F., Kievits, T., *et al.* Detection of chromosome aneuploidy in interphase nuclei from human primary breast tumors using chromosome-specific repetitive DNA probes. *Cancer Res.* 48:5825–5830, 1988.

21. Dracopoli, N. C., Houghton, A. N., and Old, L. V. Loss of polymorphism restriction fragments in malignant melanoma. Implications for tumor heterogeneity. *Proc. Natl. Acad. Sci. U.S.A.* 82:1470–1474, 1985.

22. Eisenberg, A., Silver, R., Soper, L., *et al.* The localization of breakpoints within the breakpoint cluster region (BCR) of chromosome 22 in chronic myeloid leukemia. *Leukemia* 2:642–647, 1988.

23. Engel, R., Ritterbach, J., Schwabe, D., and Lampert, F. Chromosome translocation (2;13)

(Q37;Q14) in a disseminated alveolar rhabdomyosarcoma. *Eur. J. Pediatr.* 148:69–71, 1988.

24. Gahrton, G., and Juliusson, G. Clinical implications of chromosomal aberrations in chronic B-lymphocytic leukaemia cells. *Nouv. Rev. Fr. Hematol.* 30:389–392, 1988.

25. Gerbault-Seureau, M., Vielh, P., Zafrani, B., Salmon, R., and Dutrillaux, B. Cytogenetic study of twelve human near-diploid breast cancers with chromosomal changes. *Ann. Genet.* 30:138–145, 1987.

26. Goelz, S. E., Hamilton, S. R., and Vogelstein, B. Purification of DNA from formaldehyde fixed and paraffin embedded human tissue. *Biochem. Biophys. Res. Commun.* 130:118–126, 1985.

27. Han, T., Sadamori, N., Block, A. M., *et al.* Cytogenetic studies in chronic lymphocytic leukemia, prolymphocytic leukemia and hairy cell leukemia: A progress report. *Nouv. Rev. Fr. Hematol.* 30:393–395, 1988.

28. Hecht, F. Solid tumor breakpoint update and hematologic malignancy breakpoint update. *Cancer Genet. Cytogenet.* 37:129–141, 1989.

29. Heim, S., Jin, Y., Mandahl, N., *et al.* Multiple unrelated clonal chromosome abnormalities in an *in situ* squamous cell carcinoma of the skin. *Cancer Genet. Cytogenet.* 36:149–153, 1988.

30. Heim, S., Mandahl, N., Kristoffersson, U., *et al.* Structural chromosome aberrations in a case of angioleiomyoma. *Cancer Genet. Cytogenet.* 20:325–330, 1986.

31. Heim, S., Mertens, F., Jin, Y., *et al.* Diverse chromosome abnormalities in squamous cell carcinomas of the skin. *Cancer Genet. Cytogenet.* 39:69–76, 1989.

32. Heim, S., and Mitelman, F. Cytogenetically unrelated clones in hematological neoplasms. *Leukemia* 3:6–8, 1989.

33. Hopman, A. H., Ramaekers, F. C., Raap, A. K., *et al. In situ* hybridization as a tool to study numerical chromosome aberrations in solid bladder tumors. *Histochemistry* 89:307–316, 1988.

34. Jin, Y., Heim, S., Mandahl, N., Biorklund, A., Willen, R. W., and Mitelman, F. Two unrelated clonal chromosome rearrangements in a nasal papilloma. *Cancer Genet. Cytogenet.* 39:29–34, 1989.

35. Karlsson, I., Mandahl, N., Heim, S., Rydholm, A., Willen, H., and Mitelman, F. Complex chromosome rearrangements in an extraabdominal desmoid tumor. *Cancer Genet. Cytogenet.* 34:241–245, 1988.

36. Kovacs, G., and Brusa, P. Clonal chromosome aberrations in normal kidney tissue from patients with renal cell carcinoma. *Cancer Genet. Cytogenet.* 37:289–290, 1989.

37. Lee, J. S., Pathak, S., Hopwood, V., *et al.* Involvement of chromosome 7 in primary lung cancer and nonmalignant normal lung tissue. *Cancer Res.* 47:6349–6352, 1987.

38. Lee, M. S., Chang, K. S., Cabanillas, F., Freireich, E. J., Trujillo, J. M., and Stass, S. A. Detection of minimal residual cells carrying the t(14;18) by DNA sequence amplification. *Science (Wash DC)* 237:175–178, 1987.

39. Limon, J., Mrozek, K., Nedoszytko, B., *et al.* Cytogenetic findings in two synovial sarcomas. *Cancer Genet. Cytogenet.* 38:215–222, 1989.

40. Mark, J., Havel, G., Grepp, C., Dahlenfors, R., and Wedell, B. Cytogenetical observations in human benign uterine leiomyomas. *Anticancer Res.* 8:621–626, 1988.

41. McGinnis, M., Bradley, E. L., Pretlow, T. L., *et al.* Correlation of stromal cells by morphometric analysis with metastatic behavior of human colonic cancer. *Cancer Res.,* 49:5989–5993, 1989.

42. McKeon, C., Thiele, C. J., Ross, R. A., *et al.* Indistinguishable patterns of protooncogene expression in two distinct but closely related tumors: Ewing's sarcoma and neuroepithelioma. *Cancer Res.* 48:4307–4311, 1988.

43. Mills, K. I., MacKenzie, E. D., and Birnie, G. D. The site of the breakpoint within the BCR is a prognostic factor in Philadelphia-positive CML patients. *Blood* 72:1237–1241, 1988.

44. Misawa, S., Yashige, H., Horiike, S., *et al.* Detection of karyotypic abnormalities in most

patients with acute nonlymphocytic leukemia by adding ethidium bromide to short-term cultures. *Leuk. Res.* 12:719–729, 1988.

45. Mitelman, F. *Catalog of Chromosome Aberrations in Cancer*, 3rd ed. New York, Alan R. Liss, Inc., 1988.

46. Molenaar, W. M., DeJong, B., Buist, J., et al. Chromosomal analysis and the classification of soft tissue sarcomas. *Lab. Invest.* 60:266–274, 1989.

47. Moriuchi, T., Koji, T., Nakane, P. K., Yoshida, M., Moriuchi, J., and Arimori, S. Use of non-radioactive DNA probes for the characterization of adult T-cell leukemia cells. *Nucleic Acids Symp. Ser.* 19:77–80, 1988.

48. Mugneret, F., Nacol, L. S., Volk, C., Cuisenier, J., Colin, F., and Turc-Carel, C. Association of breakpoint 14q23 with uterine leiomyoma. *Cancer Genet. Cytogenet.* 34:201–206, 1988.

49. Nilbert, M., Mandahl, N., Heim, S., Rydholm, A., Willen, H., Akerman, M., and Mitelman, F. Chromosome abnormalities in leiomyosarcomas. *Cancer Genet. Cytogenet.* 34:209–218, 1988.

50. Nilbert, M., Mandahl, N., Heim, S., Rydholm, A., Willen, H., and Mitelman, F. Cytogenetic abnormalities in an angioleiomyoma. *Cancer Genet. Cytogenet.* 37:61–64, 1989.

51. Oscier, D. G., Fitchett, M., and Hamblin, T. J. Chromosomal abnormalities in B-CLL. *Nouv. Rev. Fr. Hematol.* 30:397–398, 1988.

52. Pinkel, D., Landegent, J., Collins, C., et al. Fluorescence *in situ* hybridization with human chromosome–specific libraries: Detection of trisomy 21 and translocations of chromosome 4. *Proc. Natl. Acad. Sci. U.S.A.* 85:9138–9142, 1988.

53. Potluri, V. R., Helson, L., Elsworth, R. M., Reid, T., and Gilbert, F. Chromosomal abnormalities in human retinoblastoma, A review. *Cancer* 58:663–671, 1986.

54. Przepiorka, D. Breakpoint zone of BCR in chronic myelogenous leukemia does not correlate with disease phase or prognosis. *Cancer Genet. Cytogenet.* 36:117–122, 1988.

55. Rey, J. A., Bello, M. J., de Campos, J. M., et al. Cytogenetic clones in a recurrent neurofibroma. *Cancer Genet. Cytogenet.* 26:157–163, 1987.

56. Rigaut, J. P., Margules, S., Boysen, M., Chalumeau, M. T., and Reith, A. Karyometry of pseudostratified, metaplastic and dysplastic nasal epithelium by morphometry and stereology. 1. A general model for automated image analysis of epithelia. *Pathol. Res. Pract.* 174:342–356, 1982.

57. Rigaut, J. P., Reith, A., and El Kebir, F. Z. Karyometry by automated image analysis: Application to precancerous lesions. *Pathol. Res. Pract.* 179:216–219, 1984.

58. Rodgers, C. S., Hill, S. M., and Hulten, M. Cytogenetic analysis of human breast carcinoma: I. Nine cases in the diploid range investigated using direct preparations. *Cancer Genet. Cytogenet.* 13:95–120, 1984.

59. Saiki, R. K., Gilford, D. H., Stoffel, S., et al. Primer-directed enzymatic amplification of DNA with a thermostable DNA polymerase. *Science (Wash DC)* 239:487–491, 1988.

60. Saiki, R. K., Scharf, S., Faloona, F., et al. Enzymatic amplification of β-globin genomic sequences and restriction site analysis for diagnosis of sickle cell anemia. *Science (Wash DC)* 230:1350–1354, 1985.

61. Sandberg, A. A., Turc-Carel, C., and Gemill, R. Chromosomes in solid tumors and beyond. *Cancer Res.* 46:6019–6023, 1988.

62. Secker-Walker, L. M., Cooke, H. M., Browett, P. J., et al. Variable Philadelphia breakpoints and potential lineage restriction of BCR rearrangement in acute lymphoblastic leukemia. *Blood* 72:784–791, 1988.

63. Second MIC Cooperative Study Group. Morphologic, immunologic and cytogenetic (MIC) working classification of the acute myeloid leukaemias. *Brit. J. Haematol.* 68:487–494, 1988.

64. Seizinger, B. R., de la Monte, S., Atkins, L., Gusella, J. F., and Martuza, R. L. Molecular

genetic approach to human meningioma: Loss of genes on chromosome 22. *Proc. Natl. Acad. Sci. U.S.A.* 84:5419–5423, 1987.

65. Shapiro, J. R., Yung, W. K. A., and Shapiro, W. R. Isolation, karyotype and clonal growth of heterogeneous subpopulations of human malignant gliomas. *Cancer Res.* 41:2349–2359, 1981.

66. Showe, L. C., Croce, C. M. Chromosome translocations in B and T cell neoplasia. *Semin. Hematol.* 23:237–244, 1986.

67. Shtalrid, M., Talpaz, M., Kurzrock, R. *et al.* Analysis of breakpoints within the BCR gene and their correlation with the clinical course of Philadelphia-positive chronic myelogenous leukemia. *Blood* 72:485–490, 1988.

68. Smith, H. S., Liotta, L. A., Hancock, M. C., Wolman, S. R., and Hackett, A. J. Invasiveness and ploidy of human mammary carcinomas in short term culture. *Proc. Natl. Acad. Sci. U.S.A.* 82:1805–1809, 1985.

69. Solis, V., Pritchard, J., and Cowell, J. K. Cytogenetic changes in Wilms' tumors. *Cancer Genet. Cytogenet.* 34:223–234, 1988.

70. Stenman, G., Rorsman, F., and Betsholtz, C. Sublocalization of the human PDGF A-chain gene to chromosome 7, band Q11.23, by *in situ* hybridization. *Exp. Cell Res.* 178:180–184, 1988.

71. Teerenhovi, L., Lindholm, C., Pakkala, A., Franssila, K., Stein, H., and Knuutila, S. Unique display of a pathologic karyotype in Hodgkins' disease by Reed-Sternberg cells. *Cancer Genet. Cytogenet.* 34:305–311, 1988.

72. Teerenhovi, L., Wasenius, V-M., Franssila, K., Keinanen, M., and Knuutila, S. A method for analysis of cell morphology, banded karyotype and immunoperoxidase identification of lymphocyte subset on the same cell. *Am. J. Clin. Pathol.* 85:602–604, 1986.

73. Teyssier, J. R. Nonrandom chromosomal changes in solid tumors: Application of an improved culture method. *J. Nat. Cancer Inst.* 79:1189–1198, 1987.

74. Teyssier, J. R. The chromosomal analysis of human solid tumors. *Cancer Genet. Cytogenet.* 37:103–125, 1989.

75. Thiele, C. J., McKeon, C., Triche, T. J., Ross, R. A., Reynolds, C. P., and Israel, M. A. Differential protooncogene expression characterizes histopathologically indistinguishable tumors of the peripheral nervous system. *J. Clin. Invest.* 80:804–811, 1987.

76. Turc-Carel, C., Philip, I., Berger, M., Philip, T., and Lewdir, G.M. Chromosome study of Ewing's sarcoma (ES) cell lines: Consistency of a reciprocal translocation t(11;22)(q24;q12). *Cancer Genet. Cytogenet.* 12:1–19, 1984.

77. Vogelstein, B., Fearon, E. R., Hamilton, S. R., *et al.* Genetic alterations during colorectal-tumor development. *N. Engl. J. Med.* 319:525–532, 1988.

78. Whang-Peng, J., Triche, T. J., Knutsen, T., Miser, J., Douglass, E. C., and Israel, M. A. Chromosome translocation in peripheral neuroepithelioma. *N. Engl. J. Med.* 311:584–585, 1984.

79. Wiedemann, L. M., Karhi, K. K., Shivji, M. K. K., *et al.* The correlation of bcr rearrangement and p210 phl/abl expression with morphological analysis of Ph'-negative CML and other myeloproliferative diseases. *Blood* 71:349–353, 1988.

80. Wolman, S. R. Chromosomes in breast cancer. In: *Cellular and Molecular Biology of Experimental Mammary Cancer*, edited by D. Medina, W. Kidwell, G. Heppner, and E. Anderson. Plenum Publishing, 1987, pp. 47–65.

81. Wolman, S. R., Camuto, P. M., and Perle, M. A. Cytogenetic diversity in primary human tumors. *J. Cell Biochem.* 36:147–156, 1988.

82. Wolman, S. R., Smith, H. S., Stampfer, M., and Hackett, A. J. Growth of diploid cells from breast cancers. *Cancer Genet. Cytogenet.* 16:49–64, 1985.

83. Yokota, J., Wada, M., Shimosato, Y., Terada, M., and Sugimura, T. Loss of heterozygosity on chromosomes 3, 13, and 17 in small-cell carcinoma and on chromosome 3 in adenocarcinoma of the lung. *Proc. Natl. Acad. Sci. U.S.A.* 84:9252–9256, 1987.

84. Zhang, R., Wiley, J., Howard, S. P., Meisner, L. F., and Gould, M. N. Rare clonal karyotypic variants in primary cultures of human breast carcinoma cells. *Cancer Res.* 49:444–449, 1989.

4

Molecular Genetics of Human Disease: Clinical Applications of Genetic Linkage Analysis

Anthony A. Killeen

Harry T. Orr

Through the application of recombinant DNA technology, the study of human genetic disorders has undergone a substantial surge in activity. A number of disease-causing mutations have been characterized at the DNA level. In many cases, the ability to isolate and analyze human disease genes has provided important functional insights for specific structural features of eukaryotic genes. In the long term, the use of restriction fragment length polymorphisms (RFLPs) as genetic markers is likely to be the most important impact of recombinant DNA methods on human genetics. These markers have made it possible to follow specific disease genes within families. By using RFLPs and genetic linkage analysis, the precise chromosomal location of a disease gene may be determined. This is followed by the use of various physical and cloning strategies to isolate the disease gene. This general strategy has been designated "reverse genetics."[20]

This discussion begins with an overview of RFLPs as genetic markers in linkage analyses of human disease genes. Then the clinical applications of gene mapping are described, focusing on carrier detection and prenatal diagnosis of cystic fibrosis and 21-hydroxylase deficiency. Since this review only describes the use of DNA markers in the indirect detection of disease genes by genetic linkage analysis, the reader is referred to other excellent articles for a description of the use of DNA probes for the direct detection of disease genes (*e.g.*, reference 1).

RFLPs and Genetic Linkage

GENETIC MARKERS

A genetic marker, simply stated, is a trait whose inheritance can be followed within a family. This trait then becomes a marker for genes located within the neighboring chromosomal region. If an individual has inherited the marker from a given chromosomal region, by inference, the individual has also inherited all of the genes located within that chromosomal region. Classical genetic markers (*e.g.*, HLA antigens, protein polymorphisms, and blood groups) have a limited use in genetic linkage studies since they define a very small portion of the human genome. Recombinant DNA has introduced to human genetics a new set of markers, restriction fragment length polymorphisms (RFLPs). These genetic markers exploit the DNA sequence variations that exist between individuals and open up the entire human genome to genetic analysis.

DETECTION OF RFLPs

DNA sequence variations between individuals are routinely ascertained using restriction enzymes. Each restriction enzyme recognizes and cleaves double-stranded DNA at a specific sequence of nucleotides. Variation in the presence of restriction enzyme sites results in the generation of different-size fragments of DNA. These fragments of DNA are detected by a procedure using gel electrophoresis, transfer of the separated DNA to a filter membrane, and hybridization to a radiolabeled DNA probe. This procedure is known as Southern blotting and is described in Chapter 1.

Figure 4.1 depicts the type of variations detectable using a polymorphic DNA marker, *i.e.*, RFLPs. DNA from 7 individuals was digested with the restriction enzyme *Bcl*I, and probed with a coagulation factor XIIIa (F13A) cDNA clone.[11] The variation in size of hybridizing restriction fragments seen within this group of individuals is due to the presence or absence of *Bcl*I recognition sites in the DNA flanking the region homologous to the F13A cDNA probe. From the number of different band sizes (RFLPs) detected, it is concluded that there are multiple polymorphic *Bcl*I sites flanking the probe. In contrast to the example presented in Figure 4.1, most RFLPs that result from the presence or absence of restriction enzyme recognition sequences are simple two-allele systems. The site is either present or absent, resulting in two alleles (1 and 2) with three possible hybridization patterns (1/1, 1/2, or 2/2).

Recently, DNA markers that define multiallelic polymorphic loci have been isolated. These markers identify loci at which the polymorphism is due to a variable number of short tandem repeat sequences present between two recognition sites for a particular restriction enzyme. The number of repeat elements present determines

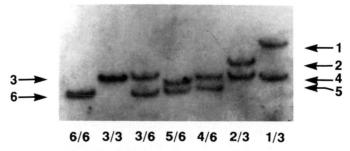

Figure 4.1.   Restriction fragment length polymorphisms detected with *Bcl* I and the F13A cDNA probe. The six variable bands are indicated by *arrows*. The F13A genotypes for each of the seven individuals analyzed are indicated below each lane.

the size of the hybridizing band. Since the number of repeats inserted can vary within the population from 0 to over 20, the number of alleles present in the population is often very large. Markers that define such highly polymorphic loci have been designated as variable number tandem repeats, or VNTRs[19] (Chapter 1).

GENETIC LINKAGE

Linkage may be defined as the nonrandom assortment of two markers within a family because of their physical proximity on the same chromosome. A measure of linkage is the recombination frequency, *i.e.*, the proportion of recombinant gametes that are detectable. The closer two markers are within the same segment of chromosomal DNA, the less likely it is that a recombination event will occur between them, and the more tightly linked they are. In any genetic linkage study, informative meioses are those that occur in an individual who is heterozygous at the marker loci. With their high level of polymorphism and their ability to detect both alleles in heterozygotes, RFLPs are particularly well suited for genetic linkage studies in humans.

Assessment of the statistical significance of linkage is made by calculating the ratio of the likelihood of linkage to the likelihood of random assortment. This calculation is the logarithm of the odds ratio and is designated the LOD score. LOD scores of 3.0 or greater (odds of 1000:1 in favor of linkage over random assortment) are considered to be statistically significant, while an LOD score of −2.0 (odds of 100:1 against linkage) statistically excludes two markers as being linked. LOD scores between −2.0 and +3.0 are considered inconclusive.

Genetic Mapping of Disease Genes Using RFLPs

LINKAGE OF RFLPs TO DISEASE GENES

The application of linkage analysis to single-gene disorders is currently one of the most active areas of human genetics. Unless the disorder is sex-linked, the location of the disease gene within the genome is unknown. Furthermore, the primary biochemical defect is also usually unknown. In such cases, the general strategy is to follow the inheritance of random DNA markers, RFLPs, with the inheritance of the disease in family studies until statistically significant data supporting linkage between one or more RFLPs and the disease are obtained.

While at first glance it may seem to be an insurmountable task to search the entire human genome with random DNA markers for linkage to a disease gene, the recent number of successes clearly establishes the feasibility of this approach. Several single-gene disorders have been linked to DNA markers over the last few years. Table 4.1 is a partial list of the genetic diseases for which there are linked DNA markers available.

This by no means presents a complete list of diseases with linked markers, nor does it include the many diseases that can be diagnosed directly, using gene-specific and mutation-specific probes. However, Table 4.1 does present some of the earliest and most notable successes in establishing disease-linked DNA markers.

Table 4.1.
A Selected List of Genetic Disorders for Which Linked DNA Markers Exist

Disorder	Chromosomal Location	Ref.
Adult polycystic kidney disease	16p12	23
Congenital adrenal hyperplasia	6p21	7, 16
Cystic fibrosis	7q22	25, 26, 28
Duchenne muscular dystrophy	Xp21	15
Hemophilia A	Xq28	10
Huntington's disease	4p16.1	12
Neurofibromatosis type I	17q11	4

USE OF LINKED MARKERS TO DETERMINE THE CHROMOSOMAL LOCATION OF A DISEASE GENE

Determining the chromosomal position of a disease gene is an important step toward isolation of the gene. Upon establishing statistically significant linkage of a DNA marker to a disease gene, there are two complementary methods that are routinely used to physically locate the disease gene. These are (*a*) mapping by chromosomal *in situ* hybridization and (*b*) mapping by analysis of somatic cell hybrids between rodent and human cells.

Chromosomal *in situ* hybridization involves the hybridization of radiolabeled probe DNA directly to stained metaphase chromosomes. Hybridization is detected by the appearance of silver grains along the chromosome in an overlaid photographic emulsion.

The second method uses a panel of somatic cell hybrids constructed between rodent and human cells.[24] The hybrid cells are obtained in a manner such that a limited number of human chromosomes are retained with a full complement of rodent chromosomes. Each somatic cell hybrid of a panel contains a different, but characterized complement of human chromosomes. Southern hybridization analysis of the entire somatic cell hybrid panel is performed, using the human DNA segment to be mapped as a probe. The chromosomal origin of the human DNA segment is determined by an analysis of the hybridization concordance and discordance with the human chromosomal makeup of each hybrid cell.

Clinical Applications of Linkage Analysis: Two Specific Examples

In the diagnostic laboratory, the goal of genetic linkage analysis using DNA markers is to identify the parental chromosomes containing the affected genes and trace their inheritance within a family. Such analyses are proving to be of use for carrier detection and prenatal diagnosis of diseases whether or not the affected gene is known. Cystic fibrosis and steroid 21-hydroxylase deficiency will be discussed to illustrate the use of DNA markers to track a disease allele.

MOLECULAR GENETICS OF CYSTIC FIBROSIS

Cystic fibrosis (CF) is an autosomal recessive disorder with incidence of about one in 2000 Caucasian births. Affected individuals have severe gastrointestinal, pulmonary, and nutritional problems. The biochemical basis of CF is unknown. In 1985, several groups obtained DNA markers linked to the CF gene.[25,26,28] These markers were used to position the CF gene on the long arm of human chromosome 7 (see Table 4.1 and Fig. 4.2). Subsequently, two additional markers, KM19 and XV2C, were found to be very tightly linked to the CF gene: only 2 recombinations have been observed between CF and KM19 and XV2C out of several thousand CF chromosomes examined to date.

The availability of a large number of markers closely linked to the CF locus has made it possible to learn much about CF mutations.[2,5,9] Linkage disequilibrium and haplotype association studies indicate that a substantial proportion of the CF genes may have descended from a single mutation. Strong linkage disequilibrium between CF and a haplotype formed with alleles of markers XV2C and KM19, designed the *b* haplotype, suggests that a single mutation may account for 85% of the CF chromosomes in the northern European and North American Caucasian

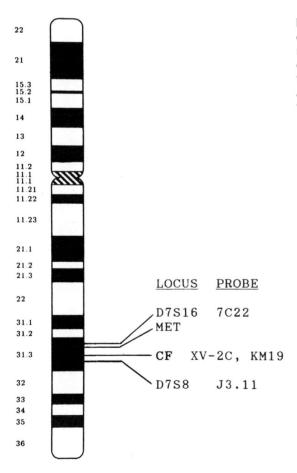

Figure 4.2. Chromosomal location of the CF locus and linked DNA markers. The karyogram of human chromosome 7 is depicted. The centromere is indicated by the *striped area*, the short arm is to the *top* and the long arm is to the *bottom*.

LOCUS	PROBE
D7S16	7C22
MET	
CF	XV-2C, KM19
D7S8	J3.11

7

populations.[2] Extended haplotype studies using the DNA markers that flank the CF locus lower this estimate to at least 50% in the Caucasian population of North America. With these analyses, it would be reasonable to assume that the estimates form upper and lower limits for the level of the most frequent CF mutation in these populations.

Information about the number of CF mutations has been obtained by similar analyses of different ethnic and clinical subgroups. Studies on an Italian population suggest that a CF mutation in addition to the one found in linkage disequilibrium with the *b* haplotype exists in the southern European population. Analyses of American blacks and the North American Hutterites indicate that there may be a few more additional CF mutations.[5,9]

CLINICAL APPLICATIONS OF MOLECULAR GENETICS TO CYSTIC FIBROSIS

An important use of the DNA markers closely linked to CF is in carrier detection and prenatal diagnosis. At the University of Minnesota nine DNA probe/restriction enzyme combinations (*i.e.*, markers) are used for carrier detection and prenatal diagnosis in families with CF (Table 4.2). Several aspects contribute to a very high reliability of diagnosis with these markers; there are several DNA markers available, these markers are located within a few centimorgans of the CF gene (Fig. 4.2), and this group of probes includes markers that flank the CF gene. In the past three years, over 40 nuclear families have been screened with these markers at the University of Minnesota, and we have yet to encounter a family that was uninformative.

To illustrate the process of CF carrier and prenatal diagnosis, Figure 4.3 presents a recent DNA marker analysis of a CF family. This family originally came to the CF clinic to find out if they were informative for prenatal testing. After genetic counseling, DNA was isolated from blood samples obtained from the parents, the CF-affected child, and the unaffected child. Their DNA was screened with the complete panel of DNA markers to establish the informative markers. Only the results of the informative markers are presented in Figure 4.3. Identification of the two parental chromosomes carrying the CF gene was determined by analyzing the parents' and affected son's hybridization patterns. For example, the father is heterozygous, T1/T2, at MetH-TaqI (*lane A1*), while the affected son is homozygous,

Table 4.2.
DNA Markers Linked to CF

Probe	Locus	RFLP (Enzyme)
pJ3.11	D7S8	4.2, 1.8 (MspI)
		6.3, 3.1 (TaqI)
pKM19	D7S23	7.8, 6.6 (PstI)
pXV2c	D7S23	2.1, 1.4 (TaqI)
pmetH	MET	5.0, 2.3, 1.8 (MspI)
		7.5, 4.0 (TaqI)
pmetD	MET	7.6, 6.8 (BanI)
		5.5, 4.3 (TaqI)
p7C22	D7S18	7.2, 5.1 (EcoRI)

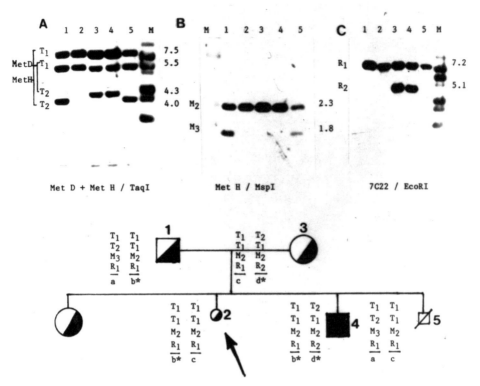

Figure 4.3. Analysis of a CF family with DNA markers. *Panels A-C* depict the Southern blot patterns obtained with the markers/enzyme combinations given. Numbering of the gel lanes corresponds to the pedigree number as shown below. Males are *squares* and females are depicted as *circles*. The *arrow* indicates the fetus.

T1/T1, at this locus (*lane A4*). Therefore, the paternal chromosome carrying the CF gene is the T1-containing chromosome. Likewise, the mother is heterozygous at MetD-TaqI, T1/T2. Since the father is homozygous, T1/T1, at MetD-TaqI, but the affected son is a T1/T2 heterozygote, he must have inherited the MetD-TaqI T2 allele from his mother. Thus, the maternal CF chromosome is marked by the MetD-TaqI T2 allele. In a similar manner, the 7C22-EcoRI R2 allele also distinguishes the CF chromosome from the mother. Analysis of the unaffected daughter's DNA indicated that she had inherited only one of the parental CF chromosomes. Therefore, she is an unaffected carrier for CF.

Once it had been determined that each of the four parental chromosome 7s could be tracked in this family, the parents were informed that their family was fully informative for prenatal diagnosis. A few months later, the laboratory was informed that the mother was in her 4th week of pregnancy and that chorionic villus sampling (CVS) was being scheduled for the 10th week. CVS tissue was divided so that a portion was used for karyotype analysis and the remainder used for DNA marker analysis. The karyotype was found to be that of a normal female. From Figure 3, *lanes 2A, 2B,* and *2C,* the chromosome 7 genotype of the fetus can be determined. Like the affected son, this fetus has inherited the MetH-TaqI T1 allele that marks the paternal CF chromosome 7. However, the fetus has not inherited the maternal MetD-TaqI T2 and 7C22-EcoRI R2 alleles that mark the CF chromosome 7 from the mother. Thus, the fetus has inherited an unaffected chromosome 7 from the mother and a CF chromosome from the father and was diagnosed as an unaffected CF carrier. After a full-term pregnancy, a female was born with no signs of CF, including a normal sweat chloride level of 25 mEq/L.

An important point worth noting on this analysis is the number of probe/enzyme combinations needed for this family to be fully informative (*i.e.*, to distinguish all four parental chromosome 7s.) Of the nine probe/enzyme combinations examined, five were found not to be informative (*i.e.*, both parents were homozygous). This illustrates the benefit of having multiple closely linked markers. The chance that a family will be fully informative is very low with a single marker and increases dramatically with multiple markers.

Another clinical application of the CF-linked DNA markers has been as an aide in resolving ambiguous CF diagnosis in siblings of known patients.[21,22] In some cases a sibling of a known CF patient presents with ambiguous sweat chloride levels, two or more sweat specimens having a chloride level in an indeterminant range between 43 and 61 mEq/L. Chromosome 7 genotyping with the linked DNA markers is then used to determine if the sibling whose CF diagnosis is ambiguous shares a chromosome 7 genotype with the affected sibling. CF-affected offspring within the same family should have an identical chromosome 7 genotype that differs from that of their unaffected siblings.

MOLECULAR GENETICS OF CONGENITAL ADRENAL HYPERPLASIA

With a disease frequency of 1 in 15,000 births, congenital adrenal hyperplasia deficiency of steroid 21-hydroxylase (21-OH) is one of the most common inborn errors of metabolism among Caucasians. The biochemical abnormalities include decreased levels of glucocorticoids and mineralocorticoids and increased levels of adrenal androgens. The latter results in virilization of affected females. Sometimes a female newborn will be mistaken for a male. The most variable characteristic of the disease is a mineralocorticoid deficiency that leads to "salt-wasting" resulting from decreased activity of the aldosterone-stimulated sodium pump in the distal renal tubule. This condition occurs in about one-half to two-thirds of individuals with 21-OH deficiency.

Since 1977 it has been known that the locus for 21-OH is closely linked to the HLA-B region on the short arm of chromosome 6.[7] Pulsed-field gel electrophoresis has been used to physically place the 21-OH locus about 600 kilobases centromeric to HLA-B.[3] In recent years the human 21-OH locus has been studied at the DNA sequence level.[13,27] There are two 21-OH genes, 21-OH A and B. The 21-OH A gene has many mutations that render it a pseudogene, while the 21-OH B gene encodes the functional enzyme. The 21-OH A and B genes are interdigitated between the complement 4A and 4B genes (Fig. 4.4).

Disease alleles can arise either from a deletion of the 21-OH B gene or from point mutations. It seems that some of the point mutations that give rise to a disease allele

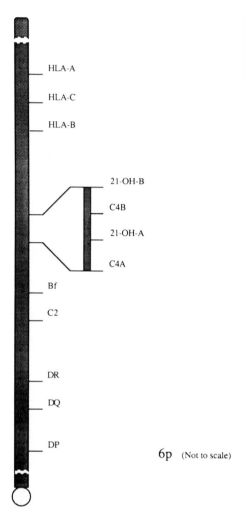

Figure 4.4.   A schematic diagram of the HLA region on human chromosome 6p. The centromere is to the *bottom*. The subregion containing the 21-OH and complement genes has been expanded.

may have resulted from gene conversions between the 21-OH A and 21-OH B genes.[17] The frequency of deletions is approximately 25%.

GENETIC LINKAGE AND PRENATAL DIAGNOSIS OF 21-OH DEFICIENCY

As described above, because of the number of different disease alleles it is not easy to identify directly the exact molecular abnormality in the 21-OH B gene in a given family with congenital adrenal hyperplasia. Therefore, DNA markers linked to the 21-OH B gene have an important role in prenatal diagnosis, even though the specific gene affected in congenital adrenal hyperplasia is known.

Prenatal diagnosis of 21-OH deficiency has important implications because of the possibility of preventing virilization of an affected female fetus by administering glucocorticoids during pregnancy.[6,8] Since the adverse effects of glucocorticoids administered during pregnancy are largely unknown, it is prudent to treat only affected female fetuses. Moreover, virilization likely occurs early during gestation, before amniocentesis can be performed. This has prompted investigators to determine whether DNA marker analysis of fetal material obtained by CVS can be used to establish a protocol for assessing the effectiveness of fetal therapy in at-risk pregnancies.

The 21-OH B gene is physically close to two polymorphic loci, the C′4B complement gene and the HLA-B gene. 21-OH lies immediately adjacent to the C′4 genes and within 600 kb of the HLA-B gene. DNA probes from both C′4 and HLA-B have been shown to detect RFLPs useful in the tracking of 21-OH B disease alleles.[14,18,29]

Figure 4.5 illustrates how DNA markers can be used to track 21-OH B disease alleles for prenatal diagnosis. *Panel A, lanes 1 and 2* depict the hybridization patterns obtained when *Taq*I-digested DNA from the parents of a 21-OH deficient daughter was probed with a DNA probe for the C′4 genes. The C′4B genes give rise to 2 RFLPs of sizes 6.0 and 5.4 kb in the mother and a single 5.4 kb band in the father. When the C′4B hybridization pattern of the affected daughter (*lane A4*) is examined, two interacting points become apparent. The affected daughter has a single C′4B fragment of 6.0 kb. Since only her mother has a 6.0 kb C′4B fragment, the maternal disease allele must be on the chromosome 6 carrying the 6.0 kb C′4B frag-

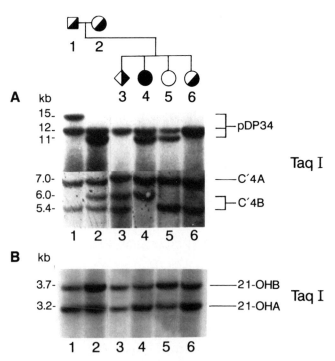

Figure 4.5. Analysis of a 21-OH family with DNA markers. The pedigree is indicated above (see legend to Fig. 4.3). Individual 3 is the fetus. Numbering of the gel lanes corresponds to the numbering in the pedigree. *Panels A and B* present the Southern blot patterns obtained with the marker/enzyme combinations indicated on the *right*.

ment. The presence of only the 6.0 kb C'4B fragment in the affected daughter's DNA also reveals that one chromosome 6 in the father carries a deletion. If the father were homozygous for the 5.4 kb C'4B fragment, he would have to pass one copy on to his daughter. Since he didn't, his disease-carrying chromosome must contain a deletion that includes that C'4B gene. That one chromosome 6 in the father carries a deletion is suggested by the relative intensity of hybridization of a 21-OH probe to the 21-OH A pseudogene and to the 21-OH B functional gene. (Compare the intensity of the 3.7 kb TaqI 21-OH B fragment to the intensity of the 3.2 kb TaqI 21-OH A fragment in *lane 1* of Fig. 4.5B.)

DNA from the fetus contains both the 6.0 kb and the 5.4 kb C'4B fragments. The 6.0 kb C'4B has to have been inherited from the mother and marks the affected maternal chromosome 6. Therefore, the 5.4 kb C'4B fragment in the fetal DNA has been inherited from the father and identifies his unaffected chromosome 6. Since the fetus has inherited markers for the affected maternal chromosome 6 and the unaffected paternal chromosome, it is an unaffected carrier of 21-OH deficiency. Glucocorticoid therapy was subsequently stopped. At the time of writing, the mother had just given birth to a clinically normal female baby.

Marker pDP34 (Fig. 4.5A) hybridizes to loci on the X and Y chromosomes and provides a useful marker for maternal contamination of the CVS. The 12.0 kb and 11.0 kb bands are from the X chromosome, and the 15.0 kb band is from the Y chromosome. The fetal pDP34 hybridization pattern (*lane 3* in Fig. 4.5A, shows that the fetus is a female homozygous for the 12.0 kb fragment. Importantly, the mother is a 12.0 kb/11.0 kb heterozygote at pDP34. Thus, the fetal CVS DNA is not contaminated by any detectable DNA from her mother.

Factors That Can Contribute to Misdiagnosis by Genetic Linkage

There are three complicating factors that are important in making an accurate diagnosis by genetic linkage. These are meiotic recombination, disease heterogeneity, and nonpaternity.

MEIOTIC RECOMBINATION

In a diagnosis made through genetic linkage, the presence of a disease allele in the DNA of an individual is inferred from the inheritance (within a family) of markers closely linked to the disease gene. The frequency of recombination between two loci is generally directly related to the physical distance between the loci on a chromosome. The greater the physical separation, the higher the recombination frequency. Thus, recombination has the potential of separating a marker from a linked disease gene in the progeny.

Use of markers located on both sides of a disease gene (*i.e.*, flanking markers) greatly reduces the error rate due to recombination. With flanking markers, for an error in diagnosis to occur, two recombination events must take place on the same chromosome. Thus, if the two flanking markers are each separated from the disease gene by recombination frequencies of 5%, the likelihood of recombination between the disease gene and both markers would be 0.05 × 0.05, or 0.25%.

DISEASE HETEROGENEITY

For linked DNA markers to provide an accurate diagnosis, the markers being used must be linked to the disease gene in question. For Huntington's and CF the data to date indicate that the genes responsible for these disorders are always linked to markers on chromosome 4 and chromosome 7, respectively. However, there are single-gene disorders where disease states that appear identical are linked to markers on different chromosomes (*e.g.*, osteogenesis imperfecta). In such cases, the specific form of the disease must be determined before genetic linkage studies are undertaken.

NONPATERNITY

To use genetic linkage analysis to track a disease gene in a family, the biological father has to be studied. Inaccurate assumptions about paternity can result in the establishment of linkage between the wrong polymorphisms and the disease gene. Therefore, it is of utmost importance that paternity issues be discussed with the family. A useful approach is to first have the mother see the genetic counselor by herself, at which time these issue are discussed. If paternity is a problem, the mother must decide if this information will be passed onto other members of the family. She should also be told that the DNA marker analysis itself may reveal instances of nonpaternity.

REFERENCES

1. Antonarakis, S. E. Diagnosis of genetic disorders at the DNA level. *N. Engl. J. Med.* 320:153–163, 1989.
2. Beaudet, A. L., Feldman, G. L., Fernbach, S. D., Buffone, G. J., and O'Brien, W. E. Linkage disequilibrium, cystic fibrosis, and genetic counseling. *Am. J. Hum. Genet.* 44:319–326, 1989.
3. Carroll, M. C., Katzman, P., Alicot, E. M., Koller, B. H., Geraghty, D. E., Orr, H. T., Strominger, J. L., and Spies, T. Linkage map of the human major histocompatibility complex including the tumor necrosis factor genes. *Proc. Natl. Acad. Sci. U.S.A.* 84: 8535–8539, 1987.
4. Collins, F. S., Ponder, B. A. J., Seizinger, B. R., and Epstein, C. J. The von Recklinghausen neurofibromatosis region on chromosome 17 genetic and physical maps come into focus. *Am. J. Hum. Genet.* 44:105, 1989.
5. Cutting, G. R., Antonarakis, S. E., Buetow, K. H., Kasch, L. M., Rosenstein, B. J., and Kazazian, H. H. Analysis of DNA polymorphism haplotypes linked to the cystic fibrosis locus in North American black and Caucasian families supports the existence of multiple mutations of the cystic fibrosis gene. *Am. J. Hum. Genet.* 44:307–318, 1989.
6. David, M., and Forest, M. G. Prenatal treatment of congenital adrenal hyperplasia resulting from 21-hydroxylase deficiency. *J. Pediatr.* 105:799–803, 1984.
7. Dupont, B., Oberfield, S. E., Smithwick, E. M., Lee, T. D., and Levine, L. S. Close genetic linkage between HLA and congenital adrenal hyperplasia (21-hydroxylase deficiency). *Lancet ii*:1309–1311, 1977.
8. Evans, M. I., Chrousos, G. P., Mann, D. W., Larsen, J. W., Green, I., McCluskey, J., Loriaux, D. L., Fletcher, J. C., Koons, G., Overpeck, J., and Schulman, J. D. Pharmacologic suppression of the fetal adrenal gland in utero: Attempted prevention of abnormal external genital masculinization in suspected congenital adrenal hyperplasia. *JAMA* 253:1015–1020, 1985.

9. Fujiwara, T. M., Morgan, K., Schwartz, R. H., Doherty, R. A., Miller, S. R., Klinger, K., Stanislovitis, P., Stuart, N., and Watkins, P. C. Genealogical analysis of cystic fibrosis families and chromosome 7q RFLP haplotypes in the Hutterite brethren. *Am. J. Hum. Genet.* 44:327–337, 1989.

10. Gitschier, J., Drayna, D., Tuddenham, E. G. D., White, R. L., and Lawn, R. M. Genetic mapping and diagnosis of haemophilia A achieved through a *BclI* polymorphism in the factor VII gene. *Nature (Lond)* 314:738–740, 1985.

11. Grundmann, U., Amann, E., Zettlemeissl, G., and Kupper, H. A. Characterization of cDNA coding for human factor XIIIa. *Proc. Natl. Acad. Sci. U.S.A.* 83:8024–8028, 1986.

12. Guesella, J. F., Wexler, N. S., Conneally, P. M., Naylor, S. L., Anderson, M. A., Tanzi, R. E., Watkins, P. C., Ottina, K., Wallace, M. R., Sakagushi, A. Y., Young, A. B., Shoulson, I., Bonilla, E., and Martin, J. B. A polymorphic DNA marker genetically linked to Huntington's disease. *Nature (Lond)* 306:234–238, 1983.

13. Higashi, Y., Yoshioka, H., Yamane, M., Gotch, O., and Fuji-Kuriyama, Y. Complete nucleotide sequence of two steroid 21-hydroxylase genes tandemly arranged in human chromosome 6: A pseudogene and a genuine gene. *Proc. Natl. Acad. Sci. U.S.A.* 83:2841–2845, 1986.

14. Killeen, A. A., Seelig, S., Ulstrom, R. A., and Orr, H. T. Diagnosis of classical steroid 21-hydroxylase deficiency using an HLA-B locus-specific DNA-probe. *Am. J. Med. Genet.* 29:703–712, 1988.

15. Koenig, M., Hoffman, E. P., Bertelson, C. J., Monaco, A. P., Feener, C., and Kunkel, L. M. Complete cloning of the Duchenne muscular dystrophy (DMD) cDNA and preliminary genomic organization of the DMD gene in normal and affected individuals. *Cell* 50:509–517, 1987.

16. Levine, L. S., Zimmerman, M., New, M. I., Prader, A., Pollack, M., O'Neil, G. J., Yang, S. Y., Oberfield, S. E., and Dupont, B. Genetic mapping of the 21-hydroxylase deficiency gene within the HLA-linkage group. *N. Engl. J. Med.* 299:911–915, 1978.

17. Miller, W. L. Gene conversion, deletions, and polymorphisms in congenital adrenal hyperplasia. *Am. J. Hum. Genet.* 42:4–7, 1988.

18. Mornet, E., Boue, J., Raux-Demay, M., Couillin, P., Oury, J. F., Demuz, Y., Dausset, J., Cohen, D., and Boue, A. First trimester diagnosis of 21-hydroxylase deficiency by linkage analysis to HLA-DNA probes and by 17-hydroxyprogesterone determination. *Hum. Genet.* 73:358–364, 1986.

19. Nakamura, Y., Leppert, M., O'Connell, P., Wolff, R., Holm, T., Culver, M., Martin, C., Fujimoto, E., Hoff, M., Kumlin, E., and White, R. Variable number tandem repeat (VNTR) markers for human gene mapping. *Science (Wash DC)* 235:1616–1622, 1987.

20. Orkin, S. H. Reverse genetics and human disease. *Cell* 47:845–850.

21. Orr, H. T., Parker, T., Wielinski, C. L., Clawson, C. C., and Warwick, W. J. Cystic fibrosis: Chromosome 7 DNA genotyping: An aide in resolving ambiguous diagnoses in siblings of known patients. *Clin. Pediatr.* 27:591–595, 1988.

22. Patton, M. A., Harris, A., Quinlan, C., and Newton, R. Use of DNA markers linked to cystic fibrosis gene to resolve equivocal sweat test results. *Lancet* ii:155–156, 1987.

23. Reeders, S. T., Breuning, M. H., Davies, K. E., Nicholl, R. D., Jarman, A. P., Higgs, D. R., Pearson, P. L., and Weatherall, D. J. A highly polymorphic DNA marker linked to adult polycystic kidney disease on chromosome 16. *Nature (Lond)* 317:542–544, 1985.

24. Ruddle, F. H. A new era in mammalian gene mapping: Somatic cell genetics and recombinant DNA methodologies. *Nature (Lond)* 294:115–120, 1981.

25. Tsui, L.-C., Buchwald, M., Barker, D., Braman, J. D., Knowlton, R., Schumm, J. W., Eiberg, H., Mohr, J., Kennedy, D., Plavsic, N., Zsiga, M., Markiewicz, D., Akots, G., Brown, V., Helms, C., Gravius, T., Parker, C., Rediker, K., and Donis-Keller, H. Cystic fibrosis locus defined by a genetically linked polymorphic DNA marker. *Science (Wash DC)* 230:1054–1057, 1985.

26. Wainwright, B. J., Scambler, P. J., Schmidtke, J., Watson, E. A., Law, H.-Y., Farrall, M., Cooke, H. J., Eiberg, H., and Williamson, R. Localization of cystic fibrosis locus to human chromosome 7cen-q22. *Nature (Lond)* 318:384–385, 1985.
27. White, P. C., New, M. I., and Dupont, B. Structure of human steroid 21-hydroxylase genes. *Proc. Natl. Acad. Sci. U.S.A.* 83:5111–5115, 1986.
28. White, R., Woodward, S., Leppert, M., O'Connell, P., Hoff, M., Herber, J., Lalouel, J.-M., Dean, M., and Vande Woude, G. A closely linked genetic marker for cystic fibrosis. *Nature (Lond)* 318:382–384, 1985.
29. Whitehead, A. S., Woods, D. E., Fleischnick, E., Chin, J. E., Yunis, E. J., Katz, A. J., Gerald, P. S., Alper, C. A., and Colten, H. R. DNA polymorphisms of the C4 genes. A new marker for analysis of the major histocompatibility complex. *N. Engl. J. Med.* 310:88–91, 1984.

5

Oncogenes: Introduction

Cecilia M. Fenoglio-Preiser
Margaret B. Listrom

Retroviral Oncogenes

Retroviruses can be divided into two general groups based on their biological activity: (*a*) *acute transforming viruses,* which efficiently transform cells and induce tumors *in vivo* within two to three weeks, and (*b*) *slow transforming viruses,* which do not appear to transform cells in culture and only induce tumors in infected animals after long latency periods.

Retroviral Replication and Oncogene Structure

Rous sarcoma virus (RSV) is the prototype of the acute transforming retroviruses. Its RNA genome carries four genes (Fig. 5.1), and except for the presence of the oncogene sequence *src,* its genetic structure is typical of all replication-competent retroviruses.

The primary translation product of the gag and env genes are protein precursors that are processed in the virion to yield multiple, functionally distinct polypeptides. The *gag* gene specifies virion core proteins, and *env* encodes viral envelope proteins. The pol sequence encodes reverse transcriptase.

The retroviral life cycle provided the first clues that cellular genes might participate in tumorigenesis. The single-stranded RNA of the diploid viral genome is transcribed into DNA by reverse transcriptase. Viral DNA then integrates into host-cell chromosomal DNA. The host cell then uses its own machinery to express the viral genes (Fig. 5.2).

During this process, retroviruses may incorporate certain genes from their hosts in a process known as "transduction." Recombination between retroviral and cellular genomes can implant cellular genes anywhere in the viral genome (Fig. 5.3), usually resulting in alterations in the sequence or structure of the cellular gene.

Transduced oncogenes are pathogenic because expression of the transduced gene is driven by potent viral signals that induce high levels of expression of the structurally altered or mutated genes.[18,19,21] Additionally, transduced genes gener-

5' —	gag	pol	env	src	— 3'

Figure 5.1. RNA genome of rous sarcoma virus (RSV) is typical of all replication-competent retroviruses: *gag* specifies virion core proteins; *pol* encodes reverse-transcription state; and *env* encodes viral envelope proteins. *src* is unique to RSV.

Figure 5.2. Integration of viral DNA into host-cell chromosomal DNA results in transduced host genes in the viral particles.

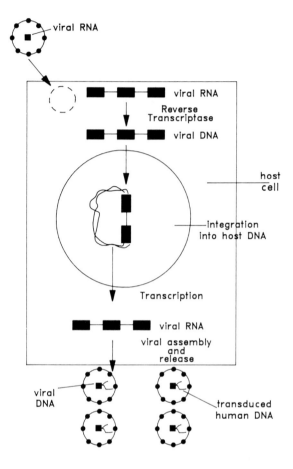

ally acquire mutations. Comparisons of retroviral oncogenes with their cellular progenitors have revealed point mutations, deletions, and genetic substitutions in the viral homologues.

In some instances retroviral oncogenes are fusions of more than one cellular oncogene. For example, avian myelocytomatosis virus (AMV) contains the *myb* oncogene fused to *ets* sequences (Table 5.1). The avian sarcoma virus MH2 also contains two fused oncogene sequences (mht and myc).

A retroviral oncogene (v-*onc*) represents the gene of an *acute* transforming retrovirus that is responsible for the malignant transformation of target cells, whereas proto-oncogenes (cellular oncogenes (c-*onc*) derived from normal cells do not usually cause cellular transformation (Table 5.1).

Expression of Proto-oncogenes

Since proto-oncogenes are conserved in sequence throughout the animal kingdom it is believed that the proteins encoded by them are important in normal cell growth and/or differentiation. In fact, oncogenes appear to be regulated in the tissue in a specific manner during development.

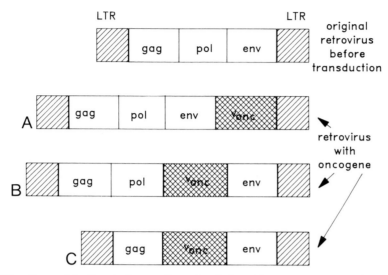

Figure 5.3. Recombination between the retroviral and host genomes can implant host genes anywhere in the viral genome. The *top* figure is the native retroviral genome sequence. The addition of v-*onc* as shown in **A** and **B** occurs without deletion of any native viral genome. In the last figure, **C,** *pol* is deleted.

For example, c-*fms* is expressed predominantly in macrophages and placental tissues, whereas c-*myb* levels are highest in hematopoietic cells. Other oncogene-encoded mRNAs, such as that for c-*myc,* are more widely expressed.

Transcription levels for different proto-oncogenes also vary at different times in development. For example, expression levels of c-*fos,* c-*int,* and c-*abl* genes vary characteristically during embryogenesis, whereas c-Ha-*ras* is continuously expressed at high levels. Other proto-oncogenes are continuously expressed at low levels or appear to be silent.

There are several explanations for how genetic damage might cause malfunction of a proto-oncogene or its product. The damage might cause constitutive expression of the oncogene or its product, which then cannot be regulated. For example, translocated genes may remove the elements that modulate their transcription. Mutations can confer constitutive activity on the products. An example is the mutated c-*ras* gene that encodes a protein with reduced GTPase activity, which normally limits the duration of its action. Mutations might also change the manner in which a protein acts. Examples include alterations in the substrate specificity of a kinase or the specificity of a transcription factor.

IDENTIFICATION OF ONCOGENES

Several experimental strategies have been utilized to identify proto-oncogenes: (*a*) The first proto-oncogenes were identified as the normal cellular homologues of acute transforming genes (viral oncogenes); (*b*) DNA transfection techniques have led to the isolation of genes that cause transformation of NIH 3T3 cells (Table 5.2)[138]; (*c*) Analysis of tumors caused by *slow* retroviruses has revealed

Table 5.1.
Retroviral Oncogenes and Cellular Proto-oncogenes

Oncogene	Location of Gene Product	Function of Gene Product	Retrovirus
fos	Nuclear	Transcription factor	FBJ murine sarcoma virus
myc	Nuclear	?	Avian myelocytomatosis virus (AMV)
myb	Nuclear	Transcription factor	Avian myeloblastosis virus
ski	Nuclear	?	SKV avian carcinoma virus
ets	Nuclear	Transcription factor	Avian myeloblastosis virus
Ha-ras	Plasma membrane	Signal transduction	Harvey murine sarcoma virus
Ki-ras	Plasma membrane	Signal transduction	Kirsten murine sarcoma virus
erb-B	Plasma membrane	Growth factor receptor	Avian erythroblastosis virus
fms	Plasma membrane	Growth factor receptor	Feline sarcoma virus
ros	Plasma membrane	Growth factor receptor	Avian sarcoma virus
kit	Plasma membrane	Growth factor receptor	Feline sarcoma virus
erb-A	Cytoplasm	Thyroid hormone receptor	Avian erythroblastosis virus
src	Membranes, cytoplasm	Protein-tyrosine kinase	Rous avian sarcoma virus
abl	Nuclear	Protein-tyrosine kinase	Abelson murine leukemia virus
yes	Membranes, cytoplasm	Protein-tyrosine kinase	Y73 avian sarcoma virus
fgr	Membranes, cytoplasm	Protein-tyrosine kinase	Feline sarcoma virus
fps/fes	Membranes, cytoplasm	Protein-tyrosine kinase	Feline sarcoma virus
mil/raf	Cytoplasm kinase	Serine-threonine kinase	MH2/murine sarcoma virus
mos	Cytoplasm kinase	Serine-threonine kinase	Moloney murine leukemia virus
rel	Cytoplasm kinase	Serine-threonine kinase	Reticuloendotheliosis virus
sis	Cytoplasm factor	Platelet-derived growth	Simian sarcoma virus

that these viruses promote tumorigenesis by inserting near cellular oncogenes and activating their expression, in a process known as "insertional mutagenesis;" (d) Other cellular oncogenes have been identified because they map to sites of consistent chromosomal rearrangements in animal tumors.

Transfection

Transfection techniques are based on the gene transfer procedure of Gram and Van der Eb.[61] In this procedure, high molecular weight DNAs are purified from donor cells that possess a phenotype of experimental interest. The purified DNA is precipitated in calcium phosphate and incubated with recipient cells, usually NIH

Table 5.2.
Cellular Oncogenes Not Transduced by Retroviruses

Oncogene	Method of Identification	Tissue Source
N-myc	Amplification	Neuroblastoma
L-myc	Amplification	Lung carcinoma
B-lym	Transfection	B cell lymphoma
met	Transfection	Osteosarcoma
neu	Transfection	Neuroblastoma
N-ras	Transfection	Neuroblastoma
mcf-2,3	Transfection	Breast carcinoma
mc-D	Transfection	Colonic carcinoma
T-lym-1,2	Transfection	T cell lymphoma
bcl-1,2	Chromosomal translocation	B cell lymphoma
bcr	Chromosomal translocation	Chronic myelogenous leukemia
tcl-1	Chromosomal translocation	T cell leukemia
int-1,2	Insertional mutagenesis	Breast carcinoma
Mlvi-1,2,3	Insertional mutagenesis	T cell lymphoma
pim-1	Insertional mutagenesis	T cell lymphoma
RMO-int-1	Insertional mutagenesis	T cell lymphoma
pvt-1	Insertional mutagenesis	Plasmacytoma

3T3 cells, in hopes that the gene controlling the property of interest will be transferred and expressed. The calcium phosphate–DNA matrix is picked up by the recipient cells, and a small proportion of the DNA is transferred to the nucleus where it is stably incorporated into the genome. In this context, genes that control cell transformation cause the recipient cell to assume an abnormal morphology and to grow without regard for density-dependent inhibition. The cells can be recognized because they grow in refractile foci that stand out against the background of normal cells. The number of foci induced by a given amount of DNA is called its transforming activity.

NIH 3T3 cells are efficiently transformed by oncogenes, probably because they have previously undergone sufficient genetic changes to allow their final transformation by an individual oncogene. Both retroviral (v-*onc*) DNA as well as c-*onc* DNA obtained from *cancer* cells can transform NIH 3T3 cells.[14,18–21,105,106,135,136,137] Transforming activity has been detected in the DNA of approximately 20% of all clinical specimens tested. The oncogenes identified are usually mutant alleles of normal cellular genes (Table 5.3).

These gene transfer experiments have revealed a diverse assortment of oncogenes, many of which were newly identified by this technique. Most of the genes were present in the tumor cells from which the DNA originated; however, some became damaged during the process of transfection.

Insertional Mutagenesis

Retroviruses that do not contain oncogenes may cause cancer by the mechanism of insertional mutagenesis. This was first shown in chicken lymphomas induced by avian leukosis virus (ALV), where the c-*myc* gene is activated by inser-

Table 5.3.
Genes Found by Transfection

ras genes
neu
k-*fgf*
mas
met

tion of retroviral DNA upstream, within, or (rarely) downstream of the gene.[65,66] Elevated c-*myc* mRNA expression occurs in 80% of these retrovirally induced B cell lymphomas.[52,53] Activation of transcription from c-*myc* is generally thought to be the first of several steps in tumor induction. The ALV long terminal repeat (LTR) (which contains sequences which activate transcription) can also integrate upstream of the c-*erb*B gene in chicken erythroblastosis cells, suggesting that erythroblastosis may be caused by activation of the c-*erb*B gene.[57,58] A mutant allele of c-*erb*B, resulting from insertion of retroviral DNA, is a remarkable facsimile of the transduced oncogene, v-*erb*B. It duplicates the amino-terminal portion of the gene product and is apparently responsible for the transforming activity. A number of cellular oncogenes may be activated by insertional mutagenesis (Table 5.4).[37,44,57,58,65,66,92,94–96,101,112,118]

Oncogene Activation

Molecular biologists have several working hypotheses to explain how cellular oncogenes become involved in cancer (Table 5.5).

The first occurs when cellular oncogenes are activated by *quantitative* changes in their expression levels. The second hypothesis postulates that the genes are altered by *qualitative structural* changes in their genes and encoded proteins (Table 5.6).

Table 5.4.
Oncogenes Activated by Insertional Mutagenesis

Virus	Oncogene Found Near the Gene Insertion Site	Tumor
Previously defined		
Avian leukosis virus (ALV)	c-*myc*	Lymphoma + erythroblastosis
Avian leukosis virus (ALV)	c-*erb*-B	Erythroblastosis
Murine leukemia virus	c-*myc*	T cell lymphoma
Moloney murine leukemia virus	c-*myb*	Lymphoma
Novel		
Mouse mammary tumor virus	*int*-1	Breast tumor
Mouse mammary tumor virus	*int*-2	Breast tumor
Murine leukemia virus (MULV)	*pra*-1	T cell lymphoma

Table 5.5.
Evidence for Oncogenes in the Induction or Maintenance of Cancer

1. Viral oncogenes are directly responsible for transformation *in vitro.*
2. Viral oncogenes are directly responsible for tumorigenesis *in vivo.*
3. Transfection of embryo fibroblast or NIH 3T3 cells with oncogenes causes transformation.
4. Transfection assays using DNA from tumors or tumor cell lines cause transformation. In some cases this DNA has been shown to contain oncogenes—often *ras.*
5. Studies with slowly transforming retroviruses that lack a viral oncogene show that they cause cancer by inserting near the site of a proto-oncogene, thereby inducing its expression.
6. The association between specific genetic abnormalities and certain human cancers often occurring at or near proto-oncogenes (best studied examples in 8;14 translocation of Burkitt's lymphoma, involving c-*myc,* or the 9;22 translocation in CML, involving c-*abl).*
7. Evidence of proto-oncogene amplification in tumors (see text). In some instances this appears to be related to the prognosis.

Table 5.6.
Mechanisms of Cellular Oncogene Activation

Point mutation
Gene deletion
Chromosomal rearrangement
Gene amplification
Insertional mutagenesis

Point Mutations

The best-characterized system demonstrating activation of oncogenes by point mutations is the *ras* gene family, where point mutations interfere with the binding of *ras* proteins to the GAP protein, described in detail below. These point mutations, observed in a wide range of human tumors, frequently recurr at specific codons (Fig. 5.4).[24,77]

Amplification and Overexpression

Gene amplification may be associated with readily recognizable chromosomal abnormalities such as double minute chromosomes (DM) or homogeneously staining regions (HSR).[15] Oncogene amplification has been documented in numerous tumors.

Some cellular oncogenes have been identified because they are consistently amplified in naturally occurring tumors (Table 5.7).[3–7,34,35,78,79,85,112,113,129] Two well-known examples include N-*myc* in neuroblastoma, and L-*myc* in small cell carcinoma of the lung (Table 5.7).

Amplification of proto-oncogenes can lead to overexpression, by increasing the amount of DNA template available for mRNA production.[5] In general, en-

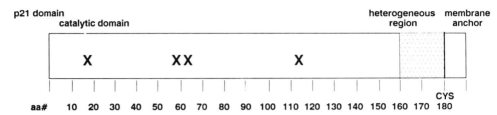

Figure 5.4. Point mutations known to occur in the p21 *ras* genome affect one of several amino acid positions (*X*).

hanced oncogene expression appears to be a prerequisite for the selective growth advantage exhibited by cells containing additional gene copies and could also be the principal contribution of gene amplification to tumorigenesis.[110,111] Furthermore, somatic amplification of specific genes has been implicated in increasing cellular adaptive responses to the environment. Selective pressures may then promote the emergence and clonal evolution of cell populations with increasingly malignant properties. It may also be that some mutagenic insults are only carcinogenic following subsequent amplification, facilitated by the action of tumor promotors or hormones on replicating cells.

Some amplified genes are also rearranged or mutated. The rearranged and amplified c-*erb*B oncogene in A431 epidermoid carcinoma cells produces an abnormal EGF receptor protein.[129] In the erythroleukemia cell line K562, an amplified DNA segment consists of portions of both the c-*abl* and the *bcr* genes resulting in an ab-

Table 5.7.
Amplified Genes

Oncogene	Cell Line	Tumor
Cell lines		
c-*myc*	HL-60	Acute promyelocytic leukemia
	COLO 320	Colon cancer
	SKBR-3	Breast carcinoma
c-*myb*	ML1-3	AML
	COLO 201/205	Colon cancer
c-*abl*	K562	CML
c-*erb*B	BA 431	Squamous carcinoma
c-Ki-*ras*	Y1	Adrenal cortical cancer
Tumor Tissues		
c-*myc*		Small cell carcinoma
		Colon cancer
N-*myc*		Neuroblastoma
		Small cell carcinoma
		Colon carcinoma
L-*myc*		Small cell carcinoma
c-*erb*B		Glioblastoma
c-*erb*B2/*neu*		Breast cancer

normal c-*abl* oncogene. It is not clear whether the structural alterations of the genes precede their amplification or whether they are acquired during amplification.

In discussing the role of gene amplification, one must keep in mind that expression of an oncogene-encoded product may have certain effects on normal cellular processes when it is subjected to normal regulatory mechanisms and expressed in a particular cell type in a defined, differentiated state. However, when the usual cell regulatory controls are lost (as in tumors), the amplified protein products may lose their routine functions or function in an abnormal way.

Defined Chromosomal Translocations of Cellular Oncogenes

Some putative cellular oncogenes have been recognized because they map to consistent chromosomal breakpoint sites in human tumors. In fact, the coincidence between chromosomal translocations and deletions characteristic of certain tumors and the known chromosomal locations of some human cellular oncogenes has been quite striking.

Translocations can affect either the expression or the biochemical function of proto-oncogenes. Effects on expression are exemplified by translocations that join c-*myc* to various immunoglobulin genes in Burkitt's lymphoma and mouse plasmacytomas. The translocation t(8;14)(q24;q32) is characteristic of Burkitt's lymphoma. The human c-*myc* gene is normally located on chromosome 8 band q24.[45] Immunoglobulin heavy (H) chain genes are located on chromosome 14 (14q32), κ light chain genes are on chromosome 2 (2p12), and the λ light chain genes are on chromosome 22 (22q11).[41,78] These tumors carry one of three specific translocations (t(8;14), t(8;2), t(8;22)) that act by juxtaposing c-*myc* to one of three immunoglobulin loci.

The consequences of the fusion of *myc* to immunoglobulin genes ultimately results in elevated *myc* expression through several mechanisms. First, transcription of c-*myc* into RNA may be released from its usual controls, allowing expression of the gene at inappropriate times. Second, regulatory influences provided by the immunoglobulin genes may drive the expression of the c-*myc* to higher levels than normal. Third, damage inflicted on c-*myc* by translocation may increase the stability of the mRNA derived from the gene. Consequently, mRNA levels increase and become less accessible to rapid modulation. Some of the structural signals governing the stability of c-*myc* mRNA are apparently located in the first (untranslated) exon of the gene, and excision of this exon by translocation (if excision occurs) could alter mRNA stability.

The Philadelphia chromosome (9q34) translocation t(9;22)(q34;q11) that typifies CML represents the second type of genetic damage. The translocation fuses a portion of the proto-oncogene c-*abl* with a region known as the breakpoint cluster region (*bcr*) on 22q11, resulting in a fusion product whose enzymatic activity is increased over that of the normal gene (See chapter 6).

Insertional Mutagenesis

As already noted, evidence exists for proto-oncogene activation by retroviral promoter or enhancer insertion. Only a small part of the retroviral genome, which contains an LTR without coding functions, is required to activate an adjacent cellular oncogene, although DNA encoding some of the viral proteins (*gag* and *env*) may also be present.[44,76,77] It is now well known that enhancers of transcription can act over longer distances than promoters,[75] and it appears that the observed effects are due to the enhancer activity of retroviral LTR sequences. The multipotential activities of slowly oncogenic viruses may be explained by their insertion into, or near, different oncogenes. Promoter insertion near a proto-oncogene, such as c-*myc* or c-*erb*B, may be necessary to immortalize the recipient cells, but insufficient to establish complete malignant transformation.

Structure and Function of Specific Cellular Oncogenes

Proteins encoded by oncogenes have been analyzed and placed into a number of functional categories based on their location in the cell or knowledge of their physiological action at the cellular level (Table 5.8). Furthermore, oncogenes may fit into a cascading hierarchy of expression in the cell, where the action of one controls another.

ONCOGENE NETWORKS

Cell proliferation and differentiation is governed by an elaborate circuitry that extends from the cell surface to the nucleus.

Proteins encoded by cellular oncogenes may function at any point in this pathway. Cellular oncogenes may encode growth factors and hormones, growth factor and hormone receptors, proteins that function as signal transducers transmitting signals from the plasma membrane to the nucleus (such as the *ras* family of G proteins and the *src* family of protein-tyrosine kinases), and proteins that bind to DNA sequences to regulate gene expression—so-called "transcriptional regulatory factors."

RELATIONSHIP OF ONCOGENES TO THE CELL CYCLE

A number of oncogenes are expressed at a precise part of the cell cycle (Table 5.9).[24,74,92]

c-*fos*, c-*myc*, c-*myb* and c-*jun* are expressed during the transition of quiescent cells from the G_0 to the G_1 phase of the cell cycle, suggesting that these genes encode proteins that then induce the transcription of other genes required for cell pro-

Table 5.8.
Functional Groups of Oncogene-Encoded Products

Nuclear proteins/transcription factors
Growth factors
Growth-factor receptors
Signal transducers

Table 5.9.
Oncogenes Expressed in Relation to the
Cell Cycle

c-*myc*
c-*fos*
c-*myb*
c-*jun*

liferation. These genes are inducible and are regulated by specific growth signals including mitogens, growth factors, drugs, hormones, TPA, and serum.[26,74]

In contrast to c-*myc* and c-*fos*, expression of the cellular protein p53 is induced later in the cell cycle, five to six hours after addition of growth factors.[114] Increases in c-Ki-*ras* mRNA have been reported to occur late in G_1 in fibroblasts and in regenerating liver.

ONCOGENES WITH A NUCLEAR FUNCTION

Several proteins encoded by cellular oncogenes are located in the nucleus and are thought to regulate gene expression.[50] Furthermore, some nuclear oncoproteins such as SV40 large T antigen, *myc*, *myb* and *fos* possess DNA-binding activity. In certain cases, these genes have been shown to encode nuclear proteins that bind regulatory regions in DNA, to regulate transcription of cellular genes. With the possible exceptions of *fos* and *ski*, the proteins have been classified as having immortalizing activity, based on their ability to rescue primary cells from senescence without concurrent tumorigenicity and their ability to cooperate with an activated *ras* gene in the transformation of primary cells.[82,83,97]

Direct evidence for the participation of a nuclear oncoprotein in transcriptional regulation is provided by the finding that *erb*A product is a nuclear receptor for thyroid hormone and by the identity of the transcription factor AP1 with the c-*jun* proto-oncogene.[8,9] c-*jun* and c-*fos* regulate gene expression by binding to DNA sequences as jun/jun homodimers and jun/fos heterodimers.

Changes in gene expression occurring during the transition from quiescence to proliferation might be modulated by families of transcription factors encoded by proto-oncogenes, either acting as single molecules or mutually interacting in an active complex.

myc Genes

The *myc* family of genes contains six functional members that include c-*myc*, N-*myc*, L-*myc*, R-*myc*, P-*myc*, and B-*myc*.[47,48] It also contains one inactive pseudogene, L-*myc*-psi. N-*myc* and L-*myc* genes were both isolated based on their frequent amplification in some types of human tumors. R-*myc* and P-*myc* were isolated based on their homology with the L-*myc* third exon. B-*myc* was isolated from a rat genomic library; it is highly homologous to c-*myc*, but it localizes to a different chromosome.

c-*myc* is the homologue of the transforming gene of avian myelocytomatosis virus involved in the pathogenesis of chicken B cell lymphomas. It is induced by the nonacute leukosis virus (RAV-2)[101] and encodes a nuclear phosphoprotein that binds to single- or double-stranded DNA.[1,22] The human c-*myc* protein has a half-life of 20–30 minutes, is found in proliferating, but not resting cells, and is expressed in almost all cell types with a high potential for proliferation.[64]

Various avian, rodent, feline, and human tumors express activated c-*myc* genes.[30,66] The bulk of the evidence suggests that aberrant expression of c-*myc* is the principal mechanism for oncogenic conversion. The possible mechanisms whereby changes in c-*myc* expression occur include: (*a*) insertion of a strong retroviral promoter in the vicinity of c-myc[66]; (*b*) amplification of c-*myc* DNA; or (*c*) chromosomal translocations that bring the c-*myc* gene in close proximity to immunoglobulin loci.[84,87]

Translocated *myc* Gene

The translocated *myc* allele is deregulated and frequently overexpressed, while the normal allele remains silent.[77] In typical translocations, the *myc* oncogene breaks at its noncoding 5′ end or at variable distances upstream of it. The coding exons of the gene are transposed to the chromosome containing immunoglobulin heavy-chain genes, head-to-head with the immunoglobulin gene. c-*myc* is then found on chromosome 14 q+, and the two gene loci are connected in a transcriptionally opposing fashion.[84] In contrast, variant translocations (those involving the immunoglobulin κ light chains from chromosome 2 or λ light chains from chromosome 22)[17] break the chromosome at the 3′ end of the gene.[41,42] The c-*myc* gene stays in its original position on chromosome 8. The constant genes and part of the variable genes of the light-chain locus are transposed to the 3′ end of the *myc* gene, giving a tail-to-head *myc* light-chain orientation. The oncogene is probably activated from the light-chain region in the downstream position.

Despite their differences, all of these translocations increase the level of expression of *myc* transcripts. The variable breakpoints upstream and downstream of the gene always leave coding exons 2 and 3 intact, suggesting that the *myc* protein plays a crucial role in transformation.

Relationship to Proliferation and Differentiation

A substantial body of evidence suggests that the *myc* gene is intimately associated with control of proliferation and differentiation (Tables 5.10 and 5.11).[26,36,49,89,62,81,139]

Many cell types that are in the exponential growth phase or that are stimulated by mitogens to enter the cell cycle, express elevated c-*myc* mRNA levels. In contrast,

Table 5.10.
Proposed Functions for c-*myc*

Promotion of DNA replication
Regulation of G_0/G_1 transition
Cell differentiation

Table 5.11.
Relationship of c-*myc* Expression to Differentiation

Quiescent 3T3 cells have lower mRNA levels than proliferating cells.

Terminal differentiation of murine teratocarcinoma cells associated with decreased mRNA levels.

Terminal differentiation of murine erythroleukemia cells associated with decreased mRNA levels.

Transfection of cells with c-*myc* inhibits differentiation.

Blocking c-*myc* expression with anti-sense oligonucleotides in HL60 cells induces differentiation without the use of a chemical inducer.

quiescent cultures express little or no c-*myc*. Significant decreases in c-*myc* levels are also found in late passages as compared to early-passage cells. Transformed cells constitutively express c-*myc* mRNA; expression rapidly diminishes with the induction of differentiation.

A strong relationship exists between the level of c-*myc* mRNA expression and the proliferative capacity of cells. Tissues with relatively high rates of cell division have substantial amounts of *myc* RNA.[74] In contrast, skeletal muscle, which has few dividing cells, has barely any detectable *myc* transcripts. In apparent contrast to somatic cells, dividing germ cells have very few *myc* transcripts and appear to proliferate, at least for a few divisions, in the absence of *myc* transcription. Regenerating rat liver cells following partial hepatectomy display rapid induction of c-*myc* mRNA.[86] Human placental cytotrophoblasts show developmental changes in c-*myc* mRNA expression that parallel their proliferative activity.[103]

myc Genes in Tumors

A causal role for *myc* in the genesis and/or progression of tumors is suggested by its frequent amplification and/or overexpression in neuroblastomas and other tumors, such as Wilms' tumor, small cell lung carcinoma, and retinoblastoma (see Chapters 8–12). Additional experimental systems that implicate c-*myc* involve the use of transgenic mice possessing a *myc* gene under the control of various promoters, which develop tumors in varying sites.

myc expression alone is insufficient for tumor formation, and few cells expressing v-*myc* will form a tumor. Furthermore, not every mouse harboring a *myc* transgene develops tumors. Furthermore, transformed cells that contain a somatically altered c-*myc* gene possess other activated oncogenes. Only certain combinations of oncogenes (*e.g.*, highly activated *myc* and mutated *ras*) effectively transform a cell, as assayed by immortalized growth *in vitro* and tumor formation in nude mice.

fos Genes

The *fos* and *jun* genes are closely related to one another, and both are considered to be immediate-early genes. (The concept of the immediate-early gene arose by analogy with viral genes that are expressed in the presence of protein synthesis inhibitors prior to gene replication following viral infection.) They are induced by growth factors and hormones and encode proteins displaying similarity with known transcription factors.[30,108,43] The proteins contain zinc fingers, a type of DNA binding domain present in several well-characterized transcription factors.

The *fos* gene encodes a nuclear phosphoprotein that forms noncovalent complexes with several other proteins including the product of another proto-oncogene, c-*jun* (Fig. 5.5). The *fos* gene products are phosphoproteins that form complexes with other nuclear proteins, bind to DNA sepharose, and appear to be associated with chromatin.[108]

Data suggest that c-*fos* represents an integral part of the intracellular messenger pathway that triggers gene expression and ultimately phenotypic alterations. In fact, c-*fos* induction is essential for the growth-promoting activities of certain growth factors and serum. It has been suggested that *fos* plays a role as the *trans* acting–transcription factor for the adipocyte gene AP2, the mouse α-1 (III) collagen gene, collagenase, some phorbol esters, and several viral promoter regions.

jun Genes

The recently discovered *jun* proto-oncogene appears to occupy a central role in the regulation of proliferation.[134,135] It was originally identified as the cellular counterpart of a retrovirally transduced oncogene, v-*jun*. The gene was identified as the oncogene of avian sarcoma virus 17, which causes fibrosarcomas in chickens. The name "*Jun*" comes from the Japanese *ju-nana*, meaning 17. Normal avian and mammalian cells also carry the *Jun* gene. Three members of the Jun/AP-1 family have been identified in mouse cDNA libraries: c-*Jun, Jun*-B, and *Jun*-D.[107] The cellular *jun* gene encodes a protein that is closely related and probably identical to the transcriptional regulator AP1. It localizes in the cell nucleus, binds to a specific DNA regulatory sequence, and can activate transcription from promoters containing the AP1 consensus sequence. The viral *jun* gene codes for similar DNA-binding protein that is also capable of activating transcription.[8,9,34] It is strongly suspected that the transcriptional regulator function of v-*jun* plays an important role in oncogenesis.

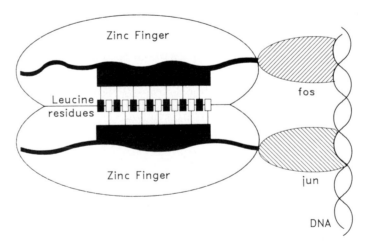

Figure 5.5. Diagram of *fos* and *jun* protein products. Their DNA-binding sites become juxtaposed when their hydrophobic leucine residues associate.

The cellular *jun* gene does not induce tumors under normal physiologic conditions of expression and of activation, and it is not yet known whether a mere quantitative up-regulation of c-*jun* expression (as can be achieved by incorporating the gene into a retroviral vector) can lead to oncogenic transformation.[134]

CELLULAR ONCOGENES THAT ENCODE PROTEIN KINASES

Protein kinases can be divided into two functional groups, those that code for intracellular protein kinases, of which the product of c-*src* gene is the prototype[27] and those that function as transmembrane growth factor receptors (Table 5.12).[68–71.]

The *src* family contains a total of 8 similar genes, all of which encode protein-tyrosine kinases. In contrast to transmembrane growth factor receptors that have an extracellular ligand-binding domain, a transmembrane region, and an intracellular tyrosine kinase domain, the *src* family kinases possess a tyrosine kinase domain, which may be associated with intracellular membranes.[33,63,68–71]

src Family of Genes

The *src* family of oncogenes includes 8 genes all of which are at least 75% identical to one another in the amino acid sequence of their terminal catalytic domains.[63,71] All of the proteins encoded by these genes exhibit tyrosine-specific protein kinase activity.[33,71,93] The protein tyrosine kinase specified by *src* does not contain hydrophobic stretches capable of spanning the plasma membrane and therefore lack extracellular ligand-binding domains. Physiologic regulators of *src*-like kinase activity in any cell type are unknown, but they are considered attractive candidates for components of a signal transduction cascade, since it is clear that mutated versions of these molecules, with increased kinase activity, can stimulate neoplastic growth in many cell types.[104]

v-*src*, the most extensively characterized viral oncogene, is capable of inducing tumors *in vivo* and transforming cells *in vitro*. The oncogene of the Rous sarcoma virus, v-*src*, arose by transduction of a cellular gene, c-*src*. c-*src* encodes a 60,000 molecular weight phosphoprotein (pp60$^{v\text{-}src}$) that catalyzes tyrosine-specific protein phosphorylation (Fig. 5.6).[68,99]

The transforming ability of v-*src* is dependent on the intrinsic tyrosine kinase activity of pp60$^{v\text{-}src}$ with several candidate *in vivo* substrates. pp60$^{c\text{-}src}$ is functionally

Table 5.12.
Protein Kinases

Transmembrane Receptors	Src-Family	Miscellaneous
CSF-1 receptor	src	abl
EGF-1 receptor	fyn	fps/fes
Insulin receptor	hck	pim-1
PDGF receptor	lck	rel
kit	lyn	ret
met	yes	sea
Her-2/neu	fgr	
ros	blk	

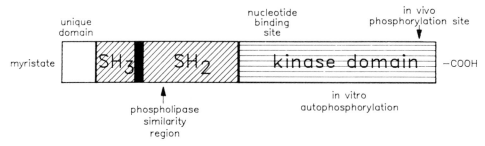

Figure 5.6. Tyrosine-specific protein kinase (pp60$^{v\text{-}src}$) encoded by v-*src* is divided into four domains. The first seven amino acids are necessary for the myristylation required for salt-resistant membrane localization. The other three include a phospholipase similarity region (SH$_2$ and SH$_3$), a nucleotide binding site, an *in vivo* phosphorylation site, and a kinase domain, which phosphorylates target proteins on tyrosine residues.

divided into four domains (Fig. 5.6).[59,100] The first seven amino acids are necessary and sufficient for myristylation. Myristylation is required for salt-resistant membrane localization.

The pp60^{src} protein phosphorylates only a relatively small percentage of given target proteins under *in vivo* conditions, for example, 1% of vinculin. Vinculin is the protein that anchors the actin microfilaments to the cell membrane and the cell substratum. Destabilizing the anchorage leads to a rounding of the cell shape (Fig. 5.7).[115–117] As a result of transformation by the *src* gene, a number of other changes occur in the cell (Table 5.13).[10–13]

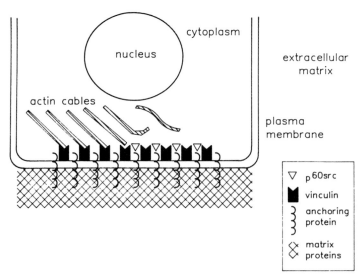

Figure 5.7. Phosphorylation of vinculin by p^{60src} results in disruption of actin binding to vinculin. (Adapted from Hunter T: The proteins of oncogenes Sci. Am. *251*:75, 1984.)

Table 5.13.
Cellular Changes Induced by *src* Genes

Altered and phosphorylated cytoplasmic proteins
Reduced epithelial adhesion
Microfilament disorganization and loss of fibronectin
Actin loss and changes in cell shape
Loss of density-dependent inhibitors of cell growth
Alteration in gap junction cell-to-cell communication
Loss of adhesion plaques and adherens junctions
Reduced junctional permeability

abl Genes

The c-*abl* gene is the cellular homologue of the transforming sequence of *abl*–murine leukemia virus, a replication-defective retrovirus capable of transforming fibroblasts in lymphoid cells and of rendering IL-3-, GM-CSF-, or IL2-requiring cells growth-factor independent. The original isolate of AMULV encodes a 160 kilodalton *gag-abl* fusion protein with tyrosine kinase activity. The cellular homologue is a tyrosine kinase family and is highly conserved. c-*abl* has the SH$_2$, SH$_3$, and kinase domains of the *src* family (Fig. 5.6) but has a unique amino terminus and a much longer carboxy terminus.

In mice, the c-*abl* gene is transcribed as two major and two minor RNAs, with the two major RNAs initiated by separate promoters and the minor transcripts arising by alternative splicing. Like the human gene, one of the alternative murine 5′ c-*abl* exons lies far upstream of the remaining exons. The major RNA that begins with this exon has 1275 nucleotides upstream of the *abl*-coding region. This unusually long upstream mRNA segment contains multiple, short, open reading frames, both in mouse and man and is highly conserved in sequence between the species. All of the four cDNA types diverge at a common position that corresponds to a splice-acceptor site in the genome, suggesting that four c-*abl* mRNAs arise by the addition of alternative 5′ exons onto a common set of 3′ exons.

The most frequent chromosomal abnormality associated with human neoplasms is the Philadelphia chromosome, which is present within leukemic cells in at least 90% of patients with CML and results from a translocation between chromosomes 9 and 22. It has been suspected for a long time that this translocation activates a cellular oncogene, which then stimulates uncontrolled proliferation of the myeloid stem cells and their progeny. This oncogene is now known to be c-*abl*, which resides at 9q34 (see Chapter 6).

erb Genes

The avian erythroblastosis virus contains two distinct oncogenes, *erb*A and *erb*B. The *erb*B oncogene, which is homologous to a portion of the epidermal growth factor receptor, is a transmembrane growth factor receptor with tyrosine kinase activity.[46,56] *erb*A potentiates the effects of *erb*B by blocking the differentiation of erythroblasts at an immature stage. Two acute transforming retroviruses, AEV-ES4 and AEV-R, were the original source of the viral *erb*B sequences, with these two avian erythroblastosis viruses probably representing independent isolates that apparently were captured during independent recombinational events. v-*erb*A is not structurally linked or related to v-*erb*B and is not oncogenic when expressed independently, although it can augment v-*erb*B-mediated transformation events.

HER2/*neu* (*erbB2*)

c-*erb*B-2 is believed to encode a growth-factor receptor similar to the EGF receptor (EGFR).[56] Sequence analysis and chromosomal-mapping studies have revealed that the three genes (*neu*, c-*erb*B2, and HER-2) are the same.[56] The *neu* transforming gene was identified as the result of transfection studies with DNA from chemically induced rat glioblastomas.[120] Another group, also using v-*erb*B as a probe, identified the same gene in a mammary carcinoma cell line, MAC 117, where it was found to be amplified 5–10-fold.

This gene encodes a new member of the tyrosine-kinase family, which is closely related to but distinct from the EGFR gene. It differs from EGFR in that it is found on chromosome 17, as compared to chromosome 7, where the EGFR is located. The HER-2/*neu* gene generates an mRNA of 4.8 kb, differing from the 5.8–10 kb transcripts of the EGFR gene. The EGFR protein HER-2/*neu* has an extracellular domain and a transmembrane domain that includes two cysteine-rich repeat clusters and an intracellular tyrosine kinase domain.[109]

The EGF receptor is the product of a proto-oncogene (*erb* B) and enables cells to undergo DNA synthesis in response to either of its known physiologic EGF receptor ligands, EGF and TGF α. The c-*erb*B-2/HER2 gene is a potent oncogene when sufficiently overexpressed in NIH 3T3 cells, even in the absence of an identifiable ligand, but its normal rat *neu* counterpart lacks similar transforming activity. On the other hand, overexpression of the EGF-R (c-*erb*B) gene causes transformation of NIH 3T3 cells only in the presence of EGF.

Tal *et al.* showed the sporadic amplification of the HER2/neu proto-oncogene in adenocarcinomas arising in various sites.[126] DNA rearrangement of the gene was noted. This rearrangement was confined to the 3′ origin of the gene and was expressed as an aberrant HER2/neu polypeptide with a molecular weight of 190,000, which is larger than the normal HER2/neu protein. It also demonstrated intrinsic protein and tyrosine kinase activity leading to cell phosphorylation.[2] Amplification of the HER2/neu proto-oncogene may play a role in transformation, since it has been shown that overexpression of the human HER2/neu proto-oncogene in NIH 3T3 cells results in transformation and tumorigenesis (see Chapter 11).

ONCOGENES ENCODING GROWTH FACTORS

int Gene

The oncogene *int*-1 was identified in mammary tumors and is activated by the integration of proviral DNA of the mouse mammary tumor virus.[94–96] *int*-1 encodes a protein in the fibroblast growth factor family. *int* proto-oncogenes are structurally related to one another and are expressed during different times in mouse embryogenesis. In adult mice *int*-1 is expressed only in the testis, but during develop-ment there is limited expression in specific regions of the neural plate and its deriv-atives, suggesting a role in the early stages of CNS development. This oncogene has also been shown to be related to pattern formation in *Drosophila*.[25]

There are currently four bona fide *int* loci identified on different mouse chro-mosomes and associated with the expression of tumor-specific RNAs from respec-tive genes. In addition, there is a suggestion of others. One of these is described in two separate cases of plaque type P mammary tumors characteristic of the GR strain of mice, and this is referred to as the *int*-P.

Sis

Simian sarcoma virus was isolated from the fibrosarcoma of a Woolly monkey and is the only known sarcoma virus of primate origin.[51] It has been shown that the v-*sis* gene is derived from genes encoding a platelet-derived growth factor. It is 92% homologous to the non-glycosylated β-chain, a potent mitogen for mesenchymal cells.

Autocrine stimulation of intracellular PDGF receptors in v-*sis*-transformed cells is the mechanism of transformation by the v-*sis* oncogene. When the PDGF re-ceptor is activated by PDGF, the receptor becomes phosphorylated on its own tyro-sine residues. Keating and Williams found that in contrast to receptor activation in normal cells, autocrine activation of PDGF receptors in v-*sis* transformed cells oc-curred in the intracellular compartments, disrupting receptor processing and divert-ing receptors and their precursors, demonstrating that intracellular activation of re-ceptors by autocrine mechanisms may play a role in cell transformation.[55,73]

ras GENE FAMILY OF SIGNAL TRANSDUCERS

ras Genes

The ras family has many members, with different degrees of homology in their effector regions (Table 5.14).

Table 5.14.
***RAS* Gene Comparisons**

Gene	Size	Chromosomal Location
Ras genes		
c-Ha-*ras*-1	< 6 kb	Chromosome 11 (11p15.1-p15.5)
c-Ki-*ras*-2	>40 kb	Chromosome 12 (12p12.1-pter)
N-*ras*	10 kb	Chromosome 1 (1p22-p32)
Ras pseudogenes		
c-Ha-*ras*-2	—	x Chromosome
c-Ki-*ras*-1	—	Chromosome 6 (6p12-p32)

Ha-*ras* and Ki-*ras* were first defined by virtue of their homology to the transforming genes of Harvey murine sarcoma virus and Kirsten murine sarcoma virus, respectively. The cellular Harvey *ras* gene, homologous to the Harvey murine sarcoma virus oncogene v-Ha-*ras,* was first isolated as the transforming gene of the EJ and T24 human bladder carcinoma cell lines.[121,122]

The N-*ras* gene was discovered in transfection assays utilizing transforming DNA from neuroblastoma cell lines seen by low-stringency hybridization to v-Ha-*ras,* named N-*ras.*[123] Each of these genes is composed of four coding exons with identical intron spacing, and each encodes a 21,000-dalton polypeptide (p21).[127] However, each polypeptide differs slightly in the carboxy terminus, suggesting that each gene product may have slightly different functions or that their activity is regulated differently.

The products of R-*ras*, *rap*-1, and *rap*-2 genes have effector regions virtually identical to the "true" *ras* proteins, whereas the *rho, ral, rab*-1, *rab*-2, *rab*-3, and *rab*-4 products differ in this region.

ras Mutations

About 15% of all human tumors including carcinomas, sarcomas, melanomas, and leukemias contain mutated c-*ras* oncogenes that can be detected by various methods.[124] *ras* point mutations may be detected by the polymerase-chain reaction using mutation-specific oligonucleotide primers. These mutations tend to preferentially affect specific amino acid positions, especially positions 12, 13, and 61 (Fig. 5.4).[24, 106,107]

ras Proteins

ras-encoded proteins are found in two states, a GTP-bound active state and a GDP-bound inactive state. Active *ras* proteins are converted to an inactive form by intrinsic guanosine triphosphatase (GTPase) activity that is stimulated by interaction with the GTPase activating protein, GAP (Fig. 5.8). Upon binding, GTP-*ras* proteins become activated and can stimulate cell proliferation through an unknown mechanism. Proteins encoded by oncogenic alleles of cellular *ras* are often mutated in a manner that decreases their intrinsic GTPase activity, by a conformational change in the guanine nucleotide binding of the protein, or by nonproductive association with the GAP protein.[60] Thus, proteins encoded by these alleles remain in the active conformation for a longer period of time.

ras proteins have been found to be enhanced in their expression in breast carcinomas, using antibodies in both immunohistochemical and immunoassay techniques. However, these data are currently controversial (see Chapter 11).

ras Activation

Both quantitative and qualitative alterations in mammalian *ras* genes seem to be involved in the activation of their oncogenic potential. Malignant transformation of established normal rat fibroblasts by mutant human *ras* oncogenes is reversible upon loss of the oncogene sequences, and moderate fluctuations in mutant *ras* gene amplification and expression have drastic consequences in the manifestation of their transformed phenotype.

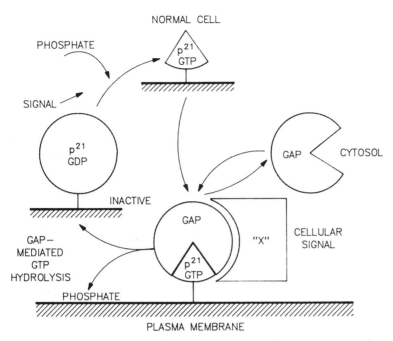

Figure 5.8. Interaction of *ras*-encoded protein p21 with quanosine triphosphatase (GTPase) and GTPase-activating protein (GAP) is illustrated. The active state, p21-GTP, is converted to the inactive p-21-GDP by GTP hydrolysis, which is stimulated by GAP.

Several lines of evidence indicate that the degree of cellular transformation by a structurally altered or mutated p21 is a function of the amount of this protein in the cell. For example, progressively higher rates of p21 synthesis controlled by an inducible promoter correlate well with the increasingly transformed phenotype of the cell. In addition, overexpression of the normal cellular Ha-*ras* gene can result in transformation of NIH 3T3 cells.

Carcinogenesis

Most carcinogens are mutagens, with a number of chemical carcinogens forming adducts with DNA bases. Some of these adducts are highly mutagenic due to their miscoding properties, whereas others can lead to mutations by the generation of apurinic sites or because of the limited fidelity of repair polymerases. Most of these mutations will have no consequence to the host; however, a number can trigger neoplastic development. Reproducible activation of *ras* oncogenes in carcinogen-induced tumors has made it possible to correlate their activating mutation with the known mutagenic effects of certain carcinogens. Each of the H-*ras*-1 oncogenes present in rat mammary carcinomas induced by NMU but not DMBA became activated by G→ A transition, which is the most common mutation induced by NMU. This has led to the proposal that NMU is directly responsible for the malignant activation of H-*ras*-1 oncogenes in mammary tumors.[140] Induction of skin carcinomas in mice by DMBA and phorbol esters involves a specific activation of H-*ras*-1 oncogenes by A→T transitions in the second base of codon 61.[23] Similarly, in hepatocarcinomas induced by treatment of B6C3F$_1$ mice with a single dose of HO-AAF generates H-*ras*-1 oncogenes activated by a C→A transversion in the first base of codon 61.

Multistep Carcinogenesis

Because viral oncogenes are the fastest-acting carcinogens known and are capable of both the initiation and the maintenance of the neoplastic phenotype, it was initially thought that "activation" of homologous cellular oncogenes might lead directly to neoplasia. However, there is no evidence that activated cellular oncogenes acting alone are necessary, much less sufficient, for carcinogenesis.[82,83,97,132] In contrast to the rapid transformation achieved by retroviruses, most naturally occurring human cancers develop from a succession of steps occurring over many years, presumably involving changes in the expression of multiple genes. Consistent with this stepwise process are data that suggest that carcinogenesis results from the concerted action of several "activated" cellular oncogenes, in addition to the loss of other cellular genes that suppress the development of the tumor cell phenotype.

Land *et al.* reported that the human bladder carcinoma EJ-Ha-*ras*1 transforming gene transformed rat embryo fibroblasts only to limited proliferation in agar. EJ-Ha-*ras*1 gene combined with v-*myc* gene, early simian virus 40, transcriptional promoter attached to c-*myc*, or DNA tumor virus genes increased the tumorigenicity of transformed rat embryo fibroblasts. These authors also observed that EJ-Ha-*ras*1 together truncated large Polyoma T antigen gene, or the E1A gene led to unlimited tumor growth, whereas EJ-Ha-*ras*1 in combination with the v-*myc* transfected into diploid rat fibroblasts produced tumors of limited size in the nude mouse. The authors concluded that the *myc* gene greatly enhanced the transformed phenotype created by the EJ-*ras* transforming gene, but did not lead to fully transformed cells obtained by cotransfection with the genes of the DNA tumor viruses. These experiments confirmed the notion that carcinogenesis is at least a two-step process in which different transforming genes act co-ordinately. It appears that transforming genes can be grouped into at least two categories. One can distinguish transforming genes necessary for immortalization or establishment of primary cells in culture from genes that are required for anchorage-independent proliferation and/or tumorigenicity.

Tumor Progression

The ability of *ras* oncogenes to induce metastatic phenotypes has been investigated. NIH 3T3 rat embryo cells transfected with *ras* oncogenes can form metastatic nodules in the lung, when injected subcutaneously into nude mice.[91] Others, however, have only observed metastatic foci when *ras*-transformed NIH 3T3 cells were re-transfected with DNAs isolated from certain metastatic human tumor cell lines or when the cells were injected intravenously.[32] In a different system, transfection of MT1C1 mouse mammary carcinoma cells with the H-*ras*-1 oncogene increased their capacity to metastasize when injected subcutaneously, but not when injected intravenously. The ability to induce metastases is independent of the levels of p21 expression, thus, induction of the metastatic phenotype is not an intrinsic property of *ras* oncogenes, a conclusion supported by the similar incidence of *ras* oncogenes in primary and metastatic tumors.

In fibroblasts in which *ras* genes have been induced *in vitro*, acquired metastatic ability is judged by the pulmonary embolism assay, and the metastatic phenotype appears almost immediately on expression of inserted *ras* genes. *Ras*-expressing cells develop new features, some of which may favour autonomous growth, such as the production of tumor growth factors and release from preexisting control by exogenous trophic stimuli. Some of these features correlate with invasive and metastatic ability, such as increase sialylation of surface glycoproteins.

Acknowledgments

The authors wish to thank Medical Media at the VA Medical Center for their expertise in preparing the illustrations and Agnes Truske for her manuscript preparation.

REFERENCES

1. Abrams, H. D., Rohrschneider, L. R., and Eisenman, R. N. Nuclear location of the putative transforming protein of avian myelocytomatosis virus. *Cell* 29:427–439, 1982.
2. Akiyama, T., Saito, T., Ogawara, H., Toyoshima, K., and Yamamoto, T. Tumor promoter and epidermal growth factor stimulate phosphorylation of the c-erbB-2 gene product in MKN-7 human adenocarcinoma cells. *Mol. Cell Biol.* 8:1019–1026, 1988.
3. Alitalo, K. Amplification of cellular oncogenes in cancer cells. *Med. Biol.* 62:304–317, 1984.
4. Alitalo, K., Koskinen, P., Makela, T. P., Saksela, K., Sistonen, L., and Winqvist, R. myc Oncogenes: Activation and amplification. *Biochem. Biophys. Acta* 907:1–32, 1987.
5. Alitalo, K., Saksela, K., Winqvist, R., Alitalo, R., Keski-Oja, J., Laiho, M., Ilvonen, M., *et al.* Acute myelogenous leukaemia with c-*myc* amplification and double minute chromosomes. *Lancet ii*:1035–1038, 1985.
6. Alitalo, K., and Schwab, M. Amplification of cellular oncogenes in tumor cells. *Adv. Cancer Res.* 11:201–215, 1983.
7. Alitalo, K., Schwab, M., Lin, C. C., Varmus, H. E., and Bishop, J. M. Homogeneously staining chromosomal regions contain amplified copies of an abundantly expressed cellular oncogene (c-*myc*) in malignant neuroendocrine cells from a human colon carcinoma. *Proc. Natl. Acad. Sci. U.S.A.* 80:1707–1711, 1983.
8. Angel, P., Allegretto, E. A., Okino, S., Hattori, K., Boyle, W. J., Hunter, T., and Karin, M. Oncogene jun encodes a sequence specific trans-activator similar to AP-1. *Nature (Lond)* 332:166–171, 1988.
9. Angel, P., Hattori, K., Smeal, T., and Karin, M. The jun proto-oncogene is positively autoregulated by its product, Jun/AP-1. *Cell* 55:875–885, 1988.
10. Azarnia, R., and Loewenstein, W. R. Intercellular communication and the control of growth: XII. Alteration of junctional permeability by Simian virus 40. Roles of the large and small T antigens. *J. Membr. Biol.* 82:213–220, 1984.
11. Azarnia, R., and Loewenstein, W. R. Intercellular communication and the control of growth: X. Alteration of junctional permeability by the *src* gene. A study with temperature-sensitive mutant Rous sarcoma virus. *J. Membr. Biol.* 82:191–205, 1984.
12. Azarnia, R., and Loewenstein, W. R. Intercellular communication and the control of growth: XI. Alteration of junctional permeability by the *src* gene in a revertant cell with normal cytoskeleton. *J. Membr. Biol.* 82:207–212, 1984.
13. Azarnia, R., Reddy, S., Kmiecik, T. E., Shalloway, D., and Loewenstein, W. R. The cellular *src* gene product regulates junctional cell-to-cell communication. *Science (Wash DC)* 239:398–401, 1988.
14. Barbacid, M. Human oncogenes. In: *Important Advances in Oncology 1986*, edited by

V.T. DeVita, S. Hellman, and S. A. Rosenberg, Philadelphia, J. B. Lippincott Co., 1986, pp. 3–22.

15. Barker, P. E. Double minutes in human tumor cells. *Cancer Genet. Cytogenet.* 5:81–94, 1982.

16. Baserga, R. *Multiplication of Animal Cells.* New York, Decker, 1976.

17. Bernheim, A., Berger, R., and Lenoir, G. Cytogenetic studies on African Burkitt's lymphoma cell lines: t(8;14), t(2;8), and t(8;22) translocation. *Cancer Genet. Cytogenet.* 3:307–315, 1981.

18. Bishop, J. M. Cellular oncogenes and retroviruses. *Annu. Rev. Biochem.* 52:301–354, 1983.

19. Bishop, J. M. Viruses, genes and cancer: II. Retroviruses and cancer genes. *Cancer* 55:2329–2333, 1985.

20. Bishop, J. M. The molecular genetics of cancer. *Science (Wash DC)* 235:305–311, 1987.

21. Bishop, J. M., and Varmus, H. E. *Functions and origins of retroviral transforming genes.* In: *RNA Tumor Viruses,* edited by R. Weiss, N. Teich, H. Varmus, and J. Coffin, Cold Spring Harbor Laboratory, Cold Spring Harbor, NY, 1984.

22. Bister, K., Lee, W.-H., and Duesberg, P. H. Phosphorylations of the nonstructural proteins encoded by three avian acute leukemia viruses and by avian fujinami sarcoma virus. *J. Virol.* 36:617–621, 1980.

23. Bizub, D., Wood, A. W., and Skalka, A. M. Mutagenesis of the Ha-ras oncogene in mouse skin tumors induced by polycyclic aromatic hydrocarbons. *Proc. Natl. Acad. Sci. U.S.A.* 83:6048–6052, 1986.

24. Bos, J. L., Toksoz, D., and Marshall, C. J. Amino-acid substitutions at codon 13 of the N-*ras* oncogene in human acute myeloid leukaemia. *Nature (Lond)* 315:726–730, 1985.

25. Cabrera, C. V., Alonso, M. C., Jonston, P., Phillips, R. G., and Lawrence, P. A. Phenocopies induced with antisense RNA identify the usingless gene. *Cell* 50:659–663, 1987.

26. Campisi, J., Gray, H. E., Pardee, A. B., Dean, M., Sonenshein, G. E. Cell-cycle control of c-*myc* but not c-*ras* expression is lost following chemical transformation. *Cell* 36:241–247, 1984.

27. Carpenter, G. Receptors for epidermal growth factor and other polypeptide mitogens. *Annu. Rev. Biochem.* 56:881–914, 1987.

28. Chambard, J. C., Franchi, A., LeCam, A., and Pouyssegur, J. Growth factor–stimulated oncogene phosphorylation in G0/G1 arrested fibroblast. Two distinct classes of growth factors with potentiating effects. *J. Biol. Chem.* 258:1706–1713, 1983.

29. Chiu, R., Boyle, W. J., Meek, J., Smeal, T., Hunter, T., and Karin, M. The c-fos protein interacts with c-Jun/AP-1 to stimulate transcription of AP-1 responsive genes. *Cell* 54:541–552, 1988.

30. Cochran, B. H., Zullo, J., Verma, I. M., and Stiles, C. D. Expression of the c-*fos* gene and of an *fos*-related gene is stimulated by platelet-derived growth factor. *Science (Wash DC)* 226:1080–1082, 1984.

31. Cohen, S., Carpenter, C., and King, L. Epidermal growth factor–receptor-protein kinase interactions. Co-purifications of receptor and epidermal growth factor-enhanced phosphorylations activity. *J. Biol. Chem.* 255:4834–4842, 1980.

32. Collard, J. G., Schijven, J. F., and Roos, E. Invasive and metastatic potential induced by *ras*-transfection into mouse BW5147 T-lymphoma cells. *Cancer Res.* 47:754–759, 1987.

33. Collett, M. S., Purchio, A. F., and Erikson, R. L. Avian sarcoma virus transforming protein p60^{src} shows protein kinase activity specific for tyrosin. *Nature (Lond)* 285:167–169, 1980.

34. Collins, S., and Groudine, M. Amplification of endogenous *myc*-related DNA sequences in human myeloid leukaemia cell line. *Nature (Lond)* 298:679–681, 1982.

35. Collins, S. J., and Groudine, M. T. Rearrangement and amplification of c-*abl* sequences

in the human chronic myelogenous leukemia cell line K-562. *Proc. Natl. Acad. Sci. U.S.A.* 80:4813–4817, 1983.

36. Coppola, J. A., and Cole, M. D. Constitutive c-myc oncogene expression blocks mouse erythroleukaemia cell differentiation but not commitment. *Nature (Lond)* 320:760–763, 1986.

37. Corcoran, L. M., Adams, J. M., Dunn, A. R., and Cory, S. Murine T lymphomas in which the cellular myc oncogene has been activated by retroviral insertion. *Cell* 37:113–122, 1984.

38. Coussens, L., Van Beveren, C., Smith, D., Chen, R. L., Mitchell, C. M., Isacke, M., Verma, I. M., and Ulrich, A. Structural alteration of viral homologue of receptor proto-oncogene *fms* at carboxyl terminus. *Nature* 320:277–280, 1986.

39. Coussens, L., Yang-Feng, T. L., Liao, Y.-C., Chen, E., Gray, A., McGrath, J., Seeburg, P. H., Libermann, T. A., Schlessinger, J., Francke, U., Levinson, A., and Ullrich, A. Tyrosine kinase receptor with extensive homology to EGF receptor shares chromosomal location with *neu* oncogene. *Science (Wash DC)* 230:1132–1139, 1985.

40. Coussens, P. M., Cooper, J. A., Hunter, T., and Shalloway, D. Restriction of the in vitro and in vivo tyrosine protein kinase activities of pp60$^{c\text{-src}}$ relative to pp60$^{v\text{-src}}$. *Mol. Cell. Biol.* 5:2753–2763, 1985.

41. Croce, C. M., Shander, M., Martinis, J., Cieurel, L., D'Ancona, G. G., Dobby, T. W., and Koprowski, H. Chromosomal location of the genes for human immunoglobulin heavy chains. *Proc. Natl. Acad. Sci. U.S.A.* 76:3416–3419, 1979.

42. Croce, C. M., Thierfdder, W., Erikson, J., Nishikura, K., Finan, J., Lenorr, G. M., and Nowell, P. C. Transcriptional activation of an unrearranged and untranslocated c-*myc* oncogene by translocation of a C lambda locus in Burkitt. *Proc. Natl. Acad. Sci. U.S.A.* 80:6922–6926, 1983.

43. Curran, T., Peters, G., Van Beveren, C., Teich, N. M., and Verma, I. M. FBJ murine osteosarcoma virus: Identification and molecular cloning of biologically active proviral DNA. *J. Virol.* 44:674–682, 1982.

44. Cuypers, H. T., Selten, G., Quint, W., Zijlstra, M., Maandag, E. R., Boelens, W., van Wezenbeck, P., and Melief, C. Murine leukemia virus–induced T cell lymphomagenesis: Integration of proviruses in a distinct chromosomal region. *Cell* 37:141–150, 1984.

45. Dalla-Favera, R., Martinotti, S., Gallo, R. C., Erikson, J., and Croce, C. M. Translocation and rearrangements of the c-*myc* oncogene locus in human undifferentiated B-cell lymphomas. *Science (Wash DC)* 219:963–967, 1983.

46. Debuire, B., Henry, C., Benaissa, M., Biserte, G., Claverie, J. M., Saule, S., Martin, P., and Stehelin, D. Sequencing the *erb*A gene of avian erythroblastosis virus reveals a new type of oncogene. *Science (Wash DC)* 224:1456–1459, 1984.

47. DePinho, R. A., Hatton, K., Ferrier, P., Zimmerman, K., Legouy, E., Tesfaye, A., Collum, R., Yancopoulos, G., Nisen, P., and Alt, F. Myc family genes: A dispersed multi-gene family. *Ann. Clin. Res.* 18:284–289, 1986.

48. DePinho, R. A., Legouy, E., Feldman, L. B., Kohl, N. E., Yancopoulos, G. D., and Alt, F. W. Structure and expression of the murine N-myc gene. *Proc. Natl. Acad. Sci. U.S.A.* 83:1827–1831, 1986.

49. Dmitrovsky, E., Kuehl, W. M., Hollis, G. F., Kirsch, I. R., Bender, T. P., and Segal, S. Expression of a transfected human c-myc oncogene inhibits differentiation of a mouse erythroleukaemia cell line. *Nature (Lond)* 322:748–750, 1986.

50. Donner, P., Greiser-Wilke, I., and Moeling, K. Nuclear localization and DNA binding of the transforming gene product of avian myelocytomatosis virus. *Nature (Lond)* 296:262–266, 1982.

51. Doolittle, R. F., Hunkapiller, M. W., Hood, L. E., Deare, S. G., Robbins, K. C., Aaronson, S. A., and Antoniades, H. N. Simian sarcoma virus onc gene v-*sis* is derived from the

gene (or genes) encoding a platelet-derived growth factor. *Science (Wash DC)* 221:275–276, 1983.

52. Duesberg, P. Retroviral transforming genes in normal cells? *Nature (Lond)* 304:219–226, 1983.

53. Duesberg, P. Activated proto-oncogenes: Sufficient or necessary for cancer? *Science (Wash DC)* 228:669–677, 1985.

54. Ebina, Y., Ellis, L., Jarnagin, K., Edery, M., Graf, L., Clauser, E., Ou, J. H., Masiarz, F., Kan, Y. W., Goldfine, I. D., Roth, R. A., and Rutter, W. J. The human insulin receptor cDNA: The structural basis for hormone-activated transmembrane signaling. *Cell* 40:747–758, 1985.

55. Ek, B., Westermark, B., Wasteson, A., and Heldin, C.-H. Stimulation of tyrosine-specific phosphorylation by platelet-derived growth factor. *Nature (Lond)* 295:419–420, 1982.

56. Fukushige, S., Matsubara, K., Yoshida, M., Sasaki, M., Suzuki, T., Semba, K., Toyoshima, K., and Yamamoto, T. Localization of a novel v-*erb*B–related gene, c-*erb*B-2, on human chromosome 17 and its amplification in a gastric cancer cell line. *Mol. Cell Biol.* 6:955–958, 1986.

57. Fung, Y.-K., Fadly, A. M., Crittenden, L. B., Kung, H.-J. On the mechanisms of retrovirus-induced avian lymphoid leukosis: Deletion and integration of the proviruses. *Proc. Natl. Acad. Sci. U.S.A.* 78:3418–3422, 1981.

58. Fung, Y.-K., Lewis, W. G., Crittenden, L. B., and Kung, H.-J. Activation of the cellular oncogene c-erb-B by LTR insertion: Molecular basis for induction of erythroblastosis by avian leukosis virus. *Cell* 33:357–368, 1983.

59. Gibbs, C. P., Tanaka, A., Anderson, S. K., Radul, J., Baar, J., Ridgway, A., Kung, H.-J., and Fujita, D. J. Isolation and structural mapping of a human c-*src* gene homologous to the transforming gene (v-*src*) of Rous sarcoma virus. *J. Virol.* 53:19–24, 1985.

60. Gibbs, J. B., Sigal, I. S., Poe, M., and Scolnick, E. M. Intrinsic GTPase activity distinguishes normal and oncogenic *ras* p21 molecules. *Proc. Natl. Acad. Sci. U.S.A.* 81:5704–5708, 1984.

61. Graham, F. L., and van der Eb, A. J. A new technique for the assay of infectivity of human adenovirus 5 DNA. *Virology* 52:456–471, 1973.

62. Griep, A. E., and Deluca, H. F. Decreased c-myc expression is an early event in retinoic acid–induced differentiation of F9 teratocarcinoma cells. *Proc. Natl. Acad. Sci. U.S.A.* 83:5539–5543, 1986.

63. Hanks, S. K., Quinn, A. M., and Hunter, T. The protein kinase family: Conserved features and deduced phylogeny of the catalytic domains. *Science (Wash DC)* 241:42–52, 1988.

64. Hann, S. R., and Eisenman, R. N. Proteins encoded by the human c-*myc* oncogene: Differential expression in neoplastic cells. *Mol. Cell Biol.* 4:2486–2497, 1984.

65. Hayward, W. S. Viral and cellular oncogenes in cancer etiology. In: *Genetics in Clinical Oncology,* edited by R. S. K. Chaganti and J. German, New York, Oxford University Press, 1985, pp. 22–38.

66. Hayward, W. S., Neel, B. G., and Astrin, S. M. Activation of cellular oncogenes by promoter insertion in ALV-induced lymphoid leukosis. *Nature (Lond)* 290:475–580, 1981.

67. Hirai, H., Maru, Y., Hagiwara, K., Nishida, J., and Takaku, F. A novel putative tyrosine kinase receptor encoded by the *eph* gene. *Science (Wash DC)* 238:1717–1720, 1987.

68. Hunter, T. The proteins of oncogenes. *Sci. Am.* 251:70–79, 1984.

69. Hunter, T. The epidermal growth factor receptor gene and its product. *Nature (Lond)* 311:414–416, 1984.

70. Hunter, T. At last the insulin receptor. *Nature (Lond)* 313:740–741, 1985.

71. Hunter, T. A thousand and one protein kinases. *Cell* 50:823–829, 1987.

72. Kasuga, M., Zick, Y., Blithe, D. L., Crettaz, M., and Kahn, C. R. Insulin stimulates tyro-

sine phosphorylation of the insulin receptor in a cell-free system. *Nature (Lond)* 298:667–669, 1982.

73. Keating, M. T., and Williams, L. T. Autocrine stimulation of intracellular PDGF receptors in v-*sis*–transformed cells. *Science (Wash DC)* 239:914–916, 1988.

74. Kelly, K., and Siebenlist, U. The role of c-*myc* in the proliferation of normal and neoplastic cells. *J. Clin. Immunol.* 5:65–77, 1985.

75. Khoury, G., and Gruss, P. Enhancer elements. *Cell* 33:313–314, 1983.

76. Klein, G. Specific chromosomal translocations and the genesis of B cell derived tumors in mice and men. *Cell* 32:311–315, 1983.

77. Klein, G., and Klein, E. Evolution of tumours and the impact of molecular biology. *Nature (Lond)* 315:190–195, 1985.

78. Kohl, N. E., Kanda, N., Schreck, R. R., Bruns, G., Latt, S. A., Gilbert, F., and Alt, F. W. Transposition and amplification of oncogene-related sequences in human neuroblastomas. *Cell* 35:359–367, 1983.

79. Kozobor, D., and Croce, C. M. Amplification of the c-*myc* oncogene in one of five human breast carcinoma cell lines. *Cancer Res.* 44:438–441, 1984.

80. Krontiris, T. G., and Cooper, G. M. Transforming activity of human tumor DNAs. *Proc. Natl. Acad. Sci. U.S.A.* 78:1181–1184, 1981.

81. Lachman, H. M., and Skoultchi, A. I. Expression of c-myc changes during differentiation of mouse erythroleukaemia cells. *Nature (Lond)* 310:592–594, 1984.

82. Land, H., Parada, L. F., and Weinberg, R. A. Cellular oncogenes and multistep carcinogenesis. *Science (Wash DC)* 222:771–778, 1983.

83. Land, H., Parada, L. F., and Weinberg, R. A. Tumorigenic conversion of primary embryo fibroblast requires at least two cooperating oncogenes. *Nature (Lond)* 304:596–602, 1983.

84. Leder, P., Battey, J., Lenoir, G., Moulding, C., Murphy, W., Potter, H., Stewart, T., and Taub, R. Translocations among antibody genes in human cancer. *Science (Wash DC)* 222:765–771, 1983.

85. Little, C. D., Nau, M. M., Carney, D. N., Gazdar, A. F., and Minna, J. D. Amplification and expression of the c-*myc* oncogene in human lung cancer cell lines. *Nature (Lond)* 306:194–196, 1983.

86. Makino, R., Hayashi, K., and Sugimura, T. c-*myc* transcript is induced in rat liver at a very early stage of regeneration or by cycloheximide treatment. *Nature (Lond)* 310:697–698, 1984.

87. Marcu, K. B. Regulation of expression of the c-myc protooncogene. *Bio-essays* 6:28–32, 1987.

88. Matsushime, H., Wang, L.-H., and Shibuya, M. Human c-*ros*-1 gene homologous to the v-*ros* sequence of UR2 sarcoma virus encodes a transmembrane receptor-like molecule. *Mol. Cell Biol.* 6:3000–3004, 1986.

89. Mulder, K. M., and Brattain, M. G. Continuous maintenance of transformed fibroblasts under reduced serum conditions: Utility as a model system for investigating growth factor–specific effects in non-quiescent cells. *J. Cell Physiol.* 138:450–458, 1989.

90. Mulder, K. M., Levine, A. E., and Hinshaw, X. H. Up-regulation of c-myc in a transformed cell line approaching stationary phase growth in culture. *Cancer Res.* 49:2320–2326, 1989.

91. Muschel, R. J., Williams, J. E., Lowy, D. R., and Liotta, L. A. Harvey *ras* induction of metastatic potential depends upon oncogene activation and the type of recipient cell. *Am. J. Pathol.* 121:1–8, 1985.

92. Mushinski, J. F., Potter, M., Bauer, S. R., and Reddy, E. P. DNA rearrangement and altered RNA expression of the c-*myb* oncogene in mouse plasmacytoid lymphosarcomas. *Science (Wash DC)* 220:795–798, 1983.

93. Naharro, G., Robbins, K. C., and Reddy, E. P. Gene product of v-*fgr* onc: Hybrid protein

containing a portion of actin and a tyrosine-specific protein kinase. *Science (Wash DC)* 223:63–66, 1984.

94. Nusse, R., van Ooyen, A., Cox, D., Fung, Y. K. T., and Varmus, H. Mode of proviral activation of a putative mammary oncogene (int-1) on mouse chromosome 15. *Nature (Lond)* 307:131–136, 1984.

95. Nusse, R., van Ooyen, A., Rijsewijk, F., van Lohuizen, M., Schuuring, E., and van't Veer, L. Retroviral insertional mutagenesis in murine mammary cancer. *Proc. R. Soc. Lond.* 226:3–13, 1985.

96. Nusse, R., and Varmus, H. E. Many tumors induced by the mouse mammary tumor virus contain a provirus integrated in the same region of the host genome. *Cell* 31:99–109, 1982.

97. Parada, L. F., Tabin, C. J., Shih, C., and Weinberg, R. A. Human EJ bladder carcinoma oncogene is homologue of Harvey sarcoma virus *ras* gene. *Nature (Lond)* 297:474–478, 1982.

98. Park, M., Dean, M., Kaul, K., Braun, M. J., Gonda, M. A., and Vande Woude, G. Sequence of MET protooncogene cDNA has features characteristic of the tyrosine kinase family of growth-factor receptors. *Proc. Natl. Acad. Sci. U.S.A.* 84:6378–6383, 1987.

99. Parker, R. C., Swanstrom, R., Varmus, H. E., and Bishop, J. M. Transduction and alteration of a cellular gene (c-*src*) created an RNA tumor virus: The genesis of Rous sarcoma virus. In: *Cold Spring Harbor Conference on Cell Proliferation and Cancer, Vol. 2: The Cancer Cell.* Cold Spring Harbor Laboratory, Cold Spring Harbor, NY, 1984, pp. 19–25.

100. Parker, R. C., Varmus, H. E., and Bishop, J. M. Cellular homology (c-*src*) of the transforming gene of Rous sarcoma virus: Isolation, mapping, and transcriptional analysis of c-*src* and flanking regions. *Proc. Natl. Acad. Sci. U.S.A.* 78:5842–5846, 1981.

101. Payne, G. S., Bishop, M. J., and Varmus, M. E. Multiple arrangements of viral DNA and an activated host oncogene in bursal lymphomas. *Nature (Lond)* 295:209–214, 1982.

102. Peters, G., Brookes, S., Smith, R., and Dickson, C. Tumorigenesis by mouse mammary tumor virus: Evidence for a common region for provirus integration in mammary tumors. *Cell* 33:369–377, 1983.

103. Pfeifer-Ohlsson, S., Gronstein, A. S., and Rydnert, J. Spatial and temporal pattern of cellular *myc* oncogene expression in developing human placenta: Implications for embryonic cell proliferation. *Cell* 38:585–592, 1984.

104. Piwnica-Worms, H., Saunders, K. B., Roberts, T. M., Smith, A. E., and Cheng, S. H. Tyrosine phosphorylation regulates the biochemical and biological properties of pp60^{c-src}. *Cell* 49:75–82, 1987.

105. Reddy, E. P. Nucleotide sequence analysis of the T24 human bladder carcinoma oncogene. *Science (Wash DC)* 220:1061–1063, 1983.

106. Reddy, E. P., Reynolds, R. K., Santos, E., and Barbacid, M. A point mutation is responsible for the acquisition of transforming properties by the T24 human bladder carcinoma oncogene. *Nature (Lond)* 300:149–152, 1982.

107. Ryder, K., Lau, L. F., and Nathans, D. A gene activated by growth factor is related to the oncogene v-jun. *Proc. Natl. Acad. Sci. U.S.A.* 85:1487–1491, 1988.

108. Sambucetti, L. C., and Curran, T. The Fos protein complex is associated with DNA in isolated nuclei and binds to DNA cellulose. *Science (Wash DC)* 234:1417–1419, 1986.

109. Schechter, A. L., Stern, D. F., Vaidyanathan, L., Decker, S. J., Drebin, J. A., Greene, M. I., and Weinberg, R. A. The *neu* oncogene: An *erb*-B related gene encoding a 185,000-M$_r$ tumour antigen. *Nature (Lond)* 312:513–516, 1984.

110. Schimke, R. T. Gene amplification in cultured animal cells. *Cell* 37:705–713, 1984.

111. Schimke, R. T., Sherwood, S. W., Hill, A. B., and Johnston, R. N. Overreplication and recombination of DNA in higher eukaryotes: Potential consequences and biological implications. *Proc. Natl. Acad. Sci. U.S.A.* 83:2157–2161, 1986.

112. Schwab, M., Ramsay, G., Alitalo, K., Varmus, H. E., Bishop, J. M., Martinsson, T.,

Levan, G., and Levan, A. Amplification and enhanced expression of the c-*myc* oncogene in mouse SEWA tumour cells. *Nature (Lond)* 315:345–347, 1985.

113. Schwab, M., Varmus, H. E., and Bishop, J. M. The human N-myc gene contributes to tumorigenic conversion of mammalian cells in culture. *Nature (Lond)* 316:160–162, 1985.

114. Schweinfest, C. W., Fujiwara, S., Lau, L. F., and Papas, T. S. c-*myc* can induce expression of G_0/G_1 transition genes. *Mol. Cell Biol.* 8:3080–3087, 1988.

115. Sefton, B. M., and Hunter, T. Vinculin: A cytoskeletal target of the transforming protein of Rous sarcoma virus. *Cell* 24:165–174, 1981.

116. Sefton, B. M., Hunter, T., Ball, E. H., and Singer, S. J. Vinculin: A cytoskeletal target of the transforming protein of Rous sarcoma virus. *Cell* 24:165–174, 1981.

117. Sefton, B. M., Trowbridge, I. S., Cooper, J. A., and Scolnick, E. M. The transforming proteins of Rous sarcoma virus, Harvey sarcoma virus and Abelson virus contain tightly bound lipid. *Cell* 31:465–474, 1982.

118. Sheng-Ong, G. L. C., Polter, M., Mushimiki, J. F., Lavu, S., and Reddy, E. P. Activation of the c-*myb* locus by viral insection mutagenesis in plasmacytoid lymphosarcoma. *Science* 226:1077–1080, 1984.

119. Sherr, C. J., Rettenmier, C. W., Sacca, R., Roussel, M. F., Look, A. T., and Stanley, E. R. The c-*fms* proto-oncogene product is related to the receptor for the mononuclear phagocyte growth factor, CSF-1. *Cell* 41:665–676, 1985.

120. Shih, C., Padhy, L. C., Murray, M., and Weinberg, R. A. Transforming genes of carcinomas and neuroblastomas introduced into mouse fibroblasts. *Nature (Lond)* 290:261–264, 1981.

121. Shih, C., and Weinberg, R. A. Isolation of a transforming sequence from a human bladder carcinoma cell line. *Cell* 29:161–169, 1984.

122. Shih, T. Y., Clanton, D. J., Hattori, S., Ulsh, L. S., and Chen, Z. Structure and function of p21 *ras* proteins: Biochemical, immunochemical, and site directed mutagenesis studies. In: *UCLA Symposium Proceedings: Growth Factors, Tumor Promotors and Cancer Genes*, 1986.

123. Shimuzu, K., Birnbaum, D., Ruley, M. A., Fasano, O., Suard, Y., Edlung, L., Taparowsky, E., Goldfarb, M., and Wigler, M. Structure of the Ki-*ras* gene of the human cell lung carcinoma cell line Calu-1. *Nature (Lond)* 304:497–500, 1983.

124. Sistonen, L., and Alitalo, K. Activation of c-*ras* oncogenes by mutations and amplification. *Ann. Clin. Res.* 18:297–303, 1986.

125. Takahashi, M., and Cooper, G. M. *ret* transforming gene encodes a fusion protein homologous to tyrosine kinases. *Mol. Cell Biol.* 7:1378–1385, 1987.

126. Tal, M., Wetzler, M., Josefberg, Z., Deutch, A., Gutman, M., Assaf, D., Kris, R., Shiloh, Y., Givol, D., and Schlessinger, J. Sporadic amplification of the HER2/neu protooncogene in adenocarcinomas of various tissues. *Cancer Res.* 48:1517–1520, 1988.

127. Taparowsky, E., Suard, Y., Fasano, O., Shimizu, K., Goldfarb, M., and Wigler, M. Activation of the T-24 bladder carcinoma transforming gene is linked to a single amino acid change. *Nature (Lond)* 300:762–765, 1982.

128. Ullrich, A., Bell, J. R., Chen, E. Y., Herrera, R., Petruzzelli, L. M., Dull, T. J., Gray, A., Coussens, L., Liao, Y.-C., Tsubokawa, M., Mason, A., Seeburg, P. H., Grunfeld, C., Rosen, O. M., and Ramachandran, J. Human insulin receptor and its relationship to the tyrosine kinase family of oncogene. *Nature (Lond)* 313:756–761, 1985.

129. Ullrich, A., Coussens, L., Hayflick, J. S., Dull, T. J., Gray, A., Tam, A. W., Lee, J., Yarden, Y., Libermann, T. A., Schlessinger, J., Downward, J., Mayes, E. L. V., Whittle, N., Waterfield, M. D., and Seeburg, P. H. Human epidermal growth factor receptor cDNA sequence and aberrant expression of the amplified gene in A431 epidermoid carcinoma cells. *Nature (Lond)* 209:418–425, 1984.

130. Ullrich, A., Gray, A., Tan, A. W., Yang-Feng, T., Tsubokawa, M., Collins, C., Henzel, W.,

Le Bon, T., Kathuria, S., Chen, E., Jacobs, S., Francke, U., Ramachandran, J., and Fujita-Yamaguchi, Y. Insulin-like growth factor I receptor primary structure: Comparison with insulin receptor suggests structural determinants that define functional specificity. *E.M.B.O. J.* 5:2503–2512, 1986.

131. Ushiro, H., and Cohen, S. Identification of phosphotyrosine as a product of epidermal growth factor–activated protein kinase in A431 cell membranes. *J. Biol. Chem.* 255:8363–8365, 1980.

132. Varmus, H. E. Molecular genetics of cellular oncogenes. *Annu. Rev. Genet.* 18:553–612, 1984.

133. Vogt, P. K., Bos, T. J., and Doolittle, R. F. Homology between the DNA binding domain of the GCN4 regulatory protein of yeast and the carboxyl-terminal region of a protein coded for by the oncogene jun. *Proc. Natl. Acad. Sci. U.S.A.* 84:3316–3319, 1987.

134. Vogt, P. K., and Tijan, R. *jun:* A transcriptional regulator turned oncogenic. *Oncogene* 3:3–7, 1988.

135. Weinberg, R. A. Alteration of the genomes of tumor cells. *Cancer* 52:1971–1975, 1983.

136. Weinberg, R. A. The action of oncogenes in the cytoplasm and nucleus. *Science (Wash DC)* 230:770–776, 1985.

137. Willecke, K., and Schafer, R. Human oncogenes. *Hum. Genet.* 66:132–142, 1984.

138. Willman, C. L., and Fenoglio-Preiser, C. M. Oncogenes, suppressor genes, and carcinogenesis. *Hum. Pathol.* 18:895–902, 1987.

139. Yokoyama, K., and Imamoto, F. Transcriptional control of the endogenous MYC protooncogene by antisense RNA. *Proc. Natl. Acad. Sci. U.S.A.* 84:7363–7367, 1987.

140. Zarbl, H., Saraswati, S., Arthur, A. V., *et al.* Direct mutagenesis of Ha-ras-1 oncogenes by N-nitrosomethylurea during initiation of mammary carcinogenesis in rats. *Nature (Lond)* 315:382–385, 1985.

6

Diagnosis of Hematopoietic Diseases of the Myeloid Lineage Utilizing Molecular Probes

Cheryl L. Willman

Introduction

Consistent cytogenetic abnormalities have been reported for many acute and chronic myeloid disorders (Table 6.1), pointing to the chromosomal loci that contain human genes which may be involved in the pathogenesis of these diseases. Cloning and sequencing these breakpoints will undoubtedly lead to the discovery of many new genes which play a critical role in the control of differentiation and proliferation in myeloid cells. Disruption of these genes by chromosomal translocation, mutation, and deletion would presumably alter their structure or expression, leading to tumorigenesis. Utilization of molecular probes derived from these genes will undoubtedly facilitate the diagnosis and classification of myeloid diseases in the future.

Identification of new genes involved in lymphoid tumorigenesis (such as BCL-1, BCL-2, TCL-1, etc.) has proceeded at a rapid rate, since the translocation breakpoints involving these genes involve known genetic loci: the immunoglobulin (Ig) or T cell receptor genes (see Chapter 7 and references 6 and 21 for a full discussion and review). Knowledge of the molecular structure of these genes, which have been cloned and at least partially sequenced, has allowed investigators to identify the "new", potentially oncogenic sequences fused to Ig or T cell receptor genes in different lymphoid malignancies. Several proto-oncogenes and other gene sequences have been mapped at the cytogenetic level near the chromosomal loci that are consistently involved in myeloid malignancies (Table 6.1; reviewed in detail in reference 2). However, more detailed molecular studies have disappointingly revealed that most of these genes do not lie near the molecular breakpoints and that their structure and expression is not altered by the cytogenetic abnormalities.[2] We have recently determined that expression of the human EVI-1 gene is induced only in human acute myeloid leukemias with karyotypic abnormalities of the long arm of human chromosome 3 (such as the inv 3; Table 6.1).[20] Since this gene maps to human chromosome 3q25-q26 and encodes a nuclear transcription factor (a potential oncogene, see Chapters 5 and 8), disruption of the structure or expression of this gene in acute leukemias with abnormalities of 3q25-26 may lead to malignancy.[20] Genomic characterization of

Table 6.1.
Consistent Chromosomal Translocations in Myeloid Disorders

Chromosomal Abnormality	Disorder	Genes Mapping to Locus[a]
t(9;22)(q34;q11)	AML-FAB M1	BCR (22), ABL (9)
t(8;21)(q22;q22)	AML-FAB M2	Hu-ETS-2 (21)
t(15;17)(q22;q21)	AML-FAB M3	MPO, G-CSF, ERB-B-2/NEU ERB A (thyroid hormone R)
inv(16)(p13q22)	AML-FAB M4e	Metallothionein cluster
del(11q); 11q23-24	AML-FAB M4	Hu-ETS-1 (11)
ins(3); inv(3); t(3q)	Thrombocytosis	Transferrin, Trfn R, Evi-1
+8	AML	MYC
−5/5q−	Secondary AML; MDS	Interleukins 3,4,5,6; CSF-1, CSF-1 R, GM-CSF
−7/7q−	Secondary AML; MDS	

[a] Numbers in parentheses refer to chromosomal locations of indicated genes.

this type of leukemia and many others is currently underway to identify the genes involved in the breakpoints in myeloid disorders, thereby revealing the molecular basis of myeloid malignancies (Table 6.1).

One chromosomal translocation recurring consistently in malignancies of the myeloid lineage has been highly characterized at the molecular level: the Philadelphia (Ph) chromosome translocation t(9;22)(q34;q11). The application of molecular probes derived from the genes involved in the Ph translocation has defined the molecular and biological basis of acute and chronic leukemias with this translocation and has refined the diagnosis of these disorders.

Molecular Genetics of the Ph Chromosome in CML, AML, and ALL

Molecular genetic analysis of the Ph chromosome translocation t(9;22) (q34;q11) has revealed the underlying molecular basis for the pathophysiology of myeloid and lymphoid diseases with this karyotypic abnormality (Fig. 6.1; see reference 13 for an excellent review). Two genes have been determined to be molecularly fused as a consequence of the Ph translocation: the ABELSON proto-oncogene (c-abl) at chromosome band 9q34 and the BCR gene mapping to chromosome 22q11 (Fig. 6.1). The Ph chromosome is found in 90–95% of patients with chronic myelogenous leukemia (CML), 2% of adults with acute myelogenous leukemia (AML), and in 20% of adults and 5% of children with acute lymphoid leukemia (ALL).[13]

STRUCTURE OF THE ABL AND BCR GENES IN THE CML Ph TRANSLOCATION

The ABELSON (c-abl) proto-oncogene was first molecularly mapped to chromosome 9 band q34 in 1983[1] and was immediately shown to be involved in the Ph chromosome translocation in CML.[1,7] c-abl was shown to be fused to a new gene on chromosome 22 band q11, referred to as the BCR gene (for "breakpoint cluster region") whose function in normal cells was then and is still unknown (Fig. 6.1). More

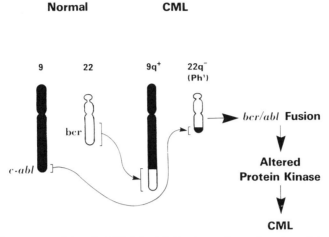

Figure 6.1. Karyogram of the t(9;22)(q34;q11). Reciprocal translocation of the distal tip of the long arm of chromosome 9 at band q34 (containing the c-*abl* gene) to the long arm of chromosome 22 at band q11 (the BCR locus) yields the 22q— or Philadelphia (Ph) chromosome translocation. The fusion of the BCR and c-*abl* genes leads to the production of a BCR-ABL fusion mRNA transcript and protein, diagnostic hallmarks of CML.

recent studies have determined that the BCR locus is very complex and that there are as many as 4 BCR-related genes in the chromosome region 22q11.[3] These genes map from the centromere to the distal portion of the long arm (q) of chromosome 22 in the following order: BCR2 → BCR4 → Immunoglobulin λ light chain gene → BCR1 → BCR3 → SIS (PDGF-β proto-oncogene). BCR1 is the gene that has been shown to be involved in the c-*abl* fusion, and all further discussion of BCR genes will refer to BCR1.[13] What role the other BCR loci play in human disease and what their functions may be in normal cells is presently unclear.

The c-*abl* proto-oncogene (see Fig. 6.2A: an exon map of the c-*abl* gene and Fig. 6.3: a schematic of the protein structure of c-*abl* in the cell) is highly conserved in evolution and encodes a protein-tyrosine kinase, an enzyme that phosphorylates tyrosine residues on cellular proteins[13] (also see Chapter 5 for a discussion of proto-oncogene protein tyrosine kinases). The BCR/*abl* tyrosine kinase is associated with the cytoplasmic face of the plasma membrane in malignant cells where it may function as a signal transducer (Fig. 6.3). However the normal c-*abl* protein is localized in the nucleus,[19] and its functional role is currently unknown. Phosphorylation of cellular proteins on tyrosine is a rare event ($< 1\%$ of total cellular phosphoamino acids) and is linked to growth control. Most polypeptide growth factor receptors are transmembrane tyrosine kinases (Fig. 6.3), and many oncogenic proteins are tyrosine kinases. The c-*abl* gene spans over 230 kb and contains 11 exons.[13] The normal c-*abl* gene undergoes alternative splicing to generate at least 2 different mRNA transcripts (6.0 kb and 7.0 kb) in normal cells. The two exons Ib may be spliced directly onto the first set of common exons II (Fig. 6.2A). Alternatively, the two exons Ia may be spliced onto the common exon II, generating an mRNA transcript with a

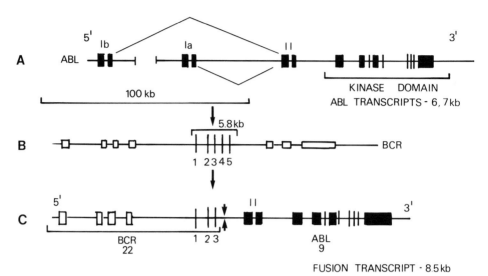

Figure 6.2. *A,* Exon map of the normal c-*abl* gene on chromosome 9q34. *B,* Exon map of the BCR gene on chromosome 22, including the small 5.8 kb breakpoint cluster in the middle of the BCR coding sequence. *C,* Exon map of the precise molecular breakpoint of the Ph translocation, with fusion of 5′ BCR sequences (in the CML 5.8 kb bcr) onto common exon II of c-*abl,* resulting in a "new" BCR-ABL fusion gene.

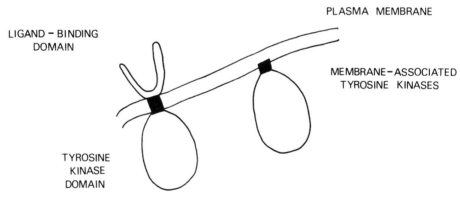

Figure 6.3. Schematic of the BCR/*abl* transforming tyrosine kinase. Unlike transmembrane tyrosine kinase growth-factor receptors that have an extracellular ligand-binding domain, a transmembrane region, and a cytoplasmic tyrosine kinase domain, the *abl* transforming tyrosine kinase lacks an extracellular domain. Transforming *abl* is associated with the inner face of the plasma membrane where it may function as a signal transducer, whereas the normal c-abl protein is localized in the nucleus.

different 5' end (Fig. 6.2A). Each mRNA has a common 3' sequence that encodes the c-abl catalytic tyrosine kinase domain. Whether these different transcripts encode proteins with different functions is unknown. Only one c-abl protein of 145kd has been identified in normal cells.[13] Sites of rearrangement within the c-abl gene in different patients with CML are very heterogeneous and may occur anywhere 100 kb upstream of the c-abl common exon II[13] (Fig. 6.2A).

In contrast, rearrangement within the BCR gene on chromosome 22 in different CML patients occurs within a very small 5.8 kb region in the BCR gene, referred to as the CML breakpoint cluster region (Fig. 6.2B). As BCR1 was studied,[7] it became clear that the BCR gene spanned at least 100 kb and that the 5.8 kb CML breakpoint cluster region was in the middle of this new gene (Fig. 6.2B). Although the function of the normal BCR gene is unknown, this gene encodes at least 2 transcripts (4.5 and 6.7 kb), and a 160kd protein has been identified.[13] In contrast to c-abl, which is expressed predominantly in hematopoietic cells, the BCR gene product is expressed at constitutive levels in many lineages.[13]

In the Ph translocation in CML, the c-abl and BCR genes fuse so that one of the five small exons in the 5.8 kb CML bcr region fuses upstream of the common exon II in c-abl, creating a "new" gene on the Ph chromosome which is a fusion between 5' BCR sequences and 3' c-abl sequences (Fig. 6.2C). This fusion gene encodes an 8.5 kb mRNA (which splices 5' BCR sequences directly onto c-abl common exon II) that encodes a 210 kd fusion protein with BCR sequences on the amino terminus and the catalytic tyrosine kinase domain of c-abl on the carboxy terminus.[13] This 210 kd fusion protein with its amino terminal BCR sequences has a greatly elevated tyrosine kinase activity compared to the 140kd protein encoded by c-abl in normal cells. Presumably, this elevated tyrosine kinase activity is somehow responsible for the increased pool of committed myeloid progenitor cells that continue to differentiate in the chronic phase of CML, resulting in elevated counts of mature, circulating myeloid cells.

With elucidation of the molecular basis of CML, DNA probes have been developed for the molecular detection of this chromosome translocation with the findings given below.

Ph⁺ CML

One hundred percent of CML patients with a demonstrable Ph chromosome on cytogenetic studies have molecular fusion of the BCR and c-abl genes. This fusion can be detected using traditional Southern blot analysis.[13]

Ph⁻ CML

A small percentage (<5%) of patients with classical clinical and morphological features of CML are Ph⁻.[13] Despite negative cytogenetic studies, many of these patients have been shown to have molecular fusion of the BCR and c-abl genes. Thus these patients do have classical CML defined by molecular technologies even though the Ph cannot be detected at the limits of cytogenetic resolution.[13] Some of these patients have complex or variant Ph translocations that mask the classic Ph. A few have been shown to have interstitial exchanges of a small amount of genetic material between chromosomes 9 and 22, leading to fusion of BCR and *abl* sequences; thus these patients have a molecular translocation not evident at the cytogenetic level.[13] The molecular rearrangement of the BCR and *abl* genes, detection of the 8.5 kb fusion transcript, or detection of the 210kd fusion protein have become the diagnostic hallmarks of CML. The absence of these features may imply that a disease is not CML, but a mimicking myelodysplastic disorder. Thus, molecular analysis of CML has become the diagnostic gold standard.

> **Case Example.** An 11-year-old Pueblo Indian female presented with a 2-week history of lethargy and decreased appetite. Physical examination revealed massive splenomegaly. WBC count was 54,000 with 5% myeloblasts, 42% immature myeloid cells, 31% bands, 16% segmented neutrophils, and 2% monocytes. No Auer rods were seen. A packed marrow had 2% blasts with granulocytic hyperplasia and decreased erythroid precursors. Although the clinical diagnosis was adult-type CML occurring in an 11-year-old child, cytogenetic studies of the peripheral blood cells revealed a karyotype of 46XX, t(2;9)(q31;q34). To determine if this translocation was a variant Ph translocation, in which the Ph translocation was masked, molecular studies were ordered.
>
> The structure of the BCR gene was first analyzed in Southern analysis using a probe for the BCR gene (Fig. 6.4). Digestion of DNA from the patient sample (Pt 1) and a liver control revealed rearrangement of the BCR gene, consistent with involvement of the BCR gene in the translocation, even though involvement of chromosome 22 was not detected at the molecular level. Northern analysis of the patient sample also revealed the presence of the 8.5 kb BCR-ABL fusion transcript, diagnostic of CML (Fig. 6.4). Presumably an initial t(9;22) underwent a secondary chromosome translocation to chromosome 2 to create the t(2;9).

Correlation of BCR Breakpoint with Prognosis in CML

CML is characterized by a triphasic clinical course. In the *chronic phase*, terminal differentiation of the increased numbers of myeloid progenitor cells is maintained, leading to marked elevations in the circulating white count. Chronic phase may last from a few weeks to several years, with a median duration of 3.5 years.[13] *Acceleration of the chronic phase* occurs when terminal differentiation is lost; additional cytogenetic abnormalities usually also occur. Finally, the patient enters *blast crisis* (which may be myeloid (65%) or lymphoid (30%)) in which the stem cells lose all ability to differentiate.[13]

Initial studies suggested that the precise location of the breakpoint within the CML breakpoint cluster might be associated with prognostic differences in CML patients. The breakpoint in most CML patients occurs either between exons 2 and 3 or 3 and 4 in the 5.8 kb CML cluster (Fig. 6.5). If the breakpoint occurs between exons 2 and 3, then only exons 1 and 2 in the CML 5.8 kb *bcr* cluster (as well as all BCR exons upstream) are involved in the fusion with c-*abl* (Fig. 6.5). This type of breakpoint has been associated with stable phase CML and a longer duration of chronic phase in 2 studies,[17,18] but these findings have not been supported by another study with fewer patients.[5] Patients with breakpoints after exon 3 have a fusion gene that contains exons 1, 2, and 3 of the 5.8 kb CML *bcr* cluster and could presumably make two proteins with alternatively spliced mRNAs (Fig. 6.5). These proteins would vary only by a small sequence encoded by exon 3. One study[17] has

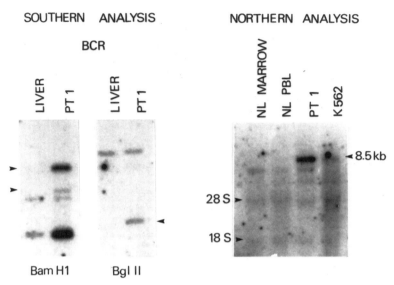

Figure 6.4. Southern and Northern analysis of a CML patient with a variant Ph translocation: t(2;9). Digestion of liver DNA used as a control and the patient sample DNA (*PT1*) with either Bam HI or Bgl II, followed by Southern hybridization with a BCR probe, revealed rearrangement of the BCR gene in the patient sample (*arrows*). Although the normal c-*abl* transcripts can barely be discerned on Northern analysis with a c-*abl* probe, an 8.5 kb BCR-ABL fusion transcript can be detected and is expressed at high levels in the patient sample (*PT1*) and in the K562 cell line (with a Ph translocation) used as a control.

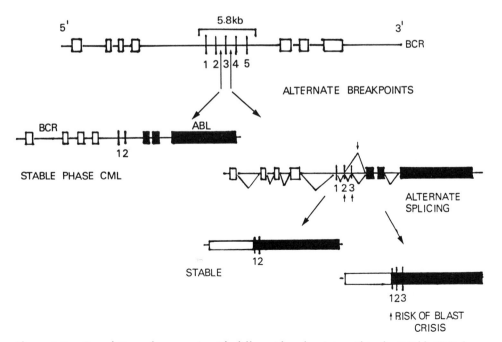

Figure 6.5. Correlation of prognosis with different breakpoints within the 5.8 kb CML *bcr* cluster in CML patients. See text for full discussion.

found that patients with this 3′ breakpoint in the BCR have a greater risk for enter-ing blast crisis (a shorter duration of chronic phase), although this issue has become highly controversial. Clearly, further clinical and biological studies must be per-formed prospectively on CML patients, and these studies must include all clinical and biological variables such as cytogenetics (further karyotypic abnormalities occur in 80% of CML patients in blast crisis[13]), to determine which features truly have predictive prognostic value.

STRUCTURE OF THE PH′ CHROMOSOME IN ACUTE LEUKEMIA

Approximately 25% of adult patients with Ph[+] acute lymphocytic leukemia (ALL) have genomic breakpoints in the 5.8 kb CML *bcr* locus. Many of these pa-tients are now assumed to be CML patients in lymphoid blast crisis who had an un-detected antecedent chronic CML phase.[11] The remaining adult Ph[+] *de novo* ALL patients and virtually all children with Ph[+] *de novo* ALL do not have fusion of the c-*abl* gene with the BCR gene in the 5.8kb CML *bcr* cluster. Rather the c-*abl* gene in these patients is fused with the first intron of the BCR gene, nearly 100 kb upstream of the 5.8 kb CML breakpoint cluster region[4,10] as shown in Fig. 6.6. Thus, although the *cytogenetic* appearance of the Ph chromosome translocation is identical in Ph[+] CML and Ph[+] ALL, these diseases have quite different molecular breakpoints and are truly different diseases at the molecular level. In ALL (and in Ph[+] acute myeloid leukemias, see below) the fusion of intron 1 of the BCR gene with the c-*abl* gene at the DNA level produces a 7.0 kb fusion mRNA containing only the first exon of BCR fused to common c-*abl* exon II (Fig. 6.7A). This fusion mRNA produces a 185–190kd BCR-ABL fusion protein with elevated tyrosine kinase activity.[11,13] In con-trast, CML patients with genomic fusion of c-*abl* into the 5.8 kb CML *bcr* locus in the BCR gene produce an 8.5 kb fusion mRNA containing all of the upstream BCR exons and some of the exons of the 5.8 kb *bcr* region (Fig. 6.5B). This CML-type BCR-ABL fusion mRNA encodes a 210kd BCR-ABL protein with elevated tyrosine kinase activity, as described above in discussions of CML. Interestingly, the trans-

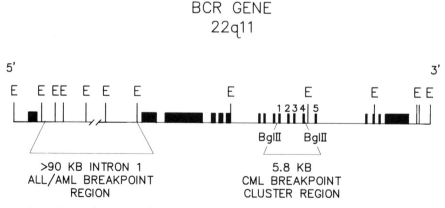

Figure 6.6. Map of the complete BCR gene. Exon map of the BCR gene reveals an up-stream exon followed by a very large intron that is the site of fusion to c-*abl* common exon II in Ph[+] *de novo* ALL and AML. Over 100 kb downstream in the BCR gene is the site of the 5.8 kb CML *bcr* cluster, the site of fusion of c-*abl* common exon II to the BCR gene in CML.

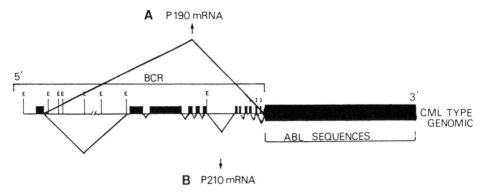

Figure 6.7. *A,* Splicing of BCR exon 1 to common exon II of c-*abl* produces a 7.0 kb BCR-ABL fusion mRNA and a P185-190 BCR-ABL fusion protein. *B,* Splicing of all BCR upstream exons including those of the 5.8 kb *bcr* cluster onto common exon II of c-*abl* produces an 8.5 kb BCR-ABL fusion mRNA and a P210 BCR-ABL protein.

forming capacity of the two different BCR-ABL fusion proteins, varying only by different BCR sequences at the amino terminus, appears to be different; the ALL type p190 BCR-ABL fusion protein is "more rapidly" transforming in *in vitro* biological systems.[14]

The Ph chromosome translocation also occurs rarely (<2%) in patients with *de novo* acute myeloid leukemia (AML).[13] Since AML cells are of the same myeloid lineage as CML cells, it was theorized that these Ph+ *de novo* AML cases would fuse c-*abl* to the BCR gene in the 5.8 kb CML breakpoint cluster. However, we[16] and others[15] have recently examined the molecular structure of the Ph translocation in Ph+ *de novo* AMLs and have determined that these AML cases fuse c-*abl* to intron 1 of the BCR gene, producing the 7.0 kb BCR-ABL fusion mRNA and the p190 BCR-ABL fusion protein. Thus, *de novo* Ph+ AML is molecularly similar to Ph+ ALL in terms of the BCR-ABL fusion. These more recent data suggest that the location of the breakpoint in the BCR gene correlates with the clinical course of the disease; Ph+ *de novo* acute leukemias (regardless of whether they have lymphoid or myeloid morphology) have the p190 BCR-ABL fusion protein, while all cases of CML (and even those in myeloid and lymphoid blast crisis[16]) have the p210 BCR-ABL fusion protein. What determines myeloid versus lymphoid morphology in the blasts of *de novo* Ph+ acute leukemia and in CML in blast crisis is unknown.

MOLECULAR DIAGNOSTICS OF DISORDERS WITH THE Ph' CHROMOSOME

PCR Diagnosis of Ph+ CML, AML, and ALL

Molecular probes for the BCR and c-*abl* genes may be used to detect rearrangements of these genes in the Ph translocation in CML using Southern blots.[13] Ph+ diseases may thus be accurately diagnosed by molecular techniques including Southern analysis (looking for rearrangements in the BCR and *abl* genes), Northern analysis (looking for fusion transcripts (8.5 kb in CML and 7.0 kb in Ph+ acute leukemia), and finally by detection of the 190 kd and the 210 kd fusion proteins. However, these approaches are not without problems. For accurate Southern analysis, 3 or more probes are necessary to examine the CML 5.8 kb *bcr* region. However, many RFLPs are being discovered in the BCR gene, indicating that a new band on Southern analysis may not reflect a true BCR rearrangement but a polymorphism.[8] At least 5 probes may be necessary to look for breakpoints in the first BCR intron,[4] and these probes may not detect a rearrangement since they do not span the entire intron.[4] It may also be impossible to distinguish the 7.0 kb fusion transcript seen in Ph[+] ALL from the normal 7.0 kb c-*abl* transcript in Northern analysis. Because of the complexity of these genes, the use of polymerase chain reaction (PCR) techniques has facilitated the diagnosis of these diseases (refer to Chapter 2 for a full discussion of PCR technology).

For PCR diagnosis of Ph[+] disorders, RNA is used as the starting material. Total RNA is isolated from a diagnostic sample and is copied into cDNA utilizing reverse transcriptase (as described in detail in reference 12). cDNA synthesis is primed with an oligonucleotide complementary to sequences present in the common exons II of c-*abl*. Thus, all mRNAs in a cell containing c-*abl* exon II sequences will be copied into cDNA. PCR is then performed in separate reactions utilizing the c-*abl* exon II primer as the 3' primer and 5' primers specific for either the CML 5.8 kb breakpoint cluster region exons or the first exon of the BCR gene (as described in reference 12). Depending upon which fusion transcripts are present in a sample (either the 7.0 kb BCR-ABL mRNA in *de novo* acute leukemia or the 8.5 kb BCR-ABL mRNA in CML), the appropriate fusion products will be amplified and may be distinguished on polyacrylamide gels.[12] After PCR, Southern analysis may also be performed utilizing probes for either BCR exon 1 or for CML *bcr* exons 2 and 3 to specifically hybridize to each of the different fusion products.[12]

> **Case Example.** A 62-year-old male with a diagnosis of CML had undergone a therapeutic bone marrow transplant two years previously. The patient was currently in clinical remission, and there was no cytogenetic evidence for the presence of the Ph translocation in the bone marrow, consistent with a cytogenetic remission. PCR analysis of the bone marrow cells was requested by the clinician to determine if the BCR-*abl* fusion transcript could be detected by the more sensitive PCR technique. Total RNA was isolated from the bone marrow cells, copied into cDNA, and amplified with a 5' primer from the CML breakpoint cluster and a 3' primer from c-*abl* exon II as described.[12] Amplification products were resolved on polyacrylamide gels and trans-

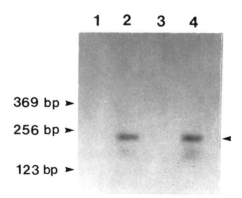

Figure 6.8.   PCR analysis of CML fusion transcripts. As described in reference 12, PCR analysis was performed on all buffer solutions (control; *lane 1*), the K562 cell line with the Ph translocation (positive control; *lane 2*), a Ph[+] *de novo* ALL (used as a negative control; *lane 3*), and the patient sample (*lane 4*). Present in the patient sample (*lane 4*) and the positive control (*lane 2*) was the 200 bp PCR-amplified product specific for the BCR-ABL CML-type fusion mRNA.[12]

Table 6.2.
BCR-ABL Rearrangements in Ph⁺ Leukemias

Disorder	BCR-ABL Genomic Rearrangements	BCR-ABL Fusion Transcripts	Proteins
Normal cells	None	6.0, 7.0 kb c-*abl* 4.5, 6.7 kb BCR	145kd ABL 160kd BCR
CML	BCR-CML 5.8 kb cluster ABL-5′ common exon II	8.5 kb BCR-ABL Fusion mRNA	210kd BCR-ABL
CML/blast crisis	BCR-CML 5.8 kb cluster ABL-5′ common exon II	8.5 kb BCR-ABL Fusion mRNA	210kd BCR-ABL
Ph⁺ *de novo* AML	BCR intron 1 ABL-5′ common exon II	7.0 kb BCR-ABL Fusion mRNA	185kd BCR-ABL
Ph⁺ ALL Adults[a]	25%: BCR-CML 5.8 kb cluster	8.5 kb BCR-ABL	210kd BCR-ABL
	75%: BCR intron 1	7.0 kb BCR-ABL Fusion mRNA	190kd BCR-ABL
Children	95%: BCR intron 1	7.0 kb BCR-ABL Fusion mRNA	190 kd BCR-ABL

[a]These cases may be lymphoid blast crisis of an undetected antecedent CML.

ferred to nitrocellulose. After hybridization of the filter with an oligonucleotide probe from the CML 5.8 kb *bcr* exon II, the products were visualized by autoradiography (Fig. 6.8). The CML-type BCR-*abl* fusion product was clearly present in the patient sample (Fig. 6.8, *lane 4*) indicating that Ph⁺ malignant cells were in the bone marrow, even though they were clinically and cytogenetically undetectable.

Summary of Ph Chromosome Diagnostics

In summary, the patterns listed in Table 6.2 may occur in Ph⁺ diseases. These diseases may thus be accurately diagnosed by the molecular techniques described above.

REFERENCES

1. Bartram, C. R., Klein, A., Hagemeijer, A., *et al.* Translocation of the c-abl oncogene correlates with the presence of the Ph′ chromosome in CML. *Nature (Lond)* 306:277–280, 1983.
2. Bitter, M., LeBeau, M. M., Rowley, J., *et al.* Associations between morphology, karyotype, and clinical features in myeloid leukemias. *Hum. Pathol. 18*:211–225, 1987.
3. Croce, C. M., Huebner, K., Isobe, M., *et al.* Mapping of four distinct BCR-related loci to chromosome 22q11: Order of BCR loci relative to CML and ALL breakpoints. *Proc. Natl. Acad. Sci. U.S.A. 84*:7174–7178, 1987.
4. Denny, C. T., Shah, N. P., Willman, C. L., *et al.* Localization of preferential sites of rearrangement within the BCR gene in Ph′+ ALL. *Proc. Natl. Acad. Sci. U.S.A. 86*:4252–4258, 1989.
5. Dreazen, D., Berman, M., Gale, R. P., *et al.* Molecular abnormalities of BCR and c-abl in CML associated with a long chronic phase. *Blood 71*:797–799, 1988.

6. Greisser, H., Tkachuk, D., Reis, M. D., *et al.* Gene rearrangements and translocations in lymphoproliferative diseases. *Blood* 73:1402, 1989.
7. Groffen, J., Stephenson, J. R., Heisterkamp, N., *et al.* Philadelphia chromosomal breakpoints are clustered within a limited region, bcr, on chromosome 22. *Cell* 36:93–99, 1984.
8. Grossman, A., Matthew, A., O'Connell, M. P., Tiso, P., Distenfield, A., and Benn, P. Multiple restriction enzyme digests are required to rule out polymorphism in the molecular diagnosis of chronic myeloid leukemia. *Leukemia* 4:63–64, 1990.
9. Heisterkamp, N., Stephenson, J. R., Groffen, J., *et al.* Localization of the c-abl oncogene adjacent to a translocation breakpoint in CML. *Nature (Lond)* 306:239–242, 1983.
10. Hermans, A., Heisterkamp, N., von Lindern, M., *et al.* Unique fusion of BCR and c-abl genes in Ph'+ ALL. *Cell* 51:33–40, 1987.
11. Hooberman, A. L., Westbrook, C. A., Davey, F., *et al.* Molecular detection of the Ph chromosome in adult acute lymphoblastic leukemia. *Blood* 74:52a, 1989.
12. Kawasaki, E., Clark, S. S., Coyne, M. Y., *et al.* Diagnosis of CML and ALL by detection of leukemia-specific mRNA sequences amplified *in vitro. Proc. Natl. Acad. Sci. U.S.A.* 85:5698–5702, 1988.
13. Kurzrock, R., Gutterman, J., and Talpaz, M. The molecular genetics of Ph'+ leukemias. *N. Engl. J. Med.* 319:990–998, 1988.
14. McLaughlin, J., Chianese, E., and Witte, O. N. Alternative forms of the BCR-ABL oncogene have quantitatively different potencies for stimulation of immature lymphoid cells. *Mol. Cell. Biol.* 9:1866–1874, 1989.
15. Najfeld, V., Cuttner, J., Figur, A., *et al.* P185 BCR-ABL in two patients with late appearing Ph+ acute nonlymphocytic leukemia. *Leukemia* 3:841–846, 1989.
16. Saikevych, I., Timson, L., Denny, C., *et al.* Philadelphia chromosome positive (Ph+) leukemias with p210 BCR-ABL and p185 BCR-ABL may be distinct disorders. *Blood* 74:79a, 1989.
17. Schaefer-Rego, K., Dudek, H., Popenoe, D., *et al.* CML patients in blast crisis have breakpoints localized to a specific region of the BCR. *Blood* 70:448–455, 1987.
18. Shtalrid, M., Talpaz, M., Kurzrock, R., *et al.* Analysis of breakpoints within the BCR gene and their correlation with the clinical course of Ph'+ CML. *Blood* 72:485–490, 1988.
19. Van Etten, R. A., Jackson, P., and Baltimore, D. The mouse type IV c-abl product is a nuclear protein, and activation of transforming ability is associated with cytoplasmic localization. *Cell* 58:669–678, 1989.
20. Whittaker, M., Morishita, K., Willman, C. L., *et al.* Expression of the Evi-1 gene, encoding a DNA binding protein, in acute myeloid leukemias with karyotypic abnormalities of chromosome 3q. *Lab. Invest.*, in press.
21. Willman, C. L., Griffith, B., and Whittaker, M. Molecular genetic approaches for the diagnosis of clonality in lymphoid neoplasms. *Clin. Lab. Med.* 10:119–149, 1990.

7

A Practical Guide to the Use of Molecular Genetic Techniques in the Diagnosis of Lymphoid Malignancies

Michael H. Whittaker
Cheryl L. Willman

Introduction

Molecular genetic studies are becoming increasingly important in the diagnosis of lymphoma and leukemia. Structural analysis of T and B lymphocyte antigen receptor genes by Southern blot, when properly interpreted, can detect monoclonal proliferations of lymphocytes. However, these data cannot be used in isolation in the diagnosis of lymphoid malignancies. The purpose of this chapter is to put molecular genetic data into perspective with clinical, morphologic, and immunophenotypic information and to provide basic guidelines for the interpretation and quality control of Southern blot analyses in the evaluation of lymphoid neoplasms. (A detailed experimental protocol for this procedure may be found in reference 26.) Practical aspects of these techniques are emphasized and illustrated with clinical cases. Several excellent reviews have been published recently, which provide interested readers with more detailed descriptions of lymphoid antigen receptor gene rearrangements and explicit summaries of gene rearrangements in various hematopoietic disorders.[3,6,26]

DEMONSTRATION OF A CLONAL PROLIFERATION OF LYMPHOCYTES IS NECESSARY (BUT NOT SUFFICIENT) FOR THE DIAGNOSIS OF LYMPHOID MALIGNANCIES

Identification of a clonal proliferation of lymphoid cells is central to the diagnosis of lymphoma and lymphocytic leukemia. This holds true for studies utilizing morphologic, immunophenotypic, or molecular genetic approaches. Each of these modalities can imply monoclonality by demonstrating shared characteristics in cells from a patient specimen. They are used in a stepwise fashion, with each method increasing in sensitivity, cost, and sophistication (Fig. 7.1). For example, when clinical circumstances suggest the differential diagnosis of reactive lymphadenopathy versus lymphoma or a nonlymphoid malignancy, laboratory evaluation usually starts

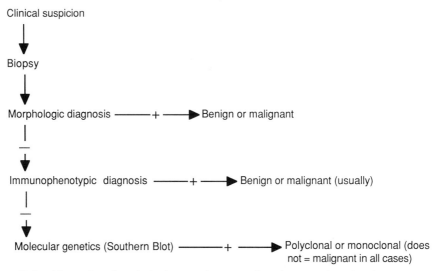

Figure 7.1. Hierarchy of analytical procedures used to diagnose lymphoid tumors. Each step increases in sensitivity, cost, and sophistication. Progression through the algorithm can stop when a definitive diagnosis is made.

with histologic and/or cytologic examination of suspicious tissue. At times, a straightforward diagnosis of malignant lymphoma can be made, if light microscopic examination reveals partial or complete replacement of normal nodal architecture by lymphoid cells sharing common morphologic characteristics. The shared morphologic features reflect the monoclonal nature of the lymphoid proliferation.

Often, however, morphologic characteristics are neither conclusively benign nor malignant. This is particularly true when the lymphoid infiltrate is extremely polymorphic instead of the usual monotonous population of cells associated with an obvious monoclonal process. Immunophenotyping may then be used to identify monoclonality. Monotypic expression of B cell markers (immunoglobulin, particularly κ or λ light chains, or the immature B cell markers CD19, 20 or 22) or aberrant expression of T cell markers may reveal the monoclonality of the lymphoid cells. These data may confirm the diagnosis of malignancy if clinical and morphologic features are supportive.[15] Note that a monoclonal proliferation defined by immunophenotypic characteristics usually, but not always, predicts a malignant course.[13] Immunophenotypic data must always be interpreted in context with clinical and morphologic information, as must molecular genetic data.

Molecular genetic techniques are usually employed when clinical, morphologic, and immunophenotypic studies are inconclusive. In order to understand how DNA from a patient specimen can be used to define clonality, the mechanism of lymphoid antigen receptor gene rearrangement must first be reviewed.

GENERATION OF DIVERSITY IN LYMPHOID ANTIGEN RECEPTORS REQUIRES GENE REARRANGEMENT AT THE LEVEL OF DNA

The immune system must recognize vast numbers of antigens. In order to accomplish this, T or B lymphocytes express unique antigen receptors on their surface membranes. B cell antigen receptors consist of two identical immunoglobulin heavy chain (IgH) molecules (IgM, IgG, IgD, IgA or IgE) paired with two identical light chain molecules (Ig κ or λ). T cell receptors (TCR) consist of paired α/β chains (on approximately 95% of T cells) or γ/δ chains on the remaining T cells. Generation of a diverse repertoire of antigen recognition sites is achieved in all of these receptors by similar means.

An antigen receptor protein contains segments encoded by different regions of the receptor gene. All of the potential coding sequences (exons) for each type of receptor are arranged in a colinear array at known chromosomal locations (Fig. 7.2). Exons are separated by noncoding sequences (introns). One segment each from the variable (V) region, diversity (D) region (not found in immunoglobulin light chains or in α or γ T cell receptors) and joining (J) regions are (apparently randomly) selected and recombined with loss of intervening DNA. This "gene rearrangement" event may be detected in the Southern blot procedure.

The rearranged VJ (for Ig κ and λ and TCR α and γ) or VDJ (for Ig heavy chains and TCR chains β and δ) DNA segment is then transcribed into mRNA with an appropriate constant (C) region segment. Introns are spliced out in the process, and the mRNA is then translated into an antigen receptor protein. The Ig or TCR receptor loci may undergo "nonfunctional" or "functional" rearrangement. A functional gene rearrangement results in an appropriate VDJ (or VJ) fusion that can be transcribed and translated into a functional protein. In contrast, nonfunctional

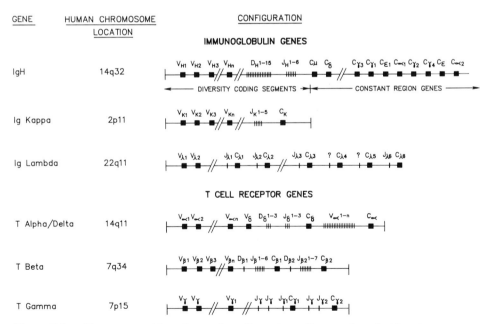

Figure 7.2. Chromosomal location and configuration of human lymphoid antigen receptor genes. Note that the Ig heavy chain, TCR β and δ loci contain V, D, J, and C regions, while Ig light chains, TCR α and γ contain all but the D regions. The entire TCR δ gene is embedded in the TCR α locus. (Reprinted with permission from Willman, C. L., Griffith, B. B., and Whittaker, M. Molecular genetic approaches for the diagnosis of clonality in lymphoid neoplasms. *Clin. Lab. Med. 10*:119–149, 1990.

rearrangements, due to out-of-frame joins or partial (DJ only) rearrangements, arrange gene segments such that they cannot be appropriately transcribed and translated to produce a functional Ig or TCR protein.

An example of an antigen receptor gene rearrangement using the Ig κ light chain gene is illustrated in Figure 7.3. Note the change in the BamHI restriction fragment size resulting from the rearrangement in this illustration. This is the phenomenon that permits detection of clonal gene rearrangements by Southern blot analysis. Many excellent, detailed reviews of this process are available, including references 24 and 25.

ANTIGEN RECEPTOR GENE REARRANGEMENTS CAN BE USED AS MARKERS OF CLONALITY

The change in restriction fragment size that occurs with lymphoid antigen receptor gene rearrangement is detectable by Southern blot analysis. Briefly, DNA is isolated from fresh or cryopreserved tissue and digested with appropriate restriction enzymes. The resulting DNA fragments are then separated by size on an agarose gel and transferred to a nitrocellulose membrane, which immobilizes the DNA in an exact image of the gel. The nitrocellulose blot is then hybridized with se-

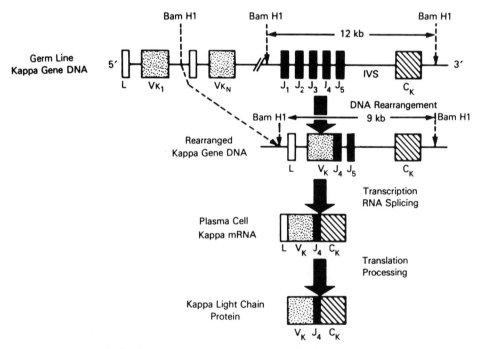

Figure 7.3. Kappa light chain gene rearrangement leading to production of a functional κ light chain protein. Note the change in restriction fragment size associated with this particular gene rearrangement. (Reprinted with permission from Waldmann, T. A. Immunoglobulin and T-cell receptor genes and lymphocyte differentiation. *In: The Molecular Basis of Blood Diseases,* edited by G. Stamatoyannopoulos, Philadelphia, W. B. Saunders Co., 1987.)

lected radiolabeled probes, washed and placed against x-ray film. The finished autoradiograph represents a one-to-one image of the nitrocellulose blot with dark bands located wherever the patient's DNA fragments match the probe (Fig. 7.4).

The distance between restriction enzyme recognition sites, and hence the size of restriction fragments in unrearranged Ig and TCR alleles is constant and predictable. This "germline" or unrearranged configuration will be seen in all nucleated, nonlymphoid cells and in any lymphoid cells that have not undergone rearrangement at the particular locus being studied. Germline restriction fragment size(s) depend upon the specific combination of restriction enzyme and probe used for analysis. Although any probe/enzyme combination could be used, only those that are least affected by artifacts have been selected for use in clinical studies. Common combinations of restriction enzymes, probes, and their resulting germline restriction fragment sizes are listed in Table 7.1. References 3 and 26 provide a detailed discussion of probes and enzymes useful in these studies.

Rearrangements of Ig or TCR genes in both polyclonal (reactive) and monoclonal (usually malignant) lymphoid proliferations produce restriction fragments differing in size from germline. In polyclonal proliferations such as reactive lymphadenopathies, each parent immunoblast and its small number of memory

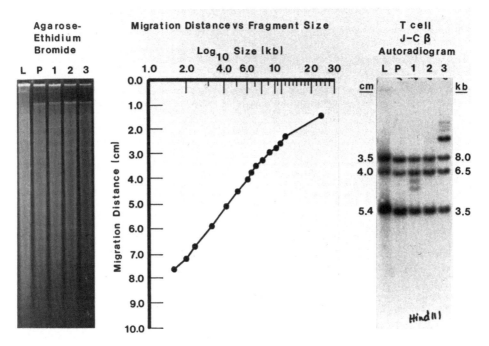

Figure 7.4. Constant-scale illustration of an ethidium bromide–stained agarose gel, containing (a) HindIII digested DNA from two controls, liver (L) and placenta (P), and 3 patients; (b) plot of migration distance vs log $_{10}$ kilobase size; and (c) finished autoradiogram hybridized with a TCR β-chain probe. Data for the plot come from DNA ladders on the same agarose gel (not shown). Kilobase sizes (*right side* of autoradiogram) are inferred from migration distances (*left side* of autoradiogram).

Table 7.1.
Immunoglobulin and T Cell Receptor Molecular Probe Descriptions[a]

Probe	Informative Restriction Enzymes	Germline Bands	Probe Description	Reference
J_H	EcoRI	16 kb	Genomic DNA J_2-J_6	19
J_κ	BamHI	8 kb	Genomic DNA of J_κ's	12
C_λ	EcoRI	16 kb 14 kb 5 kb 8 kb (polymorphic)	Genomic DNA of C region	11
T_β	EcoRI BamHI HindIII	12 kb 4.2 kb 24 kb 8 kb 6.5 kb 3.5 kb	cDNA containing some J and mostly C region	27
T_γ	EcoRI BamHI	9.5 kb 7.5 kb 6.7 kb 6.0 kb 3.8 kb 3.1 kb 24 kb 15 kb 4 kb	cDNA containing V, J and C regions	18
T_α	BamHI PvuII	6.6 kb 5.0 kb 3.5 kb 5.0 kb 4.5 kb 1.0 kb	cDNA containing 5' and 3' untranslated, V, J, and C regions	21
T_δ	EcoRI PvuII	6.6 kb 3.3 kb 1.5 kb 23 kb 6.6 kb 2.0 kb	cDNA containing V, J, and some C region	10

[a]Reprinted with permission from Willman, C. L., Griffith, B. B., and Whittaker, M. Molecular genetic approaches for the diagnosis of clonality in lymphoid neoplasms. *Clin. Lab. Med. 10*:119–149, 1990.

cell and plasma cell progeny have unique VDJ (or VJ) gene rearrangements producing a unique restriction fragment size. The multitude of antigen receptor gene restriction fragments arising in a polyclonal population of lymphoid cells produces a broad distribution down the lane when separated by size in an agarose gel. Only a few cells in a polyclonal population have identical restriction fragment sizes. Thus, the finished blot produces a faint or invisible background signal, in addition to the germline band.

In contrast, monoclonal proliferations of cells (most of which represent malignancies) share identical VDJ (or VJ) gene rearrangements. Thus restriction enzyme digestion produces large quantities of DNA at one or possibly two nongermline bands. The number and size of the novel bands depends on the relationship between restriction sites in the rearranged allele(s). Novel bands reflecting gene rearrangements may be larger, smaller, or the same size as the germline band. The germline band is usually also present, derived from unrearranged alleles in the malignant cells or unrearranged genes in admixed reactive lymphoid or nonlymphoid cells.

The standard Southern blot analysis can detect monoclonal proliferations with a sensitivity of one to five percent (*i.e.*, 1–5% of the cells in the diagnostic sample must be derived from a single clone), which is superior to morphology or immunophenotyping alone. In fact, molecular genetic techniques are so sensitive that they have identified mono- or oligoclonal proliferations in some disorders previously thought of as benign.[3,6] It is important to understand that the identification of monoclonal lymphoid proliferations by morphologic or immunophenotypic techniques usually predicts a malignant clinical course, although the grade of malignancy varies with the diagnosis. In contrast, detection of monoclonal proliferations using molecular genetic techniques may be used to confirm a malignant diagnosis, when corroborated with suspicious clinical, morphological, and immunophenotypic data. However, the diagnosis of malignancy cannot be based solely on gene rearrangement data because some clinically benign disorders may show mono- or oligoclonal gene rearrangements.[28]

Guidelines for the Interpretation of Southern Blots Used to Detect Monoclonal Lymphoid Proliferations

For any combination of restriction enzyme and Ig or TCR probe, one of three possible patterns may be seen on the finished Southern blot:

1. *All DNA is in germline configuration and no novel band(s) are present.*

In this case, no monoclonal gene rearrangement has been detected. This pattern suggests that the patient specimen consists of either (*a*) nonlymphoid tissue (including nonlymphoid neoplasms), (*b*) polyclonal lymphoid tissue, or (*c*) monoclonal cells in a quantity below the limits of detection (1 to 5%). Possible causes of a false negative study are listed below.

2. *Both germline and nongermline band(s) are present.*

This may be seen when (*a*) monoclonal lymphoid tissue (contributing the novel, rearranged band(s)) is admixed with nonlymphoid or reactive lymphoid tissue; (*b*) homogeneous monoclonal lymphoid tissue is present with one germline and one rearranged allele; or (c) some combination of these two situations is present.

3. *Only novel band(s) are present, no germline genes are detected.*

This uncommon pattern may be seen in a homogeneous population of monoclonal lymphoid cells in which both alleles have been rearranged or one allele has been rearranged and the other deleted.

CORRELATION OF GENE REARRANGEMENT STUDIES WITH DISEASES

Studies of lymphoid antigen receptor gene rearrangements in various disease states are rapidly accumulating, and generalizations must be made with caution. The following general guidelines are offered as a starting point for interpretation of Southern blot studies. References 3, 6, and 26 (from which these guidelines have been assembled) offer more detailed summaries, and current literature should be consulted for correlation with specific clinical cases.

1. Reactive lymphadenopathies usually show only the germline configuration of Ig and TCR genes. Due to the arrangement of restriction sites, reactive, polyclonal T cells may show a diminution in the TCR β/EcoRI 12.0 kb germline fragment relative to the 4.0 kb germline fragment when hybridized with a constant region probe.[3]

2. Relatively well-differentiated B cell malignancies such as B cell acute lymphocytic leukemia (ALL), B cell chronic lymphocytic leukemia (CLL), hairy cell leukemia, and well-differentiated B cell lymphomas may show both heavy and light chain Ig gene rearrangements, usually without TCR gene rearrangements. Approximately 10% of B cell lymphomas and leukemias may have nonfunctional rearrangements of TCR genes.[14]

3. Relatively well-differentiated T cell malignancies such as adult T cell leukemia/ lymphoma, T cell CLL, mycosis fungoides/Sezary's syndrome, and peripheral T cell lymphoma usually have functional TCR gene rearrangements, without Ig gene rearrangements. Occasional T cell neoplasms will have nonfunctional Ig heavy chain gene rearrangements.

4. Less-differentiated malignancies such as non-T, non-B ALL and many lymphomas (follicular or diffuse) may show rearrangements of the Ig heavy chain and TCR loci but not Ig light chain loci.

5. Cases of Hodgkin's disease may occasionally have rearrangement of either TCR, Ig, or both receptor genes. The functionality of these gene rearrangements has not yet been investigated. For this reason, Southern blot data may not be helpful in differentiating Hodgkin's versus non-Hodgkin's lymphoma.

6. "Benign clonal lymphadenopathies" are disorders in which antigen receptor gene rearrangements have been identified in patient tissue, yet clinically overt malignancy may never develop. Examples of this include lymphomatoid papulosis,[28] immunosuppression associated lymphoid proliferations (posttransplant or acquired immunodeficiency syndrome), Wiskott-Aldrich syndrome, and angioimmunoblastic lymphadenopathy with dysproteinemia (AILD).

Quality Control of Southern Blot Analysis

The most important component of quality control is correlation of the molecular genetic data with clinical, morphologic, and immunophenotypic data. A quality-control problem should be suspected if any of these tests are grossly discordant. In particular, the quantity of suspected tumor cells assessed on histologic sections and/or immunophenotype should match the relative density of germline or rearranged bands ("dosage") on Southern blots. For example, if 90% of the cells on a histologic section resemble large cell lymphoma cells, then (if both alleles are rearranged) 90% of the DNA should be present in the novel bands, and 10% in germline bands. If only one allele is rearranged, the expected dosage in the novel band is halved (45% in this example). When making this comparison, it is important to remember that some malignant appearing cells may be benign and vice versa.

One or preferably two normal germline control samples (*e.g.,* placenta or cadaveric liver) should be analyzed simultaneously with the patient sample(s) along with DNA ladders for size reference. Analysis should include probes for Ig λ and κ light chains, Ig heavy chains, and TCR β and γ. (TCR α and δ probes may also become useful as experience with them accumulates). Two or three different restriction enzyme digests should be used per probe. Comparison of multiple digests for each probe and multiple probes for each lineage are important during data interpretation. Monoclonality should not be diagnosed on the basis of a single probe/enzyme combination, since partial digests and polymorphisms may account for novel bands in a single probe/enzyme study.

The following quality control guidelines are derived from references 3, 6, and 26.

POSSIBLE CAUSES OF FALSE POSITIVE RESULTS (A NOVEL BAND IS PRESENT IN THE ABSENCE OF A CLONAL PROLIFERATION OF LYMPHOCYTES)

Partial Digests or Resistant Digestion Sites

When a restriction enzyme lacks sufficient activity, the finished blot will reflect a mixture of fully and partially digested DNA fragments. Often one or a few abnormal-sized fragments predominate, producing new nongermline fragments on the finished blot, which may be mistaken for rearranged genes. One EcoRI site in the TCR β gene is particularly resistant to digestion. This "resistant site" manifests itself as a new nongermline 8.5 kb band in the lane when hybridized with a TCR β constant region probe.

There are many possible causes of partial digestion including inappropriate reaction time, temperature, or buffer; inactive enzyme or presence of an inhibitor such as a heavy metal ion; or insufficient enzyme for the quantity of target DNA. If a partial digest is suspected, repeat digestion of patient and control DNA with the same, and perhaps other restriction enzymes under optimized conditions may resolve the problem. Use of >1U enzyme/microgram DNA is recommended. Unfortunately, some resistant sites may never completely digest, requiring consideration of other enzyme or enzyme/probe combinations to distinguish a partial digestion from a true gene rearrangement.

Restriction Fragment Length Polymorphisms (RFLPs)

RFLPs are normal variations in the size of restriction fragments that arise due to single base changes in restriction enzyme recognition sites or variations in the amount of DNA between restriction enzyme recognition sites. Naturally occurring polymorphisms in a population may cause some individuals to have normal germline bands that differ in size from the rest of the population. For example, a HindIII site in the TCR γ gene is variably present, so hybridization with a TCR γ J chain probe may produce one 4.5 kb or two (comigrating) 2.1 kb germline fragments. (Note that a person may be heterozygous for these alleles yielding both germline bands.)

Some RFLPs are caused by insertion of variable numbers of tandemly repeated sequences (VNTRs) between restriction enzyme recognition sites. This phenomenon is exploited in the "DNA fingerprint" technology described in Chapters 1 and 16 of this book. However, in antigen receptor gene rearrangement studies, VNTRs may produce novel bands, mimicking a true monoclonal gene rearrangement, often in addition to the expected germline bands. An example of this is a polymorphic site in the Ig heavy chain J region that can produce a 0.1 to 0.4 kb variation in the size of the 10.0 kb Jh/HindIII germline fragment. Also, polymorphisms in the λ light chain gene can cause replacement of the expected 8.0 kb EcoRI germline fragment with 13, 18, or 23 kb fragments.

When a polymorphism is suspected, such as when molecular genetic data are grossly discordant with clinical, morphologic, or immunophenotypic data, the Southern blot can be repeated with paired nontumor (*e.g.*, peripheral blood or skin punch biopsy) and tumor tissue from the patient. A polymorphism will be present in nontumor as well as tumor tissue from the affected individual. (Be aware that cells from a hematopoietic malignancy may infiltrate any normal appearing tissue from the patient.) Also, digestion with alternate enzymes or probe/enzyme combinations may be useful to resolve discrepancies. A true gene rearrangement is likely to yield abnormal restriction fragments with multiple enzymes and probes, whereas a polymorphism often manifests itself with only one probe/enzyme combination.

Limited Gene Rearrangements

The TCR γ gene has only a few variable regions available for recombination with joining and constant regions. Thus rearrangement of this gene in a polyclonal (reactive, not malignant) population of cells produces 7 or 8 discrete, new nongermline bands rather than the broad, often invisible, smear of rearranged bands in the other receptor genes. Sometimes, only one or two of the nongermline bands will predominate, mimicking a true gene rearrangement. For this reason, the presence of nongermline bands in any TCR γ study may not indicate monoclonality. Conversely, novel bands arising in a monoclonal T cell proliferation may become lost among the pseudoclonal bands, further diminishing the utility of TCR γ studies in clinical cases. A corresponding monoclonal gene rearrangement should be identified in the TCR β gene to confidently diagnose TCR γ rearrangements.

Administrative or Technical Errors

Like any other laboratory test, Southern blot studies are susceptible to clerical errors or technical errors such as specimen carryover. Since there are many opportunities for error in this 10- to 14-day procedure, meticulous attention must be paid to experimental technique.

POSSIBLE CAUSES OF FALSE NEGATIVE RESULTS (NO NOVEL BAND IS IDENTIFIED EVEN THOUGH DNA IS DERIVED FROM MONOCLONAL LYMPHOID TISSUE)

Sensitivity

A novel band can be identified on a Southern blot when as little as 1–5% of the cells in a specimen are derived from a single clone. When less than 1 in 100 cells are monoclonal, no novel band may be apparent.

Comigration

Conceivably, a novel restriction fragment from a monoclonal proliferation of cells may comigrate with a germline fragment and not be detected. In this setting, the rearrangement should be apparent in other restriction digests. For this reason, it is prudent to digest DNA with more than one enzyme for each probe. It is also useful to bear this situation in mind if a discrete novel band is apparent with only one enzyme in a set of digests.

Administrative or Technical Errors

Possible sampling, clerical, or technical errors must be considered when troubleshooting suspected false negative results.

Deletions of Both Alleles at an Antigen Receptor Locus

Both alleles may be deleted in a monoclonal population of cells, in which case only germline bands from nonneoplastic tissue in the specimen will be apparent on the finished blot. (If the tissue consists entirely of tumor, no bands will be present in the lane.) Each study should include probes for multiple genes from each lineage, since only one gene may be entirely deleted.

Antigen Receptor Rearrangements Occurring over Long Distances

Unusually large genes, such as the TCR α gene may require special techniques to identify novel restriction fragments. Standard electrophoresis efficiently separates DNA fragments from 0.1–25 kilobases. TCR α gene rearrangements occur over very long segments of DNA. Detection of long-distance gene rearrangements requires the use of special restriction enzymes to cut large pieces of DNA, which can only be resolved with pulsed-field gel electrophoresis. (See Chapter 1 for a discussion of pulsed-field gel electrophoresis.) Eventually, combinations of certain probes may allow resolution of TCR-α gene rearrangements on standard Southern blots.[26]

Clinical Case Studies

Clinical cases that yield conclusive diagnoses from clinical, morphologic, and immunophenotypic data usually show equally clear-cut results on Southern blot analysis. However, because of their expense and complexity, molecular genetic studies are usually reserved for the most difficult and confusing clinical cases. The examples presented in this chapter illustrate a spectrum of relatively problematic interpretations, to emphasize problem solving and quality-control procedures.

Relevant clinical, immunophenotypic, and molecular genetic data are presented for each case. The molecular genetic data illustrate examples of the following combinations of probes and restriction enzymes:

Immunoglobulin Genes

Jh/Eco-RI—a segment of the heavy chain gene joining region, is radiolabeled and hybridized to samples digested with the EcoRI restriction enzyme; Jκ/BamHI—a segment of the κ gene joining region is hybridized with BamHI digested samples; Cλ/EcoRI—a segment of the constant region of the λ gene is hybridized with EcoRI digested samples. (See Reference 26 for sources of probes.)

T Cell Receptor Genes

J-Cβ/BamHI, EcoRI, HindIII—a cDNA (reverse-transcribed messenger RNA) probe made from a TCR β transcript, prepared such that the variable region is deleted and only the constant region and a small amount of joining region is present, is hybridized with samples digested separately with each enzyme; V-J-Cγ/BamHI, EcoRI—a cDNA probe containing a variable region, joining region and constant region from a single TCR γ transcript is hybridized with sample digests. No useable TCR α or δ probes were available at the time these cases were studied. (See Reference 26 for sources of probes.)

In each example, a control sample lane (C) (*on the left*) is paired with the patient (P) sample lane. Kilobase sizes of the expected germline fragments are listed to the left of each pair of lanes. Bands from rearranged alleles are designated "R". Partial digests are marked with asterisks (*) and pseudoclonal bands are marked with "+" signs.

CASE 1

Clinical Data

A 25-year-old female presented with two painless, nodular, violaceous, cutaneous masses on her back. The 3-cm masses were nearly confluent, with smaller satellite lesions peripheral to the larger ones. The patient was otherwise asymptomatic, with a normal physical examination and no detectable adenopathy. Chest x-ray and abdominal CT scan were normal, as were laboratory data, including CBC and serum and urine immunofixation electrophoreses. Bone marrow was examined when lymphoma was initially suspected, and this also was within normal limits.

Morphology (Fig. 7.5)

The skin biopsy revealed a polymorphous lymphoid infiltrate extending deep into subcutaneous tissue. Although the infiltrate lacked well-developed follicles, a vaguely zonal pattern was present, with ill-defined aggregates of large lymphoid cells against a background of smaller, compact lymphocytes. Because of the extent and depth of invasion, and lack of definite reactive architecture, the biopsy was considered suspicious for malignancy.

Immunophenotype

Flow cytometric examination failed to identify a monoclonal population of lymphoid cells. Most cells were T cells, with a normal helper to suppressor ratio and normal coexpression of T cell antigens. No immunoglobulin light chain restriction was identified. Frozen-section immunophenotyping confirmed the zonal distribution of the cells. B cells comprised the larger aggregated cells with smaller T cells in the background. Approximately equal numbers of κ and λ light chain–bearing B cells were present.

Molecular Genetics (Fig. 7.6)

Normal germline bands are present with each combination of probe and enzyme. A 13 kb band representing a common partial digest site is seen in the TCR β/HindIII lane. No discrete novel bands are present. A strong blush that varies in intensity is present below the TCR β/BamHI germline band. The broad, diffuse bands in the TCR β/BamH1 lane are not considered diagnostic of monoclonality because corroborative novel bands are lacking in the TCR β/EcoRI, Hind III or Jh/EcoRI lanes. Partial degradation of the DNA prior to or during

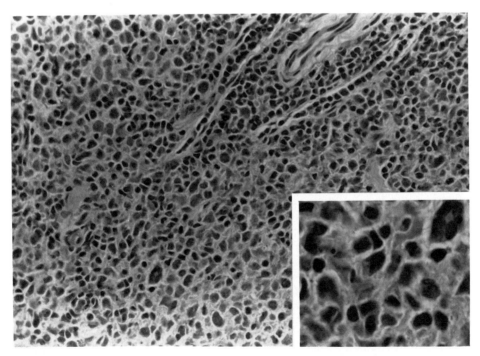

Figure 7.5. Case 1, dermis and subcutaneous tissue. Large and small lymphoid cells are admixed in a vaguely follicular pattern. Rare multinucleate giant cells are present. Neutrophils, eosinphils and basophils are not seen. The infiltrate spared the epidermis and infiltrated dermis and subcutaneous tissue. (H&E, 400X; inset 1000X.)

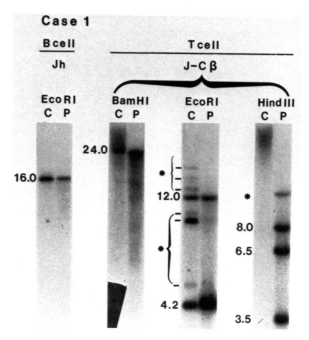

Figure 7.6. Molecular genetic results, case 1. No rearranged bands are identified in the patient lanes. The diffuse smear in TCR β/BamHI lane may be due to degraded DNA. Note partial digests in controls for TCR β/EcoRI and HindIII digests. This problem has not recurred since routine the use of >1 U restriction enzyme/microgram DNA was implemented. Appropriately digested controls and patient cases on the same blot validated data from this study. *Asterisks* mark partial digests.

isolation may produce this artifact. Tissue homogenization ("tissuemization") may contribute to this artifact and should be eliminated from the DNA isolation procedure.

The control lanes in this case illustrate some common artifacts. Appropriate germline bands are seen in other controls and clinical specimens on these blots, validating the patient results. The TCR β/EcoRI control lane shows the multiple bands that can be seen following partial digestion of the sample DNA. The control in the neighboring TCR β/HindIII lane shows minimal digestion of the sample DNA, with all of the DNA remaining in the high molecular weight region. Routine use of >1 U restriction enzyme/microgram DNA has eliminated this artifact.

Discussion

Negative data from any single laboratory test must be interpreted with caution; however, in this case we can find no clinical, immunophenotypic, or molecular genetic proof of monoclonality. We feel this case most closely resembles the "large cell lymphocytoma" described by Winkleman and colleagues,[5] and we anticipate a benign clinical course for this patient. Abnormally intense immune reactions involving the thyroid or salivary glands are known to increase a patient's risk for lymphoma.[1] By analogy, patients with exuberant cutaneous lymphoid reactions may also be at increased risk and should be watched carefully for evidence of recurrent or evolving disease. This patient's lesions spontaneously resolved, and the patient is free of recurrence 2 years after the original presentation.

CASE 2

Clinical Data

A 75-year-old female presented with cervical lymphadenopathy. An atypical lymphoid infiltrate was identified in tissue from a salivary gland biopsy 6 years earlier.

Morphology (Fig. 7.7)

Histologic sections revealed replacement of the normal architecture by diffuse islands of large, atypical lymphocytes with open chromatin and small nucleoli. Smaller lymphocytes with angulated nuclei and condensed chromatin were present peripherally. From morphology alone, the working diagnosis was probable malignant lymphoma, mixed small cleaved and large cell type.

Immunophenotype (Fig. 7.7)

Frozen section immunophenotype showed B cell markers on most of the large and many of the smaller cells. Approximately equal numbers of Ig κ and λ positive cells were present, and most of these were intermediate in size. Most of the large cells were Ig light chain negative. Many of the smaller lymphocytes stained as T cells.

Molecular Genetics (Fig. 7.8)

Gene rearrangement studies were performed because of the unusual morphology and lack of an Ig light chain predominance on immunophenotypic analysis. Novel bands are

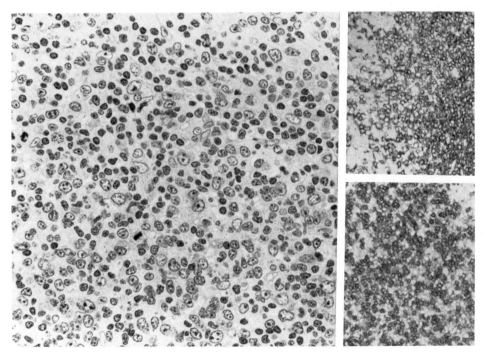

Figure 7.7.   Case 2, cervical lymph node. This lymph node was entirely replaced by abnormal appearing large and small lymphocytes. *Insets* show frozen-section immunophenotype with approximately equal numbers of κ (top) and λ (bottom) light chain positive cells. (H&E, 400X; frozen section immunophenotype, 160X.)

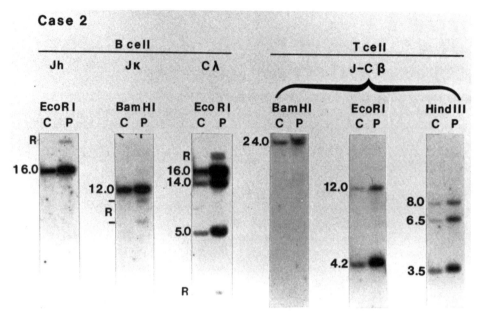

Figure 7.8. Molecular genetic data, case 2. Note the rearranged bands in Ig heavy chain (Jh) and Ig κ and possibly λ light chains and also TCR β/BamHI lanes.

present in the Ig heavy chain (Jh), and both light chain (J-κ and C-λ) lanes. The TCR β/ BamHI lane contains a possible doublet in the germline position.

Discussion

The monoclonal population of B cells identified by Southern analysis supports the diagnosis of non-Hodgkin's lymphoma. The rearrangements of both Ig heavy and light chain genes are characteristic of B cell–derived clones, which supports the follicular center cell origin for this tumor. The IgH J_h and Ig κ lanes clearly reveal the presence of rearranged bands. Note, however, that the high molecular weight doublet in the Ig λ lane is out of proportion to the novel bands present in the IgH Jh and Ig κ lanes, and the 8.5 kb Ig λ germline fragment is missing. This pattern may result from polymorphisms known to occur with this enzyme/ probe combination, yet the low molecular weight band would still be unexplained. Analysis of paired nontumor and tumor DNA from this patient would be essential to determine if these bands represent naturally occurring polymorphisms or true Ig λ gene rearrangements. The possible TCR β chain gene rearrangement in the BamHI digest is difficult to interpret, since neither the EcoRI nor the HindIII digest lanes contain novel bands, and the dosage varies relative to the rearranged IgHJ$_h$ and Ig κ bands. When TCR β gene rearrangements occur in B cell lymphomas, they are usually "nonfunctional" DJ rearrangements.

CASE 3

Clinical Data

A 57-year-old Hispanic male presented with a three-year history of recurrent skin "lumps" with no other symptoms. A previous biopsy was diagnosed as suggestive of a malignant lymphoma. Physical examination revealed a 3 × 4 cm violaceous, firm, fixed mass over the right clavicle. A total of four smaller masses were noted on the anterior and posterior thorax and posterior auricular areas. No significant abnormalities were identified on the remainder of the physical examination. No adenopathy was identified on physical or radiological examination. In order to clarify the diagnosis, the posterior thorax mass was biopsied.

Morphology (Fig. 7.9)

Histologic sections show large masses of lymphoid cells infiltrating the deep dermis and subcutaneous tissue, with sparing of the epidermis and superficial dermis. No well-defined follicles were identified, although a zonal distribution of cells was present, with large aggregates of pale-staining lymphoid cells superimposed on a background of smaller lymphocytes. A few immunoblasts were present, and acute inflammatory cells were rarely seen.

Immunophenotype

Frozen-section immunophenotype demonstrated that the larger lymphocytes were B cells and the smaller peripheral lymphocytes were T cells. No κ or λ Ig light chain predominance was identified.

Molecular Genetics (Fig. 7.10)

As in case 1, the lack of a well-defined follicular hyperplasia, and the deep and extensive invasion of subcutaneous connective tissue suggested the diagnosis of lymphoma.

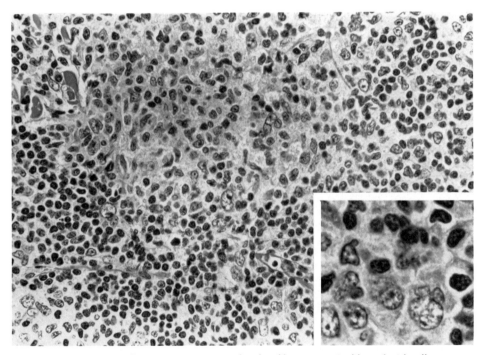

Figure 7.9.   Case 3, subcutaneous tissue. Islands of large, atypical lymphoid cells are surrounded by smaller lymphocytes. This infiltrate extends deep into the dermis and subcutaneous connective tissue. Similar to case 1, neutrophils, eosinophils, and basophils are absent. (H&E 400X, inset 1000X.)

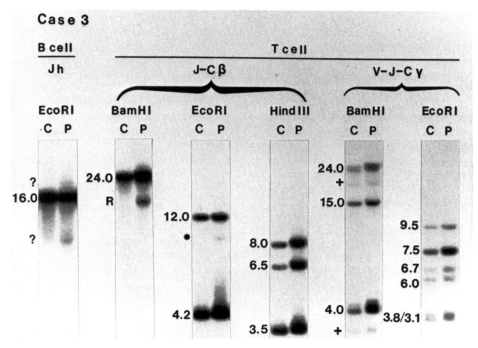

Figure 7.10. Case 3, molecular genetic data. Clonal gene rearrangements are present in the TCR β/BamHI lanes and probably the IgH Jh/EcoRI lanes. The novel TCR β gene band is not present in DNA from normal peripheral blood lymphocytes from the patient, eliminating the remote possibility that this band represents an unreported polymorphism. An *asterisk* marks the common TCR β/BamHI partial digest caused by a resistant restriction site, and the + *sign* marks pseudoclonal bands in the TCR γ/BamHI lane.

Since no Ig light chain restriction was identified, molecular genetic studies were performed. In this case, the most striking novel band is a 15 kb band identified in the TCR β/BamHI lane. No other definite gene rearrangements are identified in the T cell studies. Possible clonal gene rearrangements are seen in the IgH J$_h$ lane. A faint band resulting from partial digestion of the TCR β/EcoRI resistant site and pseudoclonal bands in the TCR γ/BamHI lane are also visible.

Discussion

These Southern blots demonstrate the presence of a monoclonal proliferation of lymphocytes, which supports (but cannot solely prove) the diagnosis of lymphoma. The actual lineage of the malignant cells is uncertain, since both Ig and TCR gene rearrangements are present. Although a novel band is present in the TCR β/BamHI analysis, no corroboration is seen in the EcoRI or HindIII lanes. To determine if the novel TCR β BamHI band truly reflects a TCR gene rearrangement, morphologically normal peripheral blood lymphocytes from this patient were analyzed by Southern blot. These cells showed no evidence of the novel band, thus increasing confidence in the diagnosis of monoclonality. It is not clear why the dosage of the rearranged IgH J$_h$ and TCR β bands do not match. No ratio of reactive to malignant cells is available from morphology or immunophenotype for comparison to gene dosage.

Staging studies in this patient showed no evidence of bone marrow involvement or serum protein abnormalities. Because the tumor was considered low grade and the patient was not symptomatic, treatment was not initiated. There has been no change in the skin lesions, nor have any new lesions, lymphadenopathy, or hepatosplenomegaly been identified with 1 year of follow-up.

CASE 4

Clinical Data

A 37-year-old white male presented with a month-long history of icterus and malaise. Bilirubin and liver function tests were abnormal, and hepatitis and HIV 1 serology were negative. Over the following two months, the patient suffered a 35 lb weight loss accompanied by severe malaise, fever, hepatosplenomegaly, and possible retroperitoneal and perisplenic lymphadenopathy. An extensive evaluation was performed, which included splenectomy and liver biopsy, both of which showed prominent atypical lymphoid infiltrates that were suggestive of lymphoma (Fig. 7.11 & 7.12). Biopsy of a mesenteric lymph node showed lymphocyte depletion with subtle atypia (Fig. 7.13). Atypical lymphoid cells were also present in the patient's bone marrow. Immunophenotype of multiple specimens showed that most of the abnormal lymphocytes were T cells. The few B cells present lacked Ig light chain restriction. Initial molecular genetic studies on splenic tissue showed very faint bands in the TCR β/BamHI lane. Since the bands were very faint and out of proportion to the number of abnormal lymphoid cells seen on tissue sections, this study was considered suggestive, but not diagnostic of monoclonality. Repeat biopsy of any new lesions was recommended.

Over the following two months, the patient developed a peripheral blood lym-

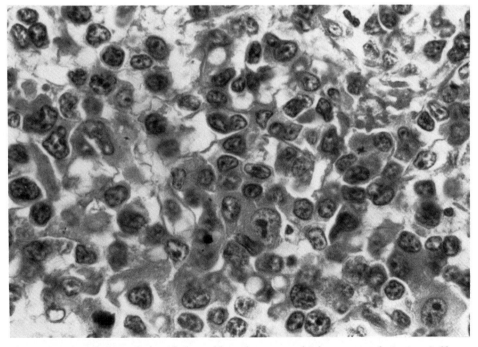

Figure 7.11. Case 4, spleen. Abnormal lymphocytes, which vary greatly in size, infiltrate the white and red pulp. Most of these cells immunophenotype as T cells. Initial molecular genetic studies were performed on this tissue, and no definite monoclonal pattern was identified. (H&E, 1000X.)

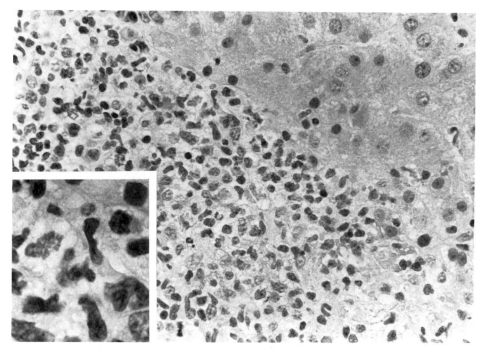

Figure 7.12. Case 4, liver. Massive infiltrates of abnormal lymphocytes are present in a periportal distribution. Similar to those from the spleen in Figure 11, the lymphocytes show wide variation in size and do not form follicles. (H&E, 400X; inset 1000X.)

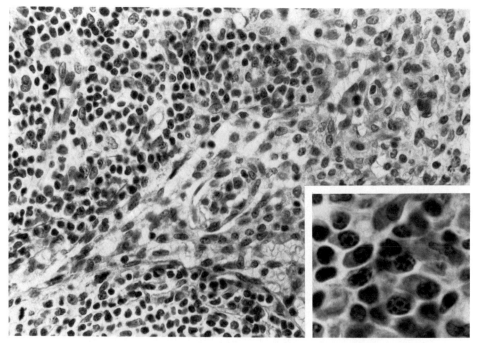

Figure 7.13. Case 4, mesenteric lymph node. Sections show lymphocyte depletion and open sinuses. Lymphocyte morphology varies from slightly atypical, small cleaved cells to normal or reactive cells including plasma cells. (H&E, 400X; inset 1000X.)

phocytosis with helper/suppressor ratio of 10:1 and generalized lymphadenopathy. A biopsy of one of the enlarged lymph nodes was performed.

Morphology (Fig. 7.14)

Histologic sections of this second lymph node biopsy revealed a confusing pattern, intermediate between lymphocyte-predominant Hodgkin's disease and malignant lymphoma. The normal architecture was replaced predominantly by small, round or cleaved lymphocytes with compact chromatin. Rarely, large mono- or binucleate cells with large nucleoli were interspersed with the small lymphoid cells.

Immunophenotyping

Immunophenotyping of formalin-fixed, embedded tissue showed that most of the smaller lymphocytes were T cells. The large binucleate cells were negative for L26 follicular center cell antigen and were variably positive for the Leu M1 and common leukocyte antigens.

Molecular Genetic Studies (Fig. 7.15)

Tissue from the patient's second lymph node biopsy reveals no novel bands with the IgH J_h probe. However, a broad, diffuse rearranged band is identified in the TCR β/BamHI lane. Nongermline bands are also present in the TCR β/EcoRI lane. In addition, numerous new bands are identified in the TCR γ/BamHI lane. As usual, the novel TCR γ bands are difficult to differentiate from pseudoclonal bands.

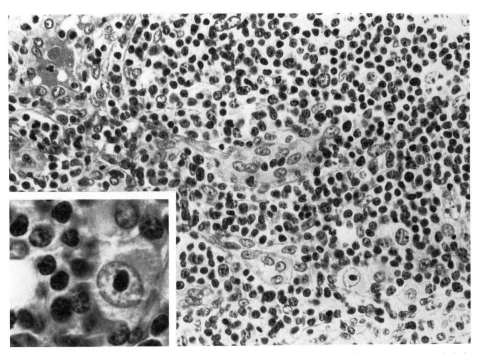

Figure 7.14. Case 4, cervical lymph node, 2 months later. Lymphocyte atypia is slightly more pronounced than in previous biopsies. The differential diagnosis includes lymphocyte-predominant Hodgkin's disease and T cell non-Hodgkin's lymphoma. (H&E, 400X; inset 1000X.)

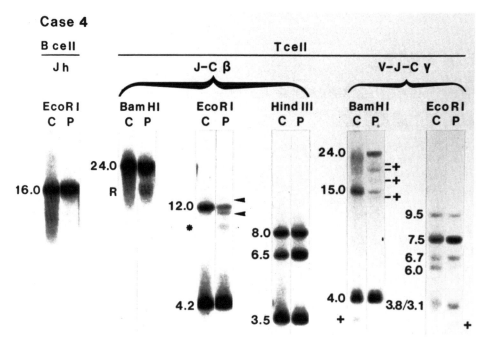

Figure 7.15.   Case 4, molecular genetic data. Monoclonal gene rearrangements are present in the TCR β/BamHI and EcoRI lanes. An *asterisk* marks a known partial digest of TCR β, and + *signs* mark pseudoclonal bands in the TCR γ gene. It is difficult to confidently differentiate pseudoclonal bands from true TCR γ monoclonal gene rearrangements.

Discussion

The molecular genetic data is considered to be diagnostic of a monoclonal T cell proliferation. Coupled with clinical, morphologic, and immunophenotypic data, the diagnosis of a malignant T cell lymphoma was made. The distinction between Hodgkin's disease and T cell non-Hodgkin's lymphoma is sometimes difficult, and particularly so in this case. Since rearrangements of TCR and/or Ig genes have been described in Hodgkin's disease, as well as T cell lymphomas, molecular genetic studies cannot be used as a sole criterion to differentiate them. While Hodgkin's disease cannot be ruled out completely in this case, morphologic, immunophenotypic, and molecular genetic evidence favor a non-Hodgkin's lymphoma, and the patient's treatment was based on this diagnosis. He is in complete remission at this time, following chemotherapy.

CASE 5

Clinical Data

A 20-year-old male presented with a left neck mass. He was otherwise asymptomatic. The mass was biopsied.

Morphology (Fig. 7.16)

Sections showed a classic pattern of nodular sclerosing Hodgkin's disease. The normal architecture was replaced by small compact lymphocytes, with histiocytes interspersed. Mononuclear variants, "mummy cells," and rare, classic Reed-Sternberg cells were present.

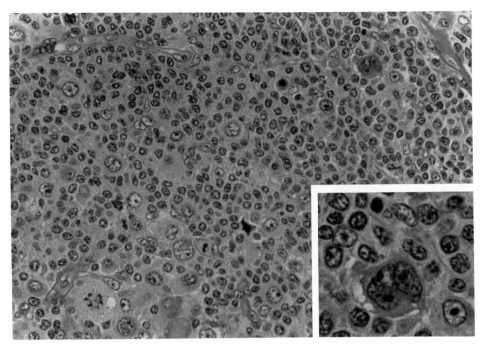

Figure 7.16.   Case 5, cervical lymph node. Normal nodal architecture is replaced by typical nodular sclerosing Hodgkin's disease. Reed-Sternberg cells, mononuclear variants, and mummy cells are admixed with small, slightly atypical lymphocytes. (H&E, 400X; inset 1000X.)

Immunophenotyping

Frozen-section immunophenotyping demonstrated Leu M1 positivity in the Reed-Sternberg cells and mononuclear variants.

Molecular Genetics (Fig. 7.17)

This case was studied as part of a survey of TCR and Ig gene rearrangements in Hodgkin's disease. Discrete, novel bands are lacking. However, broad, variably dense bands are present in the IgH J_h/EcoRI, TCR β/BamHI, and TCR γ/BamHI lanes. A typical TCR β/EcoRI partial digest band is also present.

Discussion

These results are typical of Hodgkin's disease cases that show rearranged Ig heavy chain and TCR genes. This case was included to emphasize that Hodgkin's disease cannot be differentiated from non-Hodgkin's lymphoma on the basis of gene rearrangement studies alone. The differential diagnosis of Hodgkin's versus non-Hodgkin's lymphoma must be made by combining clinical, morphologic, immunophenotypic, and molecular genetic data. This patient was staged as IIIa and is in complete remission following radiation therapy.

Modifications to Genetic Techniques Applied to Clinical Specimens

Several modifications have been introduced to make the standard Southern blot technique more amenable to clinical specimens. The use of biotinylated probes

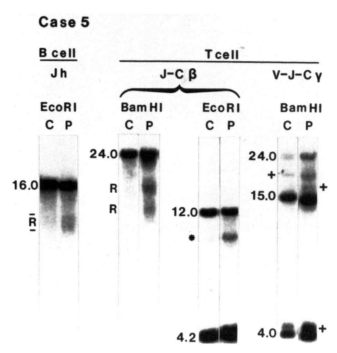

Figure 7.17. Case 5, molecular genetic data. Mono- or oligoclonal gene rearrangements are seen in the IgH Jh/EcoRI and TCR β/BamHI lanes. An *asterisk* marks the TCR β/EcoRI partial digest, and + *signs* mark pseudoclonal TCR γ bands. As usual, it is difficult to differentiate the latter from true gene rearrangements.

may eventually eliminate the need for radioactive probes; however, to date, nonradioactive probes have not been reported to be as sensitive as their radio-labeled counterparts. Nonradioactive probes will probably become more useful when examining the products of polymerase chain reaction (PCR) studies, in which tremendous amplification of the target sequence makes sensitivity less of a problem. (See Chapters 1 and 2 for a more thorough discussion of PCR.)

Theoretically, fine needle aspiration (FNA) of a lesion should be able to provide the 50 million cells needed to isolate adequate amounts of DNA for Southern blot analysis. However, this approach is still subject to the limitations of FNA such as sampling error and lack of data concerning the architecture of the involved lymph node. Attempts have also been made to isolate DNA from formalin-fixed, paraffin-embedded tissue for use in a Southern blot. In practice, so far, this has met with little success. Formalin fixation is known to denature and covalently cross-link adjacent nucleotides and proteins.[4,9] Even when a signal is obtained using DNA from formalin-fixed tissue, the sensitivity and specificity of the test will be unknown and unique to each specimen.

Polymerase chain reaction technology may be successfully applied to DNA obtained from fixed and frozen tissue. Recent reports have demonstrated the use of PCR in the detection of clonal rearrangements in both the TCR δ and TCR γ chain genes.[8,16] PCR technology has also been used to detect clonal rearrangements involving the Bcl 2 (B cell lymphoma/leukemia) oncogene that occurs in the t(14;18) of follicular center cell lymphomas.[17,23] These strategies will be extremely useful in the diagnosis and detection of occult and minimal residual disease.[20,22]

Molecular genetic techniques such as the Southern blot are becoming increasingly useful in the diagnosis of lymphoid malignancy. For this reason, it is important for pathologists to understand the indications for performance and the evaluation of this technology. With this goal in mind, this chapter outlines the interpretation and quality control of Southern blot analysis as applied to lymphoid antigen receptor gene rearrangements.

Acknowledgments

The authors thank Ms. Jerri Craven and Ms. Sheryl Curtin for assistance in manuscript preparation; Barbara Griffith, M.S.; Dan Hardy, M.D.; Kathy Foucar, M.D.; and Dan Kerrigan, M.D. for helpful advice; and Terry Evans, M.D.; Edward Abell, M.D.; Walter Forman, M.D.; Aroop Mangalik, M.D.; Robert Hilley, M.D.; Charles Haas, M.D.; and Stanley Wilson, M.D. for contribution of cases.

REFERENCES

1. Burke, J. S. The diagnosis of extranodal lymphomas and lymphoid hyperplasias ("Pseudolymphomas") other than those involving the lung. *In: Surgical Pathology of the Lymph Nodes and Related Organs*, edited by E. S. Jaffe and J. L. Bennington, Philadelphia, W.B. Saunders Co., 1985.

2. Chen-Levy, J., Nourse, J., and Cleary, M. L. The bcl-2 candidate proto-oncogene product is a 24-kilodalton integral-membrane protein highly expressed in lymphoid cell lines and lymphomas carrying the t(14;18) translocation. *Mol. Cell Biol.* 9(2):701–710, 1989.

3. Cossman, J., Uppenkamp, M., Sundeen, J., Coupland, R., and Raffeld, M. Molecular Genetics and the diagnosis of lymphoma. *Arch. Pathol. Lab. Med.* 112:117–127, 1988.

4. Dubeau, L., Chandler, L. A., Gralow, J. R., Nichols, P. W., and Jones, P.A. Southern blot analysis of DNA extracted from formalin-fixed pathology specimens. *Cancer Res.* 46:2964–2969, 1986.

5. Duncan, S. C., Evans, H. L., and Winkelmann, R. K. Large cell lymphocytoma. *Arch. Dermatol.* 116:1142–1146, 1980.

6. Griesser, H., Tkachuk, D., Reis, M. D., and Mak, T. W. Gene rearrangements and translocations in lymphoproliferative diseases. *Blood* 73(6):1402–1415, 1989.

7. Haldar, S., Beatty, C., Tsujimoto, Y., and Croce, C. M. The bcl-2 gene encodes a novel G protein. *Nature (Lond)* 342:195–198, 1989.

8. Hansen-Hagge, T., Yokota, S., and Bartram, C. R. Detection of minimal residual disease in acute lymphoblastic leukemia by in vitro amplification of rearranged T-cell receptor chain sequences. *Blood* 74(5):1762–1767, 1989.

9. Haselkorn, R. and Doty, P. The reaction of formaldehyde with polynucleotides. *J. Biol. Chem.* 236(1):2738–2745, 1961.

10. Hata, S., Brenner, M. B., and Krangel, M. S. Identification of putative human T cell receptor delta complementary DNA clones. *Science (Wash DC)* 238:678, 1987.

11. Heiter, P. A., Hollis, G. F., Korsmeyer, S. J., et al. Clustered arrangement of immunoglobulin lambda constant region genes in man. *Nature (Lond)* 294:536, 1981.

12. Hieter, P. A., Max, E. E., Seidman, J. G., et al. Cloned human and mouse kappa immunoglobulin constant and J region genes conserve homology in functional segments (Part 1). *Cell* 22:197, 1980.

13. Kamat, D., Laszewski, M. J., Kemp, J. D., et al. The diagnostic utility of immunophenotyping and immunogenotyping in the pathologic evaluation of lymphoid proliferations. *Mod. Pathol.* 3(2):105–112, 1990.

14. Knowles, D. M. Immunophenotypic and antigen receptor gene rearrangement analysis in T cell neoplasia. *Am. J. Pathol.* 134(4):761–785, 1989.
15. Little, J. V., Foucar, K., Horvath, A., and Crago, S. Flow cytometric analysis of lymphoma and lymphoma-like disorders. *Semin. Diag. Pathol.* 6(1):37–54, 1989.
16. Macintyre, E., Auriol, L. D., Amesland, F., *et al.* Analysis of junctional diversity in the preferential V_1-J 1 rearrangement of fresh T-acute lymphoblastic leukemia cells by in vitro gene amplification and direct sequencing. *Blood* 74(6);2053–2061, 1989.
17. Ngan, B-Y, Chen-Levy, Z., Weiss, L. M., Warnke, R. A., and Cleary, M. L. Expression in non-Hodgkin's lymphoma of the bcl-2 protein associated with the t(14;18) chromosomal translocation. *N. Engl. J. Med.* 318(25):1638–1644, 1988.
18. Quertermous, T., Murre, C., Dialynas, D., *et al.* Human T cell gamma chain genes: Organization. *Science (Wash DC)* 231:252, 1986.
19. Ravetch, J. V., Siebenlist, U., Korsmeyer, S., *et al.* Structure of the human immunoglobulin locus: Characterization of embryonic and rearranged J and D regions. *Cell* 27:583, 1981.
20. Shibata, D., Nichols, P., Sherrod, A., Rabinowitz, A., Bernstein-Singer, L., and Hu, E. Detection of occult CNS involvement of follicular small cleaved lymphoma by the polymerase chain reaction. *Mod. Pathol.* 3(1):71–76, 1990.
21. Sim, G. K., Yague, J., Nelson, J., *et al.* Primary structure of human T cell receptor alpha-chain. *Nature (Lond)* 312:771, 1984.
22. Stetler-Stevenson, M., Raffeld, M., Cohen, P., and Cossman, J. Detection of occult follicular lymphoma by specific DNA amplification. *Blood* 72(5):1822–1825, 1988.
23. Tsujimoto, Y., Finger, L., Yunis, J., Nowell, P., and Croce, C. M. Cloning of the chromosome breakpoint of neoplastic B cells with the t(14;18) chromosome translocation. *Science (Wash DC)* 226:1097–1099, 1984.
24. Waldmann, T. A. Immunoglobulin and T-cell receptor genes and lymphocyte differentiation. *In: The Molecular Basis of Blood Diseases*, edited by G. Stamatoyannopoulos, Philadelphia, W. B. Saunders Co., 1987.
25. Watson, J. D., Hopkins, N. H., Roberts, J. W., Steitz, J. A., and Weiner, A. M. The generation of immunological specificity: An introduction to the vertebrate immune response. *In: Molecular Biology of the Gene*, edited by J. R. Gillen, Menlo Park, CA, Benjamin/Cummings Publishing Co., 1987.
26. Willman, C. L., Griffith, B. B., and Whittaker, M. Molecular genetic approaches for the diagnosis of clonality in lymphoid neoplasms. *Clin. Lab. Med.* 10(1):119–149, 1990.
27. Yoshikai, Y., Anatoniou, D., Clark, S. P., *et al.* Sequence and expression of transcripts of the human T cell receptor B-chain genes. *Nature (Lond)* 312:521, 1984.
28. Weiss, L. M., Wood, G. S., Trela, M., Warnke, R. A., and Sklar, J. Clonal T-cell populations in lymphomatoid papulosis. *N. Engl. J. Med.* 315(8):475–479, 1986.

8

Oncogenes and Tumor Suppressor Genes in Solid Tumors: General Considerations

Cecilia M. Fenoglio-Preiser

Knudson suggests that cancer patients should fit into four groups: (*a*) those in whom spontaneous mutations produce an irreducible background level of cancer; (*b*) those who are exposed to agents that change the host genome by mutagenesis or by the addition or deletion of genetic material increasing the background incidence of cancer; (*c*) those with abnormalities that favor either spontaneous or induced mutations thereby increasing the risk of cancer; and (*d*) those who inherit an initiating mutation that would strongly predispose one to cancer.[41]

It is generally believed that the majority of cancer falls into the second group, with the causative agent being a chemical mutagen, radiation, or a virus. The third group is exemplified by patients with xeroderma pigmentosum. The somatic mutation hypothesis implies that critical genetic loci have mutations that lead to cancer. Two large classes of these genes have been identified. The first consists of oncogenes and the second of antioncogenes or tumor suppressor genes.[41] It now appears that precancerous lesions result from local phenomena that affect numerous cells, as demonstrated by the multiclonality found in polyps in polyposis coli and in the neurofibromas of neurofibromatosis. In contrast, malignant schwannomas arising on a background of neurofibromatosis and cancers arising in the colon affected by polyposis demonstrate monoclonality. Thus, each malignant tumor results from a rare second event that affects a single cell.[25,35]

Some of the proteins encoded for by oncogenes encode growth factors. Others encode growth-factor receptors. Expression of these proteins may lead to enhanced tumor cell growth. In particular, it is known that some oncogenes encoding growth-factor receptors have mutations affecting the resulting protein product in such a way that the receptor remains stimulated without the binding of its ligand. Other oncogenes encode protein kinases that phosphorylate critical subcellular components such as the cytoskeleton. Still others act as transcriptional regulators. Abnormalities in any of the above, whether it be by mutation, overexpression, dysregulated expression, or expression at the wrong time, have the ability to result in tumors.

Oncogene products form components of a signal transduction system that modulates gene activity and subsequent phenotypic manifestations (Fig. 8.1). The overexpression of a cellular or mutated oncogene may subvert normal gene function, causing transformation. Overexpression can also be achieved by transfecting cultured cells with an oncogene at high copy number or by substituting the oncogene promoter with a strong constitutive or inducible heterologous promoter. Furthermore, animals made transgenic for such constructs can be bred to express oncogenes in those somatic cells that recognize a particular promoter. Even though the oncogene products are located in different cellular compartments and have different functional properties, they may produce similar phenotypic changes.

Proto-oncogenes may be activated by various mechanisms. *ras* proto-oncogenes are frequently altered by point mutations. Chromosomal translocations involving proto-oncogenes are demonstrable in various tumors (see Chapters 3 and 6–12). Amplification of several oncogenes including c-*myc*, c-*myb*, c-Ki-*ras*, c-Ha-*ras*, c-*abl*, c-*erb*, and N-*myc* has been reported in a number of cancer cell lines and human tumors, as summarized in Chapters 8–12. Other alterations include deletions, particularly of the c-Ha-*ras* gene and c-*myb* genes. Sometimes these alterations relate to clinical behavior (see Chapters 8 and 11). It is hoped that alterations in oncogene structure or expression will be clinically useful to determine: (*a*) cancer susceptibility; (*b*) tumor etiology; (*c*) tumor diagnosis; (*d*) tumor prognosis; or (*e*) tumor therapy.[79]

Tumor Suppressor Genes

It has been recognized for a long time that certain forms of cancer cluster in families, and in these cases the genetic predisposition behaves as an autosomal dominant Mendelian trait. The most thoroughly studied example is the childhood disease retinoblastoma. Epidemiologic analyses support the hypothesis that retinoblastoma arises as the result of two separate genetic events.[40] The first event in hereditary retinoblastoma is the inheritance of a predisposing mutation from the affected parent, which by itself is insufficient to cause tumors since the neoplasms arise as discrete foci on a background of normal retinal cells, even though each cell carries the initial mutation. This implies that the predisposing mutation, although dominant in families, is recessive at the cellular level. The model also predicts that a second event will be some somatic genetic alteration that specifically affects the remaining wild-type allele at the same locus. In the sporadic form of the disease, both events are thought to take place somatically in a single cell. The detection of some tumor suppressor genes was made possible by a combination of familial studies, including cytogenetic and molecular genetic studies involving RFLP analyses.[8,10,15,44,45,50,69] Putative sites of tumor suppressor genes are listed in Table 8.1.

Additional evidence for tumor suppressor genes is provided by data demonstrating that malignant cells fused to diploid fibroblasts of the same species result in hybrid cells that are nonmalignant, as long as they retain certain specific chromosomes donated by the parent.[33,74] When these particular chromosomes are lost from the hybrid through chromosomal instability, the malignant phenotype reappears and the revertant cells can again give rise to tumors. It is also known that cancer cell growth and the transformed phenotype can be suppressed by contact with normal

Table 8.1.
Putative Sites of Tumor Suppressor Genes

Chromosome Site	Tumor	Reference Number
1p	Neuroblastoma, melanoma	6
3p	Small cell, squamous cell, adenocarcinoma of lung, testicular cancer, kidney carcinoma	43, 56, 58, 81, 82, 92
5q, 17, 18	Colon carcinoma	19, 46, 71, 85
11p	Wilms' tumor, bladder carcinoma, breast cancer, embryonal carcinoma, hepatoblastoma, rhabdomyosarcoma, testicular cancer	44, 61
13q	Retinoblastoma, ductal breast carcinoma, hepatocellular carcinoma, osteogenic sarcoma	48
22q	Bilateral acoustic neuroma, meningioma, Merkel cell	64, 67, 68
11	Multiple endocrine neoplasia type I,	5, 24, 45
1	Multiple endocrine neoplasia type 2	50
10	Multiple endocrine neoplasia type 2A, glioblastoma, papillary carcinoma of the thyroid	42
2	Melanoma	50
6p	Renal cell carcinoma, testicular teratocarcinoma	65

cells.[75] These findings are consistent with the activity of a suppressor in normal cells, which is missing in transformed cells.

Tumor suppressor genes are believed to turn off normal signals encoded for by proto-oncogenes and to only act to inhibit neoplastic growth in the presence of oncogenes. Other postulated roles include mediation of negative growth signals, enabling immune rejection, promoting senescence, and preventing angiogenesis.[12] It is presently unclear how tumor suppressor genes operate, but one would expect that they are involved in the same communication systems as proto-oncogenes and that for each positive signal produced in the network by a proto-oncogene–encoded protein there may be a stop signal sent by a tumor suppressor gene–encoded protein (Fig. 8.1, Table 8.2). For this reason, the number of tumor suppressor genes identified may equal the number of oncogenes.

Techniques Employed to Study Oncogene Expression

HYBRIDIZATION STUDIES

Many techniques have been employed to analyze tumor cell DNA, RNA, and proteins for oncogene presence or expression (see Chapters 1 and 2). Nucleic acid

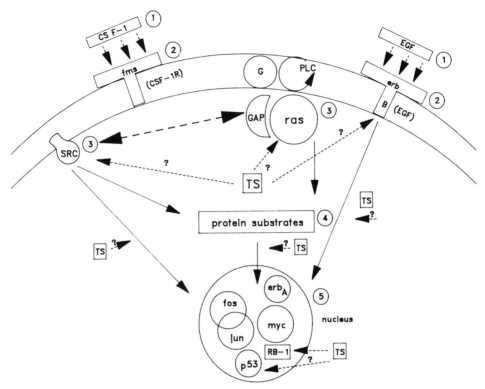

Figure 8.1. Oncogenes may disrupt normal cellular signal transduction pathways. This disruption may be modulated by tumor suppressors (*TS*). The retinoblastoma gene (*RB-1*) is a known tumor suppressor. Epidermal growth factor (*EGF*) binds to its receptor (*erb-B*) leading to tyrosine kinase activation, which may act on cytosol protein substrates or nuclear substrates. Many interactions between oncogene products are as yet unknown and only hypothetical, but most lead to phosphorylation of proteins, which promotes mitotic activity. The numbers are related to the possible sequence of the steps.

Table 8.2.
Proto-oncogenes Compared to Tumor Suppressor Genes

Proto-oncogenes	Tumor Suppressor Genes
Involved in normal growth and differentiation	Probable negative regulator of cell growth and differentiation
Functional classes: growth factors, growth-factor receptors, G proteins, nuclear-binding proteins	Probably also four major classes: Currently only evidence for nuclear binding (RB-1 & p53)
Must be activated	Must be lost or inactivated
Unclear if they play a role in hereditary cancers	Play a role in hereditary cancers

hybridization is a means of detecting homologous nucleic acid sequences and requires (*a*) the use of a radioactively or chemically labeled nucleic acid probe, (*b*) a specimen of nucleic acid to be analyzed for sequence homology with the probe, and (*c*) a means of detecting the hybridization of the labeled probe to single-stranded nucleic acids in the target. The major hybridization techniques utilized involve Northern, Southern, and dot/slot blots and *in situ* hybridization. These are summarized in reference 22 and in Chapter 1. The questions asked are (*a*) is the gene present, absent or amplified? (*b*) is the gene mutated? (*c*) is the gene expressed or overexpressed? and (*d*) are oncogene-encoded protein products present or absent?

Studies concerning gene amplification in most solid tumors are fraught with more variability than similar studies in hematological malignancies. Some tumors such as neuroblastomas, retinoblastomas, and small cell carcinomas of the lung (Fig. 8.2) are inherently intensely cellular and composed almost exclusively of neoplastic cells. Other tumors such as adenocarcinomas of the lung, carcinomas of the breast (Fig. 8.3), or carcinomas of the colon contain variable numbers of malignant cells as compared to stroma or secreted cellular components (Fig. 8.4). In addition, tumors may contain variable amounts of necrosis, hemorrhage, or inflammatory infiltrates (Fig. 8.5). Thus tremendous cellular heterogeneity exists in many

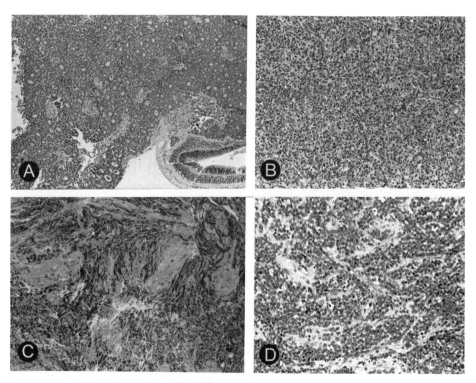

Figure 8.2. Composite of small cell neoplasms. **A,** Retinoblastoma with Flexner-Wintersteiner rosettes. The normal retina is visualized in the *lower left corner* (20x). **B,** Diffuse lymphoma (200x). **C,** Oat-cell lung carcinoma. There is significant crush artifact (200x). **D,** Neuroblastoma with fibrillary background (250x).

Figure 8.3. Invasive breast cancer. There is abundant stroma, with cancer cells comprising less than 50% of this particular field (100x).

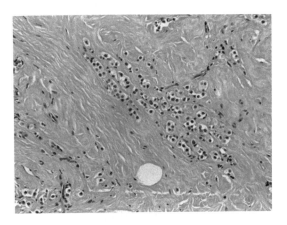

solid tumors. This leads to the dilution of tumor cell macromolecules with macro-molecules derived from normal epithelium, blood vessels, stroma, inflammatory cells, or secreted proteins. The contributions from these components can interfere with the interpretation of results obtained in all matrix blotting techniques including Southern, Western, Northern, slot and dot blots.

The more heterogeneous a tumor is, the greater the propensity to underestimate the degree of amplification or expression present. One can get around such difficulties by carefully dissecting the tissues used for extraction, first performing tissue-section analysis of the piece of tissue and then dividing it between nonneoplastic or neoplastic cells or between malignant or premalignant cells. This

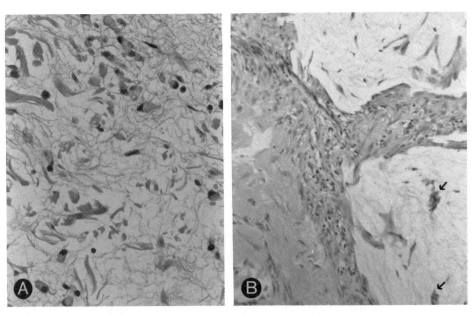

Figure 8.4. A, Signet-ring gastric carcinoma (600×). **B,** Colloid colon cancers typically have few tumor cells (*arrows*) floating in a mucinous background (400x).

Figure 8.5. The few viable lung tumor cells are difficult to find in areas with extensive hemorrhage and necrosis (200x).

was done in the study by Voglestein's group, where they carefully dissected apart the adenomatous tissue from the carcinomatous tissues on the same cases and analyzed them separately.[11] They continued to monitor this by examining frozen sections at multiple-step levels throughout the piece of frozen tissue. Similar approaches can be utilized on paraffin-embedded materials used for DNA extraction or for PCR amplification. One can also bypass these problems by using *in situ* hybridization reactions to localize specific gene products to individual cellular subsets.[21]

The *in situ* hybridization technique has been applied by many investigators to detect a variety of mRNAs encoding viral and cellular proteins[20,30,51,52] (see also Chapters 14 and 15). The ultimate usefulness of this technique depends on the sensitivity of the method and the accuracy with which the signals reflect the local concentration of target RNAs. One of the major difficulties of *in situ* hybridization methods arises from the requirement that highly labile target molecules (the RNA) must be retained in a suitable state for hybridization, while avoiding the artifacts and destruction of cellular morphology during the fixation and preparative techniques. Hybrid formation is limited by the low concentration of mRNA in the cell and any steric hindrance that may result from other intracellular structures and cell membranes. (Many of the same considerations concerning RNA detection in *in situ* hybridization reactions also apply to detection of RNA in matrix-based hybridizations such as Northern or dot blots.)

Other difficulties in studying gene structure or amplification include the integrity of the DNA, since the way tumor specimens are handled varies considerably between the time of surgical removal and DNA analysis. Small amounts of DNA degradation can lead to significant differences in signal intensity on Southern blots and can result in the inclusion of samples that incorrectly appear to have low or single-gene-copy intensity.

Early estimates on the incidence of mutated genes were probably falsely low, due to the insensitivity of the techniques utilized. This is less of a problem today, with the application of the polymerase chain reaction to tissues, followed by hybridization using synthetic oligonucleotide probes. A further discussion of this subject is provided in Chapters 1 and 2.

Special problems exist with respect to the analysis of *ras* genes (Table 8.3). These include the fact that original estimates of the presence of activated *ras* genes were based on the results of transfection assays. Since these estimates were low in early studies, they were interpreted as indicating that *ras* genes are activated in only a small percentage of human tumors. The presence of activated *ras* genes in some human tumors may have escaped detection in the NIH 3T3 transfection assay, perhaps due to the fact that DNA from the primary tumors was degraded, because of both the necrosis occurring *in vivo* and the degradation that occurs during surgery and subsequent tissue handling. For genes over 45,000 base pairs, such as the c-Ki-*ras*, this is particularly troubling; it has been estimated that the gene would only be intact in one-third of the fragments containing any part of the gene.[11] c-Ki-*ras* measures approximately 38 kilobases in length[53] and is often not intact, even if the DNA isolated is of high molecular weight. Furthermore, the presence of nonneoplastic stromal cells and inflammatory cells would dilute any positive signal from neoplastic cells in the transfection assay in the same way it dilutes it in the hybridization assays. Furthermore, the technique itself may be insensitive, since the assay is biased toward looking at *ras* effects, and many tumors contain a multiplicity of oncogene abnormalities.

IMMUNOHISTOCHEMICAL TECHNIQUES USED TO STUDY ONCOGENES: A CRITIQUE

One generally utilizes immunochemical methods to detect the proteins encoded by oncogenes; the most commonly used approaches are Western blots, immunoprecipitation reactions, and immunohistochemical staining. These methods are plagued by nonspecific reactions with other proteins, as are all immunologically based assays. The antigens are also sensitive to degradation by endogenous proteases present in the infiltrating inflammatory cells.

Immunohistochemical staining is more familiar to most pathologists than are hybridization reactions, and they are therefore particularly attractive to the diagnostic pathologist. However, this approach is fraught with difficulties, since fixation techniques can alter the subcellular localization of oncogene-encoded proteins, as has been demonstrated in the *myc* system (see below). Furthermore, many oncogene-encoded proteins are expressed in both normal and malignant tissues, and their detection may not have any significance. For example, *ras* p21 protein has been described in normal tissues with variable staining in adrenal gland, urinary

Table 8.3.
Difficulties with *ras* Assays

Early underestimates of the presence of activated *ras* genes based on transfection assays;
Assays for point mutations before use of PCR amplification and oligomer probes were
   less sensitive than other approaches;
Nonspecificity of some antibodies said to recognize *ras* p21.

bladder, bone marrow, brain, breast, heart, kidney, liver, lymph nodes, lungs, salivary glands, small intestine, spinal cord, spleen, stomach, thyroid, and trachea.[77]

As already noted, fixation may alter the subcellular distribution of oncogene-encoded proteins. c-*myc* protein is distributed in different parts of the cell, depending on the fixative used.[47] Generally speaking, frozen sections demonstrate localization of the gene product within the nucleus, whereas tissues studied after formalin-fixation and paraffin-embedding lose their nuclear staining with a concurrent appearance of cytoplasmic c-*myc* protein immunoreactivity.[36,38,47]

Another enormous problem is the specificity of the reagents used (see also Chapter 11). This has been a great dilemma with respect to the *ras*-encoded p21 proteins and has generated a large body of conflicting literature as summarized briefly here (Tables 8.4 & 8.5). Some authors report that p21 correlates with the depth of tumor invasion (Table 8.6) or tumor progression. Others have found that elevated *ras* may be important in benign tumors but that high expression is not necessary for malignancy and progression.[4,9,26,29,54,59,90] An additional, obvious difficulty is the fact that c-*ras* p21 is present in low levels in all normal tissues and could be recognized by some antibodies raised against *ras* v-onc products.[30] In addition, the majority of *ras* oncogenes that have been found to be activated are activated through point mutations, and the antibodies used may recognize proteins from both nonmutated and mutated forms of the gene. In fact, RAP-5 has been shown to recognize all p21 proteins.[49] This could account for immunohistochemical labeling of nonneoplastic as well as neoplastic cells.[89]

The peptide sequence recognized by some heteroantisera and some monoclonal antibodies to p21 may be widely distributed in proteins not related to the *ras* genes that are present in nontransformed human tissues.[89] This possibility was suggested by Robinson and associates and is consistent with the inability of Varma and associates to abolish immunoreactivity by preabsorption with stromal tissue.[83] An additional explanation could be damaged or altered p21 proteins secondary to fixation. However, in several studies, RAP-5 or 142-24EO-5 exhibited intense decoration of nonneoplastic tissues, whether or not frozen sections or sections fixed in formaldehyde were used.[18,62,72,89] In addition, protease digestion by trypsin, pepsin, pronase, ficin, or bromelain has no effect on RAP-5 immunostaining.[89]

Samowitz and associates[66] have shown that high concentrations of control monoclonal antibodies duplicated exactly the immunohistochemical staining pattern of RAP-5 on formalin-fixed tissue sections of colon, breast, and stomach. Immunoreactivity of breast lesions could not be inhibited by absorption with peptides from which the antibody was prepared, while immunoreactivity of another antisera to p21 could be inhibited by absorption to the peptide to which it was prepared. Based on studies using solid-phase radioimmunoassays, Western blots, and immunohistochemical staining, it was concluded that the RAP-5 binding to formalin-fixed tissues was nonspecific and not indicative of p21 expression.[66]

One approach used to validate RAP-5 staining was to compare it to staining with Y13 259 where similar results were obtained.[59] However, others have shown that specific tissue binding by Y13 259 is abolished by formalin fixation.[89] In addition, the staining patterns of Y13 259 on frozen sections and RAP-5 on formalin-fixed sections from the same colonic adenomas and carcinomas were not the same.[62]

Table 8.4.
ras **p21 Immunostaining Specific for Malignancy**

Antibody	Author	Site	Reference
RAP-5	Horan Hand	Breast, colon	34
RAP-5 & Y13 259	Ohuchi	Breast	59
Antibody from Triton Sciences	Tahara	Stomach	76
RAP-5	Thor	Colon	80

Table 8.5.
Antibodies to *ras* p21 Label Benign and Malignant Cells and/or Stroma

Author	Site	Reference	Antibody
Agnantis, *et al.*	Breast	1	RAP-5
Allen, *et al.*	Colon	2	RAP-5
Aoki, *et al.*	Nephroblastoma	3	RAP-5
Candlish, *et al.*	Breast	14	RAP-5
Chesa, *et al.*	Breast, colon, stomach	16	Y13-238, 24E 05, Y13 259
Embleton & Butler	Liver	18	RAP-5
Fujita, *et al.*	Stomach	27	RAP-5
Furth, *et al.*	Normal tissues	28	RAP-5
Ghosh, *et al.*	Breast	32	RAP-5
Johnson, *et al.*	Thyroid	37	RAP-5
Kerr, *et al.*	Colon	39	RAP-5
Mizukami, *et al.*	Thyroid	54	RAP-5
Robinson, *et al.*	Colon	62	RAP-5
Rodenberg, *et al.*	Ovary	63	RAP-5
Samowitz, *et al.*	Colon, breast, stomach	66	RAP-5
Tanaka, *et al.*	Stomach	78	RAP-5
Varma, *et al.*	Prostate	83	RAP-5
Walker & Wilkinson	Breast	86	Y13 259
Ward, *et al.*	Mammary tissue	88	RAP-5

Table 8.6.
p21 Correlates with Tumor Invasion or Tumor Progression

Tumor	Author, Date	Reference
Breast cancer	Clair, 1987	17
Breast cancer	Lundy, 1986	49
Stomach cancer	Tahara, 1986	76
Colon cancer	Thor, 1984	80
Urinary bladder	Viola, 1985	84
Gastrointestinal carcinoma	Yasui, 1987	91

A second means of establishing the RAP-5 specificity has been to correlate the immunohistochemical staining with the quantities of p21 as determined by liquid competition radioimmunoassay. As pointed out by Samowitz and associates,[66] one study had two cancers with more p21 by RIA and more staining by RAP-5 than benign lesions; however, the cancer with the higher percentage of positive cells by RAP-5 had a lower amount of p21 by RIA. This suggests that the RAP-5 staining was not directly related to the quantity of p21.[59,66]

Robinson and associates[62] found that RAP-5 staining of *ras* transformed cells making large quantities of p21 and nontransformed cells with very low levels of p21 were the same. In contrast, the monoclonal antibody Y13 259 stained transformed cells far more than nontransformed cells in the same study.

One approach to get around nonspecific binding is to utilize antibodies generated specifically against the mutated site in an activated *ras* protein.[14] However, one would need to first validate the specificity of the antisera, using molecular biological approaches on the same tissues before widely applying them to other tissues for analysis of their expression.

Antibodies generated to mutated *ras* proteins include those described by Sorentino and co-workers which were generated against both the normal glycine 12 and the transforming lysine 12 in p21 proteins.[73] Neither antibody preferentially recognized the normal or the activated form of p21. Two regions, amino acids 23–69 and 89–106, accounted for the epitopes of about 75% of the monoclonal antibodies raised in the study, suggesting that they encompass the most exposed and most immunogenic domains of the native *ras* p21 proteins. In addition, one of the monoclonal antibodies generated appeared to recognize molecules related to *ras* p21.[73] The reactive epitope is an unknown antigen in the cytoplasm of normal cells, perhaps a p21 precursor or an unrelated protein as seen in molecular mimicry.[60]

A comparison of a number of antibodies to *ras* p21 was conducted in selected normal fixed tissue and tumors from rats, mice, and humans.[88] Membrane staining was never apparent with the antibody to H-*ras* p21 in normal tissues, but was found in neoplastic cells.[87] It is likely that p21 levels on the cell membranes of normal cells may be present in too low a concentration to detect immunohistochemically, since it is detectable by Western blots.[28,73] A few normal tissues were immunonegative with some of the antibodies. The antibody ES-13 stained sarcoma cells on the cell membranes. Prominent granular staining, which appeared to be mitochondrial or lysosomal, was seen in many normal and neoplastic rodent tissues and in normal human colon, colon polyps, and carcinomas. This antibody may detect an epitope of mitochondrial ATP-synthetase, because of the reported homology between the c-*has*/*bos* gene product and the β subunit of bovine mitochondrial ATPase.[31] Interpretation of positive immunoreactivity with any antibody was sometimes difficult, due to the presence of nonspecific background staining. Western blots with transforming lysine 12 p21 reacted with all of the antibodies except RAP-5. In addition, immunoblots of solubilized proteins from the EJ cell line with RAP-5 indicated reactivity of this monoclonal antibody with proteins of several molecular weights. RAP-5 was also never specifically immunoreactive with cell membranes of normal or malignant cells, including EJ cells.[87]

RAP-5 is an antibody to a small peptide encompassing amino acids 10 to 17 with valine at codon 12 of Ha-*ras*. This mutation is uncommon in human or animal tumors.[7,11,23,55] Of interest is the observation that T24 and EJ cells, which do contain this mutation, produced carcinomas that did not react with RAP-5, suggesting that RAP-5 does not stain p21 in fixed tissues.[88]

One of the best studies looking at the expression of oncogene-encoded proteins involves the Her-2/*neu* proto-oncogene in human breast and ovarian cancers. In this study, Slamon and co-workers identified several potential shortcomings of the various methods used to evaluate oncogenes, including Southern, Northern, and Western blots and immunohistochemistry.[70] These were identified by carefully analyzing each of the tissues, using multiple analytical approaches. After analyzing breast cancer specimens it was found that in all 51 cases with amplification, two of the three measures of expression were concordant. The Western blot analysis was most discordant, being inconsistent with data obtained by Southern blot, Northern blot, and immunohistochemistry in 6% of cases. The authors postulated that the explanation for this was a weak signal due to dilution by a relatively large amount of stroma. Histological examination confirmed the presence of excessive amounts of stroma. DNA and RNA analyses were less sensitive to this problem, since there was evidence of amplification and increased transcript levels in these tumors. The increased susceptibility of Western blots to dilutional effects was due to the fact that large amounts of noncellular connective tissue can substantially contribute to the total protein in the sample by adding significant amounts of extracellular matrix proteins, such as collagen, to the lysate. However, since the stroma is relatively poor in cellularity, it makes only minimal contributions to the total DNA or RNA extracted from the same specimen. Immunohistochemical data were inconsistent in only 1% of cases, and Northern blot in 2%.[70]

Acknowledgments

The authors acknowledge the superb manuscript preparation by Agnes Truske and the technical expertise of the Medical Media Department at the VA Medical Center for their help with the illustrative material.

The work reported in this chapter was supported by American Cancer Society Grant #PDT-341 and also by a grant from the University of New Mexico Cancer Center.

REFERENCES

1. Agnantis, N. J., Petraki, C., Markoulatos, P., and Spandidos, D. A. Immunohistochemical study of the *ras* oncogene expression in human breast lesions. *Anticancer Res.* 6:1157–1160, 1986.
2. Allen, D. C., Foster, H., Orchin, J. C., and Biggart, J. D. Immunohistochemical staining of colorectal tissues with monocolonal antibodies to ras oncogene p21 product and carbohydrate determinant antigen 19–9. *J. Clin. Pathol.* 40:157–162, 1987.
3. Aoki, I., Yanoma, S., Misugi, K., Sasaki, Y., and Kikyo, S. ras p21 expression in nephroblastoma group tumors. *Acta Pathol. Jpn.* 37:1903–1907, 1987.
4. Augenlicht, L. H., Augeron, C., Yander, G., and Laboisse, C. Overexpression of *ras* in mucus-secreting human colon carcinoma cells of low tumorigenicity. *Cancer Res.* 47:3763–3765, 1987.
5. Bale, S. J., Bale, A. E., Stewart, K., Dachowska, L., McBride, O. W., Glaser, T., Green, J. E., Mulvihill, J. J., Brandi, M. L., Sakaguchi, K., Aurbach, G. D., and Marx, S. J. Linkage analysis of multiple endocrine neoplasia type 1 with *int*-2 and other markers on chromosome 11. *Genomics* 4:320–322, 1989.

6. Bale, S. J., Dracopoli, N. C., Tucker, M. A., Clark, W. H., Fraser, M. C., Stanger, B. Z., Green, P., Donis-Keller, H., Housman, D. E., and Green, M. H. Mapping the gene for hereditary cutaneous malignant melanoma–dysplastic nevus to chromosome 1p. *N. Engl. J. Med. 320*:1367–1372, 1989.

7. Barbacid, M. *ras* genes. *Annu. Rev. Biochem. 56*:779–827, 1987.

8. Berne, J., McColl, I., and Adinolfi, M. Recessive mutations and chromosome deletions leading to cancer. *Br. J. Surg. 76*:327–330, 1989.

9. Bizub, D., Heimer, E. P., Felix, A., Clizzonite, R., Wood, A., Skalka, A. M., Slater, D., Aldrich, T. H., and Furth, M.E. Antisera to the variable region of ras oncogene proteins and the specific detection of H-*ras* expression in an experimental model of chemical carcinogenesis. *Oncogene 1*:131–142, 1987.

10. Bodmer, W. F., Bailey, C. J., Bodmer, J., Bussey, H. J. R., Ellis, A., Gorman, P., Lucibello, F. C., Murday, V. A., Rider, S. H., Scambler, P., Sheer, D., Solomon, E., and Spurr, N. K. Localization of the gene for familial adenomatous polyposis on chromosome 5. *Nature (Lond) 328*:614–616, 1987.

11. Bos, J. L., Fearon, E. R., Hamilton, S. R., Verlaan-de Vries, M., van Boom, J. H., van der Eb, A. H., and Vogelstein, B. Prevalence of ras gene mutations in human colorectal cancers. *Nature (Lond) 327*:293–297, 1987.

12. Bouck, N. P., and Benton, B. K. Loss of cancer suppressors, a driving force in carcinogenesis. *Chem. Res. Toxicol. 2*:1–11, 1989.

13. Candlish, W., Kerr, I. B., and Simpson, H. W. Immunocytochemical demonstration and significance of p21 *ras* family oncogene product in benign and malignant breast disease. *J. Pathol. 150*:163–167, 1986.

14. Carney, W. P., Hamer, P., Petit, D., Wolfe, H., Cooper, G., Lefebore, M., and Rabin, H. A monoclonal antibody reactive to an activated *ras* protein expressing valine at position 12. *J. Cell Biochem. 32*:207–214, 1986.

15. Cavenee, W. K., Koufos, A., and Hansen, M. F. Recessive mutant genes predisposing to human cancer. *Mutat. Res. 168*:3–14, 1986.

16. Chesa, P. G., Rettig, W. J., Melamed, M. R., Old, L. J., and Niman, H. I. Expression of p21*ras* in normal and malignant human tissues. Lack of association with proliferation and malignancy. *Proc. Natl. Acad. Sci. U.S.A. 84*:3234–3238, 1987.

17. Clair, T., Miller, W. R., and Cho-Chung, Y. S. Prognostic significance of the expression of a ras protein with a molecular weight of 21,000 by human breast cancer. *Cancer Res. 47*:5290–5293, 1987.

18. Embleton, N., and Butler, P. C. Reactivity of monoclonal antibodies to oncoproteins in normal liver, carcinogen-induced tumors and premalignant liver lesions. *Br. J. Cancer 57*:48–53, 1988.

19. Fearon, E. R., Feinberg, A. P., Hamilton, S. H., and Vogelstein, B. Loss of genes on the short arm of chromosome 11 in bladder cancer. *Nature (Lond) 318*:377–380, 1985.

20. Fenoglio, C. M., Oster, M. W., LoGerfo, P., Reynolds, T., Edelson, R., Patterson, J. A. K., Madeiros, E., and McDougall, J. K. Kaposi's sarcoma following chemotherapy for testicular cancer in a homosexual man: Demonstration of cytomegalovirus RNA in sarcoma cells. *Hum. Pathol. 13*:955-959, 1982.

21. Fenoglio-Preiser, C. M., Longacre, T. A., Listrom, M. L., and Blume, P. Analysis of cellular markers in IBD: Immunohistochemistry, in situ hybridization and beyond. Proceedings of NFIC Conference on Dysplasia in Inflammatory Bowel Disease, New York, Elsevier Press, in press 1989.

22. Fenoglio-Preiser, C. M. and Willman, C. Molecular biology and the pathologist. General principles and applications. *Arch. Lab. Med. Pathol. 111*:601–619, 1987.

23. Forrester, K., Almoguera, C., Han, K., Grizzle, W. E., and Perucho, M. Detection of high incidence of K-ras oncogenes during human colon tumorigenesis. *Nature (Lond) 327*:298–303, 1987.

24. Friedman, E., Sakaguchi, K., Bale, A. E., Falchetti, A., Streeten, E., Zimering, M. B.,

Weinstein, L. S., McBride, W. O., Nakamura, Y., Brandi, M. L., Norton, J. A., Aurbach, G. D., Spiegal, A. M., and Marx, S. J. Clonality of parathyroid tumors in familial multiple endocrine neoplasia type I. *N. Engl. J. Med. 321*:21–218, 1989.

25. Friedman, J. M., Fialkow, P. J., Greene, C. L., and Weinberg, M. N. Probable clonal origin of neurofibrosarcoma in a patient with hereditary neurofibromatosis. *J. Natl. Cancer. Inst. 69*:1289–1292, 1983.

26. Fromowitz, F. B., Viola, M. V., Chao, S., Oravez, S., Mishriki, Y., Finkel, G., Grimson, R., and Lundy, J. ras p21 expression in the progression of breast cancer. *Hum. Pathol. 18*:1268–1275, 1987.

27. Fujita, J., Ohuchi, N., Yao, T., Okumura, M., Fukushima, Y., Kanakura, Y., Kitamura, Y., and Fujita, J. Frequent overexpression, but not activation by point mutation, of ras genes in primary human gastric cancers. *Gastroenterology 93*:1339–1345, 1987.

28. Furth, M. E., Aldrich, T. H., and Cordon-Cardo, C. Expression of *ras* proto-oncogene protein in normal human tissues. *Oncogene 1*:47–58, 1987.

29. Gallick, G. E., Kurzrock, R., Kloetzer, W. S., Arlinghaus, R. B., and Gutterman, G. U. Expression of p21ras in fresh primary and metastatic human colorectal tumors. *Proc. Natl. Acad. Sci. U.S.A. 82*:1795–1799, 1985.

30. Galloway, D. A., Fenoglio, C., and McDougall, J. K. Limited transcription of the herpes simplex virus genome when latent in human sensory ganglia. *J. Gen. Virol. 41*:686–691, 1982.

31. Gay, N. J., and Walker, J. E. Homology between human bladder carcinoma oncogene product and mitochondrial ATP-synthase. *Nature (Lond) 301*:262–264, 1983.

32. Ghosh, A. K., Moore, M., and Harris, M. Immunohistochemical detection of *ras* oncogene p21 product in benign and malignant mammary tissue in man. *J. Clin. Pathol. 39*:428–434, 1986.

33. Harris, H. Suppression of malignancy in hybrid cells. *J. Cell Sci. 79*:83–94, 1985.

34. Horan Hand, P., Thor, A., Wunderlich, D., Muraro, R., Caruso, A., and Schlom, J. Monoclonal antibodies of predefined specificity detect activated *ras* gene expression in human mammary and colon carcinomas. *Proc. Natl. Acad. Sci. U.S.A. 81*:5227–5231, 1984.

35. Hsu, S. H., Luk, G. D., Krush, A. J., Hamilton, S. R., and Hoover, H. H. Multiclonal origin of polyps in Gardner's syndrome. *Science (Wash DC) 221*:951–953, 1983.

36. Izumi, S. I., Moriuchi, T., Koji, T., et al. Localization of C-MYC in HL-60 cells, neoplastic and normal tissues: An immunohistochemical and in situ hybridization study. *Acta Histochem. Cytochem. 21*:327–342, 1988.

37. Johnson, T. L., Lloyd, R. V., and Thor, A. Expression of *ras* oncogene p21 antigen in normal and proliferative thyroid tissues. *Am. J. Pathol. 127*:60–65, 1987.

38. Jones, D. J., Ghosh, A. K., Moore, M., and Schofield, P. F. A critical appraisal of the immunohistochemical detection of the c-myc oncogene product in colorectal cancer. *Br. J. Cancer 56*:779–783, 1987.

39. Kerr, I. B., Lee, F. D., Quintanilla, M., and Balmain, A. Immunocytochemical demonstration of p21 *ras* family oncogene product in normal mucosa and in premalignant and malignant tumours of the colorectum. *Br. J. Cancer 52*:695–700, 1985.

40. Knudson, A. G. Mutation and cancer: Statistical study of retinoblastoma. *Proc. Natl. Acad. Sci. U.S.A. 68*:820–823, 1971.

41. Knudson, A. G. Hereditary cancer, oncogenes and antioncogenes. *Cancer Res. 45*:1437–1443, 1985.

42. Knudson, A. G., Meadows, A. T., Nichols, W. W., and Hill, R. Chromosomal deletion and retinoblastoma. *N. Engl. J. Med. 295*:1120–1123, 1976.

43. Kok, K., Osinga, J., Carritt, B., Davis, M. B., van der Hout, A. H., van der Veen, A. Y., Landsvater, R. M., deLeij, L. F., Berendsen, H. H., Postmus, P. E., Poppema, P. E., Poppema, S., and Buys, C. H. C. M. Deletion of a DNA sequence at the chromosomal region 3p21 in all major types of lung cancer. *Nature (Lond) 330*:578–581, 1987.

44. Kovacs, G., Erlandsson, R., Boldog, F., Ingvarsson, S., Muller-Brechlin, R., Klein, G., and Sumegi, J. Consistent chromosome 3p deletion and loss of heterozygosity in renal cell carcinoma. *Proc. Natl. Acad. Sci. U.S.A.* 85:1571–1575, 1988.

45. Larsson, C., Skogseid, B., Oberg, K., Nakamura, Y., and Nordenskjold, M. Multiple endocrine neoplasia type I gene maps to chromosome 11 and is lost in insulinoma. *Nature (Lond)* 332:85–87, 1988.

46. Law, D. J., Olschwang, S., Monpezat, J. P., Lefrancois, D., Jagelman, D., Petrelli, N. J., Thomas, G., and Feinberg, A. P. Concerted nonsyntenic allelic loss in human colorectal carcinoma. *Science (Wash DC)* 241:961–965, 1988.

47. Loke, S. H., Neckers, L., Scwab, G., and Jaffe, E. S. C-myc protein in normal tissue. Effects of fixation on its apparent subcellular localization. *Am. J. Pathol.* 131:29–37, 1988.

48. Lundberg, C., Skoog, L., Cavenee, W. K., and Nordenskjold, M. Loss of heterozygosity in human ductal breast tumors indicates a recessive mutation on chromosome 13. *Proc. Natl. Acad. Sci. U.S.A.* 84:2372–2376, 1987.

49. Lundy, L., Grimson, R., Mishriki, Y., Chao, S., Oravez, S., Fromowitz, F., and Viola, M. V. Elevated *ras* oncogene expression correlates with lymph node metastases in breast cancer patients. *J. Clin. Oncol.* 4:1321–1325, 1986.

50. Mathew, C. G. P., Smith, B. A., Thorpe, K., Wong, Z., Royle, N. J., Jeffreys, A. J., and Ponder, B. A. J. Deletions of genes on chromosome 1 in endocrine neoplasia. *Nature (Lond)* 328:524–528, 1987.

51. McDougall, J. K., Crum, C. P., Fenoglio, C. M., Goldstein, L. C., and Galloway, D. A. Herpesvirus-specific RNA and protein in carcinoma of the uterine cervix. *Proc. Natl. Acad. Sci. U.S.A.* 79:3853–3857, 1982.

52. McDougall, J. K. and Galloway, D. A. In situ cytological hybridization in diagnostic pathology. *Prog. Surg. Pathol.* 4:83–94, 1982.

53. McGrath, J. P., Capon, D. J., Smith, D. H., Chen, E. Y., Seeburg, P. H., Goeddel, D. V., and Levinson, A. D. Structure and organization of the human Ki-ras proto-oncogene and a related processed pseudogene. *Nature (Lond)* 304:501–506, 1983.

54. Mizukami, Y., Nonomura, A., Hashimoto, T., Terahata, S., Matsubara, F., Michigishi, T., and Noguchi, M. Immunohistochemical demonstration of ras p21 oncogene product in normal, benign and malignant human thyroid tissue. *Cancer* 61:873–880, 1988.

55. Monnat, M., Tardy, S., Saraga, P., Diggelmann, H., and Costa, J. Prognostic implications of expression of the cellular genes MYC, FOS, Ha-ras and Ki-ras in colon carcinoma. *Int. J. Cancer* 40:293–299, 1987.

56. Mori, N., Yokota, J., Oshimura, M., Cavenee, W. K., Mizoguchi, H., Noguchi, M., Shimosato, Y., Sugimara, T., and Terada, M. Concordant deletions of chromosome 3p and loss of heterozygosity for chromosomes 13 and 17 in small cell lung carcinoma. *Cancer Res.* 49:5130–5135, 1989.

57. Mukai, S. and Dryja, T. P. Loss of alleles at polymorphic loci on chromosome 2 in uveal melanoma. *Cancer Genet. Cytogenet.* 22:45–53, 1986.

58. Naylor, N. L., Johnson, B. E., Minna, J. D., and Sakaguchi, A. Y. Loss of heterozygosity of chromosome 3p markers in small cell lung cancer. *Nature (Lond)* 329:45–454, 1987.

59. Ohuchi, N., Thor, A., Page, D. L., Horan Hand, P., Haeter, S., and Schlom, J. Expression of the 21,000 molecular weight *ras* protein in a spectrum of benign and malignant human mammary tissues. *Cancer Res.* 46:2511–2519, 1986.

60. Oldstone, M. B. A. Molecular mimickry and autoimmune disease. *Cell* 50:819–820, 1987.

61. Orkin, S. H., Goldman, D. S., and Sallan, S. E. Development of homozygosity for chromosome 11 during genesis of Wilms' tumor. *Nature (Lond)* 309:172–174, 1984.

62. Robinson, A., Williams, A. R., Piris, J., Spandidos, D. A., and Wyllie, A. H. Evaluation of a monoclonal antibody to *ras* peptide, RAP-5, claimed to bind preferentially to cells of infiltrating carcinoma. *Br. J. Cancer* 54:877–883, 1986.

63. Rodenburg, C. J., Ingeborg, A. K., Nap, M., and Fleure, C. J. Immunohistochemical de-

tection of ras oncogene product p21 in advanced ovarian cancer. Lack of correlation with clinical outcome. *Arch. Pathol. Lab. Med. 112*:151–154, 1988.

64. Rouleau, G. A., Wertelecki, W., Haines, J. L., Hobbs, W. J., Trofatter, J. A., Seizinger, B. R., Martuza, R. L., Supernerau, D. W., Conneally, P. M., and Gusella, J. F. Genetic linkage of bilateral acoustic neurofibromatosis to a DNA marker on chromosome 22. *Nature (Lond) 329*:246–249, 1987.

65. Rukstalis, D. B., Bubley, G. J., Donahue, J. P., Richie, J. P., Seidman, J. G., and DeWolf, W. C. Regional loss of chromosome 6 in two urological malignancies. *Cancer Res. 49*:5087–5090, 1989.

66. Samowitz, W., Paull, G., and Hamilton, S. R. Reported binding of monoclonal antibody RAP-5 to formalin-fixed tissue sections is not indicative of ras p21 expression. *Hum. Pathol. 19*:127–132, 1988.

67. Seizinger, B. R., Martuza, R. L., and Gusella, J. F. Loss of genes on chromosome 22 in tumorigenesis of acoustic neuroma. *Nature (Lond) 322*:644–647, 1986.

68. Shabtai, F., Sternberg, A., Klar, D., Reiss, R., and Halbrecht, I. Involvement of chromosome 22 in a Merkel cell carcinoma in a patient with a previous meningioma. *Cancer Genet. Cytogenet. 38*:43–48, 1989.

69. Simpson, N. E., Kidd, K. K., Goodfellow, P. J., McDermid, H., Myers, S., Kidd, J. R., Jackson, C. E., Duncan, A. M. V., Farrer, L. A., Brasch, K., Castiglione, C., Genel, M., Gertner, J., Greenberg, C. R., Gusella, J. F., Holden, J. J. A., and White, B. N. Assignment of multiple endocrine neoplasia type 2a to chromosome 10 by linkage. *Nature (Lond) 328*:528–530, 1987.

70. Slamon, D. J., Godolphin, W., Jones, L. A., Holt, J. A., Wong, S. G., Keith, D. E., Levin, W. J., Stuart, S. G., Udove, J., Ullrich, A., and Press, M. J. Studies of the HER-2/neu proto-oncogene in human breast and ovarian cancer. *Science (Wash DC) 24*:707–713, 1989.

71. Solomon, E., Voss, R., Hall, V., Bodmer, W. F., Jass, J. R., Jeffreys, A. J., Lucibello, F. C., Patel, I., and Rider, S. H. Chromosome 5 allele loss in human colorectal carcinomas. *Nature (Lond) 328*:616–619, 1987.

72. Sorrentino, V., McKinney, M. D., Giorgi, M., Geremia, R., and Fleissner, E. Expression of cellular proto-oncogenes in the mouse male germ line: A distinctive 2.4 kilobase *pim*-1 transcript is expressed in haploid postmeiotic cells. *Proc. Natl. Acad. Sci. U.S.A. 85*:2191–2195, 1988.

73. Sorrentino, V., Nebreda, A. R., Alonso, T., and Santos, E. Preparation, characterization and properties of monoclonal antibodies against intact H-*ras* p21 proteins. *Oncogene 4*:215–221, 1989.

74. Stanbridge, E. J. A case for human tumor-suppressor genes. *Bio-essays 3*:252–255, 1986.

75. Stoker, M. Regulation of growth and orientation in hamster cells transformed by polyoma virus. *Virology 24*:165–174, 1964.

76. Tahara, E., Yasui, W., Tantyama, K., Ochiai, A., Yamamoto, T., Nakajo, S., and Yamamoto, M. Ha-*ras* oncogene product in human gastric carcinoma: Correlation with invasiveness, metastasis or prognosis. *Jpn. J. Cancer Res. 77*:517–522, 1986.

77. Tanaka, T., Ida, N., Waki, C., Shimoda, H., Slamon, D. J., and Cline, M. J. Cell type expression of ras gene products in the normal rat. *Mol. Cell. Biochem. 75*:23–32, 1987.

78. Tanaka, T., Slamon, D. J., Battifora, H., and Cline, M. J. Expression of p21/*ras* oncoproteins in human cancers. *Cancer Res. 46*:1465–1469, 1986.

79. Taylor, J. A. Oncogenes and their applications in epidemiologic studies. *Am. J. Epidemiol. 130*:6–13, 1989.

80. Thor, A., Horan Hand, P., Wunderlich, D., Caruso, A., Muraro, R., and Schlom, J. Monoclonal antibodies define differential *ras* gene expression in malignant and benign colonic disease states. *Nature (Lond) 311*:562–564, 1984.

81. Turc-Carel, C., Dal, C. P., Rao, V., Li, F. P., Zimmerman, R., Parry, D. M., Lorsom, J. M., Cowan, J. M., and Sandberg, A. A. Involvement of chromosome X in primary

cytogenetic change in human neoplasia: Nonrandom translocation in synovial sarcoma. *Proc. Natl. Acad. Sci. U.S.A. 84*:1981–1985, 1987.

82. van der Hout, A. H., Kok, K., van den Berg, A., Oosterhuis, J. W., Carritt, B., Buys, C. H. C. M. Direct molecular analysis of a deletion of 3p in tumors from patients with sporadic renal cell carcinoma. *Cancer Genet. Cytogenet. 32*:281–285, 1988.

83. Varma, V. A., Austin, G. E., and O'Connell, A. C. Antibodies to *ras* oncogene p21 proteins lack immunohistochemical specificity for neoplastic epithelium in human prostate tissue. *Arch. Pathol. Lab. Med. 113*:16–19, 1989.

84. Viola, M. V., Fromowitz, F., Oravez, S., Deb, S., and Schlom, J. *ras* oncogene p21 expression is increased in premalignant lesions and high grade bladder carcinoma. *J. Exp. Med. 161*:1213–1218, 1985.

85. Vogelstein, B., Fearon, E. R., Hamilton, S. R., Kern, S. E., Preisinger, A. C., Leppert, M., Nakamura, Y., White, R., Smits, A. M. M., and Bos, J. L. Genetic alterations during colorectal tumor development. *N. Engl. J. Med. 319*:525–532, 1988.

86. Walker, R. A., and Wilkinson, N. p21 *ras* protein expression in benign and malignant human breast. *J. Pathol. 156*:147–153, 1988.

87. Ward, J. M., Pardue, R. L., Junker, J. L., Takahashi, K., Shih, T. Y., and Weislow, O. S. Immunocytochemical localization of ras[Ha] p21 in normal and neoplastic cells in fixed tissue sections from Harvey sarcoma virus-infected mice. *Carcinogenesis 7*:645–665, 1986.

88. Ward, J. M., Perantoni, A. O., and Santos, E. Comparative immunohistochemical reactivity of monoclonal and polyclonal antibodies to H-*ras* p21 in normal and neoplastic tissues of rodents and rats. *Oncogene 4*:203–213, 1989.

89. Wick, M. R. Immunohistologic detection of *ras* oncogene products. *Arch. Pathol. Lab. Med. 113*:13–15, 1989.

90. Williams, A. R., Piris, J. M., Spandidos, D. A., and Wyllie, A. H. Immunohistochemical detection of the ras oncogene p21 product in an experimental tumour and in human colorectal neoplasms. *Br. J. Cancer 52*:687–693, 1985.

91. Yasui, W., Sumiyoshi, H., Yamamoto, T., Oda, N., Kameda, T., Tanaka, T., and Tahara, E. Expression of Ha-ras oncogene product in rat gastrointestinal carcinomas induced by chemical carcinogens. *Arch. Pathol. Jpn. 37*:1731–1741, 1987.

92. Zbar, B., Brauch, H., Talmadge, C., and Linehan, M. Loss of alleles at loci on the short arm of chromosome 3 in renal cell carcinoma. *Nature (Lond) 327*:721–724, 1987.

9

Oncogenes and Tumor Suppressor Genes in Solid Tumors: Neural Tumors

Cecilia M. Fenoglio-Preiser
Margaret B. Listrom
Teri A. Longacre

Normal Neural Tissue

Many proto-oncogenes are expressed in neural tissues, although some are only expressed in specific areas of the brain or expressed at elevated levels only during embryonal development.

src FAMILY OF GENES

High levels of the *src* protein, pp60[c-src], are found in human embryonic neural tissues and show two phases of expression.[42,64,82] In the first phase, pp60[c-src] is transiently expressed in the neural growth cone during gastrulation and neural tube formation.[87] In the second phase, pp60[c-src] immunoreactivity is found in developing neurons at the onset of terminal differentiation, when neuroblasts cease proliferating.[38,121] The protein is still detectable in adult brains, but at much lower levels.[120,126] The protein product is found in neurons and astrocytes, with the kinase activity being much higher in the neurons than in the astrocytes.[17] Kinase activity in neurons increases with their differentiation.[18] Expression of v-*src* arrests glial cell differentiation and immortalizes the cells.[136]

In contrast, pp62[c-yes], is found in relatively small amounts during embryonic development, but increases 2–4-fold in corresponding adult tissues.[125,126] In adult chicken brains, the highest levels of pp62[c-yes] and pp60[c-src] are found in the cerebellum where they concentrate in the molecular layer, and to a lesser degree, in the granular layer, particularly in cerebellar Purkinje cells and their dendritic trees.[23,125–127,141] *fyn* expression is also high in the brain.[102] Transgenic mice expressing the v-*fps* gene (not a member of the *src* family but a related tyrosine kinase) develop severe neurologic abnormalities manifest as marked trembling and occasionally marked enlargement of the trigeminal nerves, suggesting that the *fps* gene is important in normal neurogenesis.[148]

ras GENES

ras genes and other members of the *ras*-gene family are found in normal brain tissue. High levels of *ras* protein are found in embryonic and adult brains, located in the neurons of the frontal cortex, medulla, and cerebellum, as well as in peripheral nerves and dorsal root ganglia.[39,119,125,126] *ral*-A is highly expressed in brain, with two different size transcripts being detectable. The *rho* gene is also expressed.[99] The four *ras*-related genes, *rab* 1, *rab* 2, *rab* 3, and *rab* 4, were all originally isolated from a rat-brain cDNA library.[134] Expression of *rab* 1, *rab* 3, and *rab* 4 is detectable in murine brain tissue; however *rab* 2 is not found.

myc GENES

c-*myc* expression is stage and tissue specific. c-*myc* mRNA is detectable at relatively high levels throughout development, but expression declines in older adults in most tissues except spleen, thymus, heart, intestine, and adrenals.[115,126] During development of the murine cerebellum, steady-state levels of c-*myc* RNA change rapidly, with mRNA expression confined to discrete neuronal populations at specific ages. For example, on postnatal day 3, c-*myc* mRNA is found in the germinal cells of the external granular layer.[107] A week after birth, a surge of c-*myc* mRNA occurs in the Purkinje cells, coinciding with their dendritic maturation and formation of synaptic connections with the granular cells.[107]

Expression of the related genes N-*myc* and L-*myc* is highest in the newborn brain, kidney, and intestine, and rapidly declines with age.[61,62,151] *In situ* hybridization shows high levels of N-*myc* RNA in primitive neuroectoderm[29] of the human fetal cerebrum, specifically in the early embryonic germinal layer and primordial cortex;[42,53] subsequently, N-*myc* RNA levels decline. It is believed that N-*myc*, L-*myc*, and c-*myc* genes all are involved in neurogenesis, since they have restricted spatial and temporal expression in the brain, with significant elevation of N-*myc* RNA in immature neurons.[126] B-*myc* is expressed in many tissues, but the highest levels are seen in the brain.[60]

fos GENES

c-*fos* gene expression is relatively high in most postembryonic tissues, with enhanced expression in certain areas of the adult brain.[2,3,30,44] The most characteristic aspect of c-*fos* expression is its rapid and transient induction in response to various extracellular stimuli.[26,67,89,90,108] Stimulation producing partial seizures induces c-*fos* protein production in the granular cells of the dentate gyrus.[30] The seizure-inducing drug Metrazole elevates brain levels of c-*fos*.[89,90] It is postulated that, in the brain, c-*fos* plays a role in the long-term adaptive response to convulsions.[89,90]

neu GENES

c-*neu* mRNA and its gene product, p185, can be seen in embryonic neural tissues, localized to the external portions of the neural tube and dorsal root ganglia during a relatively short midgestation period.[77] Following birth, it is either present in low quantities or undetectable.

int GENES

The *int*-1 proto-oncogene is normally expressed in neural tissues during embryogenesis[63,145] and is associated with cellular migration during gastrulation, particularly with mesodermal cells migrating through the primitive streak and with migrating parietal endodermal cells. In midgestational embryos, *int*-1 mRNA is detectable in discrete regions of the neural plate and its derivatives. Following closure of the neural tube, *int*-1 mRNA is found in the dorsal walls of the brain ventricles (exclusive of the telencephalon), in walls of the midbrain and diencephalon, in spinal cord, and in the hindbrain-midbrain junction.[116,145]

Expression of the *int*-2 gene is more restricted than that of *int*-1, with *int*-2 transcripts being detected only during early gastrulation through early embryogenesis. None have been identified in adult neural tissues.[63]

OTHER PROTO-ONCOGENES

c-*mos*, c-*abl*, and c-*erb*-A have all been found in the murine brain, but their exact localization is currently unclear.[46,100,103,133,140] Transcripts for the *mas* proto-oncogene have been found in rat hippocampus and cerebral cortex.[46] c-*ets* is expressed in astrocytes in the human cortex.[5] *ret* is found in normal murine spinal cord.[117] c-*fms* is expressed at low levels,[117] and moderate expression of *jun* D is seen.[52] Table 9.1 summarizes the expression of oncogenes in nonneoplastic neural tissues.

Table 9.1.
Oncogenes in Neural Tissues

Kinases	Location	Reference
c-*abl*	Embryonic brain	94
c-*fyn*	Adult brain	102
c-*mos*	Embryonic brain	103
c-*neu*	Embryonic brain; neural tissue; dorsal root ganglia	77
c-*src*	Embryonic brain; transiently in neural growth cone, retina, developing neurons, adult brain and cerebellum, especially in astrocytes	38, 63, 120
c-*yes*	Cerebellum, particularly in adults	125, 126
Ras genes	Embryonic and adult brain, dorsal and ganglia, peripheral nerves	39, 127
Nuclear binding proteins		
c-*myc*	Embryonic brain, cerebellum, nerves	107, 126, 128
N-*myc*	Embryonic brain, cerebrum	62, 151
L-*myc*	Postnatal forebrain, hindbrain	64
c-*fos*	Entire brain, nerves	2, 3, 30, 44
jun-D	Entire brain	52
Growth factors		
int-1	Embryonic brain, dorsal wall brain, ventricular spinal cord, midbrain-hindbrain junction, neural plate	63, 116, 127
int-2	Embryonic brain	63

It is of interest that many of the neurally expressed proto-oncogenes, including *src, yes, neu, ras* and *myc*, are found in organs containing epithelial cells involved in ion transport activities such as kidney and/or intestine, leading one to speculate that they may function in ionic transport.[125,126]

Oncogenes in Neural Regeneration

It is postulatted that repair of injured neurons requires a mutual relationship between the injured neuron and its surrounding nonneural cells for successful regeneration and regrowth.[113] It has also been suggested that the proto-oncogenes *fos* and *myc* play a role in this process.[128] These genes are constitutively expressed by nonneuronal cells in the nerve. Both genes are elevated 2–4-fold above the constitutive level during regeneration.[31] c-*fos* expression is also inducible following cortical injury.[32]

int-2, which encodes a growth factor, is known to promote neurite extension and neuronal growth and survival.[91]

Neuroblastoma (NBL)

NBL represents the most common extracranial tumor seen in infants and children. It is a primitive tumor of postganglionic sympathetic neurons that arises in various sites in the body, including adrenal, head and neck, mediastinum, and pelvis. The tumor may be aggressive, particularly when it has a high stage and grade. Numerous efforts have been made to develop new prognostic tests, since the prognosis for patients with advanced-stage disease remains poor, despite aggressive multimodal therapy. Nonetheless, a great deal of progress in the past few years has advanced our knowledge of this tumor at the cellular and molecular levels. The genetic predisposition to the disease is becoming clarified, and the amplification of a specific oncogene has been found to have prognostic significance.[14,110–112,146]

By far, the most common proto-oncogene studied in NBL is N-*myc*. This gene was discovered in tumors of neuroendocrine origin such as neuroblastoma and retinoblastoma. A 2.0kb *Eco*R1 fragment of human genomic DNA homolous with the v-*myc* oncogene was cloned and found to be distinct from the classical c-*myc* gene.[61] This oncogene is commonly amplified in NBL cell lines.[25]

DM chromosomes and HSRs occur frequently in NBL tumor cells.[8,9,20,24] The most common aberration is deletion of the 1p32 on the short arm of chromosome 1.[78,104] The chromosomal deletion at this site is associated with the generation of HSRs.[84] By *in situ* hybridization, the HSRs contain multiple copies of the N-*myc* gene.[68,76] Recently, a detailed molecular analysis of chromosome 1 abnormalities in NBL has been published. The data suggest that the truncation of chromosome 1 is most likely due to a complex translocation and deletion mechanism rather than to a simple, unbalanced translocation or terminal or interstitial deletion. This conclusion is supported by the frequent removal of N-*myc* from the altered chromosome 1 to another chromosome.[104]

The presence of N-*myc* proto-oncogene amplification is recognized as an independent marker of rapidly progressive NBL.[16,75,96,97,106,110–112,114,156] Generally, am-

Table 9.2.
Neuroblastoma Stages

Stage I	:	Tumors limited to the organ of origin.
Stage II	:	Regional lymph node spread that does not cross the midline.
Stage III	:	Extension across the midline.
Stage IV	:	Metastases to distant lymph nodes, bone, brain, or lung.
Stage IV-S	:	Small primary tumors with distant metastases limited to liver, skin, and/or bone marrow (but not bone).

plification has been demonstrated by evaluation of DNA gene-copy number by Southern blot or dot blot hybridization. N-*myc* is amplified as much as 300 times in 38% of untreated NBL patients.[114] Occasionally the L-*myc* gene coamplifies with N-*myc*.[66] Amplification is usually not present in patients with stage I or II disease, whereas it is present in 50% of those with stage III or IV disease[114] (Table 9.2). These data indicate that N-*myc* amplification is a common event in untreated human NBLs and correlates with advanced stages of disease.[16] Because the N-*myc* gene-copy number appears to be constant among multiple tumor sites in patients with disseminated NBL,[16] it has been suggested that N-*myc* amplification occurs in the primary tumor before metastases arise.[15] Finally, amplification of the N-*myc* gene results in high levels of expression of structurally normal N-*myc* mRNA.[59]

Most early studies demonstrating the relationship of N-*myc* gene amplification to prognosis were performed on fresh or frozen tissues. However, these may not always be available, either because the tissue is too small to allow anything but an adequate histological evaluation or because the diagnosis was not suspected at the time the tissue was handled in the diagnostic laboratory and all of the tissue was fixed, or because the tissue arrived in the laboratory in a fixed state. However, it is possible to evaluate even these fixed tissues for the presence of gene amplification, as shown by Tsuda and co-workers who performed a retrospective study using paraffin-embedded material to determine the relationship of N-*myc* amplification to prognosis and tumor differentiation, using dot blot hybridization.[137] They studied 37 NBLs and 12 ganglioneuroblastomas (GNBLs), including 6 composite GNBLs. The N-*myc* gene was amplified 3–500-fold in 58% of cases of stage IV NBL. These authors found that survival of patients with N-*myc* gene amplification was shorter than that of patients without N-*myc* amplification ($p < 0.05$). A high incidence of N-*myc* gene amplification was seen in cases of GNBL as well as NBL but not in cases of compositte GNBL. In every case, the copy number of the N-*myc* gene was the same in both the primary site and in multiple metastatic tumors, even when the lesions showed differences in histologic subtype, *i.e.*, NBL and GNBL.[130,132,138]

However, in contrast, most studies demonstrate that morphologic subtype (Fig. 9.1) correlates with prognosis; prognosis being better in patients with GNBL and composite GNBL than in those with pure NBL.[3,13,54,57] In the Tsuda study, a relatively better prognosis existed in three patients with stage IV composite GNBL.[138] More neuronally mature histological forms of NBL (*i.e.*, GNBL) do not exhibit N-*myc* DNA amplification.[21,43,138] However the degree of heterogeneity of N-*myc* expression can be striking in some cell lines and tumors. Some tumors with predominantly primitive neuroblasts contain focal areas of early differentiation. In

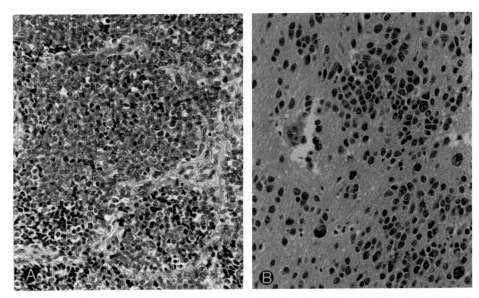

Figure 9.1.   A, Adrenal neuroblastoma composed of small cells in a fibrillary matrix (400x). **B,** Differentiating neuroblastoma with large cells and nuclear enlargement (400x).

these tumors, N-*myc* expression differs, with primitive small, dense neuroblasts expressing N-*myc*, and adjacent cells evidencing ganglioneuromatous differentiation not expressing the gene.[21]

Patients with NBL who present with the syndrome of opsoclonus and myoclonus enjoy a remarkably good prognosis, independent of their stage of disease or their age at diagnosis.[4] Four children with this syndrome were all found to contain a single copy of the N-*myc* oncogene in their tumors. All had a favorable clinical course, supporting the importance of gene copy number in predicting prognosis.[22]

To address the possibility that Southern blots may underestimate N-*myc* amplification, Cohen and co-workers studied the expression of N-*myc* genes in NBL, by *in situ* hybridization (ISH) in 28 tumors previously analyzed by Southern blot, and found that N-*myc* expression generally correlated with N-*myc* DNA amplification.[21] There were no tumors in which N-*myc* RNA expression was found by ISH in the absence of DNA amplification, nor were there tumors that had DNA amplification in the absence of RNA expression. Heterogeneity of N-*myc* expression was observed among cells within different areas of a given tumor. These findings contrast with those of others, in which N-*myc* RNA expression was detected in the absence of N-*myc* amplification.[76,82,98] Three of the 11 tumors were found to contain cells (within the tumor) that were positive for N-*myc* by ISH in the absence of amplification of the gene or RNA expression by Northern analysis,[43] confirming that ISH is a more sensitive assay for detecting isolated positive cells among a population of negative ones.

Noguchi and co-workers analyzed NBLs by ISH, using a single-step silver-enhancement method.[98] The N-*myc* gene copy numbers had been previously established by dot blot hybridization using DNA extracted from formalin-fixed, paraffin-embedded material. Silver grains were deposited above the nuclei of tumor cells, but none (or only faint deposition) were seen over infiltrating lymphocytes, stromal fibroblasts, and endothelial cells. Fourteen of 27 cases were positive for N-*myc* amplification. The results corresponded with those obtained by dot blot hybridization analysis. Eleven of 12 cases carrying less than 4 copies of the N-*myc* gene were judged to be negative by ISH. It was suggested that ISH with silver enhancement may be a useful tool in routine surgical pathology, since it can be applied to small pieces of biopsy tissue and performed without radioisotope-labeled probes.[98]

The presence and distribution of the N-*myc* gene product was examined in 13 NBLs and 5 GNBLs using an immunohistochemical technique. Nine tumors, eight NBLs and one GNBL of the composite type contained NBL cells with nuclei positive for the N-*myc* protein.[51] Histological examination showed that most of the positive NBL cells were immature, without evidence of neuronal differentiation. Nine of the 11 tumors with amplified N-*myc* genes contained positive cells, while none of the seven cases without amplified genes were immunoreactive. Survival in patients with tumors positive for N-*myc* protein was significantly lower than it was in those with immunonegative tumors. If immunohistochemical staining for the N-*myc* gene product continues to parallel gene amplification, one might expect this approach to be an alternative to slot blot hybridization in evaluating gene amplification. However, slot blots do provide a mechanism for evaluating gene copy number, something not possible utilizing antibodies to the gene product.

Case Example. A 4-year-old boy with a persistent cough was found to have a mass in the posterior mediastinum by chest x-ray. Fine needle aspiration of the mass demonstrated the presence of a highly cellular neoplasm composed of "small blue cells." Fibrillar eosinophilic material was noted among the cells, and there was a suggestion of rosette formation. Following the diagnosis of neuroblastoma, a thoracotomy with resection of the tumor was performed. At the time of surgical resection it was found that the tumor did not extend across the midline, but lymph node metastases were present. The tumor tissue was debulked and sent for pathological evaluation. Tumor, as well as grossly normal lymph nodes, were snap frozen for subsequent DNA extraction. Histologically the tumor was composed of immature neuroblastic tissue without evidence of ganglioneuromatous differentiation. Mitotic activity was high, and a suggestion of pseudo-rosettes was present (Fig. 9.2). The DNA was extracted from the tumor and from the normal lymph node tissue removed at the time of surgery, and subjected to slot blot hybridization, using a probe directed against the N-*myc* gene. The N-*myc* gene was not amplified in the tumor or nodal tissue (Fig. 9.3). The patient is alive 18 months after surgery.

OTHER ONCOGENES

Several oncogenes may play a role in the development of NBL. While amplification of the N-*myc* oncogene is closely associated with rapid tumor progression and poor clinical outcome, some NBL have been described without detectable changes in the N-*myc* gene. Furthermore, activation of the N-*ras* oncogene has been observed in the NBL cell line SK-N-SH.[35,45] In these cells the N-*ras* gene contains a mutation at position 61, with lysine substituting for glutamine.[129] However, in another study, using a rapid dot blot screening procedure based on DNA amplification and hybridization to synthetic oligonucleotide probes, 18 NBLs were investi-

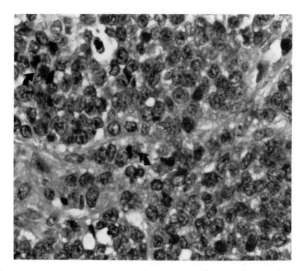

Figure 9.2.   There are numerous mitoses (*arrows*) in this mediastinal neuroblastoma.

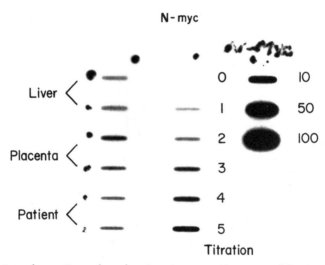

Figure 9.3.   *1st column:* Controls and patient (note: *no* N-*myc* amplification). *2nd and 3rd columns:* Titration curves of number of copies per cell of N-*myc* by dotting N-*myc* plasmid at appropriate concentrations. (Courtesy of the Center for Cellular and Molecular Diagnostics.)

gated, and no mutations were found in codons 12, 13, or 61 of the *ras* gene.[6] Thus, it is believed that mutated *ras* genes probably play a limited role in the pathogenesis of NBLs.

A transient increase in c-*ets*-1 expression is detectable in NBL, which coincides with the onset of morphologic differentiation. This gene maps to chromosome 11, in an area involved in the chromosomal translocation t(11;22) (q24;q12)[142] and is found in another pediatric neuroectodermal tumor that may be developmentally related to NBL.[131] Thus, there is the possibility that the increased expression of c-*ets*-1 occurs physiologically during a particular stage of neuroblast development.

The c-*fos* gene has been implicated in controlling processes ranging from cell proliferation[92,93] and differentiation[92,94] to nerve cell activity.[58] Induction of differentiation of human NBL cells is accompanied by rapid and transient expression of c-*fos* mRNA and down-regulation of c-*myc* RNA.[92] It is possible that decreases in N-*myc*, c-*myb*, and c-Ha-*ras* may play a role in the implementation of a neuroblast differentiation program, whereas increases in c-*ets*-1 and c-*fos* are associated with differentiation.[14]

Finally, neuroblastomas possess higher levels of pp60[c-src] activity than do glioblastomas. Cell lines of NBL do vary, however, with respect to their kinase activity.[147]

Retinoblastoma

Retinoblastoma, the most common intraocular tumor of childhood (Fig. 9.4), arises from embryonal cells and occurs in hereditary and nonhereditary forms.[95,139] It has a prevalence of about 1/20,000 live births. Nearly 60% of cases are sporadic and unilateral (nonhereditary), and approximately 40% are hereditary and bilateral. Three to five percent of patients have a constitutional deletion of chromosome 13, at 13q14.[73,150] HSRs are frequently seen *in vitro*.[41] It has been suggested that dominantly inherited tumors may result from loss or inactivation of both alleles of a regulatory or suppressor gene that, when active, prevents expression of a structural

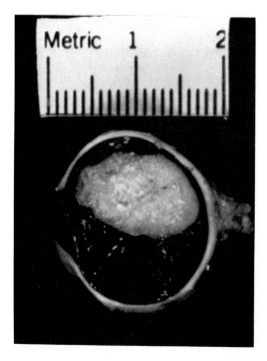

Figure 9.4. Intraocular retinoblastoma surrounded by hemorrhage (2x). (Picture courtesy of Dr. Lida Crooks, Albuquerque Veterans Administration Medical Center.)

transforming gene, possibly an oncogene, that is normally active only during embryogenesis.[72] This tumor provides evidence for the importance of tumor suppressor genes,[7,19] since they were first definitely documented in retinoblastomas.

RB GENE

It was originally postulated that the 13q14 band contained the RB gene, a mutation of which was responsible for hereditary cases of retinoblastoma that did not show a chromosomal deletion on karyotypic analysis.[71,72] Karyotypic examination of somatic cells from patients with hereditary retinoblastoma disclosed a minor subset of patients with visible deletions in the long arm of chromosome 13.[36,150] Similar deletions were also found in retinoblastoma cells.

According to the mutational theory of cancer, retinoblastoma arises via two consecutive mutations. The hereditary form develops due to an inherited first germline mutation affecting one chromosome 13 at 13q14, followed by a second somatic mutation affecting the other allele on the second chromosome 13 at the same site. The second subsequent somatic step transforms the original mutation or deletion from a heterozygous to a homozygous loss of this site.[7,19,122] Nonhereditary cases arise by two somatic mutations affecting a single retinal cell.[71,74]

The RB gene is mutated or lost in nearly all retinoblastomas, both sporadic and hereditary. In hereditary cases, every cell in the body of the affected individual carries one normal and one abnormal or nonfunctional RB allele, and when a second event occurs, the normal allele is lost, with the result that a retinoblastoma develops. It has been observed that patients with bilateral (but not unilateral) retinoblastomas have a high incidence of the development of independent second-site tumors, the majority of which are osteogenic sarcomas.[1] Importantly, survivors of nonhereditary unilateral retinoblastomas appear to have no increased risk for the development of second-site tumors.

The RB gene has been cloned, and it is now possible to detect partial or complete deletions of the RB gene by the absence of bands on Southern blots or by the presence of altered bands.[79] The gene spans approximately 200 kb of DNA within chromosome band 13q14 and contains 27 exons and 26 introns.[88] However, despite knowledge of the structure of the gene, deletions of the RB gene, as detected by Southern blots, have only been seen in 15–40% of tumors examined. In the remainder, small point mutations not detectable by Southern blots serve to create nonfunctional RB alleles.[55] Recent work has shown that the neoplastic phenotype of retinoblastoma and osteosarcoma cells can be suppressed by the introduction of a cloned RB gene,[56] supporting the hypothesis that RB gene inactivation is a critical event in tumorigenesis.

RFLPs in or near the RB gene were anticipated to be useful for genetic counseling in families segregating for mutant alleles. Bookstein and co-workers found an RFLP for the endonuclease *Bam*HI in intron 1 that was heterozygous in about 50% of individuals.[12] In another study, at least 20% of RFLPs in the RB gene were heterozygous and informative in 19 of 20 retinoblastoma kindreds.[144] RFLPs cosegregated with retinoblastoma susceptibility in all but one family.

It is anticipated that direct detection of RB gene mutations would have great utility for the following reasons: (*a*) sporadic unilateral hereditary and nonhereditary cases could be distinguished by examining the patients' fibroblasts, allowing an accurate determination of the risk of second primary cancers and of transmission to offspring; and (*b*) genetic diagnoses would be available without having to examine other family members. Antibodies to the RB protein might also have diagnostic or prognostic importance. For example, mutations in, or the absence of, the RB gene could be inferred from the absence of immunoperoxidase staining in the tumor cells, with the surrounding stroma serving as an internal positive control. Finally, if inactivation of the RB gene is important in the genesis of retinoblastomas, then restoration of this activity by RB gene transfer might provide hope of cancer therapy.[79]

The gene product of the RB gene is an 105–110 kilodalton nuclear phosphoprotein[80] associated with DNA binding, which can bind to oncogene-encoded proteins located within the nucleus.[28,55,81,85,143] It is postulated that the RB gene product P105-RB regulates transcription of specific cellular genes involved in growth control. These proteins appear to be regulated by phosphorylation.[86] They also bind to known oncogenes (*i.e.*, E1A, SV40 T, or the E7 product encoded for by human papilloma viruses). The binding inactivates the RB proteins.[33] The SV40 T antigen binds preferentially to an underphosphorylated member of the retinoblastoma susceptibility gene product family, suggesting that the growth-suppression function of RB is down-modulated by phosphorylation or T antigen binding.[86]

An aberrant RB protein was detected in J82 bladder carcinoma cells, which was not able to form a complex with E1A and was less stable than P105-RB. This defective RB protein resulted from a single point mutation within a splice-acceptor sequence eliminating a single exon of 35 amino acids from its encoded product. The binding protein encoded by the RB gene is absent from retinoblastomas lacking both RB alleles.[33]

Several groups have examined the possibility that the retinoblastoma gene product is responsible for controlling the expression of N-*myc*. The gene is under stage-specific regulation during normal development and embryogenesis. One report suggests that N-*myc* is deregulated and amplified in retinoblastomas, when compared to normal fetal or adult retinal cells.[123] However, Hansen and Cavanee[47-50] were unable to find N-*myc* expression in osteosarcomas in patients with the hereditary form of retinoblastoma, which would be at odds with a suppressing role for the RB locus. The most reasonable interpretation of these results is that expression of N-*myc* in retinoblastomas is merely a reflection of the embryonal nature of these tumors. At the present time, therefore, it is unclear which target gene(s) is (are) suppressed by the retinoblastoma gene product in neuroblastoma.

N-*myc* GENE

The N-*myc* gene is amplified 10–200-fold and is frequently overexpressed in primary retinoblastomas and in the retinoblastoma cell line Y79.[80] The N-*myc* gene product is detectable not only in Y79 cells but also in primary retinoblastomas by Western blots.[149] Immunohistochemical analysis shows nuclear staining in rosette- and fleurette-containing regions of the tumors, but none is seen in the normal retina nor in other parts of the eye. The results suggest that the level of the N-*myc* gene product may correlate inversely with both tumor differentiation and tumor aggressiveness.[149]

Brain Tumors

Not much insight has been gained on the role of proto-oncogenes in brain tumors, from the point of view of either pathogenesis or prognosis. The following highlights some of the studies to date.

Human gliomas often gain chromosome 7 or lose chromosome 10. Other abnormalities include losses of chromosome 22, deletion of 9p, and the loss of a distal region on 17p.[65] DMs are found in approximately 50% of malignant gliomas, and most tumors with DMs have demonstrable oncogene amplifications. The epidermal growth factor receptor (EGFR) gene, c-erb B, is most commonly amplified.[83] In fact, this gene was originally demonstrated by DNA transfection assays as an oncogene present in neuroblastomas and glioblastomas.[118] c-erb B can be either amplified or rearranged in these tumors. The c-raf-1 gene has also been detected by transfection of NIH 3T3 cells with DNA derived from glioblastoma.[37] Other genes that may be amplified include N-myc, c-myc, and K-ras.[10,135] The c-myc gene may also be rearranged.[135] Using a combination of immunofluorescence microscopy, flow cytometry, and immunoblot analysis to quantitate and characterize the expression of c-myc in glial tumors, Englehard and co-workers found that the c-myc oncoprotein is highly expressed in neoplastic cell lines and tumor specimens, whereas normal glial tissues are negative. In malignant cultured cells, the protein underwent an approximate 2-fold increase as the cells progressed from G_1/G_0 to G_2/M in the cell cycle.[34] High levels of N-myc expression have been found in astrocytomas, and both c-sis and c-kit are expressed in glioblastomas.[101,102]

It is of interest that even though neuroblastomas and medulloblastomas share biologic, histologic, and immunological features, the frequency of N-myc amplification varies between the two tumors. In a recent study of nine medulloblastomas, the N-myc gene was not amplified in any.[40] Our experience has been the same in the single example we examined (Fig. 9.5). In contrast, overexpression of the gene was found in 6 of 11 tumors.[40] Furthermore, a trend toward longer disease-free survival was noted in patients having low levels of N-myc protein in their tumor.[40]

The gli gene was originally identified in human glioblastoma, where it is amplified more than 50-fold.[69,105] It encodes a 4.0 kb mRNA. The gene maps to the long arm of chromosome 12 (12q13-q14.3)[69] and belongs to the DNA-binding zinc finger proteins.[70] The glioblastoma cell line 9L-5-JCK contains an activated c-raf-1 gene.[37]

Bolger and associates examined the sis proto-oncogene in patients with familial meningioma,[11] since this tumor is often characterized by the deletion of chromosome 22 or absence of the distal part of its long arm,[109] the site of the sis oncogene.[27] The sis oncogene in this family demonstrated an extra 2.3 kb band on Pst enzyme digest of DNA from two of the family members and was also seen in another subject. The variant did not segregate with the translocated chromosome in the pedigree, and it is probably located on the morphologically normal chromosome 22 in affected members. The studies of Bolger suggest, but do not prove, that a gene on

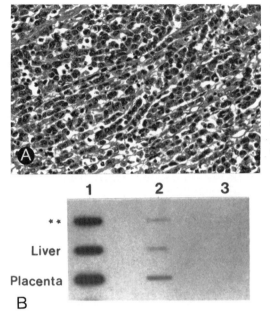

Figure 9.5. A, Anaplastic cerebellar medulloblastoma. **B,** Autoradiogram of genomic DNA isolated from the cerebellar tumor (**) and hybridized to a ³²P-labeled N-*myc* probe. *Lanes 1 through 3* represent 5 micrograms, 1.0 microgram, and 0.2 micrograms of DNA, respectively. There is no evidence of N-*myc* amplification in the brain lesion, relative to liver and placenta controls.

chromosome 22 plays a role in the pathogenesis of familial meningiomas. The relationship of the variant *sis* oncogene to the chromosomal abnormality is unknown.[11] Most recently, a human meningioma cell line has been established in which both c-*myc* and c-*fps* were amplified.[128]

Acknowledgments

The authors wish to acknowledge the helpful discussion and technical assistance of Barbara Griffith, the manuscript preparation by Agnes Truske, and the VA Medical Media Center for their expertise with the illustrative material.

The work reported in this chapter was supported by American Cancer Society Grant #PDT-341 and also by a grant from the University of New Mexico Cancer Center.

REFERENCES

1. Abramson, D. H., Ellsworth, R. M., Kitchin, E. D., and Tung, G. Second non ocular tumors in retinoblastoma survivors. Are they radiation induced? *Ophthalmology 91:* 1351–1355, 1984.
2. Adamson, E. D. Oncogenes in development. *Development 99:*449–471, 1987.
3. Adamson, E. D., Meek, J., and Edwards, S. A. Product of the cellular oncogene c-*fos* observed in mouse and human tissues using an antibody to a synthetic peptide. *E.M.B.O. J.* 4:941–947, 1985.
4. Altman, A. J., and Baehner, R. L. Favorable prognosis for survival in children with coincident opso-myoclonus and neuroblastoma. *Cancer 37:*846–852, 1976.
5. Amouyel, P., Gegonne, A., Delacourte, A., Defossez, A., and Stehelin, D. Expression of ETS proto-oncogenes in astrocytes in human cortex. *Brain Res.* 447:149–153, 1988.
6. Ballas, K., Janssen, J. W. G., and Bartram, C. R. Incidence of ras mutations in neuroblastoma. *Eur. J. Pediatr.* 147:313–314, 1988.

7. Benedict, W. F., Murphee, A. L., Banerjee, A., Spina, C. A., Sparkes, M. C., and Sparkes, R. S. Patient with 13 chromosome deletion: Evidence that the retinoblastoma gene is a recessive cancer gene. *Science (Wash DC)* 219:973–975, 1983.
8. Biedler, J. L., Helson, L., and Spengler, B. A. Morphology and growth, tumorigenicity, and cytogenetics of human neuroblastoma cells in continuous culture. *Cancer Res.* 33:2643–2652, 1973.
9. Biedler, J. L., and Spengler, B. A. A novel chromosome abnormality in human neuroblastoma and antifolate-resistant Chinese hamster cell lines in culture. *J. Natl. Cancer Inst.* 57:683–695, 1976.
10. Bigner, S. H., Vogelstein, B., and Bigner, D. D. Chromosomal abnormalities and gene amplification in malignant gliomas. *ISI Atlas of Science: Biochemistry* 1:333–336, 1988.
11. Bolger, G. B., Stamberg, J., Kirsch, I. R., Hollis, G. F., Schwarz, D. F., and Thomas, G. H. Chromosome translocation t(14;22) and oncogene (c-sis) variant in a pedigree with familial meningioma. *N. Engl. J. Med.* 312:564–567, 1985.
12. Bookstein, R., Lee, E. Y. H. P., To, H., Young, L. J., Sery, T. W., Hayes, R. C., Friedmann, T., and Lee, W. H. Human retinoblastoma susceptibility gene; Genomic organization and analysis of heterozygous intragenic deletion mutants. *Proc. Natl. Acad. Sci. U.S.A.* 85:2210–2214, 1988.
13. Bove, K. E., and McAdams, A. J. Composite ganglioneuroblastoma: An assessment of the significance of histological maturation in neuroblastoma diagnosed beyond infancy. *Arch. Pathol. Lab. Med.* 105:325–330, 1981.
14. Brodeur, G. M., and Fong, C. T. Molecular biology and genetics of human neuroblastoma: *Cancer Genet. Cytogenet.* 41:153–174, 1989.
15. Brodeur, G. M., Hayes, F. A., Green, A. A., Casper, J. T., Wasson, J., Wallach, S., and Seeger, R. C. Consistent N-myc copy number in simultaneous or consecutive neuroblastoma samples from sixty individual patients. *Cancer Res.* 47:4248–4253, 1987.
16. Brodeur, G. M., Seeger, R. C., Schwab, M., Varmus, H. E., and Bishop, J. M. Amplification of N-myc in untreated human neuroblastomas correlates with advanced disease stage. *Science (Wash DC)* 224:1121–1124, 1984.
17. Brugge, J., Cotton, P., Lustig, A., Yonemoto, W., Lipsich, L., Coussens, P., Barrett, J. N., Nonner, D., and Keane, R. W. Characterization of the altered form of the c-*src* gene product in neuronal cells. *Gene Dev.* 1:287–296, 1987.
18. Cartwright, C. A., Simantov, R., Kaplan, P. L., Hunter, T., and Eckhart, W. Alterations in pp60$^{c\text{-}src}$ accompany differentiation of neurons from rat embryo striatum. *Mol. Cell. Biol.* 7:1830–1840, 1987.
19. Cavenee, W. K., Dryja, T. P., Phillips, R. A., Benedict, W. F., Godbout, R., Gallie, B. L., Murphree, A. L., Strong, L. C., and White, R. L. Expression of recessive alleles by chromosomal mechanisms in retinoblastoma. *Nature (Lond)* 305:779–784, 1983.
20. Christiansen, H., and Lampert, F. Tumor karyotype discriminates between good and bad prognostic outcome in neuroblastoma. *Br. J. Cancer* 57:121–126, 1988.
21. Cohen, P. S., Seeger, R. C., Triche, T. J., and Israel, M. A. Detection of N-*myc* gene expression in neuroblastoma tumors by in situ hybridization. *Am. J. Pathol.* 131:391–397, 1988.
22. Cohn, S. L., Salwen, H., Herst, C. V., Maurer, H. S., Nieder, M. L., Morgan, E. R., and Rosen, S. T. Single copies of the N-*myc* oncogene in neuroblastomas from children presenting with the syndrome of opsoclonus-myoclonus. *Cancer* 62:723–726, 1988.
23. Cooper, J. A. The *src* family of protein-tyrosine kinases. In: *Peptides and Protein Phosphorylation,* edited by B. Kemp and P. F. Abewood, Boca Raton, FL, CRC Press, 1987, pp. 52–83.
24. Cox, D., Yunken, C., and Spriggs, A. I. Minute chromatin bodies in malignant tumors of childhood. *Lancet* I:55–58, 1965.
25. Csaikl, F., Mullauer, L., Csaikl, U., and Vetterlein, M. Expression of *myc* and *ras*

oncogenes in two newly established neuroblastoma cell lines. *J. Cancer Res. Clin. Oncol.* 115:242–246, 1989.

26. Curran, T., and Morgan, J. Memories of *fos*. *Bioessays* 7:255–258, 1987.
27. Dalla Favera, R., Gallo, R. C., Giallongo, A., and Croce, C. M. Chromosomal localization of the human homolog (c-*sis*) of the simian sarcoma virus *onc* gene. *Science (Wash DC)* 218:686–688, 1982.
28. DeCaprio, J. A., Ludlow, J. W., Figge, J., Shew, J.-Y., Huang, C.-M., Lee, W.-H., Marsilco, E., Paucha, E., and Livingston, D. M. SV40 large tumor antigen from a specific complex with the product of the retinoblastoma susceptibility gene. *Cell* 54:275–283, 1988.
29. Downs, K. M., Martin, G. R., and Bishop, J. M. Contrasting patterns of myc and N-myc expression during gastrulation of the mouse embryo. *Genes Dev.* 3:860–869, 1989.
30. Dragunow, M., Peterson, M. R., and Robertson, H. A. Presence of c-*fos* immunoreactivity in the adult rat brain. *Eur. J. Pharmacol.* 135:113–114, 1987.
31. Dragunow, M., and Robertson, H. A. Brain injury induces c-fos protein(s) in nerve and glial-like cells in adult mammalian brain. *Brain Res.* 455:295–299, 1988.
32. Dragunow, M., and Robertson, H. A. Localization and induction of c-fos protein-like immunoreactive material in the nuclei of adult mammalian neurons. *Brain Res.* 440:252–269, 1988.
33. Egan, C., Bayley, S. T., and Branton, P. E. Binding of the Rb1 protein to E1A products is required for adenovirus transformation. *Oncogene* 4:383–388, 1989.
34. Engelhard, H. H., III, Butler, A. B., IV, and Bauer, K. D. Quantification of the c-*myc* oncoprotein in human glioblastoma cells and tumor tissue. *J. Neurosurg.* 71:224–232, 1989.
35. Eva, A., Robbins, K. C., Anderson, P. R., Srinivasan, A., Tronick, S. R., Reddy, E. P., Ellmore, N. W., Galen, A. T., Lautenberger, J. A., Papas, T. S., Westin, E. H., Wong-Staal, F., Gallo, R., and Aaronson, S. A. Cellular genes analogous to retroviral oncogenes are transcribed in human tumor cells. *Nature (Lond)* 295:116–119, 1982.
36. Francke, U. Retinoblastoma and chromosome 13. *Birth Defects* 12:131–134, 1976.
37. Fukui, M., Yamamoto, T., Kawai, S., Maruo, K., and Toyoshima, K. Deletion of a *raf*-related and two other transforming DNA sequences in human tumor maintained in nude mice. *Proc. Natl. Acad. Sci. U.S.A.* 82:5954–5958, 1985.
38. Fults, D. W., Towle, A. C., Lauder, J. M., and Maness, P. F. pp60 c-*src* in the developing cerebellum. *Mol. Cell. Biol.* 5:27–32, 1985.
39. Furth, M. E., Aldrich, T. H., and Cordon-Cardo, C. Expression of *ras* proto-oncogene protein in normal human tissues. *Oncogene* 1:47–58, 1987.
40. Garson, J. A., Pemberton, L. F., Sheppard, P. W., Varndell, I. M., Coakham, H. B., and Kemstead, J. T. N-*myc* gene expression and oncoprotein characterisation in medulloblastoma. *Br. J. Cancer* 59:889–894, 1989.
41. Gilbert, F., Balaban, G., Breg, W. R., Gallie, B., Reid, T., and Nichols, W. Homogeneously staining region in a retinoblastoma cell line: Relevance to tumor initiation and progression. *J. Natl. Cancer Inst.* 67:301, 1981.
42. Grady, E. F., Schwab, M., and Rosenau, W. Expression of N-*myc* and c-*src* during the development of fetal human brain. *Cancer Res.* 44:2931–2936, 1984.
43. Grady-Leopardi, E. F., Schwab, M., Ablin, A. R., and Rosenau, W. Detection of N-myc oncogene expression in human neuroblastoma by in situ hybridization and blot analysis: Relationship to clinical outcome. *Cancer Res.* 46:3196–3199, 1986.
44. Gubits, R. M., Hazelton, J. L., and Simantov, R. Variations in c-*fos* gene expression during rat brain development. *Mol. Brain Res.* 3:197–202, 1988.
45. Hall, A., Marshall, C. J., Spurr, N. K., and Weiss, R. A. Identification of transforming gene in two human sarcoma cell lines as a new member of the ras gene family located on chromosome 1. *Nature (Lond)* 303:396–400, 1983.

46. Hanley, M. R. Proto-oncogenes in the nervous system. *Neuron* 1:175–182, 1988.
47. Hansen, M. F., and Cavenee, W. K. Genetics of cancer predisposition. *Cancer Res.* 47:5518–5527, 1987.
48. Hansen, M. F., and Cavenee, W. K. Tumor suppressors: Recessive mutations that lead to cancer. *Cell* 53:172–173, 1988.
49. Hansen, M. F., and Cavenee, W. K. Retinoblastoma and the progression of tumor genetics. *Trends Genet.* 4:125–128, 1988.
50. Hansen, M. F., Koufos, A., Gallie, B. L., Phillips, R. A., Fodstad, O., Brogger, A., Gedde-Dahl, T., and Cavenee, W. K. Osteosarcoma and retinoblastoma: A shared chromosomal mechanism revealing recessive predisposition. *Proc. Natl. Acad. Sci. U.S.A.* 82:6216–6220, 1985.
51. Hashimoto, H., Daimaru, Y., Enjoji, M., and Nakagawara, A. N-myc gene product expression in neuroblastoma. *J. Clin. Pathol.* 42:52–55, 1989.
52. Hirai, S. I., Ryseck, R.-P., Mechta, F., Bravo, R., and Yamiu, M. Characterization of *jun*D: A new member of the *jun* proto-oncogene family. *E.M.B.O. J.* 8:1433–1439, 1989.
53. Hirvonen, H., Sandberg, M., Kalimo, H., Hukkanen, V., Vuorio, E., Salmi, T. T., and Alitalo, K. The N-*myc* proto-oncogene and IGF-II growth factor mRNAs are expressed by distinct cells in human fetal kidney and brain. *J. Cell Biol.* 108:1093–1104, 1989.
54. Horn, R. C., Koop, C. E., and Kiesewetter, W. B. Neuroblastoma in childhood: Clinicopathologic study of forty-four cases. *Lab. Invest.* 5:106–119, 1956.
55. Horowitz, J. M., Yandell, D. W., Park, S. H., Canning, S., Whyte, P., Buchkovich, K., Harlow, E., Weinberg, R. A., and Dryja, T. P. Point mutational activation of the retinoblastoma antioncogene. *Science (Wash DC)* 243:937–941, 1989.
56. Huang, H. J. S., Yu, J. K., Shew, J. Y., Chen, P. L., Bookstein, R., Friedman, T., Lee, E. Y. P., and Lee, W.-H. Suppression of the neoplastic phenotype by replacement of the RB gene in human cancer cells. *Science (Wash DC)* 242:1563–1566, 1988.
57. Hughes, M., Marsden, H. B., and Palmer, M. K. Histologic patterns of neuroblastoma related to prognosis and clinical staging. *Cancer* 34:1706–1711, 1974.
58. Hunt, S. P., Pini, A., and Evan, G. Induction of c-*fos*-like protein in spinal cord neurons following sensory stimulation. *Nature (Lond)* 328:632–634, 1987.
59. Ibson, J. M., and Rabbitts, P. H. Sequence of a germ line N-*myc* gene and amplification as a mechanism of amplification. *Oncogene* 2:399–402, 1988.
60. Ingvarsson, S., Asker, C., Axelson, H., Klein, G., and Sumegi, J. Structure and expression of B-*myc*, a new member of the *myc* gene family. *Mol. Cell. Biol.* 8:3168–3174, 1988.
61. Izumi, S.-I., Moriuchi, T., Koji, T., *et al.* Localization of C-MYC in HL-60 cells, neoplastic and normal tissues: An immunohistochemical and in situ hybridization study. *Acta Histochem. Cytochem.* 21:327–342, 1988.
62. Jacobovits, A., Schwab, M., Bishop, J. M., and Martin, G. R. Expression of N-*myc* in teratocarcinoma stem cells and mouse embryos. *Nature (Lond)* 318:188–191, 1985.
63. Jacobovits, A., Shackleford, G. M., Varmus, H. E., and Martin, G. R. Two proto-oncogenes implicated in mammary carcinogenesis *int*-1 and *int*-2, are independently regulated during mouse development. *Proc. Natl. Acad. Sci. U.S.A.* 83:7806–7810, 1986.
64. Jacobs, C., and Rubsamen, H. Expression of pp60 c-*src* protein kinase in adult and fetal human tissue: High activities in some sarcomas and mammary carcinomas. *Cancer Res.* 43:1696–1702, 1983.
65. James, C. D., Carlbom, E., Dumanski, J. P., Hansen, M., Nordenskjold, M., Collins, V. P., and Cavenee, W. K. Conal genomic alterations in glioma malignancy stages. *Cancer Res.* 48:5546–5551, 1988.
66. Jinbo, T., Iwamura, Y., Kaneko, M., and Sawaguchi, S. Coamplification of the L-*myc* and N-*myc* oncogenes in a neuroblastoma cell line. *Jpn. J. Cancer Res.* 80:299–301, 1989.

67. Kaczmarek, L., Siedlecki, J. A., and Danysz, W. Proto-oncogene c-*fos* induction in rat hippocampus. *Mol. Brain Res.* 3:183–186, 1988.
68. Kanda, N., Schreck, R., Alt, F., Bruns, G., Baltimore, D., and Latt, S. Isolation of amplified DNA sequences from IMR-32 human neuroblastoma cells: Facilitation by fluorescence-activated flow sorting of metaphase chromosomes. *Proc. Natl. Acad. Sci. U.S.A.* 80:4069–4073, 1983.
69. Kinzler, K. W., Bigner, S. H., Bigner, D. D., Trent, J. M., Law, M. L., O'Brien, S. J., Wong, A. J., and Vogelstein, B. Identification of an amplified, highly expressed gene in human glioma. *Science (Wash DC)* 236:70–73, 1987.
70. Kinzler, K. W., Ruppert, J. M., Bigner, S. H., and Vogelstein, B. The *gli* gene is a member of the *Kruppel* family of zinc finger proteins. *Nature (Lond)* 332:371–374, 1988.
71. Knudson, A. G. Mutation and cancer: Statistical study of retinoblastoma. *Proc. Natl. Acad. Sci. U.S.A.* 68:820–823, 1971.
72. Knudson, A. G. Retinoblastoma: A prototypic hereditary neoplasm. *Semin. Oncol.* 5: 57–60, 1978.
73. Knudson, A. G., Meadows, A. T., Nichols, W. W., and Hill, R. Chromosomal deletion and retinoblastoma. *N. Engl. J. Med.* 295:1120–1123, 1976.
74. Knudson, A. G. A two-mutation model for human cancer. In: *Advances in Viral Oncology,* vol. 7, edited by G. Klein, New York, Raven Press, 1987, pp. 1–16.
75. Kohl, N. E., Gee, C. E., and Alt, F. Activated expression of the N-*myc* gene in human neuroblastomas and related tumors. *Science (Wash DC)* 226:1335–1337, 1984.
76. Kohl, N. E., Kanada, N., Schreck, R. R., Bruns, G., Latt, S. A., Gilbert, F., and Alt, F. W. Transposition and amplification of oncogene-related sequences in human neuroblastoma. *Cell* 35:359–367, 1983.
77. Kokai, Y., Cohen, J. A., Drebin, J. A., and Greene, M. I. Stage- and tissue-specific expression of the *neu* oncogene in rat development. *Proc. Natl. Acad. Sci. U.S.A.* 84:8498–8501, 1987.
78. Lampert, F., Rudolph, B., Christiansen, H., and Franke, F. Identical chromosome 1p breakpoint abnormality in both the tumor and the constitutional karyotype of a patient with neuroblastoma. *Cancer Genet. Cytogenet.* 34:235–239, 1988.
79. Lee, W.-H., Bookstein, R., and Lee, E. Y.-H. P. Studies on the human retinoblastoma susceptibility gene. *J. Cell. Biochem.* 38:213–227, 1988.
80. Lee, W.-H., Murphree, A. L., and Benedict, W. F. Expression and amplification of the N-*myc* gene in primary retinoblastoma. *Nature (Lond)* 309:458–460, 1984.
81. Lee, W.-H., Shew, J.-Y., Hong, F. D., Sery, T. W., Donoso, L. A., Young, L.-J., Bookstein, R., and Lee, E. Y.-H. P. The retinoblastoma susceptibility gene encodes a nuclear phosphoprotein associated with DNA binding activity. *Nature (Lond)* 329:642–645, 1987.
82. Levy, B. T., Sorge, L. K., Meymandi, A., and Maness, P. F. pp60 *src* kinase is in embryonic tissues of chick and human. *Dev. Biol.* 104:9–17, 1984.
83. Libermann, T. A., Nusbaum, H. R., Razon, N., Kris, R., Lax, I., Soreq, H., Whittle, N., Waterfield, M. D., and Ullrich, A. Amplification and overexpression of the EGF receptor gene in primary human glioblastomas. *Nature (Lond)* 313:144–147, 1985.
84. Longo, L., Christiansen, H., Christiansen, N. M., Cornaglia-Ferraris, P., and Lampert, F. N-myc amplification at chromosome band 1p32 in neuroblastoma cells as investigated by in situ hybridization. *J. Cancer Res. Clin. Oncol.* 114:636–640, 1988.
85. Lothe, R. A., Nakamura, Y., Woodward, S., Gedde-Dahl, T., and White, R. VNTR (variable number of tandem repeats) markers show loss of chromosome 17p sequences in human colorectal carcinomas. *Cytogenet. Cell Genet.* 48:167–169, 1988.
86. Ludlow, J. W., DeCaprio, J. A., Huang, C.-H., Lee, W.-H., Paucha, E., and Livingston, D. M. SV40 large T antigen preferentially binds to an underphosphorylated member of the retinoblastoma susceptibility gene product family. *Cell* 56:57–65, 1989.

87. Maness, P. F. pp60 c-*src* encoded by the proto-oncogene c-*src* is a product of sensory neurons. *J. Neurol. Res.* 16:127–139, 1986.
88. McGee, T. L., Yandell, D. W., and Dryja, T. P. Structure and partial genomic sequence of the human retinoblastoma susceptibility gene. *Gene* 80:119–128, 1989.
89. Morgan, J. I., Cohen, D. R., Hempstead, J. L., and Curran, T. Mapping patterns of c-*fos* expression in the central nervous system after lesion. *Science (Wash DC)* 237:192–197, 1987.
90. Morgan, J. I., and Curran, T. Role of ion flux in the control of c-*fos* expression. *Nature (Lond)* 322:552–555, 1986.
91. Morrison, R., Sharma, A., DeVellis, J., and Bradshaw, P. A. Basic fibroblast growth factor supports the survival of cerebral cortical neurons in primary culture. *Proc. Natl. Acad. Sci. U.S.A.* 83:7537–7541, 1986.
92. Muller, R. Protooncogenes and differentiation. *Trends Biochem. Sci.* 11:129–131, 1986.
93. Muller, R., Bravo, R., Burckhardt, J., and Curran, T. Induction of c-*fos* gene and protein by growth factors precedes activation of c-*myc*. *Nature (Lond)* 312:716–720, 1984.
94. Muller, R., Slamon, D. J., Tremblay, J. M., Cline, M. J., and Verma, I. M. Differential expression of cellular oncogenes during pre- and postnatal development of the mouse. *Nature (Lond)* 299:6640–6644, 1982.
95. Murphree, A. L., Benedict, W. F. Retinoblastoma: Clues to human oncogenesis. *Science (Wash DC)* 223:1028–1033, 1984.
96. Nisen, P. D., Waber, P. G., Rich, M. A., Pierce, R. S., Garvin, J. R., Gilbert, F., and Lanzkowsky, P. N-myc oncogene RNA expression neuroblastoma. *J. Natl. Cancer Inst.* 80:1633–1637, 1988.
97. Noda, M., Ko, M., Ogura, A., Liu, D.-G., Amano, T., Takano, T., and Ikawa, Y. Sarcoma viruses carrying ras oncogenes induce differentiation associated properties in a neuronal cell line. *Nature (Lond)* 318:73–75, 1985.
98. Noguchi, M., Hirohashi, S., Tsuda, H., Nakajima, T., Hara, F., and Shimosato, Y. Detection of amplified N-*myc* gene in neuroblastoma by *in situ* hybridization using the single-step silver enhancement method. *Mod. Pathol.* 1:428–432, 1988.
99. Olofsson, B., Chardin, P., Touchot, N., Zahraoui, A., and Tavitian, A. Expression of the *ras*-related *ral* A, *rho* 12 and *rab* genes in adult mouse tissues. *Oncogene* 3:231–234, 1988.
100. Oppenheimer, J. H. Thyroid hormone action at the cellular level. *Science (Wash DC)* 203:971–979, 1979.
101. Pantazis, P., Pelicci, P. G., Dalla Favera, R., and Antoniades, H. N. Synthesis and secretion of proteins resembling plate derived growth factor by human glioblastoma and fibrosarcoma cells in culture. *Proc. Natl. Acad. Sci. U.S.A.* 82:2404–2408, 1985.
102. Perlmutter, R. M., Marth, J. M., Ziegler, S. F., Garvin, A. M., Pawar, S., Cooke, M. P., and Abraham, K. M. Specialized protein tyrosine kinase proto-oncogenes in hematopoietic cells. *Biochim. Biophys. Acta* 948:245–262, 1988.
103. Propst, F., Rosenberg, M. P., Oskarsson, M. K., Russell, L. B., Nguyen-Huu, M. C., Nadeau, J., Jenkins, N. A., Copeland, N. G., and van de Woude, G. F. Genetic analysis and developmental regulation of testis-specific RNA expression of *Mos, Abl*, actin and Hox-1.4. *Oncogene* 2:227–233, 1988.
104. Ritke, M. K., Shah, R., Valentine, M., Douglass, E. C., and Tereba, A. Molecular analysis of chromosome 1 abnormalities in neuroblastoma. *Cytogenet. Cell Genet.* 50:84–90, 1989.
105. Roberts, W. M., Douglass, E. C., Peiper, S. C., Houghton, P. J., and Look, A. T. Amplification of the *gli* gene in childhood sarcomas. *Cancer Res.* 49:5407–5413, 1989.
106. Rosen, N., Reynolds, C. P., Thiele, C. J., Biedler, J. L., and Israel, M. A. Increased N-*myc* expression following progressive growth of human neuroblastoma. *Cancer Res.* 46:4139–4143, 1986.

107. Ruppert, C., Goldovitz, D., and Wille, W. Proto-oncogene c-*myc* is expressed in cerebellar neurons at different developmental stages. *E.M.B.O. J.* 5:1897–1901, 1986.

108. Sagar, S. M., Sharp, F. R., and Curran, T. Expression of c-*fos* protein in brain: Metabolic mapping at the cellular level. *Science (Wash DC)* 240:1328–1331, 1988.

109. Sandberg, A. A. *Chromosomes in Human Cancer and Leukemia.* New York, Elsevier, 1980, pp. 535–543.

110. Schwab, M., Alitalo, K., Klempnauer, K. H., Varmus, H. E., Bishop, J. M., Gilbert, F., Brodeur, G., Goldstein, M., and Trent, J. Amplified DNA with limited homology to myc cellular oncogene is shared by human neuroblastoma cell lines and a neuroblastoma tumour. *Nature (Lond)* 305:245–248, 1983.

111. Schwab, M., Alitalo, K., Varmus, H. E., Bishop, J. M., and George, D. A cellular oncogene (c-Ki-*ras*) is amplified, overexpressed, and located within karyotypic abnormalities in mouse adrenocortical tumour cells. *Nature (Lond)* 303:497–501, 1983.

112. Schwab, M., Ellison, J., Busch, M., Rosenau, W., Varmus, H. E., and Bishop, J. M. Enhanced expression of the human gene N-*myc* consequent to amplification of DNA may contribute to a malignant progression of neuroblastoma. *Proc. Natl. Acad. Sci. U.S.A.* 81:4940–4944, 1984.

113. Schwartz, M., Cohen, A., Stein-Izsak, C., and Belkin, M. Dichotomy of the glial cell response to axonal injury and regeneration. *F.A.S.E.R. J.* 3:2371–2378, 1989.

114. Seeger, R. C., Brodeur, G. M., Sather, H., Dalton, A., Siegel, S. E., Wong, K. Y., and Hammond, D. Association of multiple copies of the N-*myc* oncogene with rapid progression of neuroblastomas. *N. Engl. J. Med.* 313:1111–1115, 1986.

115. Semsei, I. M., Ma, S., and Cutler, R. G. Tissue and age specific expression of the *myc* protooncogene family throughout the life span of the C57BL/6J mouse strain. *Oncogene* 4: 465–470, 1989.

116. Shackleford, G. M., and Varmus, H. E. Expression of the proto-oncogene int-1 is restricted to posterior male germ cells and the neural tube of mid-gestational embryos. *Cell* 50:89–95, 1987.

117. Sherr, C. J., Rettenmier, C. W., Sacca, R., Roussel, M. F., Look, A. T., and Stanley, E. R. The c-*fms* proto-oncogene product is related to the receptor for the mononuclear phagocyte growth factor, CSF-1. *Cell* 41:665–676, 1985.

118. Shih, C., Padhy, L. C., Murray, M. and Weinberg, R. A. Transforming genes of carcinomas and neuroblastomas introduced into mouse fibroblasts. *Nature (Lond)* 290:261–264, 1981.

119. Slamon, D. J., and Cline, M. F. Expression of cellular oncogenes during embryonic and fetal development of the mouse. *Proc. Natl. Acad. Sci. U.S.A.* 81:7141–7145, 1984.

120. Sorge, L. K., Levy, B. T., and Maness, P. F. pp60[c-src] expression in embryonic nervous tissue: Immunocytochemical localization in the developing chicken retina. *Cancer Cells* 1:117–122, 1984.

121. Sorge, L. K., Levy, B. T., and Maness, P. F. pp60[c-src] is developmentally regulated in the neural retina. *Cell* 36:249–257, 1984.

122. Sparkes, R. S., Murphree, A. L., Lingua, R. W., Sparkes, M. C., Field, L. L., Funderburk, S. J., Benedict, W. F. Gene for hereditary retinoblastoma assigned to human chromosome 13 by linkage to esterase D. *Science (Wash DC)* 219:971–973, 1983.

123. Squire, J., Goddard, A. D., Canton, M., Becker, A., Phillips, R. A., and Gallie, B. L. Tumor induction by the retinoblastoma mutation is independent of N-*myc* expression. *Nature (Lond)* 322:555–557, 1986.

124. Stein-Izsak, C., Cohen, A., and Schwartz, M. Expression of the protooncogenes fos and myc and optic nerve regeneration. *Neurosci. Res. Commun.* 1:87–95, 1987.

125. Sudol, M. Expression of proto-oncogenes in neural tissues. *Brain Res. Rev.* 13:391–403, 1988.

126. Sudol, M., Alvarez-Buylla, A., and Hanafusa, H. Differential developmental expres-

sion of cellular *yes* and cellular *src* proteins in cerebellum. *Oncogene Res.* 2:345–355, 1988.

127. Sudol, M., Kuo, F., Shigemitsu, L., and Alvarez-Buglla, A. Expression of the *yes* proto-oncogene in cerebellar purkinge cells. *Mol. Cell. Biol.* 9:4545–4549, 1989.

128. Tanaka, K., Sato, C., Maeda, Y., Koike, M., Matsutani, M., Yamada, and Niyaki, M. Establishment of a human malignant meningioma cell line with amplified c-*myc* oncogene. *Cancer* 64:2243–2249, 1989.

129. Taprowsky, E., Shimizu, K., Goldfarb, M., and Wigler, M. Structure and activation of the human n-*ras* gene. *Cell* 34:581–586, 1983.

130. Thiele, C. J., Cohen, P. S., and Israel, M. A. Regulation of c-*myb* expression in human neuroblastoma cells during retinoic acid-induced differentiation. *Mol. Cell. Biol.* 8:1677–1683, 1988.

131. Thiele, C. J., McKeon, C., Triche, T. J., Ross, R. A., Reynolds, C. P., and Israel, M. A. Differential protooncogene expression characterizes histopathologically indistinguishable tumors of the peripheral nervous system. *J. Clin. Invest.* 80:804–811, 1987.

132. Theile, C. J., Reynolds, C. P., and Israel, M. A. Decreased expression of N-myc precedes retinoic acid-induced morphological differentiation of human neuroblastoma. *Nature (Lond)* 313:404–406, 1985.

133. Thompson, C., Weinberger, C., Lebo, R., and Evans, R. M. Identification of a novel thyroid hormone receptor expressed in the mammalian central nervous system. *Science (Wash DC)* 237:1610–1614, 1987.

134. Yeramain, P., Chardin, P., Madaule, P., and Tavitian, A. Nucleotide sequence of human rho cDNA clone 12. *Nucleic Acids Res.* 15:1869, 1987.

135. Trent, J., Meltzer, P., Rosenblum, M., Harsh, G., Kinzler, K., Mashsal, R., Feinberg, A., and Vogelstein, B. Evidence for rearrangement, amplification, and expression of c-*myc* in a human glioblastoma. *Proc. Natl. Acad. Sci. U.S.A.* 83:470–473, 1986.

136. Trotter, J., Boutter, C. A., Sontheimer, H., Schechner, M., and Wagner, E. F. Expression of v-*src* arrests murine glial cell differentiation. *Oncogene* 4:457–464, 1989.

137. Tsuda, H., Shimosato, Y., Upton, M. P., Yokata, J., Terada, M., Ohura, M., Sugimura, T., and Hirohashi, S. Retrospective study on amplification of N-myc and c-myc genes in pediatric solid tumors and its association with prognosis and tumor differentiation. *Lab. Invest.* 59:321–327, 1988.

138. Tsuda, T., Obara, M., Hirano, H., Gotoh, S., Kubomura, S., Higashi, K., Kuroiwa, A., Nakagawara, A., Nagahara, N., and Shimizu, K. Analysis of N-*myc* amplification in relation to disease stage and histologic types in human neuroblastomas. *Cancer* 60:820–826, 1987.

139. Vogel, F. Genetics of retinoblastoma. *Hum. Genet.* 52:1–54, 1979.

140. Wakai, A., Seino, S., Sakurai, A., Szilak, I., Bell, G. I., and DeGroot, L. J. Characterization of a thyroid hormone receptor expressed in human kidney and other tissues. *Proc. Natl. Acad. Sci. U.S.A.* 85:2781–2785, 1988.

141. Wang, L.-H., Ijima, S., Dorai, T., and Lin, B. Regulation of the expression of proto-oncogene c-*src* by alternative RNA splicing in chicken skeletal muscle. *Oncogene Res.* 1:43–59, 1987.

142. Whang-Peng, J., Triche, T. J., Knutsen, T., Miser, J., Douglass, E. C., and Israel, M. A. Chromosome translocation in peripheral neuroepithelioma. *N. Engl. J. Med.* 311:584–585, 1984.

143. Whyte, P., Buchkovich, K. J., Horowitz, J. M., Friend, S. H., Raybuck, M., Weinberg, R. A., and Harlow, E. Association between an oncogene and an anti-oncogene: The adenovirus E1A proteins bind to the retinoblastoma gene product. *Nature (Lond)* 334:124–129, 1988.

144. Wiedemann, H. R. Tumors and hemihypertrophy associated with Beckwith-Wiedemann syndrome. *Eur. J. Pediatr.* 141:129–132, 1983.

145. Wilkinson, D. G., Bailes, J. A., McMahan, A. P. Expression of the proto-oncogene *int*-1

is restricted to specific neural cells in the developing mouse embryo. *Cell 50*:79–88, 1987.

146. Yamada, M., Kurosawa, H., Nakagawa, Y., Nakagome, Y., Tsuda, T., and Higashi, K. Amplified allele of the human N-myc oncogene in neuroblastomas. *Jpn. J. Cancer Res.* 79:670–673, 1988.

147. Yang, X, and Walter, G. Specific kinase activity and phosphorylation state of pp60[c-src] from neuroblastomas and fibroblasts. *Oncogene 3*:237–244, 1988.

148. Yee, S.-P., Mock, D., Maltby, V., Silver, M., Rossant, J., Bernstein, A., and Pawson, T. Cardiac and neurological abnormalities in v-*fps* transgenic mice. *Proc. Natl. Acad. Sci. U.S.A. 86*:5873–5877, 1989.

149. Yokoyama, T., Tsukahara, T., Nakagawa, C., Kikuchi, T., Minoda, K., and Shimatake, H. The N-*myc* gene product in primary retinoblastomas. *Cancer 63*:2134–2138, 1989.

150. Yunis, J. J., and Ramsay, N. Retinoblastoma and subband deletion of chromosome 13. *Am. J. Dis. Child. 132*:161–163, 1978.

151. Zimmerman, K. A., Yancopoulos, G. D., Collum, R. G., Smith, R. K., Kohl, N. E., Denis, K. A., Nau, M. N., Witte, O. N., Torand-Allerand, D., Gee, C. E., Minna, J. D., and Alt, F. W. Differential expression of *myc* family genes during murine development. *Nature (Lond) 319*:780–783, 1986.

10

Oncogenes and Tumor Suppressor Genes in Solid Tumors: Gastrointestinal Tract

Cecilia M. Fenoglio-Preiser
Teri A. Longacre
Margaret B. Listrom

In the gastrointestinal tract, the greatest insights have been gained by investigation of proto-oncogene abnormalities in colorectal carcinomas and their premalignant precursors. Studies on *ras* gene mutations have been particularly rewarding, as have recent studies on p53. These studies lend insight into the pathogenesis of gastrointestinal cancers, although at this time they appear to be of limited diagnostic or prognostic value. Since work has also been carried out in organs other than colon, we will briefly summarize them as well.

Esophagus

It has been suggested that the *hst*-1 gene is amplified 2–8-fold in 42.1% of 19 esophageal squamous cell carcinomas and in metastatic tumors.[148] Esophageal carcinomas have also been shown to contain overexpressed and amplified c-*erb*B-1 proto-oncogenes. Coamplification of *hst*-1 and c-*erb*B-1 has also been demonstrated,[148] but there was no correlation between *hst*-1 gene amplification and clinical stage, histological type of cancer, or depth of invasion. Attempts to find amplified Ha-*ras*, Ki-*ras*, c-*myc*, v-*sis*, HER-2/*neu*, and v-*erb*A genes have been unsuccessful, as have efforts to find c-*myc* transcripts.[146] More recently an attempt to find *ras* gene mutations in esophageal cancers arising in a Chinese population has failed to detect their presence.

Stomach

NORMAL STOMACH

The *neu* gene product p185*neu* is identifiable in normal gastric epithelium and the lamina propria.[23] In fetuses the HER-2/*neu* product is found in the stomach as well as in the squamous esophageal epithelium.[104]

GASTRIC TUMORS

A number of oncogenes have been reported in gastric tissues, with most studies focusing on *ras* and *myc* genes (Table 10.1). Expression of proteins encoded by the *ras* gene may have prognostic significance (Table 10.2).

ras GENES

Activated mutated N-*ras* and K-*ras* genes have been detected in a transfection assay using DNA prepared directly from gastric adenocarcinomas.[16,112,163] DNA extracted from normal cells fails to contain the activated genes.

Fujita and associates extracted DNA from frozen tissues of 7 patients with advanced gastric cancer and transfected it into NIH 3T3 cells, without demonstrating transformation.[47] They also failed to demonstrate the presence of point mutations at codon 12 by restriction enzyme analysis or by altered electrophoretic mobility on Western blots, a finding recently confirmed by others.[70] Immunohistochemical analysis of *ras* p21 expression in the 11 tumors, using the monoclonal antibody RAP-5, showed more reactivity in the cancers than in the adjacent normal mucosa. The tumors ranged from well-differentiated to poorly differentiated lesions. In all instances, the percentage of cells reactive with the antibody was higher in the carcinoma than in the normal mucosa, although in 10 of 11 cases staining of the normal mucosa was evident in up to 35% of the cells.

Tahara and associates analyzed gastric carcinomas with a monoclonal antibody to the p21 protein and found high levels of c-Ha-*ras* p21 by Western blots.[137] Immunohistochemically, p21 was detected in 11% of early cancers and in 43.8% of advanced carcinomas. In the advanced carcinomas, immunoreactivity correlated with the depth of tumor invasion and was considered to be stronger in metastatic tumors than in primary ones. These authors also reported that c-Ha-*ras* p21 positive tumors were associated with a significantly worse prognosis than seen in p21 negative tumors in patients followed for up to 3 years following gastrectomy.

Czerniak and associates utilized RAP-5 to study p21 expression in cytological specimens.[28] They found that benign epithelial gastric cells were negative in 12 of 13 cases. A weakly positive reaction was only found in a few cell clusters from a smear of a benign gastric ulcer. Twenty smears of gastric carcinoma contained p21 positive cells with the degree of tumor differentiation having no impact on the staining pattern. They postulated that an immunohistochemical assay for the *ras* oncogene product may provide a useful tool for cytodiagnosis of gastric carcinomas. However, Tanaka and associates found staining of the p21 oncoprotein in both normal and neoplastic gastric tissues.[140]

More recently, Czerniak and co-workers found that *ras* p21 was overexpressed in early gastric carcinomas of the diffuse and intestinal types when compared to normal mucosa. In addition, dysplastic and metaplastic changes accompanying the cancers also overexpressed the protein.[29]

Ranzani and associates analyzed c-Ha-*ras* 1 locus polymorphisms in an Italian population with a high risk of developing gastric carcinoma.[119] This locus is characterized by an RFLP resulting from length variations in a VTR region downstream from the structural part of the gene. Thirteen different alleles were detected, but

Table 10.1.
Oncogenes in Gastric Tumors

Gene	Method of Demonstration[a]	Method of Activation[a]	Reference
N-*ras*	Transfection into NIH-3T3	Mutation codon 61	112
N-*ras*	Transfection into NIH-3T3	Mutation codon 13	108
Ha-*ras*	Western blot & IMHC with Ab to H-*ras* from Triton Biosciences	Overexpression	137
ras genes	IMHC with RAP-5	NS	28
c-*myc*	Southern blot	Amplification	110, 129
c-*myc*	Northern blot	Overexpression	129
L-*myc*	RFLP analyses (Southern blot)	NS	129
c-*myc*	ISH for mRNA	NS	69
c-*myc*	IMHC with Ab *myc*1-6E10	NS	5
c-*myc*	IMHC with Ab to c-*myc* protein (IMHC)	NS	69
AKT1	Southern blot	Amplification	135
hst	Transfection into NIH-3T3 cells	NS	125
*erb*B-2	IMHC with Ab to p185[neu]	NS	23
*erb*B-2	Southern blot	Amplification	164
*erb*B-1	Southern blot	Amplification	110
*erb*B-2[a]	Southern blot	Amplification	139

[a]Abbreviations: IMHC = immunohistochemistry; NS = not studied; RFLP = restriction fragment length polymorphism; ISH = *in situ* hybridization; [a]*erb*B-2 = (Her 2/neu).

Table 10.2.
Gastric Oncogenes in Relation to Pathological Features

Characteristic	Reference
Greater percentage of advanced cancers express p21 proteins than less advanced cases.	137
In advanced cancers, p21 expression correlates with depth of invasion.	137
p21-positive tumors are associated with worse patient prognosis than p21-negative tumors.	137
p21 expression in gastric cytology specimens may distinguish malignant cells.	28
RFLPs of L-*myc* gene in advanced cancer.	74
No correlation of p62 expression with tumor differentiation.	5
L-*myc* staining of normal, inflamed, and neoplastic gastric mucosa.	22

there was no evidence that inheritance of any allele predisposes to gastric malignancy.

Yamamoto and associates studied the expression of TGF α and c-Ha-*ras* p21 immunohistochemically in 174 gastric carcinomas comprising 27 early gastric cancers and 147 advanced cancers.[160] TFG α immunoreactivity was present in 25.9% of early carcinomas and 74.8% of advanced lesions. Of 67 p21 positive cases, 88.1% showed synchronous expression of TGF α. Expression of both markers correlated with depth of tumor invasion, presence of metastases, and prognosis.

Figure 10.1. C-*myc* DNA analysis by slot blot hybridization. Genomic DNAs were diluted serially and blotted on nitrocellulose to give 5 (*lane 1*), l.0 (*lane 2*) and 0.2 (*lane 3*) microgram amounts. The gastric carcinoma has DNA amplified for c-*myc*. Gastric mucosa adjacent to the carcinoma is not amplified.

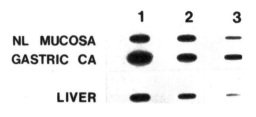

myc GENES

Shibuya and associates found that 3 of 16 human gastric adenocarcinomas maintained as solid tumors in nude mice carried amplified c-*myc* genes.[129] Two samples with a high level of c-*myc* DNA amplification (5–30-fold) contained DM chromosomes on karyotypic analysis. c-*myc* RNA was markedly elevated in a rapidly growing and poorly differentiated tumor, whereas it was only slightly elevated in a more slowly growing and better differentiated tumor. Others have noted the presence of amplified c-*myc* genes in 2–7% of gastric carcinomas (Fig. 10.1).[80,110] One tumor containing the amplified c-*myc* gene was an adenosquamous carcinoma. Variable overexpression of the c-*myc* gene occurs in about 50% of gastric cancers.[146]

Kawashima and associates demonstrated that RFLPs of the L-*myc* gene relate to the progression of human gastric carcinomas.[74] Primary tumors, as well as normal control tissues, were obtained immediately following surgery, DNA was isolated, and Southern blot analysis of the L-*myc* gene performed. All of the gastric cancers were in an advanced stage at the time of surgery. The authors found it noteworthy that almost all of the patients had either the S-S or L-S phenotype (Fig. 10.2). In normal individuals, the relative ratio of the L-L, L-S, and S-S phenotype was approximately 30%, 40%, and 30% respectively.[74]

Izumi and associates demonstrated the presence of c-*myc* protein within the nucleus of gastric carcinoma cells, as well as the presence of c-*myc* mRNA by *in situ*

Figure 10.2. Diagram of Southern blot analysis of L-*myc* RFLP. Two alleles of L-*myc* yield 6 kb and 10 kb fragments in EcoR1 digests.

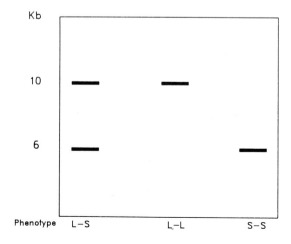

hybridization.[69] Depending on the fixative used, the c-*myc* protein could also be seen in the cytoplasm.

Allum and associates evaluated the presence of p62[c-*myc*] in a series of archival specimens of gastric cancer, and found no correlation with the degree of tumor cell differentiation.[5] Less than 40% of the tumors were positive, but there was a tendency for intestinal-type tumors[86] to stain more frequently than the diffuse type. The surface epithelium of the nonneoplastic stomach also had a tendency to stain, with staining occurring more commonly in patients with gastritis than in patients with normal mucosa. Staining was particularly marked in atrophic gastric mucosa showing intestinal metaplasia type 2b. No difference was observed between active and quiescent chronic superficial gastritis. Others also found low levels of expression in normal gastric mucosa, and increased expression was found in inflamed, metaplastic, and dysplastic epithelium.[22]

AKT1

The directly transforming retrovirus AKT8 was originally isolated from a spontaneous thymoma occurring in an AKR mouse. The viral genome was shown to contain both viral and nonviral cell-related sequences. The nonviral sequences are designated v-*akt*, the presumed viral oncogene of the AKT8 virus. This gene lacks homology with other known oncogenes. Two human homologues of the v-*akt* oncogene (AKT1 and AKT2) have been cloned. AKT1 was found to be amplified as much as 20-fold in 1 of 5 gastric adenocarcinomas detected in an effort to survey several hundred human tumors for changes involving this gene.[135]

HST

Sakamoto and associates analyzed DNA from 21 primary human gastric carcinomas, 16 metastatic gastric carcinomas in lymph nodes, and 21 apparently normal examples of gastric mucosa for their ability to induce neoplastic transformation of NIH 3T3 cells.[125] Three samples were shown to have transforming activity: one from a primary gastric carcinoma of one patient, the second from a noncancerous portion of the mucosa of the same patient, and a third from a lymph node metastasis in another patient. Transformants were tumorigenic in nude mice, and their DNA could induce secondary transformants. The transforming gene was cloned. Since it did not share homology with known transforming sequences, it was termed "*hst*" for human stomach tumor.

The authors postulated that the ability of the nonneoplastic mucosa to transform the NIH 3T3 cells could be explained in one of several ways. The first possibility was that the transforming gene was activated in the nonneoplastic gastric mucosa, but that other genetic changes are required to develop tumors. It is well established that carcinogenesis occurs as a series of multiple steps and that at least two cooperating oncogenes are required to convert a normal cell to a malignant cell.[84] Another possibility was that the sequence had acquired transforming potential by rearrangement or mutation at the time of transfection and that the changes in the other two DNAs may also have been the result of sequence changes resulting from the transfection process, since it has been shown that transfected genes can become rearranged. The final possibility was that the transforming sequences may be regulatory sequences that enhance expression of a transforming gene in NIH 3T3 cells.[125]

OTHER ONCOGENES

Other genes that are often amplified include c-erbB-1,[110] HER-2/neu (c-erbB-2),[48,162,163,164,166] c-erbA,[5,165] and raf.[130] Enhanced expression of the oncogene-encoded EGFR correlates with the depth of tumor invasion and the stage.[51] In contrast, amplification of mos, fos, and abl genes has been searched for but not found in gastric tumors.[80,110]

Kitagami and associates analyzed the expression of several oncogene-encoded proteins, utilizing immunohistochemical techniques.[79] They found rasp21, fes85, and EGFR in 65%, 65% and 40% respectively of gastric carcinomas. Multiple oncogene-encoded proteins were found in 50% of the tumors.

Intestinal Normal Tissues

c-myc, Ki-ras[106] and c-erbB2 (HER-2/neu)[104] transcripts can be identified during different stages of gastrointestinal development, with tissues showing varying expression levels. Gut epithelium is the most intense site of labeling for c-myc by *in situ* hybridization, whereas developing lymphoid follicles are devoid of transcripts. In rodents, c-myc expression is highest in the prenatal and newborn periods and decreases by six months of age. With further increases in age, a progressive pattern of increasing expression occurs in the small intestine. N-myc and L-myc expression is also highest in the prenatal and neonatal period, but there appears to be no further increases with age.

The adult colon also demonstrates c-myc expression.[22,69,72,73,136,155] When the distribution of the c-myc–encoded p62 is examined immunohistologically, its localization varies slightly according to the investigator: some investigators find that its distribution is maximal in the midzone of the crypts,[136] others find it predominantly at the free surface,[22] in both the midzone and surface but not in the basal zones,[72] or the same throughout the crypt.[73]

Recent studies demonstrate the presence of p185[neu] protein in the normal adult small intestine and colon. In the small intestine, a gradient of immunoreactivity exists, extending from the base of the crypt (which is negative) to the tip of the villus (which is strongly positive). In contrast, immunoreactivity is the same throughout the crypt length in the colon, and no gradient is seen.[23]

junD mRNA expression levels are very high in the murine intestine, and junB is also significantly expressed.[58]

c-src is present in the developing smooth muscle of the gut, a localization not surprising considering its hypothetical role in gap junction regulation.

Intestinal Tumors

Numerous oncogenes have been examined in colorectal cancers, with investigators either analyzing alterations of single oncogenes or investigating the expression and/or amplification of multiple oncogenes within a series of tumors. The following comprises a review of some of these findings.

ras GENES

One of the most extensively studied group of oncogenes has been the *ras* gene family. SW480 was the first colon carcinoma cell line shown to contain an activated transforming gene by transfection into NIH 3T3 cells;[107] the oncogene was subsequently indentified as c-Ki-*ras*-2.[93,94] Following this discovery, DNAs isolated from other cell lines and fresh tumors were found to transform NIH 3T3 cells in transfection assays[117] (Table 10.3). The majority of human colon carcinoma cell lines with an activated c-Ki-*ras* gene have mutations at amino acid positions 12, 13, 59, 61, and 113[15,20,27,45,70,93,94,161,169] (Table 10.3). Amplification of *ras* genes does not appear to play a role in the genesis of colon cancer (Fig. 10.3). Activated H-*ras* genes have also been described.[54]

An activated N-*ras* oncogene was demonstrated in a transformant derived from a rat small intestinal adenocarcinoma induced by 2-aminodipyrido[1,2-a:3',2'-D]imidazole, a compound present in fish. The activated N-*ras* oncogene

Table 10.3.
Studies on ras Genes and Colorectal Carcinoma

Study Subject	Reference
Ras expression highest in better differentiated tumors; expression decreases in metastases	8
Ki-*ras* mutated at codon 12 in >33% of colorectal cancers monitored by histologic examination and studied by oligomer hybridization *ras* mutation at 11/27 pts. 10/11 mutations were Ki-*ras* (codon 12); 1/11 N-*ras;* 9/10 mutations Ki-*ras* codon 12; 1/10 codon 61	15
Mutated *ras,* codon 12 colorectal line SW 480	20
DNA from SW 480 transforms NIH 3T3 cells	27
SW 480 and SK co-1 transform NIH 3T3 cells and contain mutated p2	30
Hypomethylation of c-Ha-*ras*-1 in 6/8 1° human cancer when compared to adjacent normal mucosa c-Ki-*ras* hypomethylated to a lesser extent	43
High incidence of mutated K-*ras* genes in colon cancer	45
Early passage cell line from differentiated cancer Ki-*ras* point mutation codon 12	54
Increased Ki and Ha-*ras* by quantitation liquid competition RIA	59
3/15 carcinomas have K-*ras* (codon 12) aspartic acid mutations	70
DNA from SW 480 transforms NIH 3T3 (it is identified as Ki-*ras*-2)	93, 94
DNA from adenocarcinoma cell lines contained NIH 3T3 cells	94
Ki-*ras*-2 isolated from mutated codon 12 with valine rather than glycine SW 4480	95
Mutated codon 13 N-*ras* in 1/35 rectal cancers	109
Increased Ki and Ha-*ras* in cancer and adenomas by blot hybridization	114
Mutated *ras,* codon 12 colorectal line SK 001	115
Mutated N-*ras,* codon 61, rectal cancer line SW 127	168
Mutated *ras,* codon 12 colorectal line KMS-4, G→T substitution	169

Figure 10.3. Southern analysis of colon carcinoma DNAs (*lanes 2 and 3*) with ³²P-labeled N-*ras* probe. Hybridization with an 8.8 kb EcoR1 fragment is present with both tumor samples, but there is no amplification. *Lane 1* represents control liver DNA.

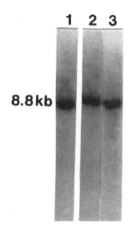

contained a G→T substitution in codon 12. The original tumor did not hybridize with the oligonucleotide probe representing the mutated allele, but did hybridize with one representing the normal allele, leading the authors to conclude that either activation of the N-*ras* oncogene had occurred during transfection or that an activated N-*ras* oncogene had been present in only a minor population of cells in the original tumor.[68]

Bos and co-workers studied 27 patients and found that 11 had tumors that contained mutated *ras* genes.[15] In 10 of 11 they were in the Ki-*ras* gene, whereas in one it was in the N-*ras* gene. The N-*ras* gene was mutated at codon 12; the Ki-*ras* gene most often was mutated at codon 12, but in one patient the mutation was at codon 61. The tumors containing *ras* mutations were located throughout the colon and rectum, were all Dukes' stages, and occurred in patients of all ages and both sexes. Amino acids present in the mutated codons included replacement of glycine (codon GGT) with aspartic acid (codon GAT), valine (codon GTT), serine (codon AGT), and cysteine (codon TGT). These authors also demonstrated that the mutated gene was present in a cancer, but not in the villous adenoma from which it arose. This prompted them to explore the association of *ras* gene activation in the adenoma-carcinoma sequence. They found that in one patient, DNA isolated from normal colonic mucosa hybridized to the normal glycine-specific oligomer but not to either serine-specific or valine-specific oligomers. In contrast, DNA isolated from fractionated cryostat sections of carcinoma or adenoma hybridized to both the glycine-specific oligomer and the serine-specific oligomer. The glycine-specific oligomer hybridized strongly to DNA prepared from the adenoma but only weakly to DNA prepared from the carcinoma. Because the DNA from both the adenoma and carcinoma was isolated from tumor regions that contained greater than 90% neoplastic cells, these results suggested to the authors that both the normal (glycine) and mutated (serine) c-Ki-*ras* alleles were present in the adenoma, and that many of the carcinoma cells had lost the normal allele. The same mutation was found in both benign and malignant regions of the tumor in 5 of 6 cancers with *ras* mutations, suggesting that the mutation preceded the development of malignancy. It is postulated that mutations at position 12 of the c-Ki-*ras* gene are more likely to occur in colon cancer than other mutations or that they impart a selective growth advantage when they do occur.

Because the adenomas studied represented a selected group, with only adenomas containing carcinoma being evaluated, the possibilities exist that either the *ras* mutations occur frequently in adenomas, but only a small fraction of those adenomas with a mutation progress to malignancy, or that *ras* mutations occur infrequently in adenomas, but those that contain them have a high probability of progression to malignancy.[15,42]

Burmer and colleagues analyzed a population of colon cancers by a combination of cell sorting, histological enrichment, polymerase chain reaction, and direct sequencing of the Ki-*ras* gene.[17] They found that mutations in codon 12 were present in both diploid and aneuploid tumor subpopulations, suggesting that these mutations precede the development of aneuploidy in tumor progression.

Overall, *ras* mutations occur in 40–50% of malignant colorectal tumors, with the incidence being higher in adenomas larger than one centimeter (58%) than in smaller adenomas (9%).[150] Furthermore, the presence of hypomethylated *ras* genes may be important in the genesis of the cancer.[43]

Spandios and associates found elevated transcripts of both c-Ki-*ras* and c-Ha-*ras* in premalignant and malignant tumors as compared to normal mucosa,[124] but could not document a direct relationship between *ras* expression levels and clinical-pathological features.[78]

As noted in the beginning of this chapter, the use of antibodies directed against *ras* p21 has generated conflicting data, and almost certainly some studies are not detecting the p21 protein.[4,28,49,77,99,100,142] Nonetheless, these antibodies, particularly RAP-5, have been extensively applied to colonic tissues; for what they are worth these studies are summarized in Table 10.4.

In considering investigations concerning *ras* genes as a whole, but ignoring the studies utilizing RAP-5, it is most likely that mutated *ras* genes arise early in the progression of colon cancer (although not necessarily as the first step, since they are uncommonly present in small adenomas) and that their presence does not correlate with the stage of disease once a cancer develops. Furthermore, their expression becomes progressively more heterogeneous as the grade and stage of the cancer increases and metastases occur. This heterogeneity can be explained by the instability of the colon cancer cell genome and the subsequent progressive accumulation of allelic deletions.[7,150,151]

myc GENES

Two human neuroendocrine tumor cell lines derived from a colonic carcinoma with either numerous DM chromosomes or HSRs (COLO 320 DM and COLO 320 HSR) contain amplified and enhanced expression of c-*myc*.[2] The HSRs are on both arms of what was once an X chromosome, and amplified c-*myc* copies are present in isolated DMs of the COLO 320 DM cells, suggesting that the amplified c-*myc* appeared first as DMs and was subsequently transposed as HSRs on an X chromosome.[88] Exon 1 and most of intron 1 have been displaced from DM*myc* by the genetic rearrangement. The incidence of amplified genes differs, depending

Table 10.4.
p21 Expression in Relationship to Prognosis

Antibody	Relationship	Reference
RAP-5	Not reliable marker for staining colonic epithelium or for malignancy	4
RAP-5	100% cancer smears positive; weakly positive cells in smears from cancer patients; degree of tumor differentiation has no impact on staining pattern	28
Y13 259	No correlation with stage or grade; elevated p21 common early, but autonomous population subsequently arises	49
Y13 259	Stained normal adult and fetal epithelium and premalignant and malignant colorectal epithelium	77, 78
RAP-5	p21 expression occurs in early stage of colon carcinogenesis, but not related to metastatic tumor progression; increase with degree of dysplasia and size of adenoma	85
RAP-5	High titers p21 worse 5-year survival; higher incidence of metastases; more advanced Duke's stage than lower titer; marker of aggressive tumors	99
RAP-5	Staining of plasma cells, histiocytes, fibroblasts, smooth muscle cells, and endothelium	100
RAP-5	No correlation with stage	123
RAP-1-5	p21 expression correlates with malignancy and depth of invasion in bowel wall	142
Y13 259	No correlation with malignancy; often found in adenomas; tumors stained with same intensity and number	157

on the underlying lesion (Fig. 10.4; Table 10.5) and the location in the bowel (Table 10.6).

Wahl proposed that episomes (submicroscopic, acentric, circular extrachromosomal DNA molecules that replicate autonomously) are the initial intermediates in the process of gene amplification. This was determined by examining the two colon cell lines HL-60 and COLO 320 that contain episomes harboring the amplified *myc* genes.[152]

Normal colonic cell cultures demonstrate a dramatic increase in the number of c-*myc* transcripts when growth is reinitiated by seeding confluent cultures at reduced density. The number of transcripts then returns to basal levels before the second cell division, suggesting that c-*myc* induction accompanies cellular release from G_0 and reentry into the cell cycle. However, c-*myc* expression in cells from colon

Figure 10.4. c-*myc* DNA analysis of two colon carcinomas with paired adjacent normal mucosa. Five micrograms (*lane 1*) and one microgram (*lane 2*) of genomic DNA were blotted on nitrocellulose and hybridized overnight. One of the colon carcinomas is amplified more than tenfold for the c-*myc* gene.

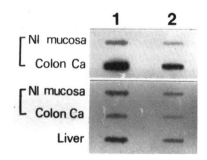

Table 10.5.
Amplified *myc* Genes

Lesion (N)	c-*myc* (%)	N-*myc* (%)	Both c-*myc* and N-*myc* (%)
Normal (16)	63	44	44
Cancer (16)	63	56	44
Inflamed (3)	66	33	33
Adenomas (3)	66	66	66
Transitional (1)	0	0	0
Hyperplastic (1)	0	0	0

Table 10.6.
Amplified Genes

Location	c-*myc* (%)	N-*myc* (%)
Left colon	40	20
Right colon	66	83

carcinoma differs. Levels of c-*myc* transcripts in confluent cells are high and essentially equivalent to those seen in normal cells during the transition from G_0 to G_1. The elevated levels do not fall significantly during subsequent growth, even after multiple population doublings, and do not correlate with increased proliferation rates.[8]

Increased c-*myc* expression parallels increased expression of two G_1-specific genes (p2A9 and ornithine decarboxylase) and an S-phase–specific gene (histone H3), suggesting that elevated levels of cell cycle–specific RNA in a tumor may not indicate overexpression of that gene, but simply reflect an increased fraction of cycling cells unless the ratio of expression between G_1 and G_1-S–phase is altered. Support for this concept is the finding that differentiation-inducing agents decrease both *myc* expression and cell proliferation; in differentiated, quiescent HT-29 cells no change occurs in *myc* expression.[8]

It has been noted that elevated expression of the c-*myc* gene occurs more frequently in tumors of the left side (rectum, sigmoid, and descending colon) than in tumors of the right side (cecum and ascending colon).[124] These findings suggest that elevated expression of the *myc* gene might be a marker for sporadic forms of colon cancer arising from adenomas, since these tend to occur on the left side of the colon. In contrast, tumors with low *myc* expression would represent the sporadic counterpart of hereditary nonpolyposis (HNP) colorectal cancer. *myc* expression may be an essential part of the neoplastic phenotype.

Recently, high levels of c-*myc* transcripts were found in 73% of colorectal carcinomas, colorectal adenomas, and metastatic colonic tumors. In adenomas, transcript levels correlate with the size and the presence of carcinoma.[66]

Elevated c-*myc* expression can occur without amplification,[33] suggesting that there may be a change in the mechanisms involved in normal gene regulation. To test this hypothesis, an osteosarcoma cell line with normal c-*myc* regulation was fused with two colon carcinoma cell lines manifesting deregulated c-*myc* expression.[34,35] The results suggested that loss of function of a *trans*-acting regulator is responsible for deregulation of c-*myc* expression in a major fraction of colorectal carcinomas. This deregulation correlates with the frequent loss of alleles on chromosome 5q. (Chromosome 5q is the region known to contain the gene for familial adenomatous polyposis (see below).) These findings, together with the earlier finding that the colonic distribution of tumors exhibiting deregulated c-*myc* expression is similar to that reported for familial polyposis, provide evidence that lost function of the familial adenomatous polyposis gene may be involved in the subset of colorectal cancers with deregulated c-*myc* expression.[35] Expression of this gene may also be modulated by growth factors such as TGF-β.

A recent study examining expression of c-*myc*, N-*myc*, and L-*myc* confirmed earlier studies that c-*myc* is overexpressed in about two-thirds of carcinomas and adenomatous tumors and that amplified levels don't correlate with expression levels. N-*myc* and L-*myc*–specific transcripts could be detected in normal mucosa and were frequently overexpressed in adenomas and carcinomas. Adenomas more frequently overexpressed L-*myc* than did carcinomas.[44]

The monoclonal antibody *myc*-1 6E10 has been raised against a synthetic peptide consisting of residues 171 to 188 of the human p62^{c-myc} sequence. This antibody was used to determine the distribution of the c-*myc* oncogene in normal colonic mucosa, colonic polyps, and carcinomas.[22,30,72,136,155] The results of these studies are summarized in Table 10.7. Almost all authors have found that poorly differentiated carcinomas exhibit lower p62^{c-myc} levels than do moderately and well-differentiated tumors.[30,136,155] The close correlation between c-*myc* levels and RNA transcript copy number validates the use of *myc*-1 6E10 to assess c-*myc* gene activity in clinical samples.

The gene product p62^{c-myc} is detectable both by immunoblotting and immunohistology, with a close correlation existing between c-*myc* mRNA copy number and p62^{c-myc} abundance. Evidence exists for progression of abnormal c-*myc* activity as one goes from normal colonic tissue to adenomas to carcinoma.[136] Two well-differentiated tumors contained high transcript levels and protein products, whereas poorly differentiated tumors have lower levels. The level of c-*myc* RNA in lymph node and hepatic metastases corresponds with that of the primary lesions, indicating that the c-*myc* RNA levels do not correlate with the degree of malignancy or metastatic potential.[146] The increased c-*myc* gene product may be due to an increase in active mRNA, secondary to enhanced transcription or increased message instability.[132]

The levels of the c-*myc* protein are consistent with mRNA levels in normal cells as well as in colon cancer. Growing populations of fibroblastic and epithelial cell lines derived from normal mucosa show small numbers of steady-state transcripts and weak nuclear immunofluorescent signals. Adenocarcinomas express 5–10-fold elevated levels of c-*myc* mRNA and a more intense fluorescent nuclear signal. Elevated expression of the c-*myc* genes at both the mRNA and protein levels occurs constitutively in colorectal carcinoma cell lines during their growth in culture, in contrast to the transiently elevated levels of expression in normal cells stimulated by a mitogenic signal.[36]

Some authors have demonstrated that L-*myc* RFLPs relate to colon cancer progression,[75] but this is controversial.[65] As noted earlier, the population is homozygous with respect to the 10 kb L-*myc* band (L-L), heterozygous with respect

Table 10.7.
myc **Oncogenes and Colon**

Study Results	Reference
HSR of COLO 320 contains amplified c-*myc*	1 2
c-*myc* expression parallels proliferative activity	18
myc-1 - 6E10 Ab showed increased expression in inflamed and dysplastic colon; correlates with degree of positivity and degree of infiltration	22
c-*myc* overexpressed in 72% of carcinomas; no amplification or rearrangement; no relationship to severe prognosis	33, 34, 35
c-*myc* overexpressed in many cancer colon cell lines; increased expression without amplification in 50%	34
Lack of correlation of L-*myc* RFLP and malignancy of colorectal cancer	65
myc-1 6E10 stained normal epithelial and terminally differentiated surface cells; 100% cancers positive, with intensity independent of differentiation; Dukes' stage ploidy independent of survival	72
Correlation of L-*myc* RFLP and colon cancer	75
Elevated c-*myc* expression seen preferentially in left colon (also Fenoglio-Preiser unpublished observation)	124
c-*myc* overexpression with amplification 12/15	132
c-*myc* oncogene product found in adenomas	136
myc-1 6E10 stained colon adenomas and carcinomas	136
Staining of normal crypts; adenomas stain intensely in dysplastic areas; cancer and undifferentiated tumors more intensely with $p62^{c-myc}$ than poorly differentiated	136
c-*myc* overexpressed in most colon cancers	146
myc-1 6E10 stains colon and adenoma nuclei; nuclei $p62^{c-myc}$ levels increase progressively from normal mucosa to adenomas and carcinomas	155
c-*myc* overexpression without amplification in colon cell line C_1	167

to the 6 kb L-*myc* band (L-S), or homozygous for the small 6 kb band (S-S) (Fig. 10.2). Patients with colon cancers have a larger proportion of L-L RFLP types than would be expected in the general population.[75]

It is conceivable that c-*myc* gene overexpression may increase the cell populations that evade terminal differentiation, allowing them to become susceptible to the secondary events, or it may bring transformed cells to a clinical stage by promoting the propagation of malignant cells. Table 10.7 summarizes studies relating to c-*myc* expression and colon cancer.

myb

The *myb* gene is both overexpressed and amplified in the colon neuroendocrine tumor cell line, COLO 320,[3] and in two cell lines, COLO 201 and 205, derived from a single colon tumor[3] (Table 10.8). In COLO 320, the amplified gene is present in either an HSR or a DM.

c-*myb* expression differs in the normal versus the neoplastic mucosa of patients with colonic cancer.[144] C-*myb* mRNA levels are low in normal mucosa, but are increased in 60% of tumors[144] (Table 10.8). In most instances, mRNA levels of c-*myc*, histone H3, and ornithine decarboxylase are also increased, suggesting that a relationship exists between elevated expression of c-*myb* and the fraction of cycling

Table 10.8.
Colon Cancer Oncogenes, Not Including *ras* and *myc*

Study Result	Reference
Overexpression p53 skin fibroblasts FAP[a] patients	81
p53 expressed in colon cancer and adenomas	149
COLO 201 and COLO 205—amplified c-*myb*, overexpression c-*myb*	3
Overexpression of c-*myc*, c-*myb*, histone protein ODC	144
pp60[c src] kinase activity elevated in lysate of colon is increased in normal colon, which does not appear to be related to overexpression	
pp60[c-src] activity regulated differently in normal and cancer cells	31
2/45 Her-2/*neu* gene amplified with gene rearrangement and expression of abnormal protein that autophosphorylates	139

[a]Abbreviation: FAP = familial adenomatous polyposis.

neoplastic cells. More interesting is the allelic deletion of the c-*myb* oncogene, which has been described in a single adenoma. This abnormality was not present in the surrounding normal mucosa nor in the carcinoma that arose from the adenoma.

src-RELATED GENES

The protein kinase activity of pp60[c-src] from colon cancer cells is generally higher than that present in normal colonic mucosa. The increased activity is due to both the presence of an increased amount of the protein in cancer cells and increased specific activity of the kinase.[14,21,31] All carcinoma cell lines with elevated pp60[c-src] *in vitro* kinase activity show increased phosphorylation on the tyrosine residues, suggesting the presence of activated protein kinase(s).[21] Analysis of the pp60[c-src] molecules from normal colon and from colon cancer cells reveals that they possess indistinguishable sites and quantities of phosphorylated serine and tyrosine residues, and both are able to complex with cellular proteins. The activation of the pp60[c-src] kinase activity in colonic tumor cells is associated with hypophosphorylation of the carboxy-terminal tyrosine residue in pp60[c-src] molecules from these tumor-derived cell lines. Once tumor cells are induced to differentiate, both pp[60src] and p56[lck] levels decrease, as do *src* and *lck* mRNA levels.

trk

One of the two colon carcinomas that Pulciani and co-workers found to be capable of transforming NIH 3T3 cells[116,117] and to contain an activated Ki-*ras* gene was also found to harbor a unique oncogene *onc*-D, now known as *trk* (tropomyosin receptor kinase).[92] This transforming sequence encodes a 70 kilodalton peptide measuring 641 amino acids, the amino terminus (1–221) of which is derived from the first seven exons of a nonmuscle tropomyosin gene, and the balance from a previously unrecognized gene with marked homology to the tyrosine kinase domains found in retroviral transforming genes.[92,101] This novel fusion product also contains a hydrophobic region characteristic of a transmembrane receptor.

The protein itself is phosphorylated on serine, threonine, and tyrosine residues. Clones of the gene which do not contain the 5' untranslated regions show diminished transforming activity.[92] This region contains a nine nucleotide sequence of unknown function, which is repeated 12 times; this recalls a similar deletion of a repeated flanking VNTR region of a mutated *ras*, which decreases *ras* transforming activity.[20]

It has been suggested that the kinase portion derives from a receptor that acts intracellularly through kinase activity in transducing a signal from an unidentified growth factor that binds to its extracellular region and that is deleted in the oncogene. Fusion to the nonmuscle tropomyosin may alter the activity, specificity, or intracellular location of the kinase, thereby imparting transforming activity.

EXPRESSION OF MULTIPLE ONCOGENES

Numerous authors have examined colorectal tumors with a battery of oncogene-related probes. The results of these studies are summarized in Table 10.9. Amplification of c-*myc* or c-*erb*-2 (HER-2/*neu*; Fig. 10.5) and allelic deletion of c-Ha-*ras* or c-*myb* are the most frequent abnormalities encountered.[98] The presence of altered oncogenes generally does not correlate with Dukes' stage, tumor progression, or patient survival following resection.[98] One study indicated that (*a*) 50% of cases showed an increase in the expression of at least one of the oncogenes studied; (*b*) overexpression is not random, with some cases overexpressing several genes; (*c*) the expression pattern of the oncogene studied varies between primary tumor and metastases; (*d*) amplification is a rare event; (*e*) tumors with high levels of mRNA in one or more oncogenes are biologically aggressive tumors; and (*f*) nonexpressors are at higher risk for local recurrence.[98]

Kitagami and associates assessed expression of a series of proto-oncogene–

Table 10.9.
Expression of Multiple Oncogenes in Colon Cancer

Author	c-*myc*	N-*myc*	c-*myb*	K-*ras*	H-*ras*	c-*fos*	c-*erbB2*	c-*fes*	N-*raf*	c-*mos*	c-*erbB1*	c-*abl*	c-*fms*	c-*sis*	p53
Dolcetti (32)	A[a] 1/44	NA NR	NA NR	NA	NS	NS	NS	NS	NS	NS	NS	NS	NS	NS	NS
Meltzer (98)	A 3/45	NS	D 4/45	NA	AD 2/45	NA	A 1/45	NA	NA	NA	NA	NA	NA	NA	NA
Trainer (145) (19 cell lines)	E	NS	E	NS	E	E	NS	E	NS	E	NS	NS	NS	NS	NS
Monnat (102)	OE 7/39 A 1/39	NS	NS	NS	OE 3/39	OE 7/39	NS	OE 9/39	NS	NS	NS	NS	NS	NS	NS
Alexander (1)	A 2/9 R 1/9	NA	NS	NA	RA	NA	NS	NA	NS	NS	NS	NS	NS	NS	NS
Imaeki (66)	E 8/11	NS	NS	NT	NT	E	NS	OE	NS	NS	NS	NS	NS	NS	NT

[a]Abbreviations: A = amplified; E = expressed; D = deleted; NA = not amplified; NR = not rearranged; NS = not studied; NT = no transcripts; OE = overexpressed; R = rearranged; RA = rare alleles.

Figure 10.5. Northern blot analysis of Her-2/ *neu* expression in nonneoplastic colorectal mucosal epithelium (*lane 2*) and adenoma (*lane 3*). Expression of the Her-2/*neu* gene in benign breast (*lane 1*), normal liver (*lane 4*), and kidney (*lane 5*) are shown for comparison. (Hybridizations performed by Rick H. Hildebrandt in the Solid Tumor Laboratory directed by Dr. Sue A. Bartow at the University of New Mexico School of Medicine.)

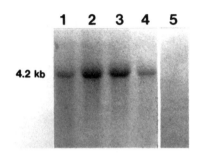

encoded products including ras p21, fes p85, and EGFR in 29 colorectal tumors.[79] *ras* p21, *fes* p85, and EGFR were found in 48%, 62%, and 62% of cases respectively. More than one oncogene was demonstrated in 18 of 29 cases.

ENDOGENOUS RETROVIRAL GENES

Human counterparts of endogenous retroviral genomes have been found in SW 116 colon carcinoma cells. These transcripts contain either LTR sequences or envelope sequences, and the former may decrease in abundance and then later increase in colon cancer.[51]

TUMOR SUPPRESSOR GENES

Lothe and associates analyzed the DNA pattern in constitutional versus tumor and polyp tissue.[89] More than half of the markers tested were of VNTR types, giving the patient panel a higher informational content, since the frequency of individuals heterozygous for a particular marker increases. Loss of alleles was demonstrated in 40% of tumors from constitutionally heterozygous patients at chromosome 17p loci. Similar losses were detected on other autosomes, but at a significantly lower frequency. These results suggest that hemi/homozygosity of 17p alleles plays a role in the development of a major subset of colorectal carcinomas.

More recently, it has been shown that deletions of specific chromosomal arms (the short arm of chromosome 17 and the long arm of 18) each provide independent information by multivariate analysis when considered individually with the Dukes' classification. Distant metastases are significantly associated with a high frequency of allelic loss and deletion of 17p and 18q. 17p deletions occur as late events associated with the transition of adenoma to carcinoma.[150,151]

ras mutations and deletions of 5q had no prognostic importance. Statistically significant associations were also found between allelic losses and family history of cancer, left-sided location, and absence of extracellular mucin.[76] Other chromosomes that demonstrate constitutional loss of heterozygosity include chromosomes 14 and 22.

Sequences related to the familial adenomatous polyposis gene are not lost in adenomas in polyposis patients but are lost in 29–35% of sporadic adenomas and carcinomas.[150] A specific region of chromosome 18 was deleted in 73% of cancers, in 47% of advanced adenomas, and only occasionally (11–13%) in early adenomas. Chromosome 17p sequences were only lost in carcinomas, where it occurred in

75% of cases. These molecular alterations accumulate in a fashion that parallels tumor progression[150] and are consistent with a model of colorectal carcinoma in which the steps required for development of cancer involve mutational activation of an oncogene, coupled with loss of genes that normally suppress tumorigenesis.[150] The region deleted on chromosome 17p is contained within bands 17p12 to 17p13.3.[9] This region contains the gene for p53.

p53

Kopelovich and DeLeo studied the expression of p53 antigen in cultured skin fibroblasts from a patient with hereditary colon carcinoma and found overexpression with respect to normal fibroblasts derived from nonpolyposis patients.[81] These investigators postulated that p53 overexpression represents an early event, possibly associated with the initiation and promotion of inherited colon cancer.[81] Others have shown that p53 is detectable in 55% of colon cancers and in 8% of adenomas. In general, p53-positive regions of polyps are histologically indistinguishable from neighboring areas (Fig. 10.6). Occasionally the p53-positive nuclei are found in foci of highly dysplastic epithelium, suggesting that p53 expression may be part of the process of neoplastic progression.

In our hands, p53 was positive in 50% of carcinomas with 50–100% of cells staining, depending on the tumor. The majority of these tumors were moderately differentiated. Two colloid carcinomas had only 5–10% of the cells positive for the p53-encoded protein. In the one poorly differentiated tumor stained with the antibody, 100% of the cells were positive. Adenomas and transitional mucosae were negative. All of the benign tissues were also negative (unpublished observations).

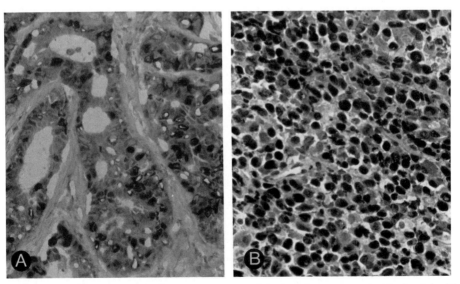

Figure 10.6.   Demonstration of p53 oncogene activity in colon cancer. **A,** Moderately differentiated carcinoma. **B,** Poorly differentiated cancer, with nearly 100% of the cells positive.

Southern and Northern blots have not demonstrated any gross alterations of the p53 gene or its surrounding sequences.[9] Two carcinomas that were subjected to detailed investigation demonstrated the loss of one 17p allele, with considerable expression of the p53 mRNA from the remaining allele. The p53 allele was mutated, with alanine substituting for valine at codon 143 in one tumor and histidine substituting for arginine at codon 175 in the other. Both mutations occurred in a highly conserved region of the p53 gene. These data suggest that mutated p53 genes may be involved in colorectal neoplasia, perhaps through the inactivation of a tumor suppressor gene.[9]

FAMILIAL ADENOMATOUS POLYPOSIS (FAP) GENE

The familial adenomatous polyposis (FAP) gene located on chromosome 5 is thought to play a role in colon carcinogenesis.[134] Initially, a single patient was observed with deletion of part of the long arm of chromosome 5 (5q). This patient had various developmental disorders including FAP.[56] Linkage analysis for RFLPs involving the long arm of chromosome 5 of patients with colon cancer and their relatives have confirmed that the FAP gene maps to 5q21 or 5q22.[13,87] Loss of heterozygosity at this locus occurs in about 20% of patients with sporadic colorectal cancer.[134]

Liver

NORMAL LIVER

*erb*A, *erb*B, Ha-*ras*, *myc*, *fos*, and *fms* expression is elevated in certain stages of fetal liver development.[153,170] In contrast, the oncogenes *rel*, *src*, *mos*, *sis*, *myb*, Ki-*ras*, and *bas* show no apparent change in their mRNA levels during embryogenesis. In rodents, c-myc expression is highest in the prenatal and newborn period and then decreases to its lowest level at age six months. With further increases in age, progressive expression of c-*myc* occurs.[121,127] In humans, peak oncogene expression occurs between four and six months of gestation.[170] Fetal liver is the organ with the largest number of c-*myc* transcripts, as detected by *in situ* hybridization. The hepatocytes label intensely, whereas islands of extramedullary hematopoiesis are negative. N-*myc* and L-*myc* are not identifiable.[127] The adult liver also expresses *neu*,[23] c-*jun*, *jun*D, *jun*B,[58] c-*fms*,[128] *rab*-1, *rab*-2,[113] and c-*rel*.[103]

The *lca* transforming DNA, which was isolated from human hepatocellular carcinoma and which has no homology with known oncogenes,[111] is expressed in fetal liver for a limited period of time (from 19 weeks through 24 weeks of gestation). No other tissues carry detectable levels of *lca* mRNA.[131]

LIVER REGENERATION

Sequential proto-oncogene expression occurs during hepatic regeneration. When growth is stimulated by partial hepatectomy, steady state levels of c-*fos*, c-*myc*, p53, and *ras* mRNAs increase sequentially during the prereplicative phase that is associated with entry into and progression through the cell cycle in the period that precedes DNA synthesis.[38–41,52,53,91,141] Levels of c-*fos* rise at least 4-fold within 15 minutes and decrease rapidly by 2 hours. c-*myc* mRNA reaches maximal levels (five times normal) between 30 minutes and 2 hours after operation. p21 levels fall, preceding a transient uncoupling of α-1 adrenergic–receptor binding from the phosphatidylinositol 4,5-biphosphate in the liver cell membranes.[26] A second transient phase of expression for both c-*fos* and c-*myc* occurs around 8 hours following partial hepatectomy. p53 mRNA levels increase between 8 and 12 hours (5-fold over normal). *ras* p21 protein levels increase much later, at a time of active DNA replication and cell division. *abl*, *mos*, and *src* genes do not appear to play a role in hepatic regeneration.

c-*myc* expression can also be induced by dietary manipulation.[60,63] When protein is withheld from the diet and then reintroduced, c-*myc* mRNA levels increase, probably reflecting reentry of the liver cells into the cell cycle.[60] Finally, increased c-*myc* expression is present in areas of cirrhosis.[143]

LIVER TUMORS

The liver has been a favorite organ for study of the successive steps of initiation, promotion, and progression of lesions from preneoplastic (hyperplastic and dysplastic) foci to benign and malignant tumors, because the histological features of each stage are well defined[38] and because activation of cellular oncogenes probably plays an important role in the initiation of hepatic tumors. Histologically, various cellular lesions are demonstrable beginning with focal areas of dysplasia, some of which progress to nodular adenomas and finally to hepatocellular carcinomas.

c-*erb*A, c-*erb*B, Ha-*ras*, c-*myc*, c-*fos*, c-*raf*, and c-*fms* are all overexpressed in hepatic carcinomas,[11,12,24,25,46,55,62,64,67,71,83,90,96,97,120,138,158,170] and cell cycle control of c-*myc* is lost following chemical transformation.[19] c-*myc* gene amplification has also been described accompanying loss of alleles on 11p in a patient with hepatocellular carcinoma,[154] but no relationship exists between ploidy status and c-*myc* expression.[133] Alterations in gap-junction function accompany the overexpression of the c-*raf* in the liver tumors, perhaps providing selective growth advantage to the neoplastic cells.[10] Study of the expression of c-*erb*B in human cirrhotic liver and hepatomas demonstrates a strong correlation between *erb*B expression and alteration of its gene structure.

Members of the *ras* gene family are frequently activated in some carcinogen-induced animal tumors,[156] and *ras* genes experimentally introduced into rat liver cells cause phenotypic changes consistent with malignant transformation.[143] In one large study, 7 of 93 hepatocellular carcinomas, 2 of 10 cholangiomas, and 2 of 137 adenomatous nodules were positive in the transfection assay. Southern blot analysis showed that the NIH 3T3 transformants induced by DNA from these tumors contained an activated Ki-*ras*, often with a G→A mutation in codon 12. The frequency of the detection of *ras* oncogene activation and the failure to observe activated oncogenes from preneoplastic lesions suggests that activation of *ras* occurs late, that it is infrequent in the evolution of some rat hepatocellular neoplasms, and that a mutation of the specific *ras* locus is not an obligatory early event in these neoplasms.[147,156] Finally, as in colon cancer, *ras* genes may be hypomethylated at the CCGG sequence.[50]

TGF-β, a multifunctional regulator of cell growth and differentiation, is a potent inhibitor of hepatocyte growth.[61,122] However, chemically transformed rat liver epithelial cells or liver cells transfected with a mutated Ha-*ras* oncogene are resistant to its inhibitory effects, supporting the idea that escape from negative growth controls might be an important step occurring during carcinogenesis.[61,122]

Models of hepatocarcinogenesis were also generated by directing the expression of SV40T-antigen, an oncogenic mutant of c-Ha-*ras*, or c-*myc* to the liver of transgenic mice, using the albumin enhancer/promoter. The majority of mice carrying the *ras* transgene were born with enlarged livers and atypical hepatic architecture and died within several days of birth. The remaining *ras* transgenic mice had lower levels of *ras* expression, exhibited mild hepatic dysplasia but no liver enlargements, and ultimately died from the development of lung tumors. The livers of mice expressing the T-antigen were relatively normal at birth, by one month displayed marked hepatic dysplasia, and by three to seven months had developed multiple hepatic adenomas and carcinomas. *myc* expression caused mild to severe hepatic dysplasia in young mice and focal hepatic adenomas in some mice over 15 years of age. Lines of mice expressing T-antigen or *myc* were established and crossed with each other to create dual transgenic mice expressing oncogene pairs. Each combination resulted in accelerated tumor development, suggesting that these oncoproteins cooperate with one another during the multistep process of carcinogenesis.[126]

The *hcr* locus is rearranged with c-*myc* in hepatocellular carcinomas. This locus is highly expressed in liver cells but not in other cell types.[105] Transformed liver cells produce a fused *hcr*/c-*myc* transcript initiated from the *hcr* promoter and extending into the c-*myc* coding sequence by differential splicing mechanisms.[37] In addition to the effect of the *hcr* locus, it is probable that abnormal levels of modified or normal *myc*-encoded proteins are produced and are involved in the malignant process.[37]

Exocrine Pancreas

NORMAL PANCREAS

In the fetal pancreas, c-*myc* transcripts[127] and c-*erb*B2 (HER-2/*neu*) gene products[104] are easily identified in the branching tubules of the gland during certain stages of embryogenesis. p185[neu] is expressed in the adult pancreatic tissues.[23]

ADENOCARCINOMAS

The only proto-oncogenes actively investigated in pancreatic adenocarcinomas belong to the *ras* gene family. As in the colon, *ras* genes appear to play an important, but as yet undefined, role in carcinogenesis. However, their alterations may contribute prognostic or diagnostic information about these tumors, which unfortunately are often diagnosed late in the disease course.

The human pancreas cell line T3M-4 contains an activated K-*ras* gene, demonstrable by transfection of the tumor DNA into NIH 3T3 cells. This gene contains an A→C mutation in the second exon of codon 61.[57] Another pancreatic cell line, Pancl, contains a G→A mutation at codon 12 of the K-*ras* gene.[112]

By utilizing PCR gene amplification and mutation detection by the RNAase A mismatch cleavage method on frozen or formalin-fixed, paraffin-embedded tissues obtained surgically or at the time of autopsy, 95% of carcinomas were found to contain c-K-*ras* genes mutated at codon 12.[6] The mutation was present in both primary tumors and their corresponding metastases. No mutations were detected in normal tissue from the same patients or in five gallbladder cancers. This led the authors to conclude that c-K-*ras* somatic mutational activation is critical in the genesis of most pancreatic cancers. Others also demonstrated a 3–6-fold amplification of c-Ki-*ras* and a 50-fold amplification of c-*myc* in a primary pancreatic cancer and its nodal metastases.[159] Because mutations are present in the majority of lesions, they were thought to represent an early event in the genesis of pancreatic cancer. The fact that the mutant gene is maintained in metastatic tumors suggests that they are also actively involved in the maintenance of the tumor phenotype and continuously contribute to tumor progression and metastasis.

Experimental evidence that supports a role for *ras* genes in pancreatic carcinogenesis is provided by the transgenic mouse model. Mutant Ha-*ras* genes placed before the elastase 1 gene promoter can induce adenocarcinomas in the pancreas of transgenic mice.[118]

Not surprisingly, p185^{neu} is found in a fraction of pancreatic adenocarcinomas,[23] since this protein can be found in normal pancreas as well as in other glandular malignancies; also, amplification of the *erb*β oncogene has been described in these tumors.[82]

Acknowledgments

The authors wish to acknowledge the helpful discussions and technical assistance of Barbara Griffith, the manuscript preparation by Agnes Truske and the VA Medical Media Center for their expertise with the illustrative material.

The work reported in this chapter was supported by American Cancer Society Grant #PDT-341 and also by a grant from the University of New Mexico Cancer Center.

REFERENCES

1. Alexander, R. J., Buxbaum, J. N., and Raicht, R. F. Oncogene alterations in primary human colon tumors. *Gastroenterology* 91:1503–1510, 1986.
2. Alitalo, K., Schwab, M., Lin, C. C., Varmus, H. E., and Bishop, J. M. Homogeneously staining chromosomal regions contain amplified copies of an abundantly expressed oncogene (c-myc) in malignant neuroendocrine cells from a human colon carcinoma. *Proc. Natl. Acad. Sci. U.S.A.* 80:1701–1711, 1983.
3. Alitalo, K., Winqvist, R., Lin, C. C., de la Chapelle, A., Schwab, M., and Bishop, J. M. Aberrant expression of an amplified c-myb oncogene in two cell lines from a colon carcinoma. *Proc. Natl. Acad. Sci. U.S.A.* 81:4534–4538, 1984.
4. Allen, D. C., Foster, H., Orchin, J. C., and Biggart, J. D. Immunohistochemical staining of colorectal tissues with monocolonal antibodies to ras oncogene p21 product and carbohydrate determinant antigen 19–9. *J. Clin. Pathol.* 40:157–162, 1987.
5. Allum, W. H., Newbold, K. M., Macdonald, F., Russell, B., and Stokes, H. Evaluation of p62 c-myc in benign and malignant gastric epithelia. *Br. J. Cancer* 56:785–786, 1987.
6. Almoguera, C., Shibata, D., Forrester, K., Martin, J., Arnheim, N., and Perucho, M. Most human carcinomas of the exocrine pancreas contain mutant c-K-ras genes. *Cell* 53:549–554, 1988.
7. Atkin, N. B., and Baker, M. C. Deficiency of all or part of chromosome 11 in several types of cancer: Significance of reduction in the normal number of chromosomes 11. *Cancer Genet. Cytogenet.* 47:106–107, 1988.

8. Augenlicht, L. H., Augeron, C., Yander, G., and Laboisse, C. Overexpression of *ras* in mucus-secreting human colon carcinoma cells of low tumorigenicity. *Cancer Res.* 47:3763–3765, 1987.

9. Baker, S. J., Fearon, E. R., Nigro, J. M., Hamilton, S. R., Preisinger, A. C., Jessup, J. M., VanTuinen, P., Ledbetter, D. H., Barker, D. F., Nakamura, Y., White, R., and Vogelstein, B. Chromosome 17 deletions and p53 gene mutations in colorectal carcinomas. *Science (Wash DC)* 244:217–221, 1989.

10. Barbacid, M. *ras* genes. *Annu. Rev. Biochem.* 56:779–827, 1987.

11. Beer, D. G., Neveu, M. J., Paul, D. L., Rapp, U. R., and Pitot, H. C. Expression of the c-*raf* protooncogene, alpha-glutamyltranspeptidase, and gap junction protein in rat liver neoplasms. *Cancer Res.* 48:1610–1617, 1988.

12. Beer, D. G., Schwartz, M., Sawada, N., and Pitot, H. C. Expression of H-*ras* and c-*myc* protooncogenes in isolated gamma-glutamyl transpeptidase positive rat hepatocytes and in hepatocellular carcinomas induced by diethylnitrosamine. *Cancer Res.* 46:2435–2441, 1986.

13. Bodmer, W. F., Bailey, C. J., Bodmer, J., Bussey, H. J. R., Ellis, A., Gorman, P., Lucibello, F. C., Murday, V. A., Rider, S. H., Scambler, P., Sheer, D., Solomon, E., and Spurr, N. K. Localization of the gene for familial adenomatous polyposis on chromosome 5. *Nature (Lond)* 328:614–616, 1987.

14. Bolen, J. B., Veillette, A., Schwartz, A. M., Deseau, V., and Rosen, N. Analysis of pp60[c-src] in human colon carcinoma and normal human colon mucosal cells. *Oncogene Res.* 1:149–186, 1987.

15. Bos, J. L., Fearon, E. R., Hamilton, S. R., Verlaan-de Vries, M., van Boom, J. H., van der Eb, A. H., and Vogelstein, B. Prevalence of ras gene mutations in human colorectal cancers. *Nature (Lond)* 327:293–297, 1987.

16. Bos, J. L., Verlaan-deVries, M., Marshall, C. J., Veeneman, G. H., van Bloom, J. H., and van der Eb, A. J. A human gastric carcinoma contains a single mutated and an amplified normal allele of the Ki-ras oncogene. *Nucleic Acids Res.* 14:1209–1217, 1986.

17. Burmer, G. C., Rabinovitch, P. S., and Loeb, L. A. Analysis of c-Ki-*ras* mutations in human colon carcinoma by cell sorting, polymerase chain reaction and DNA sequencing. *Cancer Res.* 49:2141–2146, 1989.

18. Calabretta, B., Kaczmarek, L., Min, P.-M. L., Au, F., and Ming, S. C. Expression of c-myc and other cell cycle-dependent genes in human colon neoplasia. *Cancer Res.* 45:6000–6004, 1985.

19. Campisi, J., Gray, H. E., Pardee, A. B., Dean, M., and Sonenshein, G. E. Cell-cycle control of c-*myc* but not c-*ras* expression is lost following chemical transformation. *Cell* 36:241–247, 1984.

20. Capon, D. J., Seeburg, P. H., McGrath, J. P., Hayflick, J. S., Edman, U., Levinson, A. D., and Goeddel, D. V. Activation of Ki-ras2 gene in human colon and lung carcinomas by two different point mutations. *Nature (Lond)* 304:507–513, 1983.

21. Cartwright, C. A., Kamps, M. P., Meisler, A. I., Pipas, J. M., and Eckhart, W. pp60[c-src] activation in human colon cancer. *J. Clin. Invest.* 83:2025–2033, 1989.

22. Ciclitira, P. J., Macartney, J. C., and Evan, G. Expression of c-myc in non-malignant and pre-malignant gastrointestinal disorders. *J. Pathol.* 151:293–296, 1987.

23. Cohen, J. A., Weiner, D. B., More, K. F., Kokai, Y., Williams, W. V., Maguire, H. C., LiVolsi, V. A., and Green, M. I. Expression pattern of the *neu* (NGL) gene-encoded growth factor receptor protein (p185*neu*) in normal and transformed epithelial tissues of the digestive tract. *Oncogene* 4:81–88, 1989.

24. Corcos, D., Defer, N., Raymondjcan, M., *et al.* Correlated increase of the expression of the c-*ras* genes in chemically induced hepatocarcinomas. *Biochem. Biophys. Res. Commun.* 122:259–264, 1984.

25. Cote, G. J., Lastra, B. A., Cook, J., Huang, D.-P., and Chin, J.-F. Oncogene expression in rat hepatomas and during hepatocarcinogenesis. *Cancer Lett.* 26:121–127, 1985.

26. Cruise, J. L., Muga, S. J., Lee, Y. S., and Michalopoulos, G. K. Regulaton of hepatocyte growth. Adrenergic receptor and ras p21 changes in liver regeneration. *J. Cell Physiol.* 140:195–201, 1989.

27. Cunningham, J. M., and Weinberg, R. A. Detection and analysis of oncogenes in colon carcinoma. *Prog. Cancer. Res. Ther.* 29:403–410, 1984.

28. Czerniak, B., Herz, F., Koss, L. G., and Schlom, J. ras Oncogene p21 as a tumor marker in the cytodiagnosis of gastric and colonic carcinomas. *Cancer* 60:2432–2436, 1987.

29. Czerniak, B., Herz, F., Gorczyca, W., and Koss, L. Expression of ras oncogene p21 protein in early gastric carcinoma and adjacent gastric epithelia. *Cancer* 64:1467–1473, 1989.

30. Der, C. J., and Cooper, G. M. Altered gene products are associated with activation of cellular ras^k genes in human lung and colon carcinomas. *Cell* 32:201–208, 1983.

31. DeSeau, V., Rosen, N., and Bolen, J. B. Analysis of pp60^{c-src} tyrosine kinase activity and phosphotyrosyl phosphatase activity in human colon carcinoma and normal human colon mucosal cells. *J. Cell. Biochem.* 35:113–128, 1987.

32. Dolcetti, R., de Re V, Viel, A., Pistello, M., Tavian, M., and Boiocchi, M. Nuclear oncogene amplification or rearrangement is not involved in human colorectal malignancies. *Eur. J. Cancer Oncol.* 24:1321–1328, 1988.

33. Erisman, M. D., Litwin, S., Keidan, R. D., Comis, R. L., and Astrin, S. M. Noncorrelation of the expression of the c-*myc* oncogene in colorectal carcinoma with recurrence of disease or patient survival. *Cancer Res.* 48:1350–1355, 1988.

34. Erisman, M. D., Rothberg, P. G., Diehl, R. E., Morse, C. C., Spandorfer, J. M., and Astrin, S. M. Deregulation of c-myc gene expression in human colon carcinoma is not accompanied by amplification or rearrangement of the gene. *Mol. Cell. Biol.* 5:1969–1976, 1985.

35. Erisman, M. D., Scott, J. K., and Astrin, S. M. Evidence that the familial adenomatous polyposis gene is involved in a subset of colon cancers with a complementable defect in c-*myc* regulation. *Proc. Natl. Acad. Sci. U.S.A.* 86:4264–4268, 1989.

36. Erisman, M. D., Scott, J. K., Watt, R. A., and Astrin, S. N. The c-*myc* protein is constitutively expressed at elevated levels in colorectal carcinoma cell lines. *Oncogene* 2:367–378, 1988.

37. Etiemble, J., Moroy, T., Jacquemin, E., Tiollais, P., and Buendia, M.-A. Fused transcripts of c-*myc* and a new cellular locus, *hcr* in a primary liver tumor. *Oncogene* 4:51–57, 1989.

38. Farber, E., and Cameron, R. The sequential analysis of cancer development. *Adv. Cancer Res.* 31:125–226, 1980.

39. Fasano, O., Birnbaum, D., Edlund, L., Fogh, J., and Wigler, M. New human transforming genes detected by a tumorigenic assay. *Mol. Cell. Biol.* 4:1695–1705, 1984.

40. Fausto, N., and Mead, J. E. Regulation of liver growth: Protooncogenes and transforming growth factors. *Lab. Invest.* 60:4–13, 1989.

41. Fausto, N., and Shank, P. R. Oncogene expression in liver regeneration and hepatocarcinogenesis. *Hepatology* 3:1016–1023, 1983.

42. Fearon, E. R., Hamilton, S. H., and Vogelstein, B. Clonal analysis of colorectal tumors. *Science (Wash DC)* 238:1930–1937, 1987.

43. Feinberg, A. P., and Vogelstein, B. Hypomethylation of ras oncogenes in primary human cancers. *Biochem. Biophys. Res. Commun.* 111:47–54, 1983.

44. Finley, G. G., Schultz, N. T., Hill, S. A., Geiser, J. R., Pipas, J. M., and Meisler, A. I. Expression of the *myc* gene family in different stages of human colorectal cancer. *Oncogene* 4:963–971, 1989.

45. Forrester, K., Almoguera, C., Han, K., Grizzle, W. E., and Perucho, M. Detection of high incidence of K-ras oncogenes during human colon tumorigenesis. *Nature (Lond)* 327:298–303, 1987.

46. Fox, T. R., and Watanabe, P. G. Detection of a cellular oncogene in spontaneous liver tumors of the B6C3F1 mice. *Science (Wash DC)* 228:596–597, 1985.

47. Fujita, J., Ohuchi, N., Yao, T., Okumura, M., Fukushima, Y., Kanakura, Y., Kitamura, Y., and Fujita, J. Frequent overexpression, but not activation by point mutation, of ras genes in primary human gastric cancers. *Gastroenterology 93*:1339–1345, 1987.
48. Fukushige, S., Matsubara, K., Yoshida, M., Sasaki, M., Suzuki, T., Semba, K., Toyshima, K., and Yamamoto, T. Localization of a novel v-*erb*B-related gene, c-*erb*B2 on chromosome 17 and its amplification in a gastric cancer cell line. *Mol. Cell Biol. 6*:955–958, 1986.
49. Gallick, G. E., Kurzrock, R., Kloetzer, W. S., Arlinghaus, R. B., and Gutterman, G. U. Expression of p21^ras in fresh primary and metastatic human colorectal tumors. *Proc. Natl. Acad. Sci. U.S.A. 82*:1795–1799, 1985.
50. Garcea,R., Daino, L., Pascale, R., Simile, M. M., Puddu, M., Ruggiu, M. E., Seddaiu, M. A., Satta, G., Sequenza, M. J., and Feo, F. Protooncogene methylation and expression in regenerating liver and preoplastic liver nodules induced in the rat by diethylnitrosamine: Effect of variations of S-adenosylmethionine: S-adenosylhomocysteine ratio. *Carcinogenesis 10*:1183–1192, 1989.
51. Gattoni-Celli, S., Kirsch, K., Kalled, S., and Isselbacher, K. J. Expression of type C–related endogenous retroviral sequences in human colon tumors and colon cancer cell lines. *Proc. Natl. Acad. Sci. U.S.A. 83*:6127–6131, 1986.
52. Goyette, M., Petropoulos, C. J., Shank, P. R., and Fausto, N. Expression of a cellular oncogene during liver regeneration. *Science (Wash DC) 219*:510–512, 1983.
53. Goyette, M., Petropoulos, C. J., Shank, P. R., and Fausto, N. Regulated transcription of c-Ki-*ras* and c-*myc* during compensatory growth of rat liver. *Mol. Cell. Biol. 4*:1493–1498, 1984.
54. Greenhalgh, D. A., and Kinsella, A. R. c-Ha-ras not c-Ki-ras activation in three colon tumour cell lines. *Carcinogenesis 6*:1533–1535, 1985.
55. Hayashi, K., Makino, R., and Sugimura, T. Amplification and over-expression of the c-*myc* gene in Morris hepatomas. *Gann 75*:475–478, 1984.
56. Herrea, L., Kataki, S., Gibas, L., Pietrzak, E., and Sandberg, A. A. Brief clinical report: Gardner's syndrome in a man with an interstitial deletion of 5q. *Am. J. Med. Genet. 25*:473–476, 1986.
57. Hirai, H., Okabe, T., Anraku, Y., Fujisawa, M., Urabe, A., and Takaku, F. Activation of the c-K-*ras* oncogene in a human pancreas carcinoma. *Biochem. Biophys. Res. Commun. 127*:168–174, 1985.
58. Hirai, S.-I., Ryseck, R.-P., Mechta, F., Bravo, R., and Yamiu, M. Characterization of *jun*D: A new member of the *jun* proto-oncogene family. *E.M.B.O. J. 8*:1433–1439, 1989.
59. Horan Hand, P., Vilasi, V., Thor, A., Ohuchi, N., and Schlom, J. Quantitation of Harvey *ras* p21 enhanced expression in human breast and colon carcinomas. *J. Natl. Cancer Inst. 79*:59–65, 1987.
60. Horikawa, S., Sakata, K., Hatanaka, M., and Tsukada, K. Expression of c-*myc* oncogene in rat liver by a dietary manipulation. *Biochem. Biophys. Res. Commun. 140*:574–580, 1986.
61. Houck, K. A., Michalopoulos, G. K., and Strom, S. C. Introduction of a Ha-*ras* oncogene into rat liver epithelial cells and parenchymal hepatocytes confers resistance to the growth inhibitory effects of TGF-beta. *Oncogene 4*:19–25, 1989.
62. Hsieh, L. L., Hsiao, W. L., Peraino, C., Maronpot, R. R., and Weinstein, I. B. Expression of retroviral sequences and oncogenes in rat liver tumors induced by diethylnitrosamine. *Cancer Res. 47*:3421–3424, 1987.
63. Hsieh, L. L., Wainfan, E., Hoshima, S., Dizik, M., and Weinstein, I. B. Altered expression of retrovirus-like sequences and cellular oncogene in mice fed methyl-deficient diets. *Cancer Res. 49*:3795–3799, 1989.
64. Huber, B. E., and Cordingley, M. G. Expression and phenotypic alterations caused by an inducible transforming *ras* oncogene introduced into rat liver epithelial cells. *Oncogene 3*:245–256, 1988.
65. Ikeda, I., Ishizaka, Y., Ochiai, M., Sakai, R., Itabashi, M., Onda, M., Sugimara, T., and

Nagao, M. No correlation between L-myc restriction fragment length polymorphism and malignancy of human colorectal cancers. *Jpn. J. Cancer Res.* 79:674–676, 1988.

66. Imaseki, H., Hayashi, H., Taira, M., Ito, Y., Tabata, Y., Onoda, S., Isono, K., and Tatibana, M. Expression of c-*myc* oncogenes in colorectal polyps as a biological marker for monitoring malignant potential. *Cancer* 64:704–709, 1989.

67. Ishikawa, F., Takaku, F., Nagao, M., Ochiai, M., Hayashi, K., Takayama, S., and Sugimura, T. Activated oncogenes in a rat hepatocellular carcinoma induced by 2-amino-3-methylimidazo (4,5-f) quinoline. *Gann* 76:425–428, 1985.

68. Ishizaka, Y., Ochiai, M., Ishikawa, F., Sato, S., Miura, Y., Nagao, M., and Sugimura, T. Activated N-ras oncogene in a transformant derived from a rat small intestinal adenocarcinoma induced by 2-aminodipyrido[1,2-a:3′,2′-d]imidazole. *Carcinogenesis* 8:1575–1578, 1987.

69. Izumi, S. I., Moriuchi, T., Koji, T., et al. Localization of C-MYC in HL-60 cells, neoplastic and normal tissues: An immunohistochemical and in situ hybridization study. *Acta Histochem. Cytochem.* 21:327–342, 1988.

70. Jiang, W., Kahn, S. M., Guillem, J. G., Lu, S.-H., and Weinstein, I. B. Rapid detection of *ras* oncogenes in human tumors: Applications to colon, esophageal, and gastric cancer. *Oncogene* 4:923–928, 1989.

71. Johnson, B. E., Sakaguchi, A. Y., Gazdar, A. F., Minna, J. D., Burch, D., Marshall, A., and Naylor, S. L. Restriction fragment polymorphism studies show consistent loss of chromosome 3p alleles in small cell lung cancer patients' tumors. *J. Clin. Invest.* 82:502–507, 1988.

72. Jones, D. J., Ghosh, A. K., Moore, M., and Schofield, P. F. A critical appraisal of the immunohistochemical detection of the c-myc oncogene product in colorectal cancer. *Br. J. Cancer* 56:779–783, 1987.

73. Kate, J. T., Eidelman, S., Bosman, F. T., and Damjanov, I. Expression of c-*myc* proto-oncogene in normal human intestinal epithelium. *J. Histochem. Cytochem.* 37:541–545, 1989.

74. Kawashima, K., Imoto, K., and Izawa, M. Restriction fragment length polymorphism (RFLP) of L-*myc* is related to the progression of human colon and stomach cancers. *Proc. Jpn. Acad.* 63:300–303, 1987.

75. Kawashima, K., Shikama, H., Imoto, K., Izawa, M., et al. Close correlation between restriction fragment length polymorphism of the L-myc gene and metastasis of human lung cancer to the lymph nodes and other organs. *Proc. Natl. Acad. Sci. U.S.A.* 85:2353–2356, 1988.

76. Kern, S. E., Fearon, E. R., Tersmette, K. W. F., Enterline, J. P., Leppert, M., Nakamura, Y., White, R., Vogelstein, B., and Hamilton, S. R. Allelic loss in colorectal carcinoma. *JAMA* 261:3099–3103, 1989.

77. Kerr, I. B., Lee, F. D., Quintanilla, M., and Balmain, A. Immunocytochemical demonstration of p21 *ras* family oncogene product in normal mucosa and in premalignant and malignant tumours of the colorectum. *Br. J. Cancer* 52:695–700, 1985.

78. Kerr, I. B., Spandidos, D. A., Finlay, I. G., Lee, F. D., and McArdle, C. S. The relation of *ras* family oncogene expression to conventional staging criteria and clinical outcome in colorectal carcinoma. *Br. J. Cancer* 53:231–235, 1986.

79. Kitagami, S., Itabashi, M., Hirota, T., Hyashi, I., Hojo, K., Moriya, Y., Maruyama, K., and Okabayashi, K. Immunohistochemical study of oncogene-related products in human gastrointestinal malignancies—Expression of ras p21, fes p85 and EGF receptor. *Gan No Rinsho* 32:1950–1958, 1986.

80. Koda, T., Matsushima, S., Sasaki, A., Danjo, Y., and Kakinuma, M. c-*myc* gene amplification in primary stomach cancer. *Jpn. J. Cancer Res.* 76:551–554, 1985.

81. Kopelovich, L., and DeLeo, A. B. Elevated levels of p53 antigen in cultured skin fibroblasts from patients with hereditary adenocarcinoma of the colon and rectum and its relevance to oncogenic mechanisms. *J. Natl. Cancer Inst.* 77:1241–1246, 1986.

82. Korc, M., Meltzer, P., and Trent, J. Enhanced expression of epidermal growth factor cor-

relates with alterations of chromosome #7 in human pancreatic cancer. *Proc. Natl. Acad. Sci. U.S.A. 83*:5141–5144, 1986.

83. Lafarge-Frayssinet, C., and Frayssinet, C. Overexpression of proto-oncogenes: ki-ras, fos and myc in rat liver cells treated in vitro by two liver tumor promoters: phenobarbital and biliverdin. *Cancer Lett. 44*:191–198, 1989.

84. Land, H., Parada, L. F., and Weinberg, R. A. Tumorigenic conversion of primary embryo fibroblasts requires at least two cooperating oncogenes. *Nature (Lond) 304*:596–602, 1983.

85. Lanza, G., Jr. ras p21 oncoprotein expression in human colonic neoplasia-an immunohistochemical study with monoclonal antibody RAP-5. *Histopathology 12*:595–609, 1988.

86. Lauren, P. The two histological main types of gastric carcinoma. Diffuse and so-called intestinal-type carcinoma. An attempt at histoclinical classification. *Acta Pathol. Microbiol. Immunol. Scand. 64*:31–49, 1965.

87. Leppert, M., Dobbs, M., Scrambler, P., O'Connell, P., Nakamura, Y., Stauffer, D., Woodward, S., Burt, R., Hughes, J., Gardner, E., Lathop, M., Wasmuth, J., Lalovel, J.-M., and White, R. The gene for familial polyposis coli maps to the long arm of chromosome 5. *Science (Wash DC) 238*:1411–1413, 1987.

88. Lin, C. C., Alitalo, K., Schwab, M., George, D., Varmus, H. E, and Bishop, J. M. Evolution of karyotypic abnormalities and C-MYC oncogene amplification in human colonic carcinoma cell lines. *Chromosoma 92*:11–15, 1985.

89. Lothe, R. A., Nakamura, Y., Woodward, S., Gedde-Dahl, T., and White, R. VNTR (variable number of tandem repeats) markers show loss of chromosome 17p sequence in human colorectal carcinoma. *Cytogenet. Cell Genet. 48*:167–169, 1988.

90. Makino, R., Hayashi, K., Sato, S., and Sugimura, T. Expressions of the c-Ha-*ras* and c-*myc* genes in rat liver tumors. *Biochem. Biophys. Res. Commun. 119*:1092–1102, 1984.

91. Makino, R., Hayashi, K., and Sugimura, T. c-*myc* transcript is induced in rat liver at very early stage of regeneration or by cycloheximide treatment. *Nature (Lond) 310*:697–698, 1984.

92. Martin-Zanca, D., Hughes, S. H., and Barbacid, M. A human oncogene formed by the fusion of truncated tropomyosin and protein tyrosine kinase sequences. *Nature (Lond) 319*:743–748, 1986.

93. McCoy, M., Bargmann, C. I., and Weinberg, R. A. Human colon carcinoma Ki-ras2 oncogene and its corresponding proto-oncogene. *Mol. Cell. Biol. 4*:1577–1582, 1984.

94. McCoy, M., Toole, J. J., Cunningham, J. M., Chang, E. H., Lowy, D. R., and Weinberg, R. A. Characterization of human colon/lung carcinoma oncogene. *Nature (Lond) 302*:79–81, 1983.

95. McCoy, M. S., and Weinberg, R. A. A human Ki-*ras* oncogene encodes two transforming p21 proteins. *Mol. Cell. Biol. 6*:1326–1328, 1986.

96. McMahon, G., Hanson, L., Lee, J. J., and Wogan, G. N. Identification of an activated c-Ki-*ras* oncogene in rat liver tumors induced by aflatoxin B1. *Proc. Natl. Acad. Sci. U.S.A. 83*:9418–9422, 1986.

97. McMahon, J. B., Richards, W. L., del Campo, A. A., Song, M. K., and Thorgeirsson, S. S. Differential effects of transforming growth factor-beta on proliferation of normal and malignant rat liver epithelial cells in culture. *Cancer Res. 46*:4665–4671, 1986.

98. Meltzer, S. J., Ahnen, D. J., Battifora, H., Yokota, J., and Cline, M. J. Protooncogene abnormalities in colon cancers and adenomatous polyps. *Gastroenterology 92*:1174–1180, 1987.

99. Michelassi, F., Leuthner, S., Lubienski, M., Bostwick, D., Rodgers, J., Handcock, M., and Block, G. E. Ras oncogene p21 levels parallel malignant potential of different human colonic benign conditions. *Arch. Surg. 122*:1414–1416, 1987.

100. Michelassi, F., Vannucci, L. E., Montag, A., Chappell, R., Rodgers, J., and Block, G. E. Ras oncogene expression as a prognostic indicator in rectal adenocarcinoma. *J. Surg. Res. 45*:15–20, 1988.

101. Mitra, G., Martin-Zanca, D., and Barbacid, M. Identification and biochemical characterization of p70 TRK, product of the human TRK oncogene. *Proc. Natl. Acad. Sci. U.S.A. 84*:6707–6711, 1987.

102. Monnat, M., Tardy, S., Saraga, P., Diggelmann, H., and Costa, J. Prognostic implications of expression of the cellular genes MYC, FOS, Ha-ras and Ki-ras in colon carcinoma. *Int. J. Cancer 40*:293–299, 1987.

103. Moore, B. E., and Bose, H. R. Expression of the c-*rel* and c-*myc* proto-oncogenes in avian tissue. *Oncogene 4*:845–852, 1989.

104. Mori, S., Akiyama, T., Yamada, Y., Morishita, Y., Sugawara, I., Toyoshima, K., and Yamamoto, T. C-*erb*B-2 gene product, a membrane protein commonly expressed on human fetal epithelial cells. *Lab. Invest. 61*:93–97, 1989.

105. Moroy, T., Etiemble, J., Bougueleret, L., Hadchouel, M., Tiollais, P., and Buendia, M.-A. Structure and expression of *hcr*, a locus rearranged with c-myc in a woodchuck hepatocellular carcinoma. *Oncogene 4*:59–65, 1989.

106. Muller, R., Slamon, D. J., Adamson, E. D., Tremblay, J. M., Muller, D., Cline, M. J., and Verma, I. M. Transcription of c-*onc* genes c-*ras*^{Ki} and c-*fms* during mouse development. *Mol. Cell. Biol. 3*:1062–1069, 1983.

107. Murray, M. J., Shilo, B. Z., Shih, C., Cowing, D., Hsu, H. W., and Weinberg, R. A. Three different human tumor cell lines contain different oncogenes. *Cell 25*:355–361, 1981.

108. Nishida, J., Kobayashi, Y., Hirai, H., and Takaku, F. A point mutation at codon 13 of the N-ras oncogene in a human stomach cancer. *Biochem. Biophys. Res. Commun. 146*:247–252, 1987.

109. Nitta, N., Ochiai, M., Nagao, M., and Sugimura, T. Amino-acid substitution at codon 13 of the N-ras oncogene in rectal cancer in a Japanese patient. *Jpn. J. Cancer Res. 78*:21–26, 1987.

110. Nomura, N., Yamamoto, T., Toyoshima, K., Ohami, H., Akimaru, K., Sasaki, S., Nakagami, Y., Kanauchi, H., Shoji, T., Hiraoka, Y., *et al.* DNA amplification of the c-*myc* and c-*erb*B-1 genes in a human stomach cancer. *Jpn. J. Cancer Res. 77*:1188–1192, 1986.

111. Ochiya, T., Fujiyama, A., Fukushige, S., Hatada, I., and Matsubara, K. Molecular cloning of an oncogene from a human hepatocellular carcinoma. *Proc. Natl. Acad. Sci. U.S.A. 83*:4993–4997, 1986.

112. O'Hara, B. M., Oskarsson, M., Tainsky, M. A., and Blair, D. G. Mechanism of activation of human *ras* genes cloned from a gastric adenocarcinoma and a pancreatic carcinoma cell line. *Cancer Res. 46*:4695–4700, 1986.

113. Olofsson, B., Chardin, P., Touchot, N., Zahraoui, A., and Tavitian, A. Expression of the *ras*-related *ral* A, *rho* 12 and *rab* genes in adult mouse tissues. *Oncogene 3*:231–234, 1988.

114. Parada, L. F., Tabin, C. J., Shih, C., and Weinberg, R. A. Human EJ bladder carcinoma oncogene is homologue of Harvey sarcoma virus *ras* gene. *Nature (Lond) 297*:474–478, 1982.

115. Perucho, M., Goldfarb, M., Shimuzu, K., and Lama, C. Human-tumor–derived cell lines contain common and different transforming genes. *Cell 27*:467–476, 1981.

116. Pulciani, S., Santos, E., Long, L. K., Sorrentino, V., and Barbacid, M. *ras* gene amplification and maligant transformation. *Mol. Cell. Biol. 5*:2836–2841, 1985.

117. Pulciani, S., Santos, E., Lauver, A. V., Long, L. K., Aaronson, S. A., and Barbacid, M. Oncogenes in solid human tumours. *Nature (Lond) 300*:539–542, 1982.

118. Quaife, C. J., Pinkert, C. A., Ornitz, D. M., Palmiter, R. D., and Brinster, R. L. Pancreatic neoplasia induced by *ras* expression in acinar cells of transgenic mice. *Cell 48*:1023–1034, 1987.

119. Ranzani, G. N., Salerno-Mele, P., Maltoni, M., Talarico, D., Della Valle, G., and Amadori, D. Study of the c-Ha-ras-1 locus polymorphism in an Italian population with high incidence of gastric cancer. *Mol. Biol. Med. 5*:145–153, 1988.

120. Reynolds, S. H., Stowers, S. J., Moronpot, R. R., Anderson, M. W., and Aaronson, S. A.

Detection and identification of activated oncogenes in spontaneously occurring benign and malignant hepatocellular tumors of the B6C3F1 mouse. *Proc. Natl. Acad. Sci. U.S.A.* 83:33–37, 1986.

121. Richmond, R. E., Pereira, M. A., Carter, J. H., Carter, H. W., and Long, R. E. Quantitative and qualitative immunohistochemical detection of *myc* and *src* oncogene proteins in normal, nodular, and neoplastic rat liver. *J. Histochem. Cytochem.* 36:179–184, 1988.

122. Roberts, A. B., Anzano, M. A., Wakefield, L. M., Roche, N. S., Stern, D. F., and Sporn, M. B. Type beta transforming growth factor: a bifunctional regulation of cell growth. *Proc. Natl. Acad. Sci. U.S.A.* 82:119–123, 1985.

123. Robinson, A., Williams, A. R., Piris, J., Spandidos, D. A., and Wyllie, A. H. Evaluation of a monoclonal antibody to *ras* peptide, RAP-5, claimed to bind preferentially to cells of infiltrating carcinoma. *Br. J. Cancer* 54:877–883, 1986.

124. Rothberg, P. G., Spandorfer, J. M., and Erisman, M. D. Evidence that c-myc expression defines two genetically distinct forms of colorectal adenocarcinoma. *Br. J. Cancer* 52:629–632, 1985.

125. Sakamoto, H., Mori, M., and Taira, M. Transforming gene from human stomach cancers and a noncancerous portion of stomach mucosa. *Proc. Natl. Acad. Sci. U.S.A.* 83:3997–4001, 1986.

126. Sandgren, E. P., Quaife, C. J., Pinkerf, C. A., Palmitter, R. D., and Brinsfer, R. L. Oncogene-induced neoplasia in transgenic mice. *Oncogene* 4:715–724, 1989.

127. Semsei, I. M., Ma, S., and Cutler, R. G. Tissue and age specific expression of the *myc* protooncogene family throughout the life span of the C57BL/6J mouse strain. *Oncogene* 4:465–470, 1989.

128. Sherr, C. J., Rettenmier, C. W., Sacca, R., Roussel, M. F., Look, A. T., and Stanley, E. R. The c-*fms* proto-oncogene product is related to the receptor for the mononuclear phagocyte growth factor, CSF-1. *Cell* 41:665–676, 1985.

129. Shibuya, M., Yokota, J., and Ueyama, Y. Amplification and expression of a cellular oncogene (c-*myc*) in human gastric adenocarcinoma cells. *Mol. Cell Biol.* 5:414–418, 1985.

130. Shimizu, K., Nakano, Y., Sekiguchi, M., *et al.* Molecular cloning of an activated human oncogene homologous to v-*raf* from primary stomach cancer. *Proc. Natl. Acad. Sci. U.S.A.* 82:5641–5645, 1985.

131. Shiozawa, M., Ochiya, T., Hatada, I., Imamura, T., Okudaira, Y., Hirauka, H., and Matsubara, K. The *lca* as an onco-fetal gene: Its expression in human fetal liver.

132. Sikora, K., Chan, S., Evan, G., Gabra, H., Markam, N., Stewart, J., and Watson, J. c-myc oncogene expression in colorectal cancer. *Cancer* 59:1289–1295, 1987.

133. Sinha, S., Neal, G. E., Legg, R. F., Watson, J. V., and Pearson, C. The expression of c-*myc* related to the proliferation and transformation of rat liver-derived epithelial cells. *Br. J. Cancer* 59:674–676, 1989.

134. Solomon, E., Voss, R., Hall, V., Bodmer, W. F., Jass, J. R., Jeffreys, A. J., Lucibello, F. C., Patel, I., and Rider, S. H. Chromosome 5 allele loss in human colorectal carcinomas. *Nature (Lond)* 328:616–619, 1987.

135. Staal, S. P. Molecular cloning of the *akt* oncogene and its human homologues AKT1 and AKT2: Amplification of AKT1 in a primary human gastric adenocarcinoma. *Proc. Natl. Acad. Sci. U.S.A.* 84:5034–5037, 1987.

136. Stewart, J., Evan, G., Watson, J., and Sikora, K. Detection of the c-myc oncogene product in colonic polyps and carcinomas. *Br. J. Cancer* 53:1–6, 1986.

137. Tahara, E., Yasui, W., Tantyama, K., Ochiai, A., Yamamoto, T., Nakajo, S., and Yamamoto, M. Ha-*ras* oncogene product in human gastric carcinoma: Correlation with invasiveness, metastasis or prognosis. *Jpn. J. Cancer Res.* 77:517–522, 1986.

138. Takada, S., and Koike, K. Activated N-*ras* gene was found in human hepatoma tissue but only in a small fraction of the tumor cells. *Oncogene* 4:189–193, 1989.

139. Tal, M., Wetzler, M., Josefberg, Z., Deutch, A., Gutman, M., Assaf, D., Kris, Y., Shiloh,

Y., Givolo, D., and Schlessinger, J. Sporadic amplification of the HER2/*neu* protooncogene in adenocarcinomas of various tissues. *Cancer Res.* 48:1517–1520, 1988.

140. Tanaka, T., Slamon, D. J., Battifora, H., and Cline, M. J. Expression of p21/*ras* oncoproteins in human cancers. *Cancer Res.* 46:1465–1469, 1986.

141. Thompson, N. L., Mead, J. E., Braun, L., Goyette, M., Shank, P. R., and Fausto, N. Sequential protooncogene expression during rat liver regeneration. *Cancer Res.* 46:3111–3117, 1986.

142. Thor, A., Horan Hand, P., Wunderlich, D., Caruso, A., Muraro, R., and Schlom, J. Monoclonal antibodies define differential *ras* gene expression in malignant and benign colonic disease states. *Nature (Lond)* 311:562–564, 1984.

143. Tiniakos, D., Spandidos, D. A., Kakkanas, A., Pintzas, A., Pollice, L., and Tiniakos, G. Expression of ras and myc oncogenes in human hepatocellular carcinoma and non-neoplastic liver tissues. *Anticancer Res.* 9:715–722, 1989.

144. Torelli, G., Venturelli, D., Colo, A., Zanni, C., Selleri, L., Moretti, L., Calabretta, B., Torelli, U. Expression of c-*myb* protooncogene and other cell cycle–related genes in normal and neoplastic human colonic mucosa. *Cancer Res.* 47:5266–5269, 1987.

145. Trainer, D. L., Kline, T., McCabe, F. L., Faucette, L. F., Field, J., Chaikin, M., Anzano, M., Rieman, D., Hoffstein, S., Li, D. J., *et al.* Biological characterization and oncogene expression in human colorectal carcinoma cell lines. *Int. J. Cancer* 41:287–296, 1988.

146. Tsuboi, K., Hirayoshi, K., Takeuchi, K., Sabe, H., Shimada, Y., Ohshio, G., Tobe, T., and Hatanaka, M. Expression of the c-*myc* gene in human gastrointestinal malignancies. *Biochem. Biophys. Res. Commun.* 146:699–704, 1987.

147. Tsuda, H., Hirohashi, S., Shimosato, Y., Ino, Y., Yoshida, T., and Terada, M. Low incidence of point mutation of c-Ki-*ras* and N-*ras* oncogenes in human hepatocellular carcinoma. *Jpn. J. Cancer Res.* 80:196–199, 1989.

148. Tsuda, T., Nakatani, H., Matsumura, T., Yoshida, K., Tahara, E., Nishihira, T., Sakamoto, H., Yoshida, T., Terada, M., and Sugimura, T. Amplification of the *hst*-1 gene in human esophageal carcinomas. *Jpn. J. Cancer Res.* 79:584–588, 1988.

149. van den Berg, F. M., Tigges, A. J., Schipper, M. E. I., den Hartog-Jager, F. C. A., Kroes, W. G. M. and Walboomers, J. M. M. Expression of the nuclear oncogene p53 in colon tumours. *J. Pathol.* 157:193–199, 1989.

150. Vogelstein, B., Fearon, E. R., Hamilton, S. R., Kern, S. E., Preisinger, A. C., Leppert, M., Nakamura, Y., White, R., Smits, A. M. M., and Bos, J. L. Genetic alterations during colorectal tumor development. *N. Engl. J. Med.* 319:525–532, 1988.

151. Vogelstein, B., Fearon, E. R., Kern, S. E., Hamilton, S. R., Preisinger, A. C., Nakamura, Y., and White, R. Allotype of colorectal carcinomas. *Science (Wash DC)* 244:207–211, 1989.

152. Wahl, G. M. The importance of circular DNA in mammalian gene amplification. *Cancer Res.* 49:1333–1340, 1989.

153. Wakai, A., Seino, S., Sakurai, A., Szilak, I., Bell, G. I., and DeGroot, L. J. Characterization of a thyroid hormone receptor expressed in human kidney and other tissues. *Proc. Natl. Acad. Sci. U.S.A.* 85:2781–2785, 1988.

154. Wang, H. P., and Rogler, C. E. Deletions in human chromosomal arms 11p and 13q in primary hepatocellular carcinomas. *Cytogenet. Cell Genet.* 48:72–78, 1988.

155. Watson, J. V., Stewart, J., Cox, J., Sikora, K., and Evan, G. I. Flow cytometric quantitation of the c-myc oncoprotein in archival neoplastic biopsies of the colon. *Mol. Cell. Probes* 1:151–157, 1987.

156. Watatani, M., Perantoni, A. O., Reed, C. D., Enomoto, T., Wenk, M. L., and Rice, J. M. Infrequent activation of K-*ras*, H-*ras* and other oncogenes in hepatocellular neoplasms initiated by methyl (acetoxymethyl) nitrosamine, a methylating agent, and promoted by phenobarbital in F344 rats. *Cancer Res.* 49:1103–1109, 1989.

157. Williams, A. R., Piris, J. M., Spandidos, D. A., and Wyllie, A. H. Immunohistochemical

detection of the ras oncogene p21 product in an experimental tumour and in human colorectal neoplasms. *Br. J. Cancer* 52:687–693, 1895.

158. Wiseman, R. W., Stowers, S. J., Miller, E. C., Anderson, M. W., and Miller, J. A. Activating mutations of the c-Ha-*ras* protooncogene in chemically induced hepatomas of the male B6C3 F1 mouse. *Proc. Natl. Acad. Sci. U.S.A.* 83:5825–5829, 1986.

159. Yamada, H., Sakamoto, H., Taira, M., Nishimura, S., Shimosato, Y., Terada,M., and Sugimura, T. Amplifications of both c-Ki-ras with a point mutation and c-myc in a primary pancreatic cancer and its metastatic tumors in lymph nodes. *Jpn. J. Cancer Res.* 77:370–375, 1986.

160. Yamamoto, T., Hattori, T., and Tahara, E. Interaction between transformed TGF growth factor-alpha and c-Ha-*ras* p21 in progression of human gastric carcinoma. *Pathol. Res. Pract.* 183:663–669, 1988.

161. Yanez, L., Groffen, J., and Valenzuela, D. M. c-K-*ras* mutations in human carcinomas occur preferentially in codon 12. *Oncogene* 1:315–318, 1987.

162. Yasui, W., Hata, J., Yokozaki, H., Nakatani, H., Ochiai, A., Ito, H., and Tahara, E. Interaction between epidermal growth factor and its receptor in progression of human gastric carcinoma. *Int. J. Cancer* 41:211–217, 1988.

163. Yasui, W., Sumiyoshi, H., Yamamoto, T., Oda, N., Kameda, T., Tanaka, T., and Tahara, E. Expression of Ha-ras oncogene product in rat gastrointestinal carcinomas induced by chemical carcinogens. *Arch. Pathol. Jpn.* 37:1731–1741, 1987.

164. Yokota, J., Toyoshima, K., Sugimura, T., Yamamoto, T., Terada, M., Battifora, H., and Cline, M. J. Amplification of c-erbB-2 oncogene in human adenocarcinomas *in vivo. Lancet* i:765–767, 1986.

165. Yokota, J., Tsunetsugu-Yokota, Y., Battifora, H., LeFevre, C., and Cline, M. J. Alterations of *myc, myb,* and *ras* proto-oncogenes in cancers are frequent and show clinical correlation. *Science (Wash DC)* 231:261–264, 1986.

166. Yokota, J., Yamamoto, T., Miyajima, N., Toyoshima, K., Nomura, N., Sakamoto, H., Yoshida, T., Terada, M., and Sugimura, T. Genetic alterations of the c-*erb*B-2 oncogene occur frequently in tubular adenocarcinoma of the stomach and are frequently accompanied by amplification of the v-*erb*A homologue. *Oncogene* 2:283–287, 1988.

167. Yoshimoto, K., Hirohashi, S., and Sekiya, T. Increased expression of the c-myc gene without gene amplification in human lung cancer and colon cancer cell lines. *Jpn. J. Cancer Res.* 77:540–545, 1986.

168. Yuasa, Y., Gol, R. A., Chang, A., Chiu, I.-M., Reddy, E. P., Tronick, S. R., and Aaronson, S. A. Mechanism of activation of an N-*ras* oncogene of SW-127, human lung carcinoma cells. *Proc. Natl. Acad. Sci. U.S.A.* 81:3670–3674, 1984.

169. Yuasa, Y., Oto, M., Sato, C., Miyaki, M., Iwama, T., Tinomura, A., and Namba, M. Colon carcinoma K-ras 2 oncogene of a familial polyposis coli patient. *Jpn. J. Cancer Res.* 77:901–907, 1986.

170. Zhang, X.-K., Huang, D.-P., Chiu, D.-K., and Chiu, J.-F. The expression of oncogenes in human developing liver and hepatomas. *Biochem. Biophys. Res. Commun.* 142:932–938, 1987.

11

Oncogenes and Tumor Suppressor Genes in Solid Tumors: Breast, Gynecologic, and Urologic Tumors

Cecilia M. Fenoglio-Preiser
Teri A. Longacre
Margaret B. Listrom

Breast

As in other sites of the body, oncogene expression has been detected in both the nonneoplastic as well as the neoplastic breast. Some oncogenes are of particular interest in this site. They include the *int* genes, since they were first discovered in mucinous tumors associated with the mouse mammary tumor virus. The protein encoded by this oncogene acts, presumably, in such a way as to stimulate normal growth and development. This is particularly likely since the protein is among a group of fibroblast growth factors that could stimulate the development of normal breast architecture. Also of interest are the various growth-factor receptors encoded for by the oncogenes, since their expression may well stimulate epithelial cell growth leading to autocrine-stimulated proliferation. The relationships between oncogenes, growth factors, and growth-factor receptors will undoubtedly become of even more interest as more data become available. Finally, the amplification of the Her-2/*neu* gene has some prognostic significance.

NONNEOPLASTIC BREAST

int-2 transcripts are detectable during the luteal phase of the menstrual cycle and in juvenile hyperplasia but not in pregnancy-related hyperplasia. c-*myc* expression parallels that of the *int*-2, but no *int*-1 transcripts have as yet been identified in normal human breast.[82] Low levels of H-*ras*, K-*ras*, and N-*ras* are expressed in a variety of nonmalignant breast tissues, including fibroadenomas, pregnancy-related hyperplasia, and fibrocystic disease.[255] Insulin-like growth factor II, platelet-derived growth factor, transforming growth factor α, epidermal growth factor receptor (EGFR), and c-*mos* mRNA transcripts have also been described.[177] Relatively abundant c-*myc*, EGFR, transforming growth factor β, and *erb*B-2 mRNA are pro-

219

duced in normal breast cell lines established from mammoplasties.[262] Similar studies performed on tissues obtained at autopsy have shown that the expression of the *erb*B-2 (Her-2/*neu*) gene is relatively low in breasts of pregnant, postpartum and postmenopausal women; whereas pubertal and luteal-phase breast tissues are higher expressors. Histochemical analyses with antibody to the *erb* B-2 protein localize this expression to the intermediate-sized ducts and terminal duct lobular units, with more consistent staining of the epithelium in the extralobular ductal system.

BREAST TUMORS

There have been numerous studies on proto-oncogenes in breast cancer, mostly involving *ras, myc,* HER-2/*neu*, and *int* genes. However, it remains unclear whether the observed proto-oncogene alterations are primary events in the development of human breast cancer, whether they exacerbate tumor development, or whether they represent secondary events occurring within a highly unstable and aberrant tumor cell genome, since so many genes are amplified. Unlike carcinomas of the lung, colon, and bladder, no single mutation in any oncogene is consistently associated with breast cancer. However, as will be seen, alterations of various oncogenes have been associated with breast tumor progression and metastasis; it is this latter data that constitute the strongest evidence for the involvement of oncogenes in breast cancer.

Based on their study of linkage analysis in patients at high risk for developing breast cancer, Hall and associates suggest that oncogenes are not the sites of primary lesions for inherited breast cancers, but that the primary carcinogenic lesions involve recessive alterations in tumor suppressor genes [77] in a manner similar to that which exists for retinoblastoma (see Chapter 9) or for colon cancer [18,216] (see Chapter 10). In this model, alterations in oncogenes could be associated with subsequent tumor invasion or metastasis, if abnormal oncogene expression influenced the development of existing transformed clones of breast epithelial cells.

ras Genes

Early studies demonstrated that approximately 10–15% of all human carcinomas contain *ras* gene point mutations, but these are uncommon in breast carcinomas [17,115,235] (Table 11.1). In a recent study involving breast cancer, the PCR technique was used to amplify DNA fragments containing Ki-*ras*, Ha-*ras*, and N-*ras* codons 12, 13, and 61. They were then probed on slot blots, with labeled synthetic oligomers, to detect nonconservative single-base mutations. Activating mutations were found in 1 of 40 primary tumors (Ki-*ras* codon 13), 0 of 7 lymph node and skin metastases, 1 of 9 malignant effusions (Ki-*ras* codon 12), and 2 of 5 cell lines (Ki-*ras* codons 12 and 13).[192] The rarity of mutations, even in the setting of amplified sequences, suggests that if *ras* gene activation is important in breast cancer it must occur by a mechanism other than mutation.

An alternate mechanism of *ras* activation in breast cancer involves enhanced *ras* expression,[85] and transformation of primary and established breast epithelial cells by *ras*-p21 is concentration dependent.[183] Numerous studies, utilizing

Table 11.1.
Breast Cancer and ras Genes

Study Result	References
Amplification or overexpression present	
Ki-*ras* overexpressed, MCF-7 cell line	74, 232
Variable N-*ras* amplification, MCF-7 cell line	1, 2, 85, 86
Enhanced *ras* expression in primary breast cancer	17, 215, 223
1/65 had 50x amplification c-Ha-*ras* but no overexpression	12
Expression present	
16/22 primary breast cancers, Ha-*ras*	234
Low N-*ras* expression 1° breast cancers	234
Ha-*ras* present in 65%, c-Ki-*ras* 30%, N-*ras* 15%	12
Expression of all 3 *ras* genes greater in carcinomas than normals	
No expression	
No Ki-*ras* expression in 22 primary breast cancers	233
No rearrangements or mutations	
32 primary breast cancers: no evidence of mutation codons	
12 and 61, Ha-*ras,* or codon 12, Ki-*ras*	
Mutations	
Mutated form p21 in carcinosarcoma cell line HS57AT	192
Activating mutation 1/40 primary carcinoma, Ki-ras	192
Activating mutation 1/9 malignant effusion, Ki-ras	192
Activating mutation 2/5 cell line, Ki-*ras*	192
No mutation in Ha-*ras* or Ki-*ras*	17

monoclonal antibodies reactive with the *ras* p21 protein or probes for *ras* mRNA, have demonstrated enhanced expression of the *ras* proto-oncogene in breast carcinomas as compared to benign counterparts.[1,2,74,85,160,223,235]

The two most popular antibodies used to study *ras* expression have been RAP-5, which was generated using a synthetic peptide consisting of amino acids 10–17 of mutated human Ha-*ras* p21,[85] and Y13-259, which is said to immunoprecipitate both mutated and proto-oncogenic forms of p21[ras].[39,85,160,235] Studies with these antibodies on breast tissue have generated conflicting results (Table 11.2), particularly when the antibody RAP-5 is used. As noted in Chapter 5, it is unlikely that RAP-5 specifically recognizes *ras* p21 proteins. Nonetheless, because numerous studies have been published using this antibody, they will be briefly summarized here.

Utilizing the quantitative direct-binding liquid competition radioimmunoassay, absolute levels of Ha-*ras* p21 were determined in human breast cancers, benign lesions, and/or their respective normal tissues. Enhanced Ha-*ras* expression was demonstrated in 66% of breast carcinomas as compared with their normal counterparts, with levels in breast carcinomas ranging from 10.1 to 50.4 pg *ras* p21/μ protein. Some dysplastic lesions of the breast were also associated with elevated p21 proteins.[85,86]

Enhanced p21 immunostaining was found in most carcinomas without significant reactivity of benign tissues,[85,86] although in a subsequent study, p21 *ras* expression was seen in lesions exhibiting atypical hyperplasia.[160] p21 proteins localized to

Table 11.2.
p21 in Breast Tissue

Study Result	Reference
ras p21 (RAP-5) in 63% of invasive ductal carcinomas	235
Enhanced p21 (RAP-5) staining in carcinomas; no significant staining in benign lesions; enhanced p21 expression in 66% breast cancers as compared to normal in RIA assays	85, 86,
ras p21 in 90% of breast carcinomas	229
p21 expression significantly higher in cancer cells than in normal cells. However, there was no clear cut correlation between immunoreactivity and prognosis	38
Increased p21 in patients with positive lymph nodes	130
p21 staining - atypical hyperplasia and mutiple fibroadenomas	160
p21 staining significantly greater in female than male breast cancer	130
p21 staining in invasive cancers, CIS and hyperplasia associated with fibroadenomas and cystic disease	2
Similar staining of p21 in benign and malignant tissues	24, 29, 69
p21 staining of benign breast and in 30% of tumors with high proliferative rates	252
p21 staining of normal breast and myoepithelium with occasional stronger staining in normal than cancer (Y13 259)	252
p21 staining of myoepithelial cells in normal breast, strong staining of neoplastic (Y13 259)	70

the epithelium of the terminal duct lobular unit,[160] consistent with the theory that it is the peripheral lobular portion of the breast epithelium which undergoes malignant transformation. Invasive carcinomas demonstrated the highest levels of *ras* p21 as determined by the percentage of reactive cells. Decreasing expression was seen in carcinoma *in situ*, hyperplasias with atypia, and hyperplasias without atypia, respectively.[160] The reactivity profile of lactating breast was similar to that of normal postpubertal breast without hyperplasia.[160] A heterogenous pattern of *ras* p21 expression was seen in sections of both primary and metastatic carcinomas.[160]

Immunostaining of p21 proteins, using the antibody Y13–259, in paraffin sections of tissues fixed in periodate-lysine-paraformaldehyde demonstrated strong positive staining in breast cancers and small premalignant breast lesions.[70] *In situ* and invasive cancers both showed stronger immunoreactivity than surrounding normal tissues. Staining was usually cytoplasmic, with membrane accentuation in some cases. Some nuclei also stained. These authors found that in normal breast, myoepithelial cells stained more strongly than epithelial cells.[70]

Other investigators, using the same antibodies, also identified *ras* p21 proteins in both benign and malignant tissues[24,29,69,252] (Table 11.2). Chesa and associates found staining in cells not considered to be proliferative in nature.[29] Walker and Wilkinson, utilizing the antibody Y13–259 applied to frozen sections of normal, benign, proliferative breast, fibroadenomas, and carcinomas found uniform staining of normal breast epithelium and myoepithelium, with occasional stronger staining in areas of epithelial hyperplasia and benign breast disease.[252] Contrary to other reports, decreased expression was found in half of the carcinomas. Thirty percent of carcinomas exhibited heterogeneous staining stronger than that of the normal breast, which did not relate to tumor grade or nodal status, but showed a significant correlation with proliferation rate as determined by staining with the monoclonal antibody Ki-67.[252]

Whittaker and associates examined 62 examples of benign fibrocystic disease, fibroadenomas, carcinomas, and surrounding nonmalignant tissue from 50 patients to determine the level of expression of c-*myc*, c-Ha-*ras*, c-Ki-*ras*, and N-*ras*.[255] Their results demonstrated that the frequency and relative level of expression of these oncogenes was significantly greater in carcinomas than in benign breast tissues. However, some fibrocystic disease specimens with prominent hyperplasia also exhibited enhanced oncogene expression.[255]

Liu and associates studied sequential breast cancer effusions in a single patient and found common cytogenetic abnormalities in all effusion samples, suggesting the presence of a single progenitor metastatic cell.[127] Only the last effusion exhibited a mutated c-K-*ras* gene and loss of heterozygosity of the c-H-*ras* locus. The genetic abnormalities detected in the last effusion correlated with improved *in vitro* growth of the primary cells and the ability to establish breast cell lines from the metastatic cells, as compared to cells from the primary tumor. The emergence of the mutant *ras* gene and loss of heterozygosity were associated with a more aggressive tumor phenotype occurring late in the course of the patient's disease and not with the initiation of the primary breast cancer or the establishment of metastases.[127]

RFLPs are found on Southern blot analyses with c-Ha-*ras*-1 probes.[124,208] Four common and 16 rare alleles were detected in a combined population of the cancer-bearing and normal individuals. The distribution of common and rare alleles differed significantly between the two populations. Common restriction-fragment alleles represented 91% of the allele pool in the normal control population and only 59% of the allele pool in breast carcinoma patients. The frequencies of two common alleles were diminished in breast cancer patients.[116,124,208] Concomitant increases in two rare c-Ha-*ras*-1 alleles occurred. One of the rare alleles was significantly associated with breast carcinoma.[116] By evaluating the intensity of the two constitutional c-Ha-*ras*-1–related *Bam*HI restriction fragments, 14 of 51 tumors from heterozygous patients demonstrated loss of one allele.[234] This did not alter *ras* p21 expression as determined by Western blot analysis. However, loss of the c-Ha-*ras*-1 allele did correlate significantly with histological grade III tumors, lack of estrogen and/or progesterone receptors, and a poorer clinical outcome. Thus, genotypic analyses of the c-Ha-*ras*-1 locus may be of prognostic value in determining patients at risk for developing breast carcinoma.[234]

Additional experimental evidence for the importance of *ras* genes in some phase of mammary tumorigenesis is provided by the observation that transfection of MCF-7 cells with the *ras* oncogene bypasses the dependence of breast cancer cells on exogenous estrogens for tumorigenicity in nude mice.[100] An activated c-Ha-*ras* gene has also been found in mammary carcinomas induced by N-nitroso-N-methylurea and dimethylbenzanthracene (DBA) in rats and mice.[27,39,263] The DBA-induced tumors characteristically contain an activated c-H-*ras*-1 gene with an A→T substitution in the 61st codon.[27] Furthermore, Tremblay and associates found that by constructing transgenic mice carrying an activated v-H-*ras* oncogene under the control of the mouse mammary tumor virus enhancer-promoter, several types of tumors including breast tumors were induced.[236] The mammary tumors tended to be multiple and histologically were variably differentiated adenocarcinomas or adenoacanthomas.[236] In related studies, Andres and associates demonstrated that the activated Ha-*ras* oncogene, under the control of the whey acidic protein gene, was also able to induce mammary tumors, but at a lower frequency.[6]

Her-2/*neu* (c-*erb*B-2) Genes

Amplification of the *neu* oncogene is a frequent alteration in primary breast carcinomas and breast carcinoma cell lines[89] (Table 11.3); the gene may also show structural abnormalities.[17] King and associates cloned and partially sequenced an *erb*B-related gene detected because of its amplification in mammary carcinoma. It was later shown to be identical to the *neu* oncogene originally found in neural tumors.[37] Amplification has been reported in 10–40% of primary breast carcinomas,[3,14,75,76,104,215,239,244,259,260,267] resulting in elevated expression levels of *neu* mRNA and protein.[75,76,104,108,114,191,239,240,246] Overexpression of *erb*B-2 mRNA and protein is also detectable in the absence of gene amplification, suggesting that multiple mechanisms account for the overexpression.[14,114] Several reports have linked *neu* DNA amplification to an unfavorable prognosis,[196,215,218,238,244,267] to the presence of lymph node metastases,[14,215,218,267] and to poor nuclear grade (Table 11.3).

Slamon and co-workers investigated *neu* oncogene levels in two groups of patients with breast cancers, by isolating DNA from tumor specimens and subjecting it to Southern blot analysis after digestion with restriction enzymes.[215] There was a significant increase in the incidence of *neu* gene amplification in primary tumors of patients with metastatic carcinoma in more than 3 axillary lymph nodes, as compared to those with fewer than three positive nodes or those with negative nodes. The presence of *neu* oncogene amplification correlated with positive estrogen-receptor status and with large primary tumors. More importantly, there was a statistically significant correlation between the degree of gene amplification and both survival and time to relapse. Higher copy numbers of the *neu* gene generally correlated with more aggressive tumors, in a manner similar to that observed with N-*myc* gene amplification in neuroblastoma. This led the investigators to suggest that the gene may play a role in the biologic behavior and/or pathogenesis of human breast cancer.

Although it has been suggested that HER-2/*neu* amplification correlates with poor prognosis, this association is not consistently observed (Table 11.3). Ali and associates studied the relationship between gene copy number and biological characteristics of primary human breast carcinomas and found a 2–40-fold increase in HER-2/*neu* in 12 of 122 DNAs.[3] There was no evidence in this study of an association between increased copy number and tumor behavior. The only variables important for predicting disease-free and overall survival in this study were the number of positive lymph nodes and the progesterone-receptor status,[3] the same two variables previously identified by others.[30]

Most recently, Slamon and associates have shown that the HER-2/*neu* protooncogene is amplified in 25–30% of human primary breast carcinomas and that this alteration is associated with disease progression.[218] These authors demonstrated a direct correlation between amplification and overexpression. The authors

Table 11.3.
Incidence of Her-2-*neu* Gene Amplification in Breast Cancer Patients

Reference	All Patients		Node-Positive Patients		Follow-up Comments[a]
	Number Amplified	(%)	Number Amplified	(%)	
3	12/122	(10)	8/75	(11)	No association with nodal status, hormone receptor status, grade, or age
14	13/51	(25)	7/17	(41)	Amplification associated with nodal status and grade, no association with ER
105	6/61	(10)			No association with nodal status or clinical stage
75	33/80	IBC (41)			Amplification
	27/141	NBC (19)	18/72	(25)	associated with positive LN, ER
215	53/189	(28)	34/86	(40)	Amplification of Her-2/*neu* association with LN status and early occurrence; decreased survival in LN positive patients
240	16/95	(17)	6/41	(15)	Amplification associated with overexpression
244	7/37	(19)	4/10	(40)	Amplification associated with poor prognosis
246	12/36	(33)	3/15	(20)	Amplification associated with elevated protein expression
264	52/291	(18)			Amplification associated with negative ER, PgR
267	15/86	(17)	8/37	(22)	Amplification associated with advanced stage, early recurrence, and nodal status
266	17/157	(11)		(11)	No association with size, HRS, nodal status, stage, age, or recurrence

[a]Abbreviations: ER = estrogen receptor; PgR = progesterone receptor; HRS = hormone receptor status; IBC = inflammatory breast carcinoma; NBC = noninflammatory breast carcinoma; LN = lymph node.

analyzed 668 human breast cancer specimens, 526 of which had sufficient clinical information to allow an evaluation of an association between gene amplification and disease outcome. Multivariate analysis showed that HER-2/*neu* amplification was an independent predictor of both disease relapse and overall survival, and was superior to all other known prognostic factors, with the exception of positive lymph nodes. In node-negative patients, no association existed between HER-2/*neu* amplification and prognosis.

Walker and co-workers confirmed the correlation between gene amplification and the presence of the HER-2/*neu*–encoded protein and mRNA content.[251] Both the protein and the mRNA were only detectable in patients with amplified genes.

van de Vijver produced two monoclonal antibodies directed against different epitopes of the *neu* protein and showed that immunohistochemical staining patterns correlated with *neu* gene amplification as well as several clinicopathological features of breast carcinoma.[239-241] Of 189 tumors from patients with stage II breast cancer, 14% had *neu* membrane staining. *Neu* overexpression was associated with larger tumor size, but not with lymph node involvement. *neu* immunoreactivity in nodal metastases was the same as in the primary tumors. Disease-free survival was not significantly shorter during a median follow-up of 37 months in patients with *neu* overexpression after adjustment for tumor size. Of 45 *in situ* carcinomas, 42% had *neu* membrane staining. These were all of the large cell, comedo carcinoma type. None of the 16 ductal *in situ* carcinomas of small cell, papillary, or cribriform growth type overexpressed the *neu* gene. This led the authors to conclude that *neu* overexpression may be an early step in the development of a specific subset of breast cancer.[239]

Wright and associates stained formalin-fixed, paraffin-embedded tissues from 185 primary breast cancers, using a polyclonal antibody directed against the HER-2/*neu* oncoprotein.[258] Positive staining correlated with negative estrogen receptor (ER) status and high histological grade, but not with lymph node or epidermal growth factor receptor (EGFR) status, or tumor size.

More recently, 728 primary human breast tumor specimens were analyzed for HER-2/*neu* oncogene protein using Western blot analysis. The authors found that node-positive patients with higher levels of the HER-2/*neu* protein had statistically shorter disease-free periods and overall survival times than did patients with lower levels of the protein. This relationship between HER-2/*neu* protein levels and survival was not seen in node-negative patients.[230]

van de Vijver and associates undertook a systematic study of oncogene alterations in human breast cancers and found occasional examples of c-*myc*, c-Ki-*ras*, and c-*erb*B-1 amplification.[240] The most common alteration detected by this group was coamplification of the linked oncogenes on chromosome 17, HER-2/*neu* and c-*erb*A. The presence of a normal copy number of the gene encoding the tumor antigen p53 (which is on chromosome 17p, as opposed to HER-2/*neu* which is on 17q) in all tumor DNAs demonstrates that HER-2/*neu* amplification was not due to increased ploidy of chromosome 17. It is now accepted that c-*erb*B and *ear*-1 represent a coamplification unit.[240,261] (*ear*-1 is an *erb*A-related sequence found on chromosome 17.[67,261]) Amplification of both of these genes relates to poor prognosis in patients with breast cancer.[237]

In another line of investigation, it was shown that a monoclonal antibody directed against the extracellular domain of p185$^{Her-2/neu}$ specifically inhibits growth of breast cancer cells overexpressing the HER-2/neu gene product and prevents HER-2/neu–transformed NIH 3T3 cells from forming colonies in soft agar.[88] Resistance to the cytotoxic effect of tumor necrosis factor α was also reduced by this antibody.

Her-2/neu and EGFR genes are both associated with breast cancer progression in a number of investigations.[41,117,196,197] An inverse relationship exists between the presence of estrogen receptors and EGFR in primary breast cancer, with EGFR status being a predictor of early recurrence and death from breast cancer. Primary sequence analysis indicates that the neu proto-oncogene shows about a 50% sequence homology with the EGFR gene, even though the neu proto-oncogene maps to chromosome 17 whereas the EGFR gene is located on chromosome 7.[114,175,200]

Guerin and co-workers analyzed the structure and expression of HER-2/neu and EGFR genes in inflammatory and noninflammatory breast cancers.[76] A total of 221 untreated patients were analyzed at different clinical stages. Amplification and overexpression of the HER-2/neu proto-oncogene were observed in 27% and 47% of tumors respectively, and were strongly associated with breast cancers with the most unfavorable prognosis (inflammatory breast cancer and noninflammatory breast cancer with multiple positive lymph nodes). The EGFR gene was neither amplified nor rearranged. EGFR transcripts were detected in 46% of tumors and were observed more frequently in inflammatory breast cancers than in noninflammatory breast cancers. In noninflammatory breast cancers the presence of EGFR transcripts increased linearly with lymph node involvement and was associated with estrogen receptor–negative tumors. Analysis of both genes from the same tumor samples indicated that both are independently associated with tumor aggressiveness. Furthermore, in noninflammatory breast cancers these two genes were independently activated, in contrast to inflammatory breast cancer in which activated genes were negatively correlated. This led the authors to suggest that HER-2/neu and EGFR genes play different roles in inflammatory and noninflammatory breast cancers.[76]

int and hst Genes

int genes are not detectably expressed in normal mammary tissue or in other adult tissues, except the testis of mature males, where an int-1 RNA of a slightly different size from that present in breast tumors is found.[92] However, in murine mammary tumors, transcriptional activation of the int-1 gene by the mouse mammary tumor virus (MMTV) promoter inserted near the int-1 gene is thought to play a key role in breast tumor induction.[22,43,44,157,158,171,172] Nusse proposes that the int genes act not as mitogenic growth factors, but as morphogens leading to aberrant development.[156] Normally, the breast cycles between growth, development, and regression. Tumors induced by MMTV only arise in animals going through cycles of hormonal stimulation associated with pregnancy. This stimulation, by itself, causes massive cell proliferation in the breast. During the early, hormone-dependent phases of tumorigenesis, int genes are already activated,[135,172] but the tumor cells are still completely dependent on hormonal stimuli. Activation of the morphogen then leads to aberrant growth of the tissues, with unstable cell populations. The int gene products may deregulate the interplay between different cells within the breast lobules and ducts, leading to full tumorigenesis. Another mechanism suggested by the specific biology of the breast is a delay of cell death following lactation when the gland regresses. Following a subsequent round of glandular development, cells with a prolonged lifespan could still be present, attain a specific growth advantage, and be prone to secondary tumorigenic events.[156] Transfection of int-1 into mammary cells results in transformation.[184]

An int-1 allele resembling those found in virus-induced tumors, with MMTV LTR placed 5′ to the int-1 gene in the opposite transcriptional orientation, was constructed and introduced into transgenic mice. Mice harboring this allele express int-1 RNA at high levels in both male and female breast and salivary glands and in male reproductive organs. The breasts of males and virgin females become grossly hyperplastic compared to nontransgenic littermates. Breast, and less frequently salivary gland carcinomas occur in these animals at rates indicating that activation of the int-1 gene and associated hyperplasia are initiating events in the multistep carcinogenic process involving these organs.[184]

int-2 sequences are amplified 2–20-fold in 23% of breast cancers. In addition to amplification, the EcoR1 restriction site is often lost. This loss correlates with gene amplification. These two events involve the same allele of the gene.[242]

Other genes activated by MMTV in mouse mammary tumors are int-3 and int-4.[156] Nothing is known about their normal expression. Amplification of int-2, the human homologue of the murine integration site of MMTV, has been observed in human breast carcinomas.[122,268] hst-1/int-2 represents another coamplification unit and both are frequently (17%) amplified in patients with breast cancer,[233] particularly those in a younger age group. The bcl-1 gene, which is part of the amplification unit at 11p13, is also amplified in many breast carcinomas. Amplification of these genes correlates with poor prognosis for those tumors carrying more than four copies of the gene.[238] int-p is a separate gene that has been described in the plaque-type of carcinoma characteristic of the GR strain of mice.[202]

myc Genes

The c-myc oncogene is a cell cycle–dependent gene essential for cell growth (see Chapter 5).[95,133] Its expression is activated by many peptide growth factors including platelet-derived growth factor, fibroblast growth factor, EGF, and growth hormone.[96,102,145,180]

The frequency and level of amplification of myc genes in breast cancer and breast cancer lines varies greatly in different laboratories.[55,113,119,136,217] In the line SW-613-S, clones with high levels of amplification and expression of the c-myc genes are highly tumorigenic in nude mice, whereas those with low levels are not.[119] In some carcinomas, somatic alterations of myc can also be detected.[17,53,244] The possibility that alterations of myc are involved in some stage of breast tumor development is supported by detection in one myc allele of a tumor-specific rearrangement caused by insertion of a LINE-1 sequence and accompanying amplification of the other myc allele.[139]

Escot studied the genomic organization of the c-myc locus from 121 human primary breast carcinomas and found two types of alterations: (a) the c-myc protooncogene was amplified 2–15-fold in 32% of carcinoma DNAs; and (b) a nongermline c-*myc*–related fragment of variable size was detected in 5 primary breast carcinoma DNAs.[53] With three exceptions, all tumors containing a genetic alteration of the c-*myc* locus were invasive ductal carcinomas. A significant correlation existed in patients more than 50 years old and the presence of the genetically altered c-*myc* gene. Enhanced levels of c-*myc* RNA were observed in 10 of 14 breast carcinomas. The c-*myc* gene was genetically altered in 6 of these 10 tumors. The frequency with which the gene is altered and the correlation with age may play a role in the development of breast carcinomas.

The studies of Bonilla and co-workers in 48 human primary breast tumors yielded similar results.[19] Amplification and rearrangements were observed in 56% of the tumors in this study, and the c-*myc* proto-oncogene was amplified 2–15-fold in 41% of tumors. Nongermline c-*myc*–related fragments (rearrangements) of variable size were detected in 7 primary breast tumors, one of which was benign. Four of these tumors had both rearrangements and amplification; in the other three only rearrangements were present. In this report, alterations did not correlate with aggressive tumor behavior, although they may be associated with the genesis of breast cancer. However, a more recent study suggests that amplification of the c-*myc* gene does correlate with poor prognosis.[237] It has also been suggested that c-*myc* amplification occurs preferentially in aggressive or advanced tumors,[260] but does not correlate with histologic type, grade, estrogen receptor status, or stage of the disease.[244]

Alterations in c-*myc* proto-oncogene expression occur after treatment of human breast carcinoma cells with EGF, TGF β, or estradiol.[49,50,59,199] Estradiol enhances and tamoxifen decreases c-*myc* expression in ER-positive cells. No effect of estradiol or tamoxifen is seen in ER-negative cells, suggesting that the ER-protein participates in *myc* expression. Since the action of estrogens, unlike peptide growth factors, is mediated by a nuclear estrogen receptor that is a nuclear transcriptional activator, estradiol may regulate c-*myc* oncogene expression at the transcriptional level, rather than at the posttranscriptional level.[50,59,140,199]

Insertion of the v-*myc* oncogene into mouse mammary gland alters the three-dimensional morphology of the growth patterns. Cells infected with the virus show a characteristic hyperplastic pattern, with the ducts more closely packed than usual.[52] Final evidence that the *myc* gene plays a role in breast tumor induction is provided by experimental models that fuse c-*myc* coding sequences with a retroviral long terminal repeat that leads to the formation of hormone-dependent breast cancers in transgenic mice.[120,213] The incidence of tumors in strains of transgenic mice carrying c-*myc* driven by the MMTV promoter is approximately 40%, and most tumors develop within 15 months of age.[120] In another model, transgenic mice that possess a c-*myc* gene under the control of glucocorticoid response in MTR promoter develop tumors. The tumor cell precursors are highly responsive to the glucocorticoids, and it has been suggested that the hormone-stimulated overexpression of the c-*myc* genes may be an important factor in the development of these tumors.

EGF Receptors

The MDA-468 human breast cancer cell line has a high number of EGFRs and contains an amplified EGFR gene. Its growth is inhibited *in vitro* by pharmacological doses of EGF. Several MDA-468 clones have been derived that are resistant to EGF-induced growth inhibition. The number of EGFRs in these clones is similar to that of normal fibroblasts, but the EGFR gene amplification is lost in the clones. Karyotypic analysis demonstrates that the cells have an abnormally banded region on chromosome 7p, which contains the amplified EGFR sequences as shown by ISH. Five of the six variants studied were able to generate tumors in nude mice, but their growth rate was significantly lower than that of tumors derived from the parental cells. This variant was unable to produce tumors, but was uniquely dependent on EGF for growth in soft agar.[61]

Zajchowski and co-workers studied the expression of genes that may be involved in the regulation of human mammary epithelial cell growth including TGF α, c-*myc*, *erb*B-2, EGFR, Ha-*ras*, and pS2.[262] These authors found that normal breast cells produce high levels of EGF mRNA that is translated into low-affinity EGF-binding molecules. In ER-negative cancer cell lines the EGFR gene was expressed at levels comparable to those seen in normal cells. In contrast, EGFR and TGF α mRNAs were reduced in ER-positive tumor cell lines compared to ER-negative tumor cells. These studies suggest that mRNA levels for TGF-β, *erb*B-2, c-*myc*, and Ha-*ras* cells are greater in normal cells than in tumor cells.[262]

Other Oncogenes

Breast cancer cell lines frequently express growth factors and growth factor receptors.[87,167,169,170] *erb*A-2 codes for a thyroid hormone receptor, and thyroid hormone receptors are known to be present on breast cancer cell lines.[23] c-*fms* is also found in breast cancer cell lines.[87]

The possibility that alteration of c-*myb* might occur with either tumor progression or metastasis was suggested by the loss of intensity of one *Eco*R1 allele in a metastatic tumor, compared with the primary breast tumor and normal tissue of the same patient.[260]

c-*mos* and c-*myb* are found in 20–30% of breast cancers.[17] The suggestion that *mos* is associated with human breast cancer came from the observation that a 5 kb *Eco*R1 *mos* allele appeared in tumor tissue of 6 of 75 breast cancer patients and in the normal cells of three of these patients, but did not occur among 69 unrelated control patients.[123,125] The 5 kb *mos* allele was not found in any of the 21 familial breast cancer patients studied by Hall.[77]

Biunno and co-workers described sporadic alterations in the structures of c-*erb*B-1 and c-*myb* among 65 primary human breast cancers.[17] Seventy-five percent of cases had amplified c-*fos* genes, but there was no significant correlation between the c-*fos* mRNA level and the number of positive lymph nodes.[17]

High expression of the *sis* transcript has been observed in breast cancer. All of the tumors were T3 lesions and histologically infiltrating ductal carcinomas accompanied by a marked stromal reaction. The transcripts localized to the epithelial cells of both benign and malignant lesions by *in situ* hybridization. It is possible that *sis* expression by the epithelial cells is responsible for the desmoplastic tissue response.[190]

Kasid and Lippman studied the interactions of oncogenes and estrogen receptors in MCF-7, an ER-positive cell line dependent on estradiol for DNA synthesis, elevation of the thymidine-labeling index, nucleoside incorporation into DNA, and cell growth.[99] Estradiol seems to exert its control of replication by transcriptional regulation of two "second messenger" proteins—thymidine kinase and dihydrofolate reductase. Antiestrogens reverse these effects. In addition, estradiol induces the secretion of three polypeptide growth factors: IGF-1 (insulin growth factor–1), TGF-α, and platelet-derived growth factor. MCF-7 cells transfected with v-Ha-*ras* oncogene become independent of estradiol for growth and tumorigenicity, and refractory to control by antiestrogens; but they continue to constitutively secrete high levels of IGF-1, TGF-α, and PDGF.[167] These data indicate that activated oncogenes may mimic or bypass estrogen function and that tumor cells may acquire autocrine capabilities by the coexpression of growth factors and their receptors.

TUMOR SUPPRESSOR GENES

p53

Crawford and co-workers utilized antibodies reacting to the protein encoded by p53 and found that p53 was present in the sera of 9% of patients with localized or metastatic breast cancer.[37] The location of the first metastasis in patients whose sera was positive was unusual, with more lung metastases and fewer bone metastases than expected. The presence of circulating anti-p53 antibodies indicates that p53 is altered in amount, type, or presentation, so that it becomes immunogenic in breast cancer patients.

More recently, Cattoretti and co-workers studied 200 primary breast cancers with the anti-p53 mouse monoclonal antibody and found that 15.5% of tumors were positive.[28] The positivity was significantly associated with EGFR-positive (67.7%; $p < 0.001$) and ER-negative (73.3%; $p < 0.001$) cases. Low values for progesterone receptor (mean 67.20 ± 25.2 fmol/mg; $p < 0.05$) and a high number of cells positive for the proliferation-associated antigen Ki67 (log mean 6.88 ± 0.33; $p < .01$) were found in p53 positive tumors, as well as a high number of grade 3 infiltrating duct carcinomas. The authors concluded that p53 is associated with ER-negative, growth factor receptor–positive, Ki-67 positive, high-grade tumors and is a promising new parameter for evaluating the cellular biology and prognosis of breast cancer.

Retinoblastoma and Other Genes

Lundburg and co-workers showed that 4 of 10 breast cancers of differing histological types had similar specific losses of chromosome 13 heterozygosity, indirectly suggesting involvement of the RB gene.[129] These tumors occurred in premenopausal women and were poorly differentiated. More recently, Lee and co-workers found partial DNA deletions and absent or altered RB mRNA in two breast cancer cell lines.[121]

Structural aberrations of the retinoblastoma locus are present in 25% of breast tumor cell lines and 7% of primary tumors.[231] These changes include homozygous internal deletions, total deletion of the retinoblastoma locus, and a duplication of an exon in one case. In all cases, structural changes result in either the absence or truncation of the RB1 transcript.

Analysis of primary and metastatic breast carcinomas, as well as benign breast tissues, reveals structural abnormalities of RB-1 in 19% of primary breast cancers. The same alteration is seen in metastatic lesions, but not in benign tissues.[243]

Finally, partial deletion of chromosome 11p correlates with the primary tumor size and estrogen receptor levels. In these tumors one Ha-*ras* allele was deleted 22% of the time, as found by RFLP analysis (see above). This reduction to hemi- or homozygosity of alleles suggests the presence of a tumor suppressor gene that is important in the early genesis of breast cancer.[131] Ali had previously demonstrated the somatic loss of heterozygosity in 20% of breast cancer patients, particularly in those tumors lacking estrogen or progesterone receptors.[4]

Placenta

The early villous placenta expresses c-*myc*, c-*sis*, and c-*erb*B genes, with peak expression occurring 5 weeks following conception, followed by a decline by the end of the first trimester.[47,73,173,205] *In situ* hybridization shows a strong correlation between *myc* transcript abundance and the presence of trophoblastic proliferation.[173] c-*fos* expression is 3–100-fold greater in term placenta than it is in any other normal human cells or tissues.[142-144] The levels in human amniotic and chorionic cells are close to the level of v-*fos* expression that results in the induction of osteosarcomas in mice or transformation of fibroblasts *in vitro*.[144]

c-*fms* expression is restricted to cells of the monocyte/macrophage lineage, placenta,[144,145] and tumors of trophoblastic origin.[87,90,144,209,249] Since uterine CSF-1 levels are regulated by the female sex steroids and are elevated 10,000-fold during pregnancy,[174] it is postulated that CSF-1 may have a role in regulating placental trophoblast proliferation, differentiation, and ion transport during pregnancy.[209] It has been suggested that both the c-*fos* and the c-*fms* encoded proteins may be associated with embryo-derived cells whose primary functions are the protection and nourishment of the human fetus.[145] It has been shown that exon 1 of the c-*fms* gene is expressed differently in trophoblastic tissue than it is in hematopoietic cells.[249]

The *int*-1 gene is highly conserved during evolution. A gene showing 38% homology with the *int* gene, called *irp*, located close to the cystic fibrosis locus on chromosome 7,[228] is expressed in fetal tissue and placenta.[156] c-*mos* is also expressed at low levels.[177]

myc, *fos*, and *ras* gene expression has been demonstrated by *in situ* hybridization in hydatidiform moles and in the BeWo choriocarcinoma cell line.[145,174]

Cervix

As in other systems, *ras* and *myc* genes are the most actively investigated genes in the female genital tract.

ras Genes

When tumor DNA preparations from cervical cancers at different clinical stages were incubated with *Msp*I restriction enzyme, analyzed by Southern blot, and hybridized with a c-Ha-*ras* VNTR probe, several electrophoretic patterns were obtained.[26,57,71] The basis of this RFLP is a region consisting of variable number tandem repeats (VNTRs) of 28 base pairs located 3′ to exon 4 of the c-Ha-*ras* locus. [185–189] One or two DNA fragments were present in cancer DNA preparations from tumors of homozygous and heterozygous patients, respectively. DNA prepared from the lymphocytes of the normal population displayed similar electrophoretic patterns. The size and frequency of the alleles occurred with similar frequencies in both cervical cancer (7.5–11%) and in normal patients (6.5%),[188] indicating that the presence of variant alleles at the c-Ha-*ras* locus is not predictive of a genetic predisposition to cervical cancer. In contrast, the loss of one c-Ha-*ras* allele (determined by comparing the c-Ha-*ras* VNTR hybridization patterns of tumor versus lymphocyte DNAs from patients homozygous for the c-Ha-*ras* alleles) is present in a relatively high proportion of patients. Heterozygosity was found in about 60% of the tumor-lymphocyte pairs analyzed. Loss of heterozygosity did not correlate with the clinical stage of disease.[188] It has been suggested that somatic loss of heterozygosity results in homozygosity for a recessive mutant allele on these chromosomes, which may represent a critical step in the development of the malignancies by the loss of an unknown tumor suppressor gene, in a manner analogous to the loss of the RB-1 gene in retinoblastoma.

In cervical cancers, mutations at codon 12 of the c-Ha-*ras* gene were detected in 2% of stage I and II tumors and in 24% of stage III and IV tumors.[188] Furthermore, the mutant oncogene was observed in 40% of carcinomas exhibiting a deletion of the c-Ha-*ras* gene on the other allele. It is unclear whether the allelic loss is the precipitating event or whether the lost allele initially contained a mutated c-Ha-*ras* gene. c-Ha-*ras* allele deletion could contribute to activation of the mutated c-Ha-*ras* gene present on the retained allele. It is interesting to note that all tumors with the mutated c-Ha-*ras* gene also contained an amplified and/or overexpressed c-*myc* gene. c-*myc* gene activation was found in 100% and 70% of tumors containing the *ras* mutation and deletions, respectively. This observation led Riou and coworkers to suggest that these two proto-oncogenes cooperate in the progression of cervical cancers.[188] In addition, more than 90% of the tumors contained human papillomavirus (HPV) sequences, suggesting that these too may have an effect in the genesis and progression of cervical cancers, in a manner similar to that described for other sites.[118] Finally, cervical squamous cell carcinomas can be induced by sequential transfection with HPV 16 DNA and V-Ha-*ras*.[45]

c-*myc* Genes

Elevated c-*myc* transcripts, without genetic rearrangements, are found in 49% of the tumors.[188] The c-*myc* gene in normal cervical epithelium was neither amplified nor rearranged. Others have detected c-*myc* gene amplification in 48% of tumors, but they also detected a high proportion of c-*myc* rearrangements.[159] Since 46% of cervical cancers contained an overexpressed gene with only a single DNA copy, it was suggested that amplification is not the only mechanism by which c-*myc* proto-oncogenes are overexpressed.[159] The proportion of tumors with amplified and overexpressed c-*myc* genes dramatically increases in advanced cancers with poor prognosis. Some early-stage cancers with overexpressed c-*myc* genes had early relapses.[187,188] This prompted the authors to study the relative prognostic importance of c-*myc* gene expression in predicting early relapse. Carcinoma samples were obtained from 72 untreated patients with either stage I or II tumors. Some patients had received external radiotherapy. High levels of c-*myc* RNA were observed in 35% of cases. c-*myc* overexpression did not significantly associate with stage, nodal status, age, or geographic origin of the patients, but a significant association existed between elevated levels of c-*myc* transcripts and the risk of relapse, irrespective of other prognostic factors.[188]

Hendy-Ibbs compared flow cytometric quantitation of DNA and c-*myc* oncoproteins in archival biopsies of uterine and cervical neoplasia, and showed that normal biopsies did contain higher oncoprotein levels than did carcinomas,[78] contrasting with the work of Riou.[188]

Sowani and co-workers analyzed p62^{c-myc} in cervical lesions ranging from cervical intraepithelial neoplasia to advanced grade 4 tumors using a monoclonal antibody, and found no correlation between the clinical stage of disease at presentation and the c-*myc* expression levels.[222] However, patients with a c-*myc* immunonegative tumor had a better disease-free and total survival time than did patients with tumors positive for this marker. The pattern of recurrence also differed between the two groups of patients. c-*myc*–positive tumors were more likely to develop extrapelvic metastatic disease, suggesting that the identification of the c-*myc* oncoprotein in cervical tissue could have potential prognostic value.

Perhaps most intriguing is the fact that HPV sequences integrate near cellular oncogenes in some cervical cancers.[51] Two HPV 16 flanking sequences derived from cervical carcinoma localize to chromosome regions 20ptr>20q13 and 3p25>3qter, which also contain the proto-oncogenes c-*src*-1 and c-*raf*-1, respectively. The HPV 16 integration site in the SiHa cervical carcinoma–derived cell line is chromosome region 13q14>13q32. In the two cervical cell lines HeLa and C4-I, HPV 18 DNA integrates in chromosome 8, and in HeLa cells it occurs within 40 kb 5' of the c-*myc* gene inside the HL60 amplification unit surrounding and including the c-*myc* gene. Additionally, steady-state levels of c-*myc* mRNA are elevated in HeLa and C4-I cells, relative to those in other cervical carcinoma cell lines, leading Durst and co-workers to postulate that insertional *cis*-acting promoter-enhancer mutagens that could activate nearby cellular proto-oncogenes may be involved in the malignant transformation of cervical cells.[51]

The role of HPV sequences in the genesis of cervical cancer is still unknown, but in addition to its effect on the *myc* gene, type 16 viral DNA has been shown to cooperate with the mutated c-Ha-*ras* gene in transforming primary cells.[134] HPV 18 E6 and E7 are amplified in Hela cells. It is of interest that the E7 product of HPV 16 and 18 binds to the RB protein product. This binding might lead to the inhibition of tumor suppressor gene functions in cervical cells infected with the virus.

Endometrium

The influence of estrogen on the induction of N-*myc* and c-*myc* proto-oncogenes in rat uteri has been examined, since it has been shown that the *myc* proto-oncogene is rapidly activated in quiescent cells when exposed to various mitogens and hormones. Estradiol induced increased c-*myc* mRNA expression three hours following administration. The N-*myc* gene responds more rapidly, within 15 minutes, suggesting that estradiol has a direct effect on the N-*myc* expression, as compared to the c-*myc* response. The latter may be mediated via autocrine, paracrine, or circulating estrogen dependent growth factors.[146] Estrogen induces epidermal growth factor receptors in uterine cells.[140]

Kacinski and associates have described high expression levels of the *fms* proto-oncogene in clinically aggressive endometrial carcinomas.[94] Occasionally, the c-*erb*B-1 gene is rearranged in uterine adenocarcinomas. The rearrangement resembles that found in the chicken v-*erb*-B oncogene that produces a truncated form of the EGFR.[265]

The chromosome segment 12q 13–15 is consistently rearranged in uterine leiomyomas. The rearrangement is mostly a reciprocal translocation between chromosomes 12 and 14 t(12;14)(q14–15;q23–24).[152] The oncogenes *int*-1 and *gli*-1 map to the segment 12q 13–14[8,9]; however, neither of these genes are amplified or rearranged when the chromosomal translocation occurs.[8]

Ovary

NORMAL OVARY

The c-*mos* gene is a member of the cellular protein kinases, which function in signal transduction patterns to regulate cell growth and differentiation. Since it is expressed in reproductive tissues, including the testis and ovary,[147] the gene is believed to play a regulatory role in germ cell development. Oocyte maturation, fertilization, and the early stages of embryonic development are controlled by maternal proteins and mRNAs that are synthesized and stored in growing oocytes.[40] The maternal RNA pool supports new protein synthesis and development of embryos to the two-cell stage.[20] Approximately 50% of the bulk mRNA of the oocyte is retained in both mature and fertilized eggs.[31] Ninety percent of the maternal mRNA is degraded by the mid to late two-cell embryo, when active transcription of the embryonic genes becomes permanent.[31] c-*mos* RNA is among the mRNAs that undergo polyadenylation in concert with oocyte maturation, correlating with the recruitment of stored mRNAs for translation.[31]

The c-*mos* mRNA is retained in mature ovulated eggs but is degraded by the late two-cell embryo stage.[72] *In situ* hybridization demonstrates that in the ovary, oocytes are the predominant, if not exclusive, source of the c-*mos* transcripts.[103] They accumulate in growing oocytes, increasing 40–90-fold during oocyte and follicular development, suggesting that developmentally regulated expression of c-*mos* exists in oocytes and may function in normal germ cell differentiation early in embryogenesis. c-*mos* levels substantially decrease in oocytes undergoing meiotic maturation. At the two-cell stage, levels of c-*mos* drop below detection levels and remain undetectable through the blastocyst stage of embryogenesis.[148]

c-*mos* is considered to be the candidate "initiator" for oocyte maturation[164,195] since c-*mos* product pp39^{mos} is required for progesterone-induced meiotic maturation in *Xenopus* oocytes. Treatment of oocytes with progesterone induces a rapid increase in pp39^{mos} that precedes both activation of maturation promoting factor and germinal vesicle breakdown. It may serve as a trigger for the G2→M transition.[195]

The *ras* gene product p21 may also be functionally important for oocyte maturation. Microinjection of a monoclonal antibody directed at a highly conserved region of p21 (amino acids 29–44) required for p21 function inhibits *Xenopus* oocyte maturation induced by insulin. The p21 appears to be active in mediating insulin-induced maturation of oocytes in this system and p21 is found to be a substrate of the insulin receptor kinase.[110] The *ras*-related genes, *ral* 2, *rho* 12, *rab* 1, and *rab* 2, are all expressed in the ovary.[161]

A series of *jun* genes are also expressed in the normal ovary. Moderate levels of *jun*D, high levels of c-*jun* and even higher levels of *jun*B are seen.[83] In addition, the c-*rel* gene is expressed in oocytes.[137]

OVARIAN TUMORS

ras and *myc* Genes

Early passages of the human teratocarcinoma cell line, PA1, are not tumorigenic in nude mice, whereas late passages are, due to the presence of a mutated N-*ras* gene at codon 12 which seems to have arisen during cell culture *in vitro*.[227]

Transcript levels of c-Ki-*ras*, c-*myc*, and c-*fos* are significantly higher in the PCC4 cells and in undifferentiated and malignant (PCC4) teratocarcinoma cells, whereas transcripts of c-Ha-*ras* are unchanged. Southern blot analysis does not reveal any structural alterations of c-*myc* or c-*fos*. The Ki-*ras* proto-oncogene is amplified 10–20-fold. N-*myc* and c-*myc* are differentially regulated during the transition of murine embryonal carcinoma cells to a proliferating state.[204]

DNA isolated from the tumor of a patient with a serous cystadenocarcinoma contained an activated *ras* gene, as detected by transfection into NIH 3T3 cells. Abnormalities in the electrophoretic mobility of the transforming gene product suggested the presence of an undescribed mutation.[56] DNA isolated from normal cells of the same patient lacked transforming activity.

However, other workers, who demonstrated the presence of *ras* gene activation in a smaller percentage of tumors, were unable to detect any mutations in codons 12 or 61 in the activated *ras* genes, using PCR techniques. The *ras* genes were activated through amplification.[242]

Zhou and associates demonstrated a unique pattern of altered *ras* proto-oncogenes in ovarian carcinomas.[268] Amplification of Ki-*ras* was found in 3 of 7 ovarian tumors; amplification of Ha-*ras* was found in 1 of 12. Other proto-oncogenes altered in ovarian adenocarcinomas included c-*myc* and c-*erb*B-2 (HER-2/*neu*).[268] Proto-oncogene abnormalities are found more frequently in aggressive high-grade tumors. Three of five high-grade tumors had detectable proto-oncogene abnormalities, as did one of four low-grade tumors. Distinctive patterns of proto-oncogene alteration had the following features: (*a*) a high incidence of *ras* oncogene amplification relative to other tumors (33% versus <3% of others); (*b*) a relatively high frequency of c-*myc* amplification (33% versus 2–12% in other tumors); and (*c*) a high frequency of multiple proto-oncogene alterations in a single tumor.

The stability of c-K-*ras* amplification was studied in a patient with a progressive serous cystadenocarcinoma.[60] Ten to twenty fold amplification of the cellular oncogene K-*ras* was present in the tumor, whereas in normal cells purified from the malignant ascites it was not amplified. Five consecutive samples were obtained by paracentesis over a nine-month period, during which time the patient received chemotherapy and underwent clinical progression of her disease. The level of c-K-*ras* amplification did not change despite progression of the tumor and induction of chemotherapy, suggesting either that the amplification was stable or that a constant selection pressure over a homogeneously amplified tumor cell population was present during this time period.

Fukomoto and associates used a modified in-gel DNA denaturation technique to detect amplified DNA sequences on surgical specimens of primary and metastatic ovarian cancers.[66] Amplified DNA sequences were found in 2 of 8 tumors. Hybridization with different oncogene probes revealed that both tumors contained amplified Ki-*ras*, which in one case was coamplified with c-*myc*.

Rodenburg and associates studied 57 advanced ovarian cancers and 28 normal ovaries, utilizing the antibody RAP5.[193] Although the pattern of staining of tumor specimens was similar to that of the germinal epithelium of normal ovaries, the intensity of the staining was more enhanced in the carcinomas. There was no correlation between staining intensity, histological type, histological grade, ploidy, or clinical outcome. Tanaka and associates demonstrated p21 protein in 4 of 4 tumors studied.[229]

c-*erb*B Proto-oncogenes

c-*erb*B-2 genes are occasionally amplified in ovarian cancers,[218,265] but c-*erb*B-1 is not amplified.[265]

Recently Slamon and associates analyzed a series of ovarian tumors for changes in Her-2/*neu*, since ovarian tumors, like breast cancers, often contain steroid hormone receptors, and epidemiologic studies suggest that they share etiologies.[218] DNA from 120 primary ovarian malignancies were evaluated by Southern blot analysis. Twenty-six percent of the cases showed HER-2/*neu* amplification. Intact RNA was available in 72 of 120 cases; these were grouped into four expression categories. Additional material was available to perform Western blot analysis using the anti-HER-2/*neu* (c-*erb*B-2) antibody, and frozen tissue was available for 75 of 120 cases. There was complete correlation between gene amplification and overexpression. Every ovarian cancer judged to be amplified had evidence of overexpression. HER-2/*neu* amplification and expression in immunohistochemical reactions correlated. Twelve percent of tumors had a single gene copy associated with gene overexpression. These authors found a statistically significant correlation between gene amplification and survival, with median survival of 1879, 959, and 243 days for patients having one copy, two copies, and five copies or more of HER-2/*neu*, respectively. Additionally, in the 72 cases for which follow-up data and expression analysis based on the immunohistochemistry were available, a significant association between HER-2/*neu* expression and survival was found. Thus, amplification of the gene occurred in 25–30% of ovarian tumors, with amplification uniformly associated with overexpression. Approximately 10% had evidence of a single gene copy with overexpression, and there was an association between amplification and clinical outcome.

fms Genes

Slamon and associates observed *fms* transcripts in RNA extracted from ovarian tumors in a dot blot assay.[217] However, it is difficult to say whether they represent transcripts present in the tumor tissue itself or whether they derive from infiltrating macrophages. However, since c-*fms* transcripts have also been described in ovarian cancer cell lines,[87] it is likely that ovarian epithelial tumors do produce these transcripts.

int Genes

As seen in an earlier section, *int* genes are transcribed at modest levels in developing murine embryos at a number of distinct stages that show both spatial and temporal regulation.[92] Thus, it is not surprising that both *int*-2[21] and *int*-1[203] are oncogenes expressed in teratocarcinoma cells. *int*-4 is transiently expressed in embryonal carcinoma cell lines.[156] The related gene k-*fgf*'s expression is repressed during differentiation in embryonic carcinoma cells.

Testis

NORMAL TESTIS

The proto-oncogene c-*myc* may play a role in normal somatic cell proliferation and differentiation, since c-*myc* RNA and protein are expressed in a cell cycle–dependent manner in spermatogonia, but not in spermatocytes or spermatids.[108] However, even though the c-*myc* oncogene relates to cell proliferation, testes that are actively proliferating have low c-*myc* RNA levels.[224] c-*myc* expression can be transiently elevated in Leydig cells by exposure to the hormone human chorionic gonadotropin.[126]

int-4 is normally expressed in the adult mouse testis.[92] Meiotic round spermatids have the highest expression of *int*-1,[257] suggesting that this proto-oncogene functions specifically in sperm maturation.[206] c-*mos* transcripts are detectable in germ cells and become more abundant after the cells have entered the haploid stage of spermatogenesis.[147,176–178] A 43kDa c-*mos* protein localizes specifically to the seminiferous tubules.[81] Testis also contains transcripts for c-*erbA*,[12] *junD*, and *junB*.[80]

Hu-*ets*-2,[16] *erg*,[181] and *elk*[181] sequences appear to be transcriptionally active in testis. The testis-specific expression of these genes suggests that *elk* and related sequences may play a role in the differentiation of male germ cell lineage. A distinctive 2.4 kb *pim*-1 transcript is expressed in haploid, postmeiotic testicular cells.[221] Finally, the *ras*-related genes, *ral* A, *rho* 12, *rab* 1, and *rab* 2, are all expressed in testicular tissue.[161]

TESTICULAR TUMORS

c-*myc* is often amplified in seminomas, but its expression declines as the tumor becomes more poorly differentiated and is associated with more rapid growth.[212] Both c-*myc* and N-*myc* are expressed in embryonal carcinomas.[204] *ras* p21 protein is not detectable,[229] but the gene is mutated in 43% of seminomas.[141] Most embryonal carcinoma cell lines do not express int-1,[156] although once they are induced to differentiate the gene may become expressed.[220]

Tumor suppressor genes may also play a role in testicular teratocarcinomas, as suggested by the nonrandom loss of chromosome 6p DNA centromeric to the HLA locus.[194] Other alleles that are lost include 3p and 11p.[128]

Urinary Bladder

ras Genes

Transfection experiments proving that cancer cells contain transforming genes were first carried out in the T24 and EJ bladder carcinoma cell lines.[179,210,211] Comparative analysis of the bladder carcinoma oncogene with retroviral transforming genes revealed an internal fragment of the T24 oncogene that was closely related to the onc genes of Harvey and Kirsten murine sarcoma viruses.[42,162,198,232] ras is the most frequently identified oncogene detected in transfection assays.[11,46] Reddy and co-workers demonstrated that the genetic change leading to the activation of the oncogene in the T24 human bladder carcinoma cells is a single point mutation of guanosine to thymidine, resulting in the incorporation of valine instead of glycine at the twelfth amino acid residue of the T24 oncogene–encoded p21 protein.[182] The single amino acid substitution appears to be sufficient to confer transforming properties on the gene product of the T24 human bladder carcinoma oncogene. Tabin and co-workers performed similar studies on the EJ bladder carcinoma cell line and detected an altered p21 in the absence of altered expression levels.[226]

Fujita and co-workers systematically evaluated a large series of urinary tract tumors and cell lines for ras oncogenes by DNA transfection and by molecular genetic analysis.[64,65] Ha-ras genes were detected in 2 of 38 tumors by transfection and were shown to contain a single base mutation at codon 61, leading to substitutions of arginine and leucine, respectively, for glutamine at this position. An additional Ha-ras oncogene was identified in a bladder carcinoma by restriction polymorphisms at codon 12. In 1 of 21 tumors, a 40-fold amplification of the Ki-ras gene was seen. No amplification of other ras genes was detected in any of the tumors analyzed. These data support the conclusion that codons 12 and 61 are major hotspots of ras oncogene activation and suggest that quantitative alterations in expression due to gene amplification may provide an alternative mechanism for ras gene activation in primary urinary tract infections.[58]

Capon and co-workers analyzed the complete nucleotide sequences of the T24 human bladder carcinoma oncogene and its normal homologue and found two alleles of its normal cellular progenitor, c-Ha-ras-1, indicating that the genes encompass at least four exons and that a single point mutation residing within the first exon distinguishes the coding region of both alleles of the normal gene from their activated counterpart.[25] Nucleotide sequencing determined that the T24 oncogene and the normal c-Ha-ras gene encode a p21 highly homologous to the product of the v-Ha-ras viral oncogene. The human and viral p21 differ by only 3 amino acid residues, despite extensive nucleotide divergence in the coding regions. Others have also shown that ras genes are frequently mutated in urothelial tumors.[93]

Viola and co-workers studied the expression of the *ras* oncogene p21 in premalignant lesions of the bladder as well as in high-grade bladder carcinomas, utilizing immunohistochemical techniques.[248] In contrast to other studies showing *ras* activation and mutations in less than 10% of bladder carcinomas,[57,65,225] they demonstrated increased *ras* p21 expression in all high-grade bladder carcinomas and further demonstrated that this increased expression might be an indicator of malignant potential in premalignant lesions of the bladder. However, the authors used the antibody RAP-5 for these studies, making it unlikely that the *ras* protein was being detected.

Geiser and co-workers studied the human tumor cell line EJ, which expresses the activated c-Ha-*ras* oncogene, and fused it with a normal human fibroblast cell line.[68] The fusion resulted in hybrids that behaved as transformed cells in culture, but failed to form tumors in nude mice. After repeated cell passage, two tumorigenic segregants of the hybrids arose in culture. The expression levels of activated c-Ha-*ras* mRNA and its protein product, p21, were similar in the EJ cell line, the nontumorigenic hybrids, and the tumorigenic segregants. DNA transfections of the hybrids were performed with activated c-Ha-*ras* plasmid constructs, and transfectants expressing a 2-fold level of c-Ha-*ras* relative to the hybrid cells were found to maintain the nontumorigenic phenotype. This suggests that the active c-Ha-*ras* oncogene is insufficient to maintain malignant transformation of these human cells. These results also demonstrate that the *ras*-induced tumorigenic phenotype can be suppressed in human cell lines that contain a dominantly acting oncogene. These EJ fibroblast hybrids clearly show that a tumor cell expressing an activated oncogene can be suppressed for tumorigenecity and that the expression appears to be independent of the expression of the activated oncogene.[68]

A *ras*-related oncogene has been found in the urine of patients with bladder cancer.[225]

Activated *ras* genes are often found in the benign and malignant tumors experimentally induced by chemicals in the urinary bladder, but not usually in the hyperplastic or preneoplastic lesions preceding tumor development.[253]

Other Proto-oncogenes

Skouv and co-workers studied the effect of 12-0-tetradecanoylphorbol-13-acetate (TPA) on the induction of transcription in c-*myc* and c-*fos* genes by mortal and immortal human urothelial cells *in vitro*.[214] A single treatment of TPA transiently increased the transcription of c-*fos* and c-*myc* proto-oncogenes at least 20-fold in the mortal cell line HU 1752 and induced a transient change in cellular morphology. When immortalized cell lines were treated with TPA, a similar rapid and transient morphologic response was observed, but the TPA treatment only increased the level of c-*fos* mRNA, suggesting that normal regulation of c-*myc* transcription is altered in immortalized cells, irrespective of their tumorigenic properties.

Overexpression of the EGF RNA or protein product has been associated with bladder carcinomas.[13,151] The EGFR is normally expressed on urothelium, but the concentration increases as cancer develops, especially as depth and grade increase.[151] The *FGF*-related *gene* HST is expressed in 7% of bladder carcinomas.[234] Finally, human bladder epithelial cells can be transformed by transfection with the v-*raf* oncogene.

Tumor Suppressor Genes

Fearon found the loss of 11p alleles in 40% of urinary bladder carcinomas in a small series of patients, suggesting that a tumor suppressor gene may be present at this site.[54]

Prostate

myc Genes

myc genes may be expressed in benign and neoplastic prostate, but amplification of these genes has not been observed (Fig. 11.1).

Fleming and co-workers examined the level of c-*myc* transcripts in prostatic tissue from patients with both benign prostatic hyperplasia and adenocarcinoma.[62] A significantly higher level of c-*myc* transcripts was observed in patients with adenocarcinoma. In addition, a subset of patients with adenocarcinoma had c-*myc* transcript levels twice that of the group mean. Levels of prostatic acid phosphatase at the time of diagnosis did not correlate with c-*myc* levels. Histological classification of the adenocarcinoma indicated that 2 patients with high levels of c-*myc* expression had grade 1 lesions, whereas 2 patients had grade 2 lesions without evidence of systemic disease.

ras Genes

The c-Ki-*ras* gene has also been found in human prostatic cancer.[165,166] Viola *et al.* utilizing RAP-5, suggest that the expression of c-*ras* may arise in conjunction with increasing malignancy of prostatic carcinoma cells and that the expression correlates with tumor grade.[247] However, Varma and co-workers were unable to confirm their results.[245]

Increased expression of Ha-*ras* is associated with increased tumorigenicity in

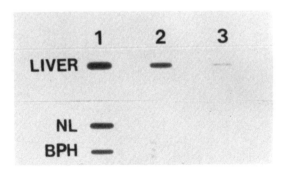

Figure 11.1. Autoradiograph of DNA from normal adult prostate and benign prostatic hyperplasia hybridized to c-*myc*. Lanes *1 through 3* represent 5.0, 1.0, and 0.2 micrograms of genomic DNA. c-*myc* is not amplified in these tissues.

one lineage of the Dunning R3327 system.[34,35] p21 expression also increases in parallel with mRNA expression.

Other Oncogenes

Interestingly, the detection of proto-oncogene products in body fluids may be useful diagnostically. Niman and co-workers recently reported increased levels of c-*ras*, c-*sis*, or c-*fes* proteins in urines of patients with different types of urologic tumors, including prostatic carcinomas.[153]

Cooke and co-workers measured steady-state levels of *myc, fos,* p53, *sis,* and *neu* mRNAs in 8 variants of the Dunning 3327 rat prostate adenocarcinoma and compared the levels to normal prostate.[34,35] *neu* and *myb* expression were below detection levels, but *myc*, p53, and *sis* mRNA were all elevated. *fos* mRNA levels were below control levels for 4 of 5 anaplastic tumors and above control levels in the remaining tumors. Increased oncogene expression did not correlate with tumor progression. Specific and high-affinity EGF receptors have been demonstrated in prostate cancers and have been shown to correlate with histological degrees of differentiation. Well-differentiated tumors express more receptors than do poorly differentiated tumors.[132]

Kidney

NORMAL KIDNEY

Kokai has demonstrated staining of the normal proximal renal tubules with an antibody to the *neu*-encoded protein.[109] The c-*erb*B2 gene product is present in transitional cells of the renal pelvis and ureters and renal tubules in the human fetus. The protein is also expressed on adult tubular epithelium.[138] Transcripts for c-*erb*A are also present.[250]

myc genes are expressed in the fetal kidney, where they are confined first to the metanephric bud and later to the renal tubular glomerular epithelium. In fact, N-*myc* is preferentially expressed in this fetal organ.[269] N-*myc* transcripts are found primarily in epithelially differentiating mesenchyme. Stromal cells are largely negative.[84] The glomeruli do not express *myc* proto-oncogenes.[201] c-*mos*,[177] *rho* 12, *rob* 1, and *rab* 2[161] are also expressed.

REGENERATING KIDNEY

The c-*myc* gene is activated in kidney tubule cells that are induced to proliferate as a consequence of folic acid–induced renal injury *in vivo*. Transcriptional initiation is induced in a coordinate fashion, with a strong block to transcriptional elongation in the gene's first exon. Therefore, transcription is not effectively induced throughout the c-*myc* gene in renal tubular cells.

In contrast, initiation of the c-*fos* gene transcription is induced while the degree of transcriptional blockage within the gene's first exon remains constant. Steady-state levels of *myc* transcripts originate from the genes P1 and P2. Start sites accumulate to high levels within 4 to 6 hours and are maintained for at least 24 hours after folic acid treatment.[10]

KIDNEY TUMORS

ras Genes

ras mutations are found in 3–17% of renal cell carcinomas.[63] Nanus and associates introduced v-Ki-*ras* and v-Ha-*ras* into primary cultures of human proximal renal tubules and found that this procedure initiates a series of events that eventually results in cells with a transformed phenotype.[150] The cells pass through two distinct phases. In the first phase, *ras*-infected proximal tubule cells manifest distinct morphological manifestations and undergo a burst of proliferation and outgrowth. Subsequently, the cells no longer senesce and become immortalized. The *ras*-infected cells also produce a series of TGFs that may function as autocrine growth factors to further stimulate cellular growth.

c-Ha-*ras* 1 is located at 11p15, a site commonly duplicated in patients with the Beckwith-Wiedemann syndrome. This syndrome is associated with an increased number of tumors, including nephroblastoma, adrenocortical carcinoma, hepatoblastoma, and rhabdomyosarcoma.[256] However, despite these associations, the *ras* gene, which is located on chromosome 11, does not appear to play a role in the pathogenesis of the tumors, since the gene for H-*ras* 1 does not cosegregate with the syndrome.[79] Nonetheless, tumors do stain with antibodies to *ras* p21.[7]

myc Genes

N-*myc* expression has been found by some investigators in Wilm's tumor[5,155,207] and renal cell carcinoma,[91,154,155] but evidence of N-*myc* or c-*myc* amplification has not been identified (Fig. 11.2).[207,238] By *in situ* hybridization, N-*myc* RNA transcripts are present in the blastemic elements of the Wilm's tumor.[207] However, in one unusual case in which an intrarenal neuroblastoma was present, an am-

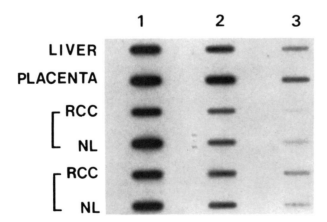

Figure 11.2. DNA slot blot evaluating c-*myc* in two renal cell carcinomas with paired, nonneoplastic kidney. Serial dilutions of genomic DNA were blotted onto nitrocellulose membranes to give final concentrations of 5.0, 1.0, and 0.2 micrograms (*lanes 1–3,* respectively).

plified N-*myc* gene was identified and its presence correlated with a poor prognosis,[154] as is true for NBLs arising elsewhere in the body.

Kakehi and associates analyzed RFLPs of the L-*myc* gene in 50 patients with the sporadic form of renal cell carcinoma and found that there were no significant differences between normal and tumor tissues digested with *Eco*R1.[97] Three genotypes were detected in both the tumor patient and normal patient populations. However, of 16 patients who demonstrated distant metastases at the time of surgery, only one was a 10 kb fragment homozygote. The incidence of distant metastases in 10 kb homozygotes was significantly lower than that seen in 6.6 homozygotes or in heterozygotes. These results are similar to those described for lung cancer patients[101] and suggest that L-*myc* RFLPs may be used to predict prognosis in cancer patients.

Other Oncogenes

c-*fms* has been found to be elevated in renal cell carcinoma,[217] but as in the ovarian tumors, it is not clear whether the oncogene derives from tumor cells or infiltrating macrophages.

Tal and associates found no evidence of HER-2/*neu* amplification,[228] whereas Yokota did find c-*erb*B-2 to be amplified in these tumors.[259] The mRNA levels of the EGFR was 2–3 times greater than those seen in the surrounding normal tissues. In addition, these tissues had high levels of TGF-α whereas none was detected in the normal tissues.[149] Most recently, high levels of EGF receptors have been found in renal cell carcinoma, as compared to surrounding normal kidney.[168]

Tumor Suppressor Genes

Although retinoblastoma is the prototypic cancer for demonstrating the role of tumor suppressor genes, Wilms' tumors also typify the role of these genes.

Hereditary cases are not as numerous in Wilms' tumor as they are in retinoblastoma, and about 50% of persons with the characteristic mutation do not develop the tumor.[106] The deletions were discovered because of the association with aniridia in 1–3% of cases. This observation led Knudson and Strong[107] to postulate that these cases might result from deletions of neighboring genes, one for aniridia and one for Wilms' tumor. The deletion is now known to involve the region 11p13>11p14.1, to which c-Ha-*ras*-1 has been mapped.[219] Heterozygosity for this locus precedes the development of homozygosity. Homozygosity results from a second somatic event such as mutation, chromosomal loss, or genetic recombination.[55,98,106,111] Finally, it has been shown that a normal chromosome 11p can suppress the tumorigenicity of Wilms' tumor cells.[254]

Renal cell carcinoma usually occurs in a sporadic form. However, hereditary cases occur at an early age and are frequently bilateral. Deletions of chromosome 3p and the short arm of chromosome 11 occur in both the sporadic and hereditary forms of the tumor.[15,33,48,112,163] Inherited translocations t(3;8)(p21;q24) involving the c-*myc* gene and t(3;11)(p13;p15) have also been associated with hereditary renal cell carcinoma.[32] A more recently recognized abnormality is a on chromosome 6 in a region centromeric to the HLA locus in some renal cell carcinomas.[194]

Acknowledgments

The authors wish to acknowledge the helpful discussions and technical assistance of Barbara Griffith, the manuscript preparation by Agnes Truske, and the VA Medical Media Center for their expertise with the illustrative material. The work reported in this chapter was supported by American Cancer Society Grant #PDT-341 and also by a grant from the University of New Mexico Cancer Center.

REFERENCES

1. Agnantis, N. J., Pariessi, P., Angnostakis, D., and Spandidos, D. A. Comparative study of Harvey-*ras* oncogene expression with conventional clinicopathologic parameters of breast cancer. *Oncology* 43:36–39, 1986.
2. Agnantis, N. J., Petraki, C., Markoulatos, P., and Spandidos, D. A. Immunohistochemical study of the *ras* oncogene expression in human breast lesions. *Anticancer Res.* 6:1157–1160, 1986.
3. Ali, I. U., Campbell, G., Lidereau, R., and Callahan, R. Amplification of c-erbB-2 and aggressive human breast tumors? *Science (Wash DC)* 240:1975–1796, 1988.
4. Ali, I. U., Lidereau, R., Theillet, C., and Callahan, R. Reduction to homozygosity of genes on chromosome 11 in human breast neoplasia. *Science (Wash DC)* 238:185–188, 1987.
5. Alt, F. W., DePinho, R., Zimmerman, K., Legouy, E., Hatlon, K., Ferrier, P., Tesfaye, A., Yancopoulos, G., and Nisen, P. The human myc gene family. *Cold Spring Harbor Symp. Quant. Biol.* 81:931–941, 1986.
6. Andres, A. C., Schonenberger, C. A., Grower, B., Hennighausen, L., Lemur, M., and Gerlinger, P. Ha-ras oncogene expression directly by a milk protein gene promoter: Tissue specificity, hormonal regulation and tumor induction in transgenic mice. *Proc. Natl. Acad. Sci.U.S.A.* 84:1299–1303, 1987.
7. Aoki, I., Yanoma, S., Misugi, K., Sasaki, Y., and Kikyo, S. ras p21 expression in nephroblastoma group tumors. *Acta Pathol. Jpn.* 37:1903–1907, 1987.
8. Arheden, K., Mandahl, N., Strombeck, B., Isakson, M., and Mitelman, F. Chromosome localization of the human oncogene INT1 to 12q13 by in situ hybridization. *Cytogenet. Cell. Genet.* 47:86–87, 1988.
9. Arheden, K., Nilbert, M., and Mitelman, F. No amplification or rearrangement of INT1, GLI or COL2A1 in uterine leiomyomas with t(12;14)(q14-15;q23-24). *Cancer Genet. Cytogenet.* 39:195–201, 1989.
10. Asselin, C., and Marcu, K. B. Mode of c-*myc* regulation in folic acid-induced kidney regeneration. *Oncogene Res.* 5:67–72, 1989.
11. Balmain, A. Transforming *ras* oncogenes and multistage carcinogenesis. *Br. J. Cancer* 51:1–7, 1985.
12. Benbrook, D., and Pfahl, D. A novel thyroid hormone receptor encoded by cDNA clones from a human testis library. *Science (Wash DC)* 238:788–794, 1987.
13. Berger, M. S., Greenfield, C., Gullick, W. J., Haley, J., Downward, J., Neal, D. E., Harris, A. L., and Waterfield, M. D. Evaluation of epidermal growth factor receptors in bladder tumours. *Br. J. Cancer* 56:533–537, 1987.
14. Berger, M. S., Locher, G. W., Saurer, S., Gullick, W. J., Waterfield, M. D., Groner, B., and Hynes, N. E. Correlation of c-*erbB-2* gene amplification and protein expression in human breast carcinoma with nodal status and nuclear grading. *Cancer Res.* 48:1238–1243, 1988.
15. Bergerheim, U., Nordenskjold, M., and Collins, V. P. Deletion mapping in human renal cell carcinoma. *Cancer Res.* 49:1390–1396, 1989.
16. Bhat, N. K., Fisher, R. J., Fujiwara, S., Ascione, R., and Papas, T. S. Temporal and tissue-

specific expression of mouse *ets* genes. *Proc. Natl. Acad. Sci. U.S.A.* 84:3161–3165, 1987.

17. Biunno, I., Pozzi, M. R., Pierotti, M. A., Pilotti, S., Cattoretti, G., and Della Porta, G. Structure and expression of oncogenes in surgical specimens of human breast carcinomas. *Br. J. Cancer* 57:464–468, 1988.

18. Bodmer, W. F., Bailey, C. J., Bodmer, J., Bussey, H. J. R., Ellis, A., Gorman, P., Lucibello, F. C., Murday, V. A., Rider, S. H., Scambler, P., Sheer, D., Solomon, E., and Spurr, N. K. Localization of the gene for familial adenomatous polyposis on chromosome 5. *Nature (Lond)* 328:614–616, 1987.

19. Bonilla, M., Ramirez, M., Lopez-Cueto, J., and Gariglio, P. In vivo amplification and rearrangement of c-myc oncogene in human breast tumors. *J. Natl. Cancer Inst.* 80:665–671, 1988.

20. Braude, P., Pelham, H., Flach, G., and Lobatto, R. Post-transcriptional control in the early mouse embryo. *Nature (Lond)* 282:102–105, 1979.

21. Brookes, S., Smith, R., Casey, G., Dickson, C., and Peters, G. Sequence organization of the human *int*-2 gene and its expression in teratocarcinoma cells. *Oncogene* 4:429–436, 1989.

22. Brown, A. M., Wildin, R. S., Prendergast, T. J., and Varmus, H. E. A retrovirus vector expressing the putative mammary oncogene *int*-1 causes partial transformation of a mammary epithelial cell line. *Cell* 46:1001–1009, 1986.

23. Burke, R., and McGuire, W. L. Nuclear thyroid hormone receptors in a human breast cancer cell line. *Cancer Res.* 38:3769–3773, 1978.

24. Candlish, W., Kerr, I. B., and Simpson, H. W. Immunocytochemical demonstration and significance of p21 *ras* family oncogene product in benign and malignant breast disease. *J. Pathol.* 150:163–167, 1986.

25. Capon, D. J., Chen, E. Y., Levinson, A. D., Seeburg, P. H., and Goeddel, D. V. Complete nucleotide sequences of the T24 human bladder carcinoma oncogene and its normal homologue. *Nature (Lond)* 302:33–37, 1983.

26. Capon, D. J., Seeburg, P. H., McGrath, J. P., Hayflick, J. S., Edman, U., Levinson, A. D., and Goeddel, D. V. Activation of Ki-ras2 gene in human colon and lung carcinomas by two different point mutations. *Nature (Lond)* 304:507–513, 1983.

27. Cardiff, R. D., Gumerlock, P. H., Song, M.-M., Dandekar, S., Barry, P. A., Young, L. J. T., and Meyers, E. J. c-H-*ras*-1 expression in 7, 12 dimethylbenzanthracene-induced Balb/c mouse mammary hyperplasias and their tumors. *Oncogene* 3:205–213, 1988.

28. Cattoretti, G., Rilke, F., Andreola, S., D'Amato, L., and Delia, D. p53 expression in breast cancer. *Int. J. Cancer* 41:178–183, 1988.

29. Chesa, P. G., Rettig, W. J., Melamed, M. R., Old, L. J., and Niman, H. I. Expression of p21*ras* in normal and malignant human tissues. Lack of association with proliferation and malignancy. *Proc. Natl. Acad. Sci. U.S.A.* 84:3234–3238, 1987.

30. Clark, G. M., McGuire, W. L., Hubay, C. A., Pearson, O. H., and Marshall, J. S. Progesterone receptors as a prognostic factor in stage II breast cancer. *N. Engl. J. Med.* 309:1343–1347, 1983.

31. Clegg, K. B., and Piko, L. Poly (A) length, cytoplasmic adenylation and synthesis of poly (A)+ RNA in early mouse embryos. *Dev. Biol.* 95:331–341, 1983.

32. Cohen, A. J., Li, F. P., Berg, S., Machetto, D. J., Tsai, S., Jacobs, S. C., and Brown, R. S. Hereditary renal cell carcinoma associated with a chromosomal translocation. *N Engl. J. Med.* 301:592–595, 1979.

33. Comings, D. E. A general theory of carcinogenesis. *Proc. Natl. Acad. Sci. U.S.A.* 70:3324–3328, 1973.

34. Cooke, D. B., Quarmby, V. E., Mickey, D. D., Isaacs, J. T., and French, F. S. Oncogene expression in prostate cancer: Dunning R3327 rat dorsal prostate adenocarcinoma system. *Prostate* 13:263–272, 1988.

35. Cooke, D. B., Quarmby, V. E., Petrusz, P., Mickey, D. D., Der, C. J., Isaacs, J. T., and

French, F. S. Expression of ras proto-oncogenes in the Dunning R3327 rat prostatic adenocarcinoma system. *Prostate* 13:273–287, 1988.

36. Coussens, L., Yang-Feng, T. L., Liao, Y.-C., Chen, E., Grey, A., McGrath, J., Seeburg, P. H., Lieberman, T. A., Schlessinger, J., Franke, U., Levinson, A., and Ullrich, A. Tyrosine kinase receptor with extensive homology to EGF receptor shares chromosomal location with *neu* oncogene. *Science* 230:1132–1139, 1985.

37. Crawford, L. V., Pim, D. C., and Bulbrook, R. D. Detection of antibodies against the cellular protein p53 in sera from patients with breast cancer. *Int. J. Cancer* 30:403–408, 1982.

38. Czerniak, B., Chen, R., Tuziak, T., Markiewski, M., Kram, A., Gorczyca, W., Deitch, D., Herz, F., and Koss, L. G. Expression of ras oncogene p21 protein in relation to regional spread of human breast carcinomas. *Cancer* 63:2008–2013, 1989.

39. Dandekar, S., Sukumar, S., Zarbl, H., Young, L. J., and Cardiff, R. D. Specific activation of the cellular Harvey-ras oncogene in dimethylbenzanthrene-induced mouse mammary tumors. *Mol. Cell. Biol.* 6:4104–4108, 1986.

40. Davidson, E. H. Gene Activity in Early Development. Orlando, FL, Academic Press, 1986.

41. Delarue, J. C., Friedman, S., Mouriesse, H., May-Levin, F., Sancho-Garnier, H., and Contesso, G. Epidermal growth factor receptor in human breast cancers: Correlation with estrogen and progesterone receptors. *Breast. Cancer. Res. Treat* 11:173–178, 1988.

42. Der, C. J., Krontiris, T. G., and Cooper, G. M. Transforming genes of human bladder and lung carcinoma cell lines are homologous to the *ras* genes of Harvey and Kirsten sarcoma viruses. *Proc. Natl. Acad. Sci. U.S.A.* 79:3637–3640, 1982.

43. Dickson, C., and Peters, G. Potential oncogene product related to growth factors. *Nature (Lond)* 326:833, 1987.

44. Dickson, C., Smith, R., Brooke, S., and Peters, G. Tumorigenesis by mouse mammary tumor virus: Proviral activation of a cellular gene in the common integration region *int*-1. *Cell* 37:529–536, 1984.

45. DiPaolo, J. A., Woodworth, C. D., Popescu, N. C., Notario, V., and Doniger, J. Induction of human cervical squamous cell carcinoma by sequential transfection with human papillomavirus 16 DNA and viral Harvey *ras. Oncogene* 4:395–399, 1989.

46. Dolnick, B. J., and Reznikoff, C. A. Amplification of the Ha-*ras* oncogene correlates with the malignant properties of several rodent cell lines in vitro and is associated with a clinical metastatic bladder carcinoma. *Proc. Am. Assoc. Cancer Res.* 25:65, 1984.

47. Downs, K. M., Martin, G. R., and Bishop, J. M. Contrasting patterns of myc and N-myc expression during gastrulation of the mouse embryo. *Gene Dev.* 3:860–869, 1989.

48. Drabkin, H. A., Bradley, C., Hart, I., Bleskan, J., Li, F. P., and Patterson, D. Translocation of c-*myc* in the hereditary renal cell carcinoma associated with a t(3:8)(p 14.2:q 24.13) chromosomal translocation. *Proc. Natl. Acad. Sci. U.S.A.* 82:6980–6984, 1985.

49. Dubik, D., Dembinski, T. C., and Shiu, R. P. Stimulation of c-myc oncogene expression associated with estrogen induced proliferation of human breast carcinoma cells. *Cancer Res.* 47:6517–6521, 1987.

50. Dubik, D., and Shiu, R. P. C. Transcriptional regulation of c-myc oncogene expression by estrogen in hormone-responsive human breast cancer cells. *J. Biol. Chem.* 263:12705–12708, 1988.

51. Durst, M., Croce, C. M., Gissmann, L., Schwarz, E., and Huebner, K. Papillomavirus sequences integrate near cellular oncogenes in some cervical carcinomas. *Proc. Natl. Acad. Sci. U.S.A.* 84:1070–1074, 1987.

52. Edwards, P. A. W., Ward, J. C., and Brabbery, J. M. Alteration of morphogenes by the v-myc oncogene in transplants of mammary glands. *Oncogene* 2:407–412, 1988.

53. Escot, C., Theillet, C., Lidereau, R., Spyratos, F., Champeme, M. H., Gest, J., and Callahan, J. Genetic alteration of the c-*myc* protooncogene (MYC) in human primary breast carcinomas. *Proc. Natl. Acad. Sci. U.S.A.* 83:4834–4838, 1986.

54. Fearon, E. R., Feinberg, A. P., Hamilton, S. H., and Vogelstein, B. Loss of genes on the short arm of chromosome 11 in bladder cancer. *Nature (Lond)* 318:377–380, 1985.

55. Fearon, E. R., Vogelstein, B., and Feinberg, A. P. Somatic deletion and duplication of gene on chromosome 11 in Wilms' tumor. *Nature (Lond)* 309:176–178, 1984.

56. Feig, L. A., Bast, R. C., Knapp, R. C., and Cooper, G. M. Somatic activation of ras^K gene in a human ovarian carcinoma. *Science (Wash DC)* 223:698–701, 1984.

57. Feinberg, A. P., and Vogelstein, B. Hypomethylation of ras oncogenes in primary human cancers. *Biochem. Biophys. Res. Commun.* 111:47–54, 1983.

58. Feinberg, A., Vogelstein, B., Droller, M., Baylin, A., and Nelkin, B. Mutations affecting the 12th amino acid of c-Ha-*ras* oncogene product occur infrequently in bladder cancer. *Science (Wash DC)* 220:1175–1177, 1983.

59. Fernandez-Pol, J. A., Talkad, V. D., Klos, D. J., and Hamilton, P. D. Suppression of the EGF-dependent induction of c-myc proto-oncogene expression by transforming growth factor B in a human breast carcinoma cell line. *Biochem. Biophys. Res. Commun.* 144:1197–1205, 1987.

60. Filmus, J. E., and Buick, R. N. Stability of c-K-*ras* amplification during progression in a patient with adenocarcinoma of the ovary. *Cancer Res.* 45:4468–4472, 1985.

61. Filmus, J., Trent, J. M., Pollak, M. N., and Buick, R. N. Epidermal growth factor receptor gene-amplified MDA-468 breast cancer cell line and its nonamplified variants. *Mol. Cell. Biol.* 7:251–257, 1987.

62. Fleming, W. H., Hamel, A., MacDonald, R., Ramsey, E., Pettigrew, N. M., Johnston, B., Dodd, J. G., and Matusik, R. J. Expression of the c-*myc* proto-oncogene in human prostate carcinoma and benign prostatic hyperplasia. *Cancer Res.* 46:1535–1538, 1986.

63. Fujita, J., Kraus, M. H., Onoue, H., Srivasta, S. K., Ebi, Y., Kitamura, Y., and Rhim, J. S. Activated H-*ras* oncogenes in human kidney tumors. *Cancer Res.* 48:5251–5255, 1988.

64. Fujita, J., Srivastava, S. K., Kraus, M. H., Rhim, J. S., Tronick, S. R., and Aaronson, S. A. Frequency of molecular alterations affecting *ras* protooncogenes in human urinary tract tumors. *Proc. Natl. Acad. Sci. U.S.A.* 82:3849–3853, 1985.

65. Fujita, J., Yoshida, O., Yuasa, Y., Rhim, J., Hatanaka, and Aaronson, S. Ha-*ras* oncogenes are activated by somatic alterations in human urinary tract tumours. *Nature (Lond)* 309:464–466, 1984.

66. Fukumoto, M., Estensen, R. D., Sha, L., Oakley, G. J., Twiggs, L. B., Adcock, L. L., Carson, L. F., and Roninson, I. B. Association of Ki-*ras* with amplified DNA sequences, detected in human ovarian carcinomas by a modified in-gel renaturation assay. *Cancer Res.* 49:1693–1697, 1989.

67. Fukushige, S., Matsubara, K., Yoshida, M., Sasaki, M., Suzuki, T., Semba, K., Toyshima, K., and Yamamoto, T. Localization of a novel v-*erb*B-related gene, c-*erb*B2 on chromosome 17 and its amplification in a gastric cancer cell line. *Mol. Cell. Biol.* 6:955–958, 1986.

68. Geiser, A. G., Der, C. J., Marshall, C. J., and Stanbridge, E. J. Suppression of tumorigenicity with continued expression of the c-Ha-*ras* oncogene in EJ bladder carcinoma–human fibroblast hybrid cells. *Proc. Natl. Acad. Sci. U.S.A.* 83:5209–5213, 1986.

69. Ghosh, A. K., Moore, M., and Harris, M. Immunohistochemical detection of *ras* oncogene p21 product in benign and malignant mammary tissue in man. *J. Clin. Pathol.* 39:428–434, 1986.

70. Going, J. J., Williams, R. W., Wyllie, A. H., Anderson, T. J., and Piris, J. Optimal preservation of ras p21 immunoreactivity and morphology in paraffin-embedded tissue. *J. Pathol.* 155:185–190, 1988.

71. Goldfarb, M., Shimizu, K., Perucho, M., and Wigler, M. Isolation and preliminary characterization of a human transforming gene from T24 bladder carcinoma cells. *Nature (Lond)* 296:404–409, 1982.

72. Golman, D. S., Kiessling, A. A., and Cooper, G. M. Post-transcriptional processing sug-

gests that c-*mos* functions as a maternal message in mouse eggs. *Oncogene* 3:159–162, 1988.

73. Goustin, A. S., Betsholtz, C., Pfeifer-Ohlsson, S., Persson, H., Rydnert, J., Bywater, M., Holmgren, G., Heldin, C-H, Westermark, B., and Ohlsson, R. Coexpression of the *sis* and *myc* proto-oncogenes in developing human placenta suggests autocrine control of trophoblast growth. *Cell* 41:301–312, 1985.

74. Graham, K. A., Richardson, C. L., Minden, M. D., Trent, J. M., and Buick, R. N. Varying degrees of amplification of the N-*ras* oncogene in the human breast cancer cell line MCF-7. *Cancer Res.* 45:2201–2205, 1985.

75. Guerin, M., Barrois, M., Terrier, M. J., Spielmann, M., and Riou, G. Over-expression of either c-myc or c-erbB-2/neu proto-oncogenes in human breast carcinomas: Correlation with poor prognosis. *Oncogene Res.* 3:21–31, 1988.

76. Guerin, M., Gabillot, M., Mathieu, M.-C., Travagli, J.-P., Spielmann, M., Andrieu, N., and Riou, G. Structure and expression of c-erbB-2 and EGF receptor genes in inflammatory and non-inflammatory breast cancer: Prognostic significance. *Int. J. Cancer* 43:201–208, 1989.

77. Hall, J. M., Zuppan, P. J., Anderson, L. A., Huey, B., Carter, C., and King, M.-C. Oncogenes and human breast cancer. *Am. J. Hum. Genet.* 44:577–584, 1989.

78. Hendy-Ibbs, P., Cox, H., Evan, G. I., and Watson, J. V. Flow cytometric quantitation of DNA and c-myc oncoprotein in archival biopsies of uterine cervix neoplasia. *Br. J. Cancer* 55:275–282, 1987.

79. Henry, I., Jeanpierre, M., Barichard, F., Serre, J. L., Mallet, J., Turleau, C., deGrouchy, J., and Junien, C. Duplication of HRAS1, INS, and IGF2 is not a common event in Beckwith-Wiedemann syndrome. *Ann. Genet.* 31:216–220, 1988.

80. Herrea, L., Kataki, S., Gibas, L., Pietrzak, E., and Sandberg, A. A. Brief clinical report: Gardner's syndrome in a man with a interstitial deletion of 5q. *Am. J. Med. Genet.* 25:473–476, 1986.

81. Herzog, N. B., Singh, B., Elder, J., Lipkin, I., Trauger, R. J., Millette, C. F., Goldman, D. S., Wolfes, H., Cooper, G. M., and Arlinghaus, R. B. Identification of the protein product of c-*mos* proto-oncogene in mouse testes. *Oncogene* 3:225–229, 1988.

82. Hildebrandt, R., Longacre, T., Willman, C., Hoffman, M. S., and Bartow, S. Nonmalignant human breast tissue expresses the int-2 proto-oncogene. *Mod. Pathol.* 2:41A, 1989.

83. Hirai, S.-I., Ryseck, R.-P., Mechta, F., Bravo, R., and Yamiu, M. Characterization of junD: A new member of the *jun* proto-oncogene family. *EMBO J.* 8:1433–1439, 1989.

84. Hirvonen, H., Sandberg, M., Kalimo, H., Hukkanen, V., Vuorio, E., Salmi, T. T., and Alitalo, K. The N-*myc* proto-oncogene and IGF-II growth factor mRNAs are expressed by distinct cells in human fetal kidney and brain. *J. Cell. Biol.* 108:1093–1104, 1989.

85. Horan Hand, P., Thor, A., Wunderlich, D., Muraro, R., Caruso, A., and Schlom, J. Monoclonal antibodies of predefined specificity detect activated *ras* gene expression in human mammary and colon carcinomas. *Proc. Natl. Acad. Sci. U.S.A.* 81:5227–5231, 1984.

86. Horan Hand, P., Vilasi, V., Thor, A., Ohuchi, N., and Schlom, J. Quantitation of Harvey *ras* p21 enhanced expression in human breast and colon carcinomas. *J. Natl. Cancer Inst.* 79:59–65, 1987.

87. Horiguchi, J., Sherman, M. L., Sampson-Johannes, A., Weber, B. L., and Kufe, D. W. CSF-1 and c-fms expression in human carcinoma cell lines. *Biochem. Biophys. Res. Commun.* 157:395–401, 1988.

88. Hudziak, R. M., Lewis, G. D., Winget, M., Fendly, B. M., Shepard, H. M., and Ullrich, A. p185[her2] monoclonal antibody has antiproliferative effects in vitro and sensitizes human breast tumor cells to tumor necrosis factor. *Mol. Cell. Biol.* 9:1165–1172, 1989.

89. Hynes, N. E., Gerber, H. A., Saurer, S., and Groner, B. Overexpression of the c-erbB-2 protein in human breast tumor cell lines. *J. Cell. Biochem.* 39:167–173, 1989.

90. Izhar, M., Siebert, P. D., Oshima, R. G., DeWolf, W. C., and Fukuda, M. N. Trophoblastic differentiation of human teratocarcinoma cell line HT-H. *Dev. Biol.* 116:510–518, 1986.
91. Jacobovits, A., Schwab, M., Bishop, J. M., and Martin, G. R. Expression of N-*myc* in teratocarcinoma stem cells and mouse embryos. *Nature (Lond)* 318:188–191, 1985.
92. Jacobovits, A., Shackleford, G. M., Varmus, H. E., and Martin, G. R. Two proto-oncogenes implicated in mammary carcinogenesis *int*-1 and *int*-2, are independently regulated during mouse development. *Proc. Natl. Acad. Sci. U.S.A.* 83:7806–7810, 1986.
93. Joyce, A. D., D'Emilia, J. C., Steele, G., Libertino, J. A., Silverman, M. L., and Summerhayes, I. C. Detection of altered H-*ras* proteins in human tumors using Western blot analysis. *Lab. Invest.* 61:212–218, 1989.
94. Kacinski, B. M., Carter, D., Mittal, K., Kohorn, E. I., Bloodgood, R. S., Donahue, J., Donofrio, L., Edwards, R., Schwartz, P. E., Chambers, J. T., and Chambers, S. K. High level expression of *fms* proto-oncogene mRNA is observed in clinically aggressive human endometrial adenocarcinomas. *Int. J. Radiat. Oncol. Biol. Phys.* 15:823–829, 1988.
95. Kaczmarek, L., Hyland, R., Watt, R., Rosenberg, M., and Baserga, R. Microinjected c-*myc* as competence factor. *Science (Wash DC)* 228:1313–1315, 1985.
96. Kaibuchi, K., Tsuda, T., Kikuchi, A., Tanimoto, T., Yamashita, T., and Takai, Y. Possible involvement of protein kinase C and calcium in growth factor-induced expression of c-*myc* oncogene in Swiss 3T3 fibroblasts. *J. Biol. Chem.* 261:1187–1192, 1986.
97. Kakehi, Y., and Yoshida, O. Restriction fragment length polymorphism of the L-myc gene and susceptibility to metastasis in renal cancer patients. *Int. J. Cancer* 43:391–394, 1989.
98. Karfas, Hausen, M. F., Lampkin, B. C., Wakman, M. L., Copeland, N. G., Jenkins, N. A., and Cavenee, W. K. Loss of alleles at loci on human chromosome 11 during genesis of Wilms' tumor. *Nature (Lond)* 309:170–172, 1984.
99. Kasid, A., and Lippmann, M. E. Estrogen and oncogene mediated growth regulation of human breast cancer cells. *J. Steroid Biochem.* 27:465–470, 1987.
100. Kasid, A., Lippman, M. E., Papageorge, A. G., Lowry, D. R., and Gelmann, E. P. Harvey murine sarcoma virus DNA transfected into MCF-7 human breast cancer cells bypasses their dependence on estrogen for tumorogenicity. *Science (Wash DC)* 228:725–728, 1985.
101. Kawashima, K., Shikama, H., Imoto, K., Izawa, M., *et al.* Close correlation between restriction fragment length polymorphism of the L-myc gene and metastasis of human lung cancer to the lymph nodes and other organs. *Proc. Natl. Acad. Sci. U.S.A.* 85:2353–2356, 1988.
102. Kelly, K., Cochran, B. H., Stiles, C. D., and Leder, P. Cell-specific regulation of the c-*myc* gene by lymphocyte mitogens and platelet-derived growth factor. *Cell* 35:603–610, 1983.
103. Keshet, E., Rosenburg, M. P., Mercer, J. A., Propst, F., Van de Woude, G. F., Jenkins, N. A., and Copelmand, N. G. Developmental regulation of ovarian-specific *ras* expression. *Oncogene* 2:235–240, 1988.
104. King, C. R., Kraus, M. H., and Aaronson, S. A. Amplification of a novel v-*erb*B-related gene in human mammary carcinoma. *Science (Wash DC)* 229:974–976, 1985.
105. King, C. R., Swain, S. M., Porter, L., Steinberg, S. M., Lippman, M. E., and Gelmann, E. P. Heterogeneous expression of *erb*B-2 messenger RNA in human breast cancer. *Cancer Res.* 49:4185–4191, 1989.
106. Knudson, A. G. Hereditary cancer, oncogenes and antioncogenes. *Cancer Res.* 45:1437–1443, 1985.
107. Knudson, A. G., and Strong, L. C. Mutation and cancer. A model for Wilms' tumor of the kidney. *J. Natl. Cancer Inst.* 48:313–324, 1972.
108. Koji, T., Izumi, S., Tanno, M., Moriuchi, T., and Nakane, P. K. Localization *in situ* of c-*myc* mRNA and c-*myc* protein in adult mouse testis. *Histochem. J.* 20:551–557, 1988.

109. Kokai, Y., Cohen, J. A., Drebin, J. A., and Greene, M. I. Stage- and tissue-specific expression of the *neu* oncogene in rat development. *Proc. Natl. Acad. Sci. U.S.A.* 84:8498–8501, 1987.

110. Korn, L. J., Siebel, C. W., McCormick, F., and Roth, R. A. Ras p21 as a potential mediator of insulin action in *Xenopus* oocytes. *Science (Wash DC)* 236:840–843, 1987.

111. Koufos, A., Hansen, M. F., Lampkin, B. C., Workman, M. L., Copeland, N. G., Jenkins, N. A., and Cavenee, W. K. Loss of alleles at loci on human chromosome 11 during genesis of Wilms' tumour. *Nature (Lond)* 309:170–172, 1984.

112. Kovacs, G., Erlandsson, R., Boldog, F., Ingvarsson, S., Muller-Brechlin, R., Klein, G., and Sumegi, J. Consistent chromosome 3p deletion and loss of heterozygosity in renal cell carcinoma. *Proc. Natl. Acad. Sci. U.S.A.* 85:1571–1575, 1988.

113. Kozbor, D., and Croce, C. M. Amplification of the c-*myc* oncogene in one of five human breast carcinoma cell lines. *Cancer* 44:438–441, 1984.

114. Kraus, M. H., Popescu, N. C., Amsbaugh, S. C., and King, C. R. Overexpression of the EGF receptor related protooncogene *erb*B-2 in human mammary tumor cell lines by different molecular mechanisms. *EMBO J.* 6:605–620, 1987.

115. Kraus, M. H., Yuasa, Y., and Aaronson, S. A. A position 12-activated H-*ras* oncogene in all HS578T mammary carcinosarcoma cells but not normal mammary cells of the same patient. *Proc. Natl. Acad. Sci. U.S.A.* 81:5384–5388, 1984.

116. Krontiris, T. G., DeMartino, N. A., Colb, M., and Parkinson, D. R. Unique allelic restriction fragments of the human Ha-*ras* locus in leukocyte and tumour DNAs of cancer patients. *Nature (Lond)* 313:369–374, 1985.

117. Lacroix, H., Iglehart, J. D., Skinner, M. A., and Kraus, M. H. Overexpression of erbB-2 or EGF receptor proteins present in early state mammary carcinoma is detected simultaneously in matched primary tumors and regional metastases. *Nature (Lond)* 333:87–90, 1988.

118. Land, H., Parada, L. F., and Weinberg, R. A. Tumorigenic conversion of primary embryo fibroblasts requries at least two cooperating oncogenes. *Nature (Lond)* 304:596–602, 1983.

119. Laviale, C., Modjtahedi, N., Cassingena, R., and Brinson, O. High c-myc amplification level contributes to the tumorgenic phenotype of the human breast carcinoma cell line SW G13-S. *Oncogene* 3:335–339, 1988.

120. Leder, A., Pattengale, P. K., Kuo, A., Steward, T. A., and Leder, P. Consequences of widespread deregulation of the c-*myc* gene in transgenic mice: Multiple neoplasms and normal development. *Cell* 45:485–495, 1986.

121. Lee, W.-H., Bookstein, R., and Lee, E. Y.-H. P. Studies on the human retinoblastoma susceptibility gene. *J. Cell. Biochem.* 38:213–227, 1988.

122. Lidereau, R., Callahan, R., Dickson, C., Peters, G, Escot, C., and Ali, I. U. Amplification of the int-2 gene in primary human breast tumors. *Oncogene Res.* 2:285–291, 1988.

123. Lidereau, R., Cole, S. T., Larsen, C. J., and Mathieu-Mahul, D. A single point mutation responsible for c-*mos* polymorphism in cancer patients. *Oncogene* 1:235–237, 1987.

124. Lidereau, R., Escot, C., Theillet, C., Champeme, M. H., Brunet, M., Gest, J., and Callahan, R. High frequency of rare alleles of the human c-Ha-*ras*-1 protooncogene in breast cancer patients. *J. Natl. Cancer Inst.* 77:697–701, 1986.

125. Lidereau, R., Mathieu-Mahul, D., Theillet, C., Renaud, M., Mauchauffe, M., Gest, J., and Larsen, C. J. Presence of an allelic EcoR1 restriction fragment of the c-*mos* locus in leukocyte and tumor cell DNAs of breast cancer patients. *Proc. Natl. Acad. Sci. U.S.A.* 82:7068–7070, 1985.

126. Lin, T., Blaisdell, J., Barbour, K. W., and Thompson, E. A. Transient activation of c-myc protooncogene expression in Leydig cells by human chorionic gonadotropin. *Biochem. Biophys. Res. Commun.* 157:121–126, 1988.

127. Liu, E., Dollbaum, C., Scott, G., Rochlitz, C., Benz, C., and Smith, H. S. Molecular lesions involved in the progression of a human breast cancer. *Oncogene* 3:323–327, 1988.

128. Lothe, R. A., Fossa, S. D., Stenwig, A. E., Nakamura, Y., White, R., Borresen, A. L., and

Brogger, A. Loss of 3p or 11p alleles is associated with testicular cancer tumors. *Genomics* 5:134–138, 1989.

129. Lundberg, C., Skoog, L., Cavenee, W. K., and Nordenskjold, M. Loss of heterozygosity in human ductal breast tumors indicates a recessive mutation on chromosome 13. *Proc. Natl. Acad. Sci. U.S.A.* 84:2372–2376, 1987.

130. Lundy, L., Grimson, R., Mishriki, Y., Chao, S., Oravez, S., Fromowitz, F., and Viola, M. V. Elevated *ras* oncogene expression correlates with lymph node metastases in breast cancer patients. *J. Clin. Oncol.* 4:1321–1325, 1986.

131. Mackay, J., Elder, P. A., Porteous, D. J., Steel, C. M., Hawkins, R. A., Going, J. J., and Chetty, U. Partial deletions of chromosome 11p in breast cancer correlates with size of primary tumor and oestrogen receptor level. *Br. J. Cancer* 58:710–714, 1988.

132. Maddy, S. Q., Chisholm, G. D., Busuttil, A., and Habib, F. K. Epidermal growth factor receptors in human prostate cancer: Correlation with histological differentiation of the tumor. *Br. J. Cancer* 60:41–44, 1989.

133. Mariani-Costantini, R., Escot, C., Theillet, C., Gentile, A., Merlo, G., Lidereau, R., and Callahan, R. In situ c-myc expression and genomic status of the c-myc locus in infiltrating ductal carcinomas of the breast. *Cancer Res.* 48:199–205, 1988.

134. Matlashewski, G., Schneider, J., Banks, L., Jones, N., Murray, A., and Crawford, L. Human papillomavirus type 16 DNA cooperates with activated ras in transforming primary cells. *EMBO J.* 6:1741–1746, 1987.

135. Mester, J., Wagenaar, E., Sluyser, M., and Nusse, R. Activation of *int*-1 and *int*-2 mammary oncogenes in hormone-dependent and -independent mammary tumors of GR mice. *J. Virol.* 61:1073–1078, 1987.

136. Modjtahedi, N., Lavialle, C., Poupon, M.-F., Landin, R. M., Cassingena, R., and Monier, R. Increased level of amplification of the c-myc oncogene in tumors induced in nude mice by a human breast cancer line. *Cancer Res.* 45:4372–4379, 1985.

137. Monnat, M., Tardy, S., Saraga, P., Diggelmann, H., and Costa, J. Prognostic implications of expression of the cellular genes MYC, FOS, Ha-ras and Ki-ras in colon carcinoma. *Int. J. Cancer* 40:293–299, 1987.

138. Mori, S., Akiyama, T., Yamada, Y., Morishita, Y., Sugawara, I., Toyoshima, K., and Yamamoto, T. C-*erb*B-2 gene product, a membrane protein commonly expressed on human fetal epithelial cells. *Lab. Invest.* 61:93–97, 1989.

139. Morse, B., Rotherg, P. G., South, V. J., Spandorfer, J. M., and Astrin, S. M. Insertional mutagenesis of the *myc* locus by a LINE-1 sequence in a human breast carcinoma. *Nature (Lond)* 333:87–90, 1988.

140. Mukku, V. R., and Stancel, G. M. Regulation of epidermal growth factor by estrogen. *J. Biol. Chem.* 260:9820–9824, 1985.

141. Mulder, M. P., Keijzer, W., Boot, A. J. M., Verkerk, T., Prins, E., Splinter, T., and Bos, J. L. Activated *ras* genes in human seminoma: evidence of tumor heterogeneity. *Oncogene* 4:1345–1351, 1989.

142. Muller, R. Protooncogenes and differentiation. *Trends Biochem. Sci.* 11:129–131, 1986.

143. Muller, R., Bravo,R., Burckhardt, J., and Curran, T. Induction of c-*fos* gene and protein by growth factors precedes activation of c-*myc*. *Nature (Lond)* 312:716–720, 1984.

144. Muller, R., Tremblay, J. M., Adamson, E. D., and Verma, I. M. Tissue and cell type-specific expression of two human c-oncogenes. *Nature (Lond)* 304:484–486, 1983.

145. Murphy, L. J., Bell, G. I., and Friesen, H. G. Growth hormone stimulates sequential induction of c-*myc* and insulin-like growth factor I expression *in vivo*. *Endocrinology* 120:1806–1812, 1987.

146. Murphy, L. J., Murphy, L. C., and Friesen, H. G. Estrogen induction of N-*myc* and c-*myc* proto-oncogene expression in the rat uterus. *Endocrinology* 120:1882–1888, 1987.

147. Mutter, G. L., Grills, G. S., and Wolgemuth, D. L. Evidence for the involvement of the

proto-oncogene c-*mos* in mammalian meiotic maturation and possibly early embryogenesis. *EMBO J.* 7:683–689, 1988.

148. Mutter, G. L., and Wolgemuth, D. J. Distinct developmental patterns of c-mos proto-oncogene expression in female and male mouse germ cells. *Proc. Natl. Acad. Sci. U.S.A.* 84:5301–5305, 1987.

149. Mydlo, J. H., Michaeli, J., Cordon-Cardo, C., Goldenberg, A. S., Heston, W. D., and Fair, W. R. Expression of transforming growth factor alpha and epidermal growth factor receptor messenger RNA in neoplastic and nonneoplastic kidney tissue. *Cancer Res.* 49:3407–3411, 1989.

150. Nanus, D. M., Ebrahim, S. A. D., Bander, N. H., Real, F. X., Pfeffer, L. M., Shapiro, J. R., and Albino, A. P. Transformation of human kidney proximal tubule cells by ras-containing retroviruses. Implications for tumor progression. *J. Exp. Med.* 169:953–972, 1989.

151. Neal, D. E., Bennett, M. K., Hall, R. R., Marsh, C., Abel, P. D., Sainsbury, J. R. C., and Harris, A. L. Epidermal growth factor receptors in human bladder cancer: Comparison of invasive and superficial tumours. *Lancet I*:366–368, 1985.

152. Nilbert, M., Heim, S., Mandahl, N., Floderus, U.-M., Willen, H., Akerman, M., and Mitelman, F. Ring formation and structural rearrangements of chromosome 1 as secondary changes in uterine leiomyomas with t(12;14)(q14–15; q23–24). *Cancer. Genet. Cytogenet.* 36:183–190, 1988.

153. Niman, H. L., Thompson, A. M. H., Yu, A., Markman, M., Willems, J. J., Herwig, K. R., Habib, N. A., Wood, C. B., Houghten, R. A., and Lerner, R. A. Anti-peptide antibodies detect oncogene-related proteins in urine. *Proc. Natl. Acad. Sci. U.S.A.* 82:7924–7928, 1985.

154. Nisen, P. D., Rich, M. A., Gloster, E., Valderrama, E., Saric, O., Shende, A., Lanzkowsky, P., and Alt, F. W. N-myc oncogene expression in histopathologically unrelated bilateral pediatric renal tumors. *Cancer* 61:1821–1826, 1987.

155. Nisen, P. D., Zimmerman, K. A., Cotter, S. V., Gilbert, F., and Alt, F. Enhanced expression in the N-*myc* gene in Wilms' tumors. *Cancer Res.* 46:6217–6222, 1986.

156. Nusse, R. The *int* genes in mammary tumorigenesis and in normal development. *Trends Genet.* 4:291–295, 1988.

157. Nusse, R., van Ooyen, A., Cox, D., Fung, Y. K. T., and Varmus, H. Mode of proviral activation of a putative mammary oncogene (*int*-1) on mouse chromosome 15. *Nature (Lond)* 307:131–136, 1984.

158. Nusse, R., and Varmus, H. E. Many tumors induced by the mouse mammary tumor virus contain a provirus integrated in the same region of the host genome. *Cell* 31:99–109, 1982.

159. Ocadiz, R., Sauceda, R., Cruz, M., Graef, A. M., and Gariglio, P. High correlation between molecular alterations of the c-myc oncogene and carcinoma of the uterine cervix. *Cancer Res.* 47:4173–4177, 1987.

160. Ohuchi, N., Thor, A., Page, D. L., Horan Hand, P., Haeter, S., and Schlom, J. Expression of the 21,000 molecular weight *ras* protein in a spectrum of benign and malignant human mammary tissues. *Cancer Res.* 46:2511–2519, 1986.

161. Olofsson, B., Chardin, P., Touchot, N., Zahraoui, A., and Tavitian, A. Expression of the *ras*-related *ral* A, *rho* 12 and *rab* genes in adult mouse tissues. *Oncogene* 3:231–234, 1988 .

162. Parada, L. F., Tabin, C. J., Shih, C., and Weinberg, R. A. Human EJ bladder carcinoma oncogene is homologue of Harvey sarcoma virus *ras* gene. *Nature (Lond)* 297:474–478, 1982.

163. Pathak, S., Strong, L. C., Ferrell, R. E., and Trindade, A. A familial renal cell carcinoma with a 3:11 chromosome translocation limited to tumor cells. *Science* 217:939–941, 1982.

164. Paules, R. S., Buccione, R., Moschel, R. C., VandeWoude, G. F., and Eppig, J. J. Mouse *mos* protooncogene product is present and functions during oogenesis. *Proc. Natl. Acad. Sci. U.S.A. 86*:5395–5401, 1989.

165. Peehl, D. M. Molecular biology of proto-oncogenes in genitourinary cancers. *Cancer 60*:645–649, 1987.

166. Peehl, D. M., Wehner, N., and Stamey, T. A. Activated c-Ki-*ras* proto-oncogene in human prostatic adenocarcinoma. *Prostate 10*:281–289, 1987.

167. Pekonen, F., Partanen, S., Makinen, T., and Rutanen, E.-M. Receptors for epidermal growth factor and insulin-like growth factor 1 and their relation to steroid receptors in human breast cancer. *Cancer Res. 48*:1343–1347, 1988.

168. Pekonen, F., Partanen, S., and Rutanen, E.-M. Binding of epidermal growth factor and insulin-like growth factor I in renal carcinoma and adjacent normal kidney tissue. *Int. J. Cancer 43*:1029–1033, 1989.

169. Perez, R., Betsholtz, C., Westermark, B., and Heldin, C. Frequent expression of growth factors for mesenchymal cells in human mammary carcinoma cell lines. *Cancer Res. 47*:3425–3429, 1987.

170. Perez, R., Pascual, M., Macias, A., and Lage, A. Epidermal growth factor receptors in human breast cancer. *Breast Cancer Res. 4*:189–192, 1984.

171. Peters, G., Brookes, S., Smith, R., and Dickson, C. Tumorigenesis by mouse mammary tumor virus: Evidence for a common region for provirus integration in mammary tumors. *Cell 33*:369–377, 1983.

172. Peters, G., Lee, A. E., and Dickson, C. Concerted activation of two potential proto-oncogenes in carcinomas induced by mouse mammary tumor virus. *Nature (Lond) 320*:628–631, 1986.

173. Pfeifer-Ohlsson, S., Goustin, A. S., Rydnert, J., Wahlstrom, T., Bjersing, L., Stehlin, D., and Ohlsson, R. Spatial and temporal pattern of cellular *myc* oncogene expression in developing human placenta: Implications for embryonic cell proliferation. *Cell 38*:585–596, 1984.

174. Pollard, J. W., Bartocci, A., Arcecis, R., Orlofsky, A., Ladner, M. D., and Stanley, E. R. Apparent role of the macrophage growth factor, CSF-1, in placental development. *Nature (Lond) 330*:484–486, 1987.

175. Popescu, N. C., King, C. R., and Kraus, M. H. Localization of the human c-*erb*B-2 gene in normal and rearranged chromosomes 17 to bands q12–21.3. *Genomics 4*:362–366, 1989.

176. Propst, F., Rosenberg, M. P., Iyer, A., Kaul, K., and Van de Woude, G. F. C-*mos* proto-oncogene transforms mouse tissues: Structural features, developmental regulation, and localization in specific cell types. *Mol. Cell. Biol. 7*:1629–1637, 1987.

177. Propst, F., Rosenberg, M. P., Oskarsson, M. K., Russell, L. B., Nguyen-Huu, M. C., Nadeau, J., Jenkins, N. A., Copeland, N. G., and Van de Woude, G. F. Genetic analysis and developmental regulation of testis-specific RNA expression of *Mos, Abl*, actin and Hox-1.4. *Oncogene 2*:227–233, 1988.

178. Propst, F., and Van de Woude, G. F. Expression of c-*mos* proto-oncogene transcripts in mouse tissues. *Nature (Lond) 315*:516–518, 1985.

179. Pulciani, S., Santos, E., Long, L. K., Sorrentino, V., and Barbacid, M. *ras* gene amplification and malignant transformation. *Mol. Cell. Biol. 5*:2836–2841, 1985.

180. Ran, W., Dean, M., Levine, R. A., Henkle, C., and Campisi, J. Induction of c-*fos* and c-*myc* mRNA by epidermal growth factor or calcium ionophore is cAMP dependent. *Proc. Natl. Acad. Sci. U.S.A. 83*:8216–8220, 1986.

181. Rao, V. N., Huebner, K., Isobe, M., Ar-Rushdi, A., Croce, C., Shyam, E., and Reddy, P. *elk* tissue specific *ets*-related genes on chromosome X and 14 near translocation breakpoints. *Science (Wash DC) 244*:66–70, 1989.

182. Reddy, E. P., Reynolds, R. K., Santos, E., and Barbacid, M. A point mutation is responsible for the acquisition of transforming properties by the T24 human bladder carcinoma oncogene. *Nature (Lond) 300*:149–152, 1982.

183. Redmond, S. M. S., Reichman, E., Muller, R. G., Friis, R. R., Groner, B., and Hynes, N. E. The transformation of primary and established mouse mammary epithelial cells by p21-*ras* is concentration dependent. *Oncogene* 2:259–265, 1988.

184. Rijsewijk, F., van Deemter, L., Wagenaar, E., Sonnenberg, A., and Nusse, R. Transfection of the *int*-1 mammary oncogene in cuboidal RAC mammary cell line results in morphological transformation and tumorigenicity. *EMBO J.* 6:127–131, 1987.

185. Riou, G. F. Proto-oncogenes and prognosis in early carcinoma of the uterine cervix. *Cancer Surv.* 7:441–456, 1988.

186. Riou, G., Barrois, M., Dutronquay, V., and Orth, G. Presence of papillomavirus DNA sequences, amplification of c-myc and c-Ha-ras oncogenes and enhanced expression of c-myc in carcinomas of the uterine cervix. In: *Papillomaviruses: Molecular and Clinical Aspects*, edited by Broker, T.R., and Howley, P.M. New York, Alan R. Liss, 1985, vol. 32, pp. 47–56.

187. Riou, G., Barrois, M., Le, M. G., George, M., Le Doussal, V., and Haie, C. C-*myc* proto-oncogene expression and prognosis in early carcinoma of the uterine cervix. *Lancet* 1:761–763, 1987.

188. Riou, G., Barrois, M., Sheng, Z.-M., Duvillard, P., and Lhomme, C. Somatic deletions and mutations of c-Ha-*ras* gene in human cervical cancers. *Oncogene* 3:329–333, 1988.

189. Riou, G., Barrois, M., Tordjman, I., Dutronquay, V., and Orth, G. Presence de genomes de papillomavirus et amplification des oncogenes c-myc et c-Ha-ras dans des cancers envahissants du col de l'uterus. *C. R. Acad. Sci. (Paris)* 299:575–580, 1984.

190. Ro, J., Bressle, J., Ro, J. Y., Brasfield, F., Hortobagyi, G., and Blick, M. Sis/PDGF-β expression in benign and malignant breast lesions. *Oncogene* 4:351–354, 1989.

191. Robinson, A., Williams, A. R., Piris, J., Spandidos, D. A., and Wyllie, A. H. Evaluation of a monoclonal antibody to *ras* peptide, RAP-5, claimed to bind preferentially to cells of infiltrating carcinoma. *Br. J. Cancer* 54:877–883, 1986.

192. Rochlitz, C. F., Scott, G. K., Dodson, J. M., Liu, E., Dollbaum, C., Smith, H. S., and Benz, C. C. Incidence of activating *ras* oncogene mutations associated with primary and metastatic human breast cancer. *Cancer Res.* 49:357–360, 1989.

193. Rodenburg, C. J., Koelma, I. A., Nap, M., and Fleuren, G. J. Immunohistochemical detection of the *ras* oncogene product p21 in advanced ovarian cancer. *Photochem. Photobiol.* 41S:110s, 1985.

194. Rukstalis, D. B., Bubley, G. J., Donahue, J. P., Richie, J. P., Seidman, J. G., and DeWolf, W. C. Regional loss of chromosome 6 in two urological malignancies. *Cancer Res.* 49:5087–5090, 1989.

195. Sagata, N., Daar, I., Oskarsson, M., Showalter, S. D., and Van de Woude, G. F. The product of the *mos* proto-oncogene as a candidate "initiator" for oocyte maturation. *Science (Wash DC)* 245:643–645, 1989.

196. Sainsbury, J. R. C., Farndon, J. R., Needham, G. K., Malcolm, A. J., and Harris, A. L. Epidermal-growth-factor receptor status as predictor of early recurrence of and death from breast cancer. *Lancet* I:1398–1402, 1987.

197. Sainsbury, J. R. C., Sherbet, G. V., Farndon, J. R., and Harris, A. C. Epidermal growth factor receptors and oestrogen receptors in human breast cancer. *Lancet* 1:364–366, 1985.

198. Santos, E., Tronick, S. R., Aaronson, S. A., Pulciani, S., and Barbacid, M. T24 human bladder carcinoma oncogene is an activated form of the normal human homologue of BALB- and Harvey-MSV transforming genes. *Nature (Lond)* 298:343–347, 1982.

199. Santos, G. F., Scott, G. K., Lee, W. M. F., Liu, E., and Benz, C. Estrogen-induced post-transcriptional modulation of c-myc proto-oncogene expression in human breast cancer cells. *J. Biol. Chem.* 263:9565–9568, 1988.

200. Schechter, A. L., Hung, M.-C., Vaidyanathan, L., Weinberg, R. A., Yang-Feng, T. L., Francke, U., Ullrich, A., and Coussens, L. The *neu* gene: An *erb*B-homologous gene distinct from and unlinked to the gene encoding the EGF receptor. *Science (Wash DC)* 229:976–978, 1985.

201. Schmid, P., Schultz, W. A., and Hameister, H. Dynamic expression pattern of the *myc* protooncogene in midgestation mouse embryos. *Science (Wash DC)* 243:226–229, 1989.
202. Schuermann, M., and Michalides, R. A rare common integration site of provirus of the mouse mammary tumor virus in P-type mammary tumors of mouse strain GR. *Virology 1556*:229–237, 1987.
203. Schuuring, E., VanDeemter, L., Roelink, H., and Nusse, R. Transient expression of the proto-oncogene int-1 during differentiation of P19 embryonal carcinoma cells. *Mol. Cell. Biol. 9*:1357–1361, 1989.
204. Sejersen, T., Rahm, M., Szabo, G., Ingvarsson, S., and Sumegi, J. Similarities and differences in the regulation of N-myc and c-myc in murine embryonal cells. *Exp. Cell Res. 172*:304–317, 1987.
205. Semsei, I. M., Ma, S., and Cutler, R. G. Tissue and age specific expression of the *myc* protooncogene family throughout the life span of the C57BL/6J mouse strain. *Oncogene 4*:465–470, 1989.
206. Shackleford, G. M., and Varmus, H. E. Expression of the proto-oncogene int-1 is restricted to posterior male germ cells and the neural tube of mid-gestational embryos. *Cell 50*:89–95, 1987.
207. Shaw, A. P. W., Poirier, V., Tyler, S., Mott, M., Berry, J., and Maitland, N. J. Expression of the N-*myc* oncogenes in Wilms' tumor and related tissues. *Oncogene 3*:143–149, 1988.
208. Sheng, Z. M., Guerin, M., Gabillot, M., Spielmann, M., and Riou, G. c-Ha-*ras*-1 polymorphism in human breast carcinomas: Evidence for a normal distribution of alleles. *Oncogene Res. 2*:245–250, 1988.
209. Sherr, C. J., and Rettenmier, C. W. The *fms* gene and the CSF-1 receptor. *Cancer Surv. 5*:221–232, 1986.
210. Shih, C., and Weinberg, R. A. Isolation of a transforming sequence from a human bladder carcinoma cell line. *Cell 29*:161–169, 1984.
211. Shih, T. Y., Clanton, D. J., Hattori, S., Ulsh, L. S., and Chen, Z. Structure and function of p21 *ras* proteins: Biochemical, immunochemical, and site directed mutagenesis studies. In: *UCLA Symposium Proceedings: Growth Factors, Tumor Promoters and Cancer Genes*, 1986. p. 124.
212. Sikora, K., Evan, G., Stewart, J., and Watson, J. V. Detection of the c-*myc* oncogene product in testicular cancer. *Br. J. Cancer 52*:171–176, 1985.
213. Sinn, E., Muller, W., Pattengale, P., Tepler, I., Wallace, R., and Leder, P. Coexpression of MMTV/v-Ha-ras and MMTV/c-myc genes in transgenic mice: Synergistic action of oncogenes in vivo. *Cell 49*:465–475, 1987.
214. Skouv, J., Cristensen, B., and Autrup, H. Differential induction of transcription of c-*myc* and c-*fos* proto-oncogenes by 12-0-tetradecanoylphorbol-13-acetate in mortal and immortal human urothelial cells. *J. Cell. Biochem. 34*:71–79, 1987.
215. Slamon, D. J., Clark, G. M., Wong, S. G., Levin, W. J., Ulrich, A., and McGuire, W. L. Human breast cancer: Correlation of relapse and survival with amplification of the HER-2/*neu* oncogene. *Science (Wash DC)* 235:177–182, 1987.
216. Slamon, D. J., and Cline, M. F. Expression of cellular oncogenes during embryonic and fetal development of the mouse. *Proc. Natl. Acad. Sci. U.S.A. 81*:7141–7145, 1984.
217. Slamon, D. J., deKernion, J. B., Verma, I. M., and Cline, M. J. Expression of cellular oncogenes in human malignancies. *Science (Wash DC)* 224:256–262, 1984.
218. Slamon, D. J., Godolphin, W., Jones, L. A., Holt, J. A., Wong, S. G., Keith, D. E., Levin, W. J., Stuart, S. G., Udove, J., Ullrich, A., and Press, M. J. Studies of the HER-2/*neu* proto-oncogene in human breast and ovarian cancer. *Science (Wash DC)* 24:707–713, 1989.
219. Slater, R. M., and de Kraker, J. Chromosome number 11 and Wilms' tumor. *Cancer Genet. Cytogenet. 5*:237–245, 1982.
220. Smith, R., Peters, G., and Dickson, C. Multiple RNAs expressed from the *int-2* gene in mouse embryonal carcinoma cell lines encode a protein with homology to fibroblast growth factors. *EMBO J. 7*:1013–1022, 1988.

221. Sorrentino, V., McKinney, M. D., Giorgi, M., Geremia, R., and Fleissner, E. Expression of cellular proto-oncogenes in the mouse male germ line: A distinctive 2.4 kilobase *pim*-1 transcript is expressed in haploid postmeiotic cells. *Proc. Natl. Acad. Sci. U.S.A.* 85:2191–2195, 1988.

222. Sowani, A., Ong, G., Dische, S., Quinn, C., White, J., Soutter, P., Waxman, J., and Sikora, K. c-*myc* oncogene expression and clinical outcome in carcinoma of the cervix. *Mol. Cell. Probes* 3:117–123, 1989.

223. Spandidos, D. A., and Agnantis, N. J. Human malignant tumors of the breast, as compared to their respective normal tissue, have elevated expression of the Harvey-*ras* oncogene. *Anticancer Res.* 4:269–272, 1984.

224. Stewart, T. A., Bellve, A. R., and Leder, P. Transcription and promoter usage of the *myc* gene in normal somatic and spermatogenic cells. *Science (Wash DC)* 226:707–710, 1984.

225. Stock, L. M., Brosman, S. A., Fahey, J. L., and Liu, B. C. S. *Ras* related protein as a marker in transitional cell carcinoma of the bladder. *J. Urol.* 137:789–791, 1987.

226. Tabin, C. J., Bradley, S. M., Bargmann, C. I., Weinburg, R. A., Papageorge, A. G., Scolnick, E. M., Dhar, R., Lowy, D. R., and Chang, E. H. Mechanism of activation of a human oncogene. *Nature (Lond)* 300:143–149, 1982.

227. Tainsky, M. A., Cooper, C. S., Giovanella, B. C., and Vande Woude, G. F. An activated *ras*^N gene: Detected in late but not early passage human PA1 teratocarcinoma cells. *Science (Wash DC)* 225:643–645, 1984.

228. Tal, M., Wetzler, M., Josefberg, Z., Deutch, A., Gutman, M., Assaf, D., Kris, Y., Shiloh, Y., Givolo, D., and Schlessinger, J. Sporadic amplification of the HER2/*neu* protooncogene in adenocarcinomas of various tissues. *Cancer Res.* 48:1517–1520, 1988.

229. Tanaka, T., Slamon, D. J., Battifora, H., and Cline, M. J. Expression of p21/*ras* oncoproteins in human cancers. *Cancer Res.* 46:1465–1469, 1986.

230. Tandon, A. K., Clark, G. M., Chamness, G. C., Ullrich, A., and McGuire, W. L. HER-2/*neu* oncogene protein and prognosis in breast cancer. *J. Clin. Oncol.* 7:1120–1128, 1989.

231. T'Ang, A., Varley, J. M., Chakraborty, S., Murphree, A. L., and Fung, Y.-K. T. Structural rearrangement of the retinoblastoma gene in human breast carcinoma. *Science (Wash DC)* 242:263–242, 1988.

232. Taprowsky, E., Shimizu, K., Goldfarb, M., and Wigler, M. Structure and activation of the human N-*ras* gene. *Cell* 34:581–586, 1983.

233. Theillet, C., Le Roy, X., DeLapyriere, O., Grosgeorges, J., Adnane, J., Raynaud, S. D., Simony-Lafontaine, J., Goldfarb, M., Escot, C., Birnbaum, B., and Gaudry, P. Amplification of *FGF*-related genes in human tumors: Possible involvement of *HST* in breast carcinomas. *Oncogene* 4:915–922, 1989.

234. Theillet, C., Lidereau, R., Escot, C., Hutzell, P., Brunet, M., Gest, J., Schlom, J., and Callahan, R. Loss of a c-Ha-*ras*-1 allele and aggressive human primary breast carcinomas. *Cancer Res.* 46:4776–4781, 1986.

235. Thor, A., Ohuchi, N., Horan, P. H., Caalahan, R., Weeks, M. O., Theillet, C., Lidereau, R., Escot, C., Page, D., Vilasi, V., and Schlom, J. *ras* gene alterations and enhanced levels of *ras* p21 expression in a spectrum of benign and malignant human mammary tissues. *Lab. Invest.* 55:603–615, 1986.

236. Tremblay, P. J., Pothier, F., Hoang, T., Tremblay, G., Brownstein, S., Liszauer, A., and Jolicouer, P. Transgenic mice carrying the mouse mammary tumor virus *ras* fusion gene: Distinct effects on various tissues. *Mol. Cell. Biol.* 9:854–859, 1989.

237. Tsuda, H., Hirohashi, S., Shimosato, Y., Hirota, T., Tsugane, S., Yamamoto, H., Miyajima, N., Toyoshima, K., Yamamoto, T., Yoshida, T., Sakamoto, H., Terada, M., and Sugimura, T. Correlation between long-term survival in breast cancer patients and amplification of two putative oncogene-coamplification units: *hst*-1/*int*-2 and c-*erb*B-2/*ear*-1. *Cancer Res.* 49:3104–3108, 1989.

238. Tsuda, H., Shimosato, Y., Upton, M. P., Yokata, J., Terada, M., Ohura, M., Sugimura, T., and Hirohashi, S. Retrospective study on amplification of N-myc and c-myc genes in

pediatric solid tumors and its association with prognosis and tumor differentiation. *Lab. Invest.* 59:321–327, 1988.

239. van de Vijver, M. J., Mooi, W. J., Wisman, P., Peterse, J. L., and Nusse, R. Immunohistochemical detection of the *neu* protein in tissue sections of human breast tumors with amplified *neu* DNA. *Oncogene* 2:175–178, 1988.

240. van de Vijver, M., van de Bersselaar, R., Devilee, P., Cornelisse, C., Petersen, J. L., and Nusse, R. Amplification of the *neu* (c-*erb*-B-2) oncogene in human mammary tumors is relatively frequent and is often accompanied by amplification of the linked c-*erb*A oncogene. *Mol. Cell. Biol.* 7:2019–2023, 1987.

241. van de Vijver, M. J., Peterse, J. L., Mooi, W. J., Wisman, P., Lomans, J., Dalesio, O., and Nusse, R. *Neu*-protein overexpression in breast cancer. Association with comedo-type ductal carcinoma in situ and limited prognostic value in stage II breast cancer. *N. Engl. J. Med.* 319:1239–1245, 1988.

242. Van't Veer, L. J., Hermens, R., Vanden Berg-Baker, L. A. M., Cheng, N. C., Fleuren, G. J., Bos, J. L., Cleton, E. J., and Schrier, P. I. Oncogene activation in human ovarian carcinomas. *Oncogene* 2:157–165, 1988.

243. Varley, J. M., Armour, J., Swallow, J. E., Jeffrey, A. J., Powder, B. A., T'Anga, Fung, Y-K., Brammar, W. J., and Walker, R. A. The retinoblastoma gene is frequently altered leading to loss of expression in primary breast tumors. *Oncogene* 4:725–729, 1989.

244. Varley, J. M., Swallow, J. E., Brammar, W. J., Whittaker, J. L., and Walker, R. A. Alterations to either c-erbB-2 (*neu*) or c-*myc* proto-oncogenes in breast carcinomas correlate with poor short-term prognosis. *Oncogene* 1:423–430, 1987.

245. Varma, V. A., Austin, G. E., and O'Connell, A. C. Antibodies to *ras* oncogene p21 proteins lack immunohistochemical specificity for neoplastic epithelium in human prostate tissue. *Arch. Pathol. Lab. Med.* 113:16–19, 1989.

246. Venter, D. J., Tuzi, N. L., Kumar, S., and Gullick, W. J. Overexpression of the c-*erb*B-2 oncoprotein in human breast carcinomas: Immunohistological assessment correlated with gene amplification. *Lancet* 2:69–72, 1987.

247. Viola, M. V., Fromowitz, F., Oravez, S., Deb, S., Finkel, G., Lundy, J., Hand, P., Thor, A., and Schlom, J. Expression of *ras* oncogene p21 in human prostate cancer. *N. Engl. J. Med.* 314:133–137, 1986.

248. Viola, M. V., Fromowitz, F., Oravez, S., Deb, S., and Schlom, J. *ras* oncogene p21 expression is increased in premalignant lesions and high grade bladder carcinoma. *J. Exp. Med.* 161:1213–1218, 1985.

249. Visvader, J., and Verma, I. M. Differential transcription of exon 1 of the human c-*fms* gene in placental trophoblasts and monocytes. *Mol. Cell. Biol.* 9:1336–1341, 1989.

250. Wakai, A., Seino, S., Sakurai, A., Szilak, I., Bell, G. I., and DeGroot, L. J. Characterization of a thyroid hormone receptor expressed in human kidney and other tissues. *Proc. Natl. Acad. Sci. U.S.A.* 85:2781–2785, 1988.

251. Walker, R. A., Senior, P. V., Jones, J. L., Critchley, D. R., and Varley, J. M. An immunohistochemical and in situ hybridization study of c-*myc* and c-erbB-2 expression in primary human breast carcinomas. *J. Pathol.* 158:97–105, 1989.

252. Walker, R. A., and Wilkinson, N. p21 *ras* protein expression in benign and malignant human breast. *J. Pathol.* 156:147–153, 1988.

253. Ward, J. M., Hagiwara, A., Tsuda, H., Tatematsu, M., and Ito, N. H-*ras* p21 and peanut lectin immunoreactivity of hyperplastic, preneoplastic and neoplastic urinary bladder lesions in rats. *Jpn. J. Cancer Res.* 79:152–155, 1988.

254. Weissman, B. E., Saxon, P. J., Pasquale, S. R., Jones, G. R., Geiser, A. G., and Stanbridge, S. J. Introduction of a normal human chromosome 11 into a Wilms' tumor cell line controls its tumorigenic expression. *Science (Wash DC)* 236:175–180, 1987.

255. Whittaker, J. L., Walker, R. A., and Varley, J. M. Differential expression of cellular oncogenes in benign and malignant human breast tissue. *Int. J. Cancer.* 38:651–655, 1986.

256. Wiedemann, H. R. Tumors and hemihypertrophy associated with Beckwith-Wiedemann syndrome. *Eur. J. Pediatr. 141*:129–132, 1983.

257. Wilkinson, D. G., Bailes, J. A., and McMahan, A. P. Expression of the proto-oncogene *int*-1 is restricted to specific neural cells in the developing mouse embryo. *Cell 50*:79–88, 1987.

258. Wright, C., Angus, B., Nicholson, S., Sainsbury, J. R. C., Cairns, J., Gullick, W. J., Kelly, P., Harris, A. L., and Horne, C. H. W. Expression of c-*erb*B-2 oncoprotein: A prognostic indicator in human breast cancer. *Cancer Res. 49*:2087–2090, 1989.

259. Yokota, J., Toyoshima, K., Sugimura, T., Yamamoto, T., Terada, M., Battifora, H., and Cline, M. J. Amplification of c-erbB-2 oncogene in human adenocarcinomas *in vivo*. *Lancet i*:765–767, 1986.

260. Yokota, J., Tsunetsugu-Yokota, Y., Battifora, H., LeFevre, C., and Cline, M. J. Alterations of *myc, myb*, and *ras* proto-oncogenes in cancers are frequent and show clinical correlation. *Science (Wash DC) 231*:261–264, 1986.

261. Yokota, J., Yamamoto, T., Miyajima, N., Toyoshima, K., Nomura, N., Sakamoto, H., Yoshida, T., Terada, M., and Sugimura, T. Genetic alterations of the c-*erb*B-2 oncogene occur frequently in tubular adenocarcinoma of the stomach and are frequently accompanied by amplification of the v-*erb*A homologue. *Oncogene 2*:283–287, 1988.

262. Zajchowski, J., Band, V., Pauzie, N., Tager, A., Stampfer, M., and Sager, R. Expression of growth factors and oncogenes in normal and tumor-derived human mammary epithelial cells. *Cancer Res. 48*:7041–7047, 1988.

263. Zarbl, H., Sarawati, S., Arthur, A. V., Martin-Zanca, D., and Barbacid, M. Direct mutagenesis of Ha-ras-1 oncogenes by N-nitroso-N-methylurea during initiation of mammary carcinogenesis in rats. *Nature (Lond) 315*:382–385, 1985.

264. Zeillinger, R., Kury, F., Czerwenka, K., Kubista, E., Sliutz, G., Knogler, W., Huber, J., Zielinski, C., Reiner, G., Jakesz, R., Staffen, A., Reiner, A., Wrba, F., and Spona, J. HER-2 amplification, steroid receptors and epidermal growth factor receptor in primary breast cancer. *Oncogene 4*:109–114, 1989.

265. Zhang, X., Silva, E., Gersherson, D., and Hung, M.-C. Amplification and rearrangements of c-erb β protooncogenes in cancer of human female genital tract. *Oncogene 4*:985–989, 1989.

266. Zhou, D. J., Ahuja, H., and Cline, M. J. Proto-oncogene abnormalities in human breast cancer: c-erbB2 amplification does not correlate with recurrence of disease. *Oncogene 4*:105–108, 1989.

267. Zhou, D., Battifora, H., Yokota, J., Yamamoto, T., Cline, M. J. Association of multiple copies of the c-*erb*B-2 oncogene with spread of breast cancer. *Cancer Res. 47*:6123–6125, 1987.

268. Zhou, D. J., Casey, G., and Cline, M. J. Amplification of human int-2 in breast cancers and squamous carcinomas. *Oncogene 2*:279–282, 1988.

269. Zimmerman, K. A., Yancopoulos, G. D., Collum, R. G., Smith, R. K., Kohl, N. E., Denis, K. A., Nau, M. N., Witte, O. N., Torand-Allerand, D., Gee, C. E., Minna, J. D., and Alt, F. W. Differential expression of *myc* family genes during murine development. *Nature (Lond) 319*:780–783, 1986.

12

Oncogenes and Tumor Suppressor Genes in Solid Tumors: Lung, Endocrine, Skin, Salivary Gland, Head and Neck, and Soft Tissue Tumors

Cecilia M. Fenoglio-Preiser
Margaret B. Listrom
Teri A. Longacre

Lung

Lung tissues and lung tumors have not been studied as frequently as gastrointestinal, breast, and neural tissues and tumors. However, one subset of pulmonary neoplasms has been actively investigated, small cell tumors of the lung, and several aspects of oncogene and tumor suppressor gene functions have emerged. For example, in these tumors, amplification of the *myc* genes, particularly L-*myc*, has prognostic significance. Furthermore, because a consistent chromosomal deletion (3p14) is present in small cell carcinomas, it is quite likely that a tumor suppressor gene is deleted, giving rise to the propensity to develop tumors.

NORMAL LUNG

The protein encoded by HER-2/*neu* (c-*erb*B2) is found lining normal fetal bronchioles, [76] usually on the basolateral aspects of the cells.[100] In adults, the protein is expressed at low levels. L-*myc* is also found in fetal lung.[192] c-*jun* is expressed at high levels, and intermediate levels of *jun*D are present in normal lung. [60]

elk-related sequences appear to be transcriptionally active in lung tissue, but they have not been actively investigated in lung tumors.[121] *lyn* and other members of the *src* family have also been found to be expressed in lung tissue.[115]

LUNG TUMORS

There are four major histologic subtypes of lung cancer—small cell lung cancer (SCLC), squamous cell carcinoma, adenocarcinoma, and large cell carcinoma (the latter three referred to as non-SCLC or NSCLC), and therapeutic decisions and the overall prognosis depend on the histologic subtype present (*i.e.*, SCLC versus NSCLC). While there have been advances in the clinical management of all types of lung cancer, these have not had a significant impact on overall survival.[9,66]

Any breakthrough in the treatment or prevention of this disease that emerges from a more complete understanding of its molecular pathogenesis should allow the development of strategies to target therapy more rationally and to develop ways to prevent lung cancer by identifying persons with genetic susceptibilities to the disease. Activation of proto-oncogenes is likely to be important in establishing and maintaining the transforming properties of human lung cancer. Loss of tumor suppressor genes should predispose to the subsequent development of cancer. The oncogenes that appear to play a role in human lung cancer belong predominantly to the *myc* and *ras* families, with *myc* genes being most important in SCLC and *ras* genes in adenocarcinomas.

ras Genes

Early studies demonstrated that DNA obtained from the lung cancer cell lines Calu 1, Lx-1, and Sk-Lu-1 are able to transform NIH 3T3 cells *in vitro*.[34,35,116] These transforming genes, now known to be the mutated Harvey and Kirsten cellular *ras* genes, are also present in DNA isolated from bladder and colon carcinoma cell lines.[21,35,92,93,106,114,119,144–147]

Adenocarcinomas show a high incidence of mutationally activated K-*ras* genes,[26,97,126,127] which may be an early event since they are found in small primary tumors. *ras* mutations are also present in squamous cell carcinoma. The K-*ras* mutations found in ⅓ of pulmonary adenocarcinomas may impart a selective growth advantage to subpopulations of cells, instead of being essential for malignant transformation. The mutations that are identified may be the direct result of one or more ingredients of tobacco smoke,[127] since chemical carcinogens are notorious for inducing *ras* gene mutations.

ras mutations in lung cancers are summarized in Table 12.1. The exact mechanism of transformation by altered *ras* genes is elusive, but chromosomal analysis reveals frequent breaks and gaps, suggesting that *ras* mutations exert a destabilizing effect on the genome or that the mutations result from an unstable genome.[182]

Additional experimental evidence for the role of *ras* genes in the genesis of lung cancer comes from two sources, transfection assays and the construction of transgenic mice carrying an activated *ras* gene. Transfection of normal bronchial epithelial cells with a plasmid containing the Harvey *ras* oncogene changes the growth requirements, terminal differentiation, and tumorigenicity of the recipient cells. In one study, cell lines isolated after transfection were shown to contain v-Ha-*ras* DNA. The cells expressed the phosphorylated v-Ha-*ras* p21 and became aneuploid. As the transfectants become transplantable tumors, they acquire indefinite life spans, anchorage independence, and show many phenotypic abnormalities.[6,18,87,182]

Transgenic mice carrying the v-Ha-*ras* oncogene under the control of the MMTV LTR develop a number of tumors including bronchoalveolar carcinomas of the lung.[164] This was unexpected, since lung tumors were not reported by Sinn and

Table 12.1.
ras Mutations in Lung Cancer

Experimental Material	Cancer Type	Reference
Cell lines		
HS 242, Ha-*ras*, mutation codon 61, leucine incorporated	?	189
Q656, mutation codon 61	Squamous	70
SW1251, N-*ras* activation, mutated codon 61, G→A mutation	?	42, 188
Cancer line Lu 65, amplified c-Ki-*ras*, mutated codon 12	Giant cell	160
PR 371, mutated codon 12, cysteine substituted	Adenocarcinoma	106
Sk-Lu-1, not specified	?	116
LX1, mutation ?	Squamous	34
LC 10, mutation codon 12, ki-*ras*, arginine substituted	Squamous	132
A549, mutation?	Adenocarcinoma	93
Tumors		
Mutation codon 12, adenocarcinoma		126
Mutation codon 12, squamous		126
Mutation codon 12, metastases		126
Mutation codon 12, carcinoid		126
Mutation codon 12, large cell		

associates despite the fact that the v-Ha-*ras* transgene was expressed in the lungs of their transgenic mice.[148] This difference in tumor induction may reflect strain susceptibility differences in the animals used in the two studies.

RFLPs in the expression of some oncogenes, particularly the *myc* genes and *ras* genes, have provided potential information about tumor susceptibility. In an effort to extend such studies to the lung, Heighway and co-workers investigated the frequency of RFLPs involving the *ras* gene in 132 patients with lung cancer. They found four common alleles, one of which (a4) occurs more frequently in patients with non–small cell lung carcinoma than in either SCLC or normal control tissue. However, no difference in the a4 allele frequency was noted between tumor and normal tissue from the same patients, suggesting that this change is not restricted to tumor cells.[59]

myc Genes

Members of the *myc* gene family are amplified in all types of lung cancer. c-*myc* gene amplification is commonly present in human tumor cell lines that contain DM chromosomes and chromosomal HSRs.[5,73,85,140,168,176] In some cell lines the amplified sequences are present in the DMs, in others they are present in both DMs and HSRs. In one cell line they were dispersed throughout the genome.[168]

Either N-*myc* or L-*myc* was amplified in 24 SCLC, while c-*myc* was amplified in 25 squamous cell tumors. In another squamous cell carcinoma, amplification was found in the primary tumor and hilar and pleural metastasis but not in liver or paraaortic lymph node metastases. In one squamous cell, c-*myc* was amplified in a pleural and lymph node metastasis but not in the primary tumor. In two cases of SCLC, amplification or rearrangement of c-*myc* was detected in cell lines, but not in the original tumors from which the lines were derived. These data led the authors to suggest that lung tumors are heterogenous for amplification and rearrangement of *myc* oncogenes and that these oncogenes are activated during progression rather than transformation.[185]

c-*myc* RNA transcripts are also elevated, paralleling the level of DNA amplification.[49,130] However, because abundant c-*myc* mRNA can also be seen in the absence of c-*myc* amplification, c-*myc* expression might be deregulated in this type of carcinoma.[56,95,130,187] c-*myc* amplification was present in 44% of tumors from patients who had relapsed after chemotherapy, as compared to 11% of newly diagnosed tumors. Patients whose tumors exhibited c-*myc* amplification had shorter survival times than those who did not,[66,67] in a manner similar to that of N-*myc* and NBL.

SCLC fall into two groups: classic and variant phenotypes. Classic SCLC have a typical morphology and produce various gene products. Variant tumors are highly malignant, grow as larger cells, contain prominent nucleoli, grow faster, are more metastatic, and fail to express classical neuroendocrine properties such as L-dopadecarboxylase activity. A number of investigators have noted a relationship between the aggressive behavior of small cell tumors and their ability to be placed in a tissue culture.[50,66]

Small cell lung cancer lines, both classic and variant, have been found to exhibit N-*myc* gene amplification,[45,140] contrasting with the absence of N-*myc* gene amplification in non-SCLC lines.[130] Amplification of N-*myc* in biopsies from untreated patients with SCLC is associated with poor response to chemotherapy, rapid tumor growth, and short survival times.[47,61]

The relationship of the copy numbers of c-*myc* and N-*myc* to tumor formation or progression was studied in SCLC. Ninety-six tumors from 45 patients were examined and could be grouped into three categories: high copy number (tumors with greater than 3 copies of N-*myc* or c-*myc* gene/haploid genome), middle copy number (with 1.5–3 copies/genome), and normal copy number.[131] Fourteen of the patients had middle copy numbers, almost always resulting from chromosomal duplication rather than from amplification of a small genetic locus. This contrasted with 5 patients who had high copy numbers due to gene amplification. The amplification did not occur in a heterogeneous fashion within individual patients, since all the metastatic lesions from patients with high copy numbers also had a high copy number. None of the 41 metastatic lesions from the other patients had high copy numbers. These data suggest that gene amplification is an important step in the neoplastic growth in a subset of patients with SCLC and that this genetic event occurs relatively early, *i.e.*, before metastases.

A related gene, L-*myc*, is amplified in SCLC lines. Its amplification was not present in cell lines containing extra copies of the c-*myc* or N-*myc* gene, suggesting that only one member of the *myc* family can be activated in a given tumor.

Gene amplification is generally unstable,[134,135] as is shown in the drug resistance gene system. By analogy, myc genes present in SCLC should be rapidly lost during their growth and development unless they provide the tumors with some selective growth advantage.[49] The amplification of *myc* in SCLC may represent an autocrine process whereby a cell stimulates its own growth by secreting a factor to which it can respond.[151] SCLC cells secrete a variety of neural peptides, including bombesin, which induces proliferation of SCLC lines and is mitogenic for cultured normal bronchial epithelium.[173] Antibodies to bombesin block SCLC proliferation and induce regression of tumors transplanted into nude mice.[29] Furthermore, bombesin increases c-*myc* and c-*fos* gene expression.[29]

Not all SCLC contain amplified levels of *myc* genes. Possible reasons for this include: (*a*) tumors without *myc* gene amplification may be expressing these genes at a high level through a mechanism other than amplification; (*b*) the effect of the elevated *myc* gene expression may be replaced by activation of other oncogenes acting in a similar fashion (such as p53 or *fos* genes); and (*c*) it is possible that *myc* amplification only provides a selective growth advantage to tumors under well-defined circumstances, such as those mediated by immune responses or the microenvironment.

Bepler and associates identified three subclasses of SCLC on the basis of p64c-*myc* expression: (*a*) slow-growth group, with neuroendocrine differentiation and absent or very little c-*myc* expression; (*b*) a variant group with fast growth, absent to low neuroendocrine differentiation, and high c-*myc* expression; and (*c*) a transitional group with moderate neuroendocrine differentiation, moderate growth, and high p64 c-*myc* expression.[16]

RFLP patterns of the L-*myc* gene correlate with the extent of metastasis, particularly to lymph nodes at the time of surgery.[72] Patients with the L band (10 kb) have few lymph node metastases, whereas patients with the S (6 kb) or S + L bands almost always have lymph node metastases. This correlation is strongest in lung adenocarcinomas and may provide a useful preoperative marker predictive of the presence of metastases.

> **Case Example.** A 66-year-old man presented to the hospital with increased shortness of breath and dyspnea. A chest x-ray revealed a right pleural effusion and a radiodense lesion in the posterior upper lobe. The effusion was drained and no malignant cells were found on cytological examination. Bronchial brushings and biopsy were positive for SCLC (Fig. 12.1). The patient underwent extensive chemotherapy, but died 18 months later of widely disseminated metastatic disease. Portions of the tumor obtained at the time of autopsy were snap frozen in liquid nitrogen for future studies. DNA was extracted from the frozen tumor after monitoring the tissue for the presence of tumor cells and the absence of significant necrosis (Fig. 12.2). DNA was blotted onto nitrocellulose and hybridized to probes for c-*myc* and N-*myc* genes. Neither gene was amplified (Fig. 12.3).

Other Proto-oncogenes

Simultaneous amplification of c-*myb* and N-*myc* has been identified in SCLC lines.[54] Of interest is the observation that in two cell lines with low neuroendocrine differentiation (but without c-*myc* expression) c-*myb* expression was observed, suggesting that p75[c-*myb*] may be able to substitute for the p64[c-*myc*] protein.[16] Other proto-oncogenes found in NSCLC include c-*fur*, c-*fms*, *lck*, c-*raf*, c-*erb*B-1, c-*erb*B-4,

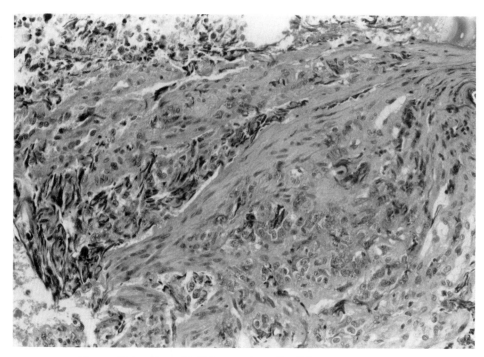

Figure 12.1. Bronchial biopsy of small cell lung cancer (250×).

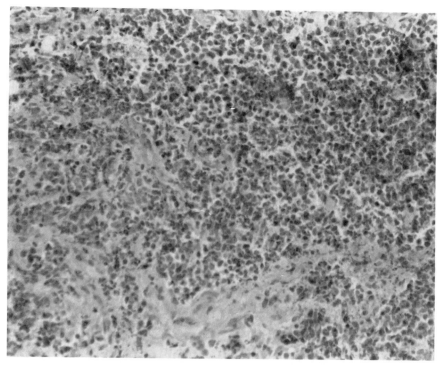

Figure 12.2. Frozen section of small cell carcinoma utilized for DNA analysis illustrated in Figure 12.3.

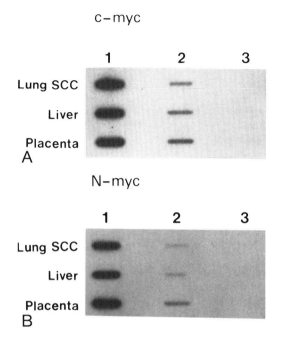

c-myc

N-myc

Figure 12.3. Autoradiograms of DNA from lung small cell carcinoma (SCC) hybridized to c-*myc* (**A**) and N-*myc* (**B**) ³²P-labeled probes. Five, 1.0, and 0.2 micrograms of genomic DNA were blotted on nitrocellulose membranes (*lanes 1–3*), hybridized overnight, and exposed to film at −70°C. DNA is not amplified for either gene.

and HER-2/*neu*.[54,63,115,138,157,183] Amplification of c-*erb*B-1 and HER-2/*neu* may provide epithelial cells with a proliferative advantage.

The *fur* gene encodes a membrane-associated protein with receptor-like characteristics. Its expression has been described recently in adenocarcinoma and squamous cell carcinoma,[133] but not in SCLCs. Perhaps this differential expression can be used as a discriminating marker in future studies on human lung cancer.

Finally, mink lung cells can be transformed with the oncogene v-*mos*. The viral genome integrates into the lung cells, causing cellular transformation. Both v-*mos* amplification and overexpression are required for transformation to occur.[47]

Tumor Suppressor Genes

A small deletion of the short arm of chromosome 3 at 3p14 is the most consistent finding in lung cancer of all types, is seen in nearly 10% of SCLC,[20,95,99,175,186] and may predispose individuals to develop lung cancer. This region coincides with the breakpoint in translocations associated with renal cell carcinoma.[3;8 and 3;11] It also corresponds to the approximate location of the *raf* oncogene.[19] The SCLC-associated deletion arises by a somatic mutation, since it is not present in normal cells of these patients. The deletion is only visible in one of two of the homologous chromosomes in heterozygous individuals, as revealed by RFLP analysis. Individuals at risk would be heterozygous for this recessive gene, carrying one mutated and one functional gene in each cell (Fig. 12.4). In these individuals, environmental factors, especially those present in tobacco smoke, may produce somatic changes such as deletions or mutations of the normal gene, thereby unmasking the effects of the abnormal allele. This might then deregulate another gene, such as a member of the *myc* family or a gene that encodes for a growth factor or its receptor. The presence of

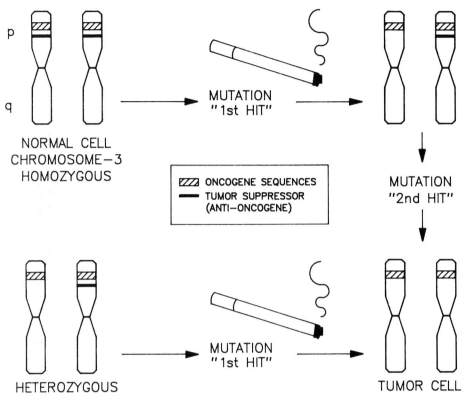

Figure 12.4. Simple diagram of lung cancer pathogenesis. A change in the genome of somatic cells occurs as a result of smoking. In this example loss of cancer suppressor genes is postulated.

these genetic deletions in concert with *myc* gene amplification, suggests an analogy to the retinoblastoma system, in which there is loss of the RB-1 gene and N-*myc* amplification, lending support to the possible role of tumor suppressor genes in this system. Loss of heterozygosity for 3p,[68] 15q, and 17p occurs in nearly 100% of SCLC,[16,99,184] although no common cytogenetic abnormalities of chromosomes 15 and 17 have been observed. Sithanandan and coauthors identified new c-*raf* polymorphisms and used RFLP analysis to determine the frequency of loss of this locus.[149] They found that one allele of the c-*raf*-1 locus located at 3p25 is consistently deleted in SCLC. It is possible that this locus contains genes acting as suppressor genes. p74[raf-1] expression does not appear to correlate with growth or differentiation.[16] Recent data suggest that deletion occurs frequently on 3p and that mitotic recombination or mitotic nondisjunction is more frequent than deletion for 13q and 17p or SCLC.[99]

p53 may be rearranged in some lung tumors, and overproduction of the p53 antigen makes established cells highly tumorigenic.[40] Additional evidence for a role of p53 in the genesis of lung tumors is provided by the transgenic mouse model in which animals carrying mutated p53 genomic fragments under the control of their own promoters exhibit a high incidence of lung adenocarcinoma.[80]

Endocrine Tissues

THYROID

Normal Thyroid

The expression of c-*myc* and c-*ras* proto-oncogenes has been investigated using Northern blot analysis during the establishment of thyroid cells in primary cell culture,[61] a time when major changes occur in thyroid cell function. In intact thyroids, c-*myc* and c-*ras* mRNA are detectable only at low levels, since c-*myc* expression relates to the state of cellular differentiation and whether the cell is primed for division as well as its stage in the cell cycle.[79,162] However, c-*myc* mRNA expression increases in cultured cells to 430%, 670%, and 330% of initial tissue levels on the first, second, and third days of culture, respectively. In contrast, *ras* oncogene mRNA expression does not change significantly during the 3-day culture period. Southern blot analysis indicates that gene amplification does not account for the increase in c-*myc* mRNA levels. These data suggest that c-*myc* (but not c-*ras*) oncogene expression is enhanced during the transition of thyroid follicle cells into a monolayer culture, which consistent with a role for c-*myc* expression in the regulation of normal, differentiated thyroid cell function.

Thyroid-stimulating hormone acutely stimulates both c-*myc* and β-actin mRNA levels in rat thyroid cells.[37,163] Elevated levels of c-*fos* RNA can also be seen in nonneoplastic thyroid tissue. In thyroid cells, c-*myc* and c-*ras* mRNA levels are stimulated by cAMP.[36,37]

Thyroid Tumors

A series of differentiated rat thyroid epithelial cell lines, either uninfected or infected with Kirsten murine sarcoma virus, were studied to determine the levels of p21 of the v-Ki-*ras* oncogene.[43] All cell lines showed a significant increase in p21 levels after transformation, confirming the potential ability of this oncogene product to transform epithelial cells.

A high frequency of *ras* activation (all members of the *ras* family) has been demonstrated in all stages of human thyroid carcinogenesis.[83,154] *ras* oncogene activation was detected in 33% of all thyroid adenomas (in 50% of microfollicular adenomas), in 53% of differentiated follicular carcinomas, and in 60% of undifferentiated carcinomas. Activation is most often due to the presence of mutation, which occurs in 0–60% of tumors, depending on the type.[46,82,83] *ras* mutations occur most often in follicular or undifferentiated tumors, but not in papillary lesions. The predominant amino acid substitution is glutamine to arginine at position 61 of Ha-*ras* or N-*ras*.[82,83]

Seventy-three thyroid lesions were stained with the antibody RAP-5.[69] Normal thyroid tissue showed the least reactivity. Papillary carcinomas, Hurthle cell carcinomas, and follicular carcinomas had slightly more intense staining than Hurthle cell adenomas or follicular adenomas. Anaplastic carcinomas and medullary carcinomas had much less staining. Since inflammatory thyroid lesions also showed moderate to intense immunoreactivity (as did multinodular goiters), the antibody is not useful in distinguishing benign from malignant thyroid lesions.

Another group analyzed thyroid lesions with the same antibody in Western blots and immunohistochemical preparations and found that elevation of *ras* p21 was common in thyroid lesions, with immunoreactivity localizing to apical and cytoplasmic areas. The authors suggest that this apical staining pattern is most characteristic of papillary carcinomas and that this pattern of staining may even be useful in differentiating papillary from nonpapillary neoplasms.

Insertion and expression of the v-*ras* gene in TT cells (cells derived from a patient with the sporadic form of medullary thyroid cancer), suggest that thyroid endocrine tumors induce cellular differentiation, as shown by slowed cellular growth, increased transcription of the calcitonin gene, and alterations of the calcitonin gene mRNA splicing events toward more mature cytoplasmic secretory granules.[86,105,107]

Final evidence that *ras* genes are involved in thyroid tumorigenesis is provided by the fact that injection of the retroviruses carrying the v-ki-*ras* oncogenes into the thyroid gland of rats induces thyroid carcinomas when the animals are treated with a goitrogenic agent.[117]

Yamashita and associates studied the expression of the *myc* cellular protooncogene in human thyroid tissue.[180] Four thyroid adenomas and three thyroid carcinomas expressed c-*myc* protooncogene mRNA transcripts, but they were not expressed in the normal thyroid tissues.

c-*erb*B-2 (HER-2/*neu*) and c-*erb*B are amplified and/or overexpressed in human thyroid tumors. Two to 3 fold higher levels of c-*erb*B-2 RNA were present in papillary carcinomas and lymph node metastases, as well as in one adenoma, when compared to nonneoplastic tissue.[1] Thyroid tumors also express EGF receptors.[89] c-*myc* and c-*fos* are variably expressed in both nontumor and tumor tissue. RNA transcripts specific for the platelet-derived growth factor A and B chains and the N-*ras* gene were detected in one anaplastic carcinoma.[1] Neither rearrangements nor amplifications of oncogenes were observed.

Medullary thyroid carcinomas express Ha-*ras*, *fos*, c-*myc*, and N-*myc* in primary and metastatic lesions, as shown by *in situ* hybridization and Northern blot analysis. Significant overexpression of Ha-*ras*, c-*myc*, and N-*myc* RNA are seen, as compared to normal thyroid. Expression of the proto-oncogenes *sis*, *fms*, and *abl* is not detectable.[75]

Tumor Suppressor Genes

Genetic linkage analyses have indicated an association between a region of chromosome 10 that may contain a tumor suppressor gene and multiple endocrine neoplasia type II. Medullary carcinomas constitute one component of this syndrome. However, medullary carcinomas have a low incidence of loss of chromosome 10 markers.[108]

PANCREATIC ENDOCRINE TISSUES

Normal Tissues

Normal islet cells, particularly a subset, strongly express p185^{neu}.[27]

Islet Cell Tumors

The mRNA expression of the proto-oncogenes Ha-*ras*, Ki-*ras*, *fos*, c-*myc*, N-*myc*, and *sis* was studied in five pancreatic endocrine tumors and in nonneoplastic pancreas by *in situ* hybridization and Northern blot analysis. Ha-*ras*, Ki-*ras*, *fos*, and c-*myc* (but not N-*myc* or *sis*) were detectable in all of the tumors, as well as in nonneoplastic pancreatic tissues.[25,62] Compared with the nonneoplastic components, Ha-*ras* mRNA was overexpressed up to 42 times more in all tumors. Metastasizing tumors showed 2 to 6 times higher Ha-*ras* mRNA levels than benign tumors. Ki-*ras*, *fos*, and c-*myc* mRNA expression did not correlate with any histological or biological properties of the tumors or with clinical outcome.[62] In contrast, c-*myc* RNA levels were higher in normal tissue than in tumors, and c-*fos* mRNA levels did not differ significantly between tumors and normal tissues.

ADRENAL AND PITUITARY

The normal adrenal medulla contains high levels of the c-*src* gene product. It is present in cell fractions enriched for chromaffin cell membranes and is stably associated with a 38 kilodalton protein through disulfide bonds. The subcellular localization suggests that p38 may be involved in anchoring the p60^{c-src} to the chromaffin granule membrane. Evidence exists that v-*src* is capable of inducing differentiation in pheochromocytoma cells.[4] This gene product is also abundant in pituitary tissue, again complexed to a 38 kilodalton protein.[52]

myc, *ras*, and *src* viral oncogenes cooperate in the transformation of nonestablished adrenal cortical cells.[88] This is analogous to its effects on medullary carcinoma of thyroid cell lines. Schwab and associates demonstrated the presence of amplified N-*ras* in murine adrenocortical tumor cells that contained karyotypic abnormalities in the form of DM chromosomes and HSRs.[139] Amplification and enhanced expression of cellular oncogenes may contribute to the maintenance of some of these tumors.[74] The amplified oncogene was c-Ki-*ras*.

Although activated *ras* genes generally promote cell growth, a rat pheochromocytoma cell line has been shown to stop proliferating and to differentiate when transfected with Kirsten murine sarcoma virus.[109] Mutated p21 promotes the morphological differentiation of PC12, a rat pheochromocytoma cell line.[12,15]

MULTIPLE ENDOCRINE NEOPLASIA

Multiple endocrine neoplasia is inducible by the promiscuous expression of the simian virus 40 large tumor antigen. When the gene is placed under the control of a major histocompatibility complex class I gene enhancer, the mice develop multiple endocrine neoplasms with tumors involving the pancreas, pituitary, thyroid, adrenals, and testis.[125]

Skin

NORMAL SKIN

In mice, c-*myc* expression is highest in the prenatal and newborn age groups and then declines until about age 6 months. Subsequently c-*myc* levels rise.[143] The continued expression in the adult may reflect the fact that the skin regularly undergoes cellular renewal.

When *myc* genes are introduced into keratinocytes, the cells increase their sensitivity to growth factors such as EGF, as demonstrated by an increased mitogenic activity.[123]

Nonneoplastic skin fibroblasts derived from patients with the Li-Fraumeni syndrome have been shown to demonstrate elevated expression of c-*myc* and an apparent activation of c-*raf*-1.[23] These individuals have an autosomal dominant disorder in which there is an increased incidence of tumors of all sites including leukemias, carcinomas, and sarcomas. The increased proto-oncogene expression may provide some insight into the pathogenesis of these tumors.

SQUAMOUS LESIONS

Ogiso and associates studied the expression of c-*fos*, c-*myc*, Ha-*ras*, N-*ras*, EGFR, and actin genes in 7 examples of normal epidermis, 3 cellular nevi, and 8 skin tumors, including 6 malignant and 2 benign lesions.[110] The genes were transcribed in most normal and tumor tissues, although no tumor-specific expression of c-*fos*, c-*myc*, Ha-*ras*, and N-*ras* was detected. However, there was a characteristic parallelism between expression of c-*fos* and c-*myc* in the normal epidermis, which was lost in tumors. The ratio of c-*fos* to c-*myc* transcripts in the normal epidermis was constant, compared with the expression of other genes. The data suggest that c-*fos* and c-*myc* are expressed in all normal skin tissues and that a constant c-*fos*/c-*myc* ratio maintains orderly tissue growth. Ha-*ras* and N-*ras* expression was not found, in keeping with the findings reported by Slamon[150] and Tatosyan.[159] Expression of the EGF-receptor gene was detected in 2 of 5 cases of normal epidermis and in a basal cell carcinoma (1 of 1), but not in cellular nevi (0 of 2) or malignant melanomas (0 of 4).

Ananthaswamy and associates analyzed the DNA extracted from 8 fresh human skin cancers occurring on sun-exposed sites for transformability of NIH 3T3 cells.[7] DNA from 6 of the human skin cancers induced tumors in nude mice. Four of the tumors contained *ras* oncogenes, and the DNAs from the other 2 did not contain any known oncogene. In one squamous cell carcinoma there were highly amplified copies of the Ha-*ras* oncogene. Some ultraviolet radiation (UV) induced murine skin cancers express an activated Ki-*ras* oncogene. UV irradiation also causes a transient increase in the m-RNAs for c-*myc* and c-Ha-*ras* in an SV-40 transformed human keratinocyte cell line, SVK-14.[128]

Mutational activation of the Ha-*ras* gene occurs in chemically induced benign skin papillomas and keratoacanthomas,[10,84] and the mutations found are consistent with the known reactivity of the carcinogens used. The *ras* mutations occur early in the process.[10] Furthermore, v-*ras* induces a cellular phenotype similar to that of carcinogen-treated keratinocytes.[190] *ras* transformation is also associated with decreased gap junctional communication.

MELANOMAS

Trent and associates reported a recurring translocation involving the long arm of chromosome 6 (6q) in malignant melanoma.[165] These authors found a t(1;6) involving band region 6q11-13 with two different regions of chromosome 1 (p22,q12-21). Chromosome 6 contains several oncogenes, including *ros, myb*, and *mas-1*. These findings suggest that any or all of these oncogenes may be biologically important in the biology of malignant melanoma.

Activated or mutated *ras* genes have been found in melanomas,[141] but they are uncommon.[122] Of five cell lines originating from separate metastatic deposits of a single patient, only one contained an activated *ras* gene, indicating heterogeneity in *ras* activation and suggesting that *ras* activation was not involved in tumor initiation or maintenance.[3,170]

It is unclear whether *ras* mutations in melanoma occur early or late. Sometimes it is concluded that the mutation occurs late,[3] other times it is thought to be present earlier.[170] Some tumors contain clones of cells with different *ras* mutations.[170] Of interest is the fact that when one compares melanomas with and without *ras* mutations, one finds that *ras* mutations are more common in tumors arising in sun-exposed areas of the body.[170]

Differences in the pattern of *ras* p21 have been noted in melanomas and melanocytic nevi. Nodular melanomas, epithelioid melanomas, and deeply invasive tumors have higher p21 levels than other types of tumors. The degree of expression parallels the degree of biologic aggressiveness.[181]

An analysis of the DNA of peripheral blood leukocytes from 55 patients with malignant melanomas and 53 normal healthy volunteers failed to show any significant association between melanomas and rare H-*ras*-1 alleles as defined by MSp I/Hpa II digestion. However, analysis of the same DNAs for a different polymorphism, based on the presence of additional Taq1 sites in the VTR region of H-*ras*-1, showed that the total frequency of a group of allelic variants called TP was significantly higher in melanoma patients than in normal donors.[120]

sis expression has also been noted,[174] and the EGFR is expressed in human melanomas containing an extra copy of chromosome 7.[78] The expression of EGFR may be related to the late steps in the malignant process.[78]

Eight percent of melanomas demonstrate amplification of the *hst* oncogene,[161] a gene encoding fibroblast growth factor–related protein. The *hst* gene is coamplified with the *int*-2 oncogene that maps to the same chromosomal location.[2] Transfection of polyoma middle T oncogene into an immortal murine melanocyte line, Mel-ab, which is dependent on TPA for *in vitro* growth, induces cellular transformation in the absence of TPA.[38]

Head and Neck Tumors

Gene amplification and overexpression of EGFR occurs in 19% of squamous cell carcinomas of the head and neck region, particularly in well-differentiated tumors, but no correlation exists between amplification and/or overexpression and the clinical stage or tumor site.[64,65,179]

However, the expression of this oncogene and others can be altered by administration of γ-interferon. The compound results in rapid morphologic changes; elevated *ras*, EGFR, and keratin expression; and cell death. Cell death appears to result from the induction of terminal differentiation of the tumor cells.[24]

Repeated applications of dimethylbenzanthracene (DMBA) induces squamous cell carcinomas in the cheek pouch of the hamster. During the process, the onco-gene c-*erb*B becomes amplified, and increased levels of c-*erb*B mRNA appear. The expression does not appear during the early hyperplastic events in the mucosa, but can be seen once the tumor develops and starts to invade the underlying connective tissue.[177]

H-*ras* mutations at codon 13, amplification of c-*erb*B-1, and c-myc amplifica-tion have all been described in oral cavity carcinomas.[155] Nasal carcinomas experi-mentally induced by a carcinogenic acylating agent contain mutated *ras* genes, with an A→T mutation at codon 61.[48] Similarly, chemically transformed normal mucosal epithelium shows evidence of transcription of the K-*ras* gene.[179]

The tumor cell line SQ-20B was established from a squamous cell carcinoma of the larynx in a patient whose tumor progressed while undergoing radiotherapy. This tumor remains radioresistant *in vitro* and is associated with the production of abnormal c-*raf* transcripts.[71] The cell line is also associated with chromosome 3 ab-normalities; chromosome 3 is the site of the *raf* gene.[19]

The *bcl*-1 locus is located on chromosome 11 at band q13, and chromosome 11 abnormalities occur frequently in head and neck tumors.[101] In addition, *bcl*-1 is am-plified 2–8-fold in squamous head and neck tumors. The amplification occurs more frequently in poorly differentiated tumors.[17]

The *int*-2 gene is amplified in 25% of head and neck squamous cell cancers. All of the involved tumors had metastasized to regional lymph nodes, and many were clinically aggressive tumors.[191]

Normal Thymus and Thymomas

The normal thymus expresses the *ras*-related genes *rho*-12, *rab*-1, and *rab*-2.[111]

Thymomas induced by x-rays in BALB/c mice were investigated for their oncogene expression; H-*ras*, K-*ras*, N-*ras*, c-*myc*, c-*myb*, and c-*abl* genes were con-sistently expressed in both normal thymus and thymomas. A 1.5–3-fold elevated c-*myc* expression was found in 42% of thymomas; an altered ratio of two normal promoters, P1 and P2, was found in some of the tumors. In one instance, expression from only the P2 promoter was present. In this instance, an alteration in the DNA sequence to the 5′ side of the promoter may have disrupted transcription from the P1 promoter.[11] From the study, it is unclear whether this alteration derived from the epithelial or lymphoid component of the tissue.

Salivary Gland

NORMAL SALIVARY GLAND

Fetal submandibular gland expresses high levels of the c-*myc* proto-oncogene at certain times in its organogenesis.[136] The protein encoded by the *neu* gene is ex-pressed in normal adult glands. Staining with an antibody directed at this protein shows that its expression is restricted to the intercalated and interlobular ducts; acinar cells are negative for the protein.[27] The adult salivary gland also expresses high levels of *rab* 1 and *rab* 2.

SALIVARY GLAND TUMORS

Stenman *et al.* studied the relationship between chromosomal patterns and proto-oncogenes in various human salivary gland cancers and found that comparisons with known localization of proto-oncogenes revealed that almost 60% of the abnormal stemlines of salivary gland tumors showed abnormalities that would fit with oncogene activation.[152] However, specific studies analyzing individual oncogenes were not undertaken.

sis-1 induces salivary gland tumors in mice. c-*erb*B-2 (HER-2/*neu*) is often amplified.[142,183]

Recently, Tsukamoto and associates constructed a transgenic mouse with the MMTV long terminal repeat placed 5' to the *int*-1 gene, in the opposite transcriptional orientation. Mice harboring this allele expressed *int*-1 mRNA at high level in their salivary glands. Salivary gland carcinomas also developed in some of the animals.[167] Other transgenic mouse constructs that result in the genesis of mammary tumors also often produce associated salivary gland tumors.[164] This is of interest also because some salivary gland tissues express gp52, a protein encoded by the mouse mammary tumor virus.[94]

Soft Tissues and Bone

NORMAL SOFT TISSUES

Fetal mesodermal tissues have high c-*myc* expression levels, particularly fetal cartilage.[136] It has also been shown that c-*myc* expression inhibits the differentiation of myoblasts cultured *in vitro*.[33,137] Transfection of myoblasts with mutationally activated *ras* genes blocks morphologic and molecular events associated with myogenic differentiation,[77,113] perhaps by preventing accumulation of regulatory factors required for transcriptional induction of muscle-specific genes.[153] The ability of the *ras* oncogene to interfere with developmental regulation of actin and intermediate filament mRNAs is dependent upon mutational activation and is not seen with the proto-oncogenic form of the gene. The oncogenic form of the *ras* gene prevents up-regulation of muscle-specific products during myogenesis and interferes with down-regulation of genes whose expression declines during myoblast fusion.[112]

The *trk* oncogene was isolated from an ascending colon cancer using DNA transfection techniques.[91] Its general structural features show a resemblance to the *bcr*/*abl* fusion proteins,[55] which arose by recombination. The gene may be activated by alternative splicing in muscle and nonmuscle tropomyosin sequences.[13]

v-*fos* represents the oncogene associated with the FBJ murine osteogenic sarcoma virus, and expression levels are markedly increased in virus-induced tumors in mice. However, increased expression may be insufficient for its carcinogenic potential, since similar expression levels are found in human placenta. This suggests that transformation and tumor induction may be tissue specific or may result from expression at the wrong time.[104] Of interest is the fact that neonatal bone expresses high levels of the c-*fos* gene.[102,103] Normal endothelial cells express the c-*sis* gene.[14]

c-*raf* and A-*raf* expression increases following initiation of differentiation of preadipocyte fibroblasts into adipocytes.[201] c-*rel* is expressed in normal muscle cells.[98]

SOFT TISSUE AND BONE TUMORS

The *fos* gene is responsible for the induction of osteogenic sarcomas in mice infected with the FBJ murine sarcoma virus.[28] Recently it has been shown that c-*fos* expression induces bone tumors, especially chondrosarcomas, in transgenic mice. The tumors have a strong tendency to develop in male animals.[129] Osteosarcomas express c-*sis*.[42] The *met* proto-oncogene was identified by DNA transfection from a human osteogenic sarcoma line transformed *in vitro* with N-nitronitrosoguanidine. Ewing's sarcomas are associated with expression of the *dbl* proto-oncogene.[171]

Rhabdomyosarcomas contain activated *ras* genes, and fibrosarcomas express N-*ras* and c-*sis*.[53] Embryonal cell rhabdomyosarcoma DNA transforms NIH 3T3 cells, and this cell line was shown to contain the c-Ki-*ras* gene and an elevated p21 protein level.[118,119] The c-*myc* gene was amplified 8-fold in one case of rhabdomyosarcoma.[166] The N-*myc* gene may also be amplified.[96]

A novel, rapidly migrating *ras* p21 was described in a malignant fibrous histiocytoma by Tanaka *et al.*, suggesting that mutational change had occurred in the *ras* gene.[158]

Viral and cellular *ras* genes are directly involved in the control of tumorigenic potential in two human sarcomas cell lines and in their revertants. Integration and loss of a single v-Ki-*ras* gene controls the malignant potential of the human osteosarcoma cell line TE 85, changing its ability to proliferate within normal tissues in an organ culture invasion assay. In flat revertants derived from the human HT 1080 fibrosarcoma cell lines, decreased expression of an activated N-*ras* oncogene affects the tumorgenic phenotype.[22,90]

Recent cytogenetic studies on synovial sarcoma have demonstrated a characteristic chromosomal translocation t(18;X)(q11;q11); the breakpoint on chromosome X has been assigned to Xp11.2. The *elk*-1 gene may be a candidate for involvement in this translocation. *elk*-2 also maps to chromosome 14q32.[121]

Retroviruses are known to be activated during the course of radiation-induced carcinogenesis. Occasionally, but not always, the c-*myc* gene is amplified during this process.[169]

Mesotheliomas strongly express the c-*sis* oncogene, which is barely detectable in normal mesothelial cells.[51,172]

A unique oncogene was isolated by transfecting the DNA from Kaposi's sarcoma and transforming NIH 3T3 cells.[31] These genes encode for a member of the fibroblast growth factor family.[32] This oncogene, termed K-*fgf*,[129] is probably identical to the *hst* gene isolated by Taira.[156]

Metastasizing hibernomas have been induced in transgenic mice expressing an α-amylase–SV-40 T antigen hybrid gene.[44]

Tumor Suppressor Genes

Hansen and associates[58] have shown that survivors of the heritable form of retinoblastomas subsequently develop second primary osteosarcomas at a substantially greater frequency than either the general population or survivors of nonheritable retinoblastoma. This involves specific somatic loss of the constitutional heterozygosity for the region of chromosome 13 that includes the RB locus.[57] Six osteosarcoma cell lines contained abnormalities in the RB genes, including DNA deletions and shortened mRNA transcripts. Similar events occur during the genesis of nonheritable osteosarcoma, but not in several other embryonal tumors or sarcomas.

In addition, survey of other tumors indicates that synovial sarcoma may have an alteration of the RB mRNA[81] and that alveolar rhabdomyosarcoma is associated with a unique translocation between chromosomes 2 and 13. The breakpoint for the translocation is at 13q14, the region containing the RB-1 locus.[39]

Recently, a series of unselected human sarcomas was studied in patients without an antecedent history for retinoblastoma. Three of nine osteosarcomas and 4 of 29 soft-tissue sarcomas, including two malignant fibrous histiocytomas, one leiomyoblastoma, and one liposarcoma had complete or partial deletion of the RB locus. These data suggest that the RB locus may be inactivated in human sarcomas unrelated to retinoblastoma.[124]

Transgenic mice carrying mutated p53 genomic fragments under the control of their own promoters have a high incidence of osteogenic sarcomas.[80] This gene had been previously reported to be rearranged in these tumors.

The meth A cell line, derived from a methylcholanthrene-induced fibrosarcoma, produces high levels of p53,[30] which represent a mutated p53.[8] Two mutant p53 alleles exist, representing independently occurring mutational events. Furthermore, some of the p53 mRNA species undergo alternative splicing events.[8] These mutations appear to play a critical role in p53-mutated transformation.[41]

Acknowledgments

The authors wish to acknowledge the helpful discussions and technical assistance of Barbara Griffith, the manuscript preparation by Agnes Truske, and the VA Medical Media Center for their expertise with the illustrative material.

The work reported in this chapter was supported by American Cancer Society Grant #PDT-341 and also by a grant from the University of New Mexico Cancer Center.

REFERENCES

1. Aasland, R., Lillehaug, J. R., Male, R., Josendal, O., Varhaug, J. E., and Kleppe, K. Expression of oncogenes in thyroid tumours: Coexpression of c-erbB2/neu and c-erbB. Br. J. Cancer 57:358–363, 1988.
2. Adelaide, J., Mattei, M.-G., Marics, I., Rayband, F., Planche, J., deLapeyriere, O., and Birnbaum, D. Chromosomal localization of the hst oncogene and its co-amplification with the int-2 oncogene in a human melanoma. Oncogene 2:413–416, 1988.
3. Albino, A. P., LeStrange, R., Oliff, A. I., Furth, M. E., and Old, L. J. Transforming ras genes from human melanoma: A manifestation of tumour heterogeneity? Nature (Lond) 308:69–71, 1984.

4. Alema, S., Casalbore, P., Agnostini, E., and Tato, F. Differentiation of PC12 pheochromocytoma cells induced by v-src oncogene. *Nature (Lond)* 316:557–559, 1985.
5. Alitalo, K., Schwab, M., Lin, C. C., Varmus, H. E., and Bishop, J. M. Homogeneously staining chromosomal regions contain amplified copies of an abundantly expressed oncogene (c-myc) in malignant neuroendocrine cells from a human colon carcinoma. *Proc. Natl. Acad. Sci. U.S.A.* 80:1701–1711, 1983.
6. Amstad, P., Reddel, R. R., Pfeiffer, A., Alan, M., Sibley, L., Mark, G. E., and Harris, C. L. Neoplastic transformation of a human bronchial epithelial cell line by a recombinant retrovirus encoding viral Harvey ras. *Mol. Carcinog.* 1:151–160, 1988.
7. Ananthaswamy, H. N., Price, J. E., Goldberg, L. H., and Bales, E. S. Detection and identification of activated oncogenes in human skin cancers occurring in sun-exposed body sites. *Cancer Res.* 48:3341–3346, 1988.
8. Arai, N., Nomura, D., Yokata, K., Wolf, D., Brill, E., Shohat, O., and Rotter, V. Immunologically, distinct p53 molecules generated by alternative splicing. *Mol. Cell. Biol.* 6:3232–3239, 1986.
9. Bailar, J. C. and Smith, E. M. Progress against cancer? *N. Engl. J. Med.* 314:1226–1232, 1986.
10. Bailleul, B., Brown, K., Ramsden, M., Akhurst, R. J., Fee, F., and Balmain, A. Chemical induction of oncogene mutations and growth factor activity in mouse skin carcinogenesis. *Environ. Health Perspect.* 81:23–27, 1989.
11. Bandyopadhyay, S. K., D'Andrea, E., and Fleissner, E. Expression of cellular oncogenes: Unrearranged c-myc gene but altered promoter usage in radiation-induced thymoma. *Oncogene Res.* 4:311–318, 1989.
12. Barbacid, M. *ras* genes. *Annu. Rev. Biochem.* 56:779–827, 1987.
13. Barnes, D., Clayton, L., Chumbley, G., and MacLeod, A. R. Activation of the *trk* oncogene by alternatively spliced muscle and non-muscle tropomyosin sequences. *Oncogene* 4:259–262, 1989.
14. Barrett, T. B., Gajdusek, C. M., Schwartz, S. M., McDougall, J. K., and Benditt, E. P. Expression of the sis gene by endothelial cells in culture and in vivo. *Proc. Natl. Acad. Sci. U.S.A.* 81:6772–6774, 1984.
15. Bar-sagi, D. and Feramisco, J. R. Microinjection of the ras oncogene protein into PC12 cells induces morphologic differentiation. *Cell* 42:841–848, 1985.
16. Bepler, G., Bading, H., Heimann, B., Kiefer, P., Havemann, K., and Moelling, K. Expression of p64c-myc and neuroendocrine properties define three subclasses of small cell lung cancer. *Oncogene* 4:45–50, 1989.
17. Berenson, J. R., Yang, J., and Mickel, R. A. Frequent amplification of the *bcl*-1 locus in head and neck squamous cell carcinomas. *Oncogene* 4:1111–1116, 1989.
18. Bonfil, R. D., Reddel, R. R., Ura, H., Reich, R., Fridman, Harris, C. C., and Klein-Szanto, A. J. P. Invasive and metastatic potential of a v-Ha-ras transformed human bronchial epithelial cell line. *J. Natl. Cancer Inst.* 81:587–594, 1989.
19. Bonner, T. I., O'Brien, J. J., Nash, W. G., and Morton, C. The human homologs of the raf/mil oncogene are located on human chromosomes 3 and 4. *Science (Wash DC)* 223:71–74, 1984.
20. Brauch, H., Johnson, B., Hovis, J., Yano, T., Gazdar, A., Pettingill, O. S., Graziano, S., Sorenson, G. D., Poiesz, B. J., Minna, J., Linehan, M., and Zbar, B. Molecular analysis of the short arm of chromosome 3 in small cell and non small cell carcinoma of the lung. *N. Engl. J. Med.* 317:1109–1113, 1987.
21. Capon, D. J., Seeburg, P. H., McGrath, J. P., Hayflick, J. S., Edman, U., Levinson, A. D., and Goeddel, D. V. Activation of Ki-ras2 gene in human colon and lung carcinomas by two different point mutations. *Nature (Lond)* 304:507–513, 1983.
22. Carloni, G., Paterson, H., Mareel, M., Augey-Bourgit, Y., Sabraoui, J., Reibiustein, E., Siearez, H., and Azzarone, B. N-*ras* dependent revertant phenotyping in human HT1080 fibrosarcoma cells is associated with loss of proliferation within normal tis-

sues and expressions of an adult membrane antigenic phenotype. *Oncogene* 4:873–880, 1989.

23. Chang, E. H., Pirollo, K. F., Zou, Z. Q., Cheunng, H. Y., Lawler, E. L., Garner, R., White, E., Bernstein, W. B., Fraumeni, J. W., and Blattner, W. A. Oncogene in radioresistant, noncancerous skin fibroblasts from a cancer prone family. *Science (Wash DC)* 237:1036–1039, 1987.

24. Chang, E. H., Ridge, J., Yu, Z., Richtsmeir, W. J., Harford, J. B., and Black, R. Induction of altered oncogene expression and differentiation of squamous cell carcinoma *in vitro*. In: *The Success of Differentiation Therapy of Cancer*. Second S Publications from Raven Press Vol 45; S. Waxman, G. B. Rossi, F. Takuda, eds., Raven Press NY, pp 59–77, 1988.

25. Chesa, P. G., Rettig, W. J., Melamed, M. R., Old, L. J., and Niman, H. I. Expression of p21*ras* in normal and malignant human tissues. Lack of association with proliferation and malignancy. *Proc. Natl. Acad. Sci. U.S.A.* 84:3234–3238, 1987.

26. Cline, M. J. and Battifora, H. Abnormalities of protooncogenes in non small cell lung cancer. Correlations with tumor type and clinical characteristics. *Cancer* 60:2669–2674, 1987.

27. Cohen, J. A., Weiner, D. B., More, K. F., Kokai, Y., Williams, W. V., Maguire, H. C., LiVolsi, V. A., and Green, M. I. Expression pattern of the *neu* (NGL) gene-encoded growth factor receptor protein (p185*neu*) in normal and transformed epithelial tissues of the digestive tract. *Oncogene* 4:81–88, 1989.

28. Curran, T. and Morgan, J. Memories of *fos*. *Bioessays* 7:255–258, 1987.

29. Cuttitta, F., Carney, D. N., Mulshine, J., Moody, T. N., Fedorko, J., Fischler, A., and Minna, J. D. Bombesin-like peptides can function as autocrine growth factors in human small cell lung cancer. *Nature (Lond)* 316:823–826, 1985.

30. DeLeo, A. B., Jay, G., Appella, E., Duboids, G. C., Law, L. W., and Old, L. J. Detection of a transformation-related antigen in chemically induced sarcomas and other trans-formed cells of the mouse. *Proc. Natl. Acad. Sci. U.S.A.* 76:2420–2424, 1979.

31. Delli Bovi, P. and Basilico, C. Isolation of a rearranged human transforming gene fol-lowing transfection of Kaposi's sarcoma DNA. *Proc. Natl. Acad. Sci.* 84:5660–5664, 1987.

32. Delli Bovi, P., Curatola, A. M., Kern, F. G., Greco, A., Ittmann, M., and Basilico, C. An oncogene isolated by transfection of Kaposi's sarcoma DNA encodes a growth factor that is a member of the FGF family. *Cell* 50:729–737, 1987.

33. Denis, N., Blanc, S., Leibovitch, M. P., Nicolaiew, N., Dautry, F., Raymondjean, Kruh, J., and Kitzis, A. c-*myc* oncogene expression inhibits initiation of myogenic differentia-tion. *Exp. Cell Res.* 172:21–217, 1987.

34. Der, C. J. and Cooper, G. M. Altered gene products are associated with activation of cel-lular ras^k genes in human lung and colon carcinomas. *Cell* 32:201–208, 1983.

35. Der, C. J., Krontiris, T. G., and Cooper, G. M. Transforming genes of human bladder and lung carcinoma cell lines are homologous to the *ras* genes of Harvey and Kirsten sarcoma viruses. *Proc. Natl. Acad. Sci. U.S.A.* 79:3637–3640, 1982.

36. Dere, W. H., Hirayu, H., and Rapoport, B. TSH and cAMP enhance expression of the *myc* proto-oncogene in cultured thyroid cells. *Endocrinology* 117:2249–2251, 1985.

37. Dere, W. H., Hirayu, H., and Rapoport, B. Thyrotropin and cyclic AMP regulation of *ras* proto-oncogene expression in cultured thyroid cells. *FEBS Lett.* 196:305–308, 1986.

38. Dooley, T. P., Wilson, R. E., Jones, N. C., and Hart, I. R. Polyoma middle T abrogates TPA requirement of murine melanocytes and induces malignant melanoma. *Oncogene* 4:531–535, 1989.

39. Douglas, E. C., Valentine, M., Etcubanes, E., Parhan, D., Webber, B. L., Houghton, P. J., Houghton, J. A., and Green, A. A. A specific chromosomal abnormality in rhabdomyo-sarcoma. *Cytogenet. Cell Genet.* 45:148–155, 1987.

40. Eliyahu, D. A., Raz, A., Gruss, P., Givol, D., and Oren, M. Participation of p53 tumor antigen in transformation of normal embryonic cells. *Nature (Lond)* 312:646–649, 1984.

41. Emanuel, B. S. Chromosomal in situ hybridization and the molecular cytogenetics of cancer. *Surv. Synth. Pathol. Res.* 4:269–281, 1985.

42. Eva, A., Robbins, K. C., Anderson, P. R., Srinivasan, A., Tronick, S. R., Reddy, E. P., Ellmore, N. W., Galen, A. T., Lautenberger, J. A., Papas, T. S., Westin, E. H., Wong-Staal, F., Gallo, R., and Aaronson, S. A. Cellular genes analogous to retroviral oncogenes are transcribed in human tumor cells. *Nature (Lond)* 295:116–119, 1982.

43. Ferrentino, M., Di Fiore, P. P., Fusco, A., Colletta, G., Pinto, A., and Vecchio, G. Expression of the *onc* gene of the Kirsten murine sarcoma virus in differentiated rat thyroid epithelial cell lines. *J. Gen. Virol.* 65:1955-1961, 1984.

44. Fox, N., Crooke, R., Hwang, L. H. S., Schibler, U., Knowles, B. B., and Solter, D. Metastatic hibernomas in transgenic mice expressing an alpha-amylase-SV 40 T antigen hybrid gene. *Science (Wash DC)* 244:460–463, 1989.

45. Funa, K., Steinholtz, L., Nov, E., and Beigh, J. Increased expression of N-myc in human small cell lung cancer biopsies predicts lack of response to chemotherapy and poor prognosis. *Am. J. Clin. Pathol.* 88:216–220, 1987.

46. Fusco, A., Grieco, M., Santoro, M., Berlingieri, M. J., Pilotti, S., Pierotti, M. A., Della Porta, G., and Vecchio, G. A new oncogene in human thyroid papillary carcinomas and their lymph node metastases. *Nature (Lond)* 328:170–172, 1987.

47. Gao, C., Wang, L.-C., Vass, W. C., Seth, A., and Chang, K. S. S. The role of v-*mos* in transformation, oncogenicity and metastatic potential of mink lung cells. *Oncogene* 3:267–273, 1988.

48. Garte, S. J. and Hochwalt, A. E. Oncogene activation in experimental carcinogenesis. The role of carcinogens and tissue specificity. *Environ. Health Perspect.* 81:29–31, 1989.

49. Gazdar, A. F., Carney, D. N., Nau, M. M., and Minna, J. D. Characterization of variant subclasses of cell lines derived from small cell lung cancer having distinctive biochemical, morphological, and growth properties. *Cancer Res.* 45:2924–2930, 1985.

50. Gazdar, A. F. and Oie, H. K. Cell culture methods for human lung cancer. *Cancer Genet. Cytogenet.* 19:5–10, 1986.

51. Gerwin, B. I., Lechner, J. F., Reddel, R., Betsholtz, C., Roberts, A., and Harris, C. C. Comparison of production of transforming growth factor-beta and platelet-derived growth factor by normal human mesothelial cells and mesothelioma cell lines. *Eur. J. Cancer Clin. Oncol.* 23:1752, 1987.

52. Grandori, C. and Hanafusa, H. p60[c-src] is complexed with a cellular protein in subcellular compartments involved in exocytosis. *J. Cell Biol.* 107:2125–2135, 1988.

53. Graves, D. T., Owen, A. J., Barth, R. K., Tempst, P., Winoto, A., Fors, L., Hood, L. E., and Antoniades, H. N. Detection of c-sis transcripts and synthesis of PDGF-like proteins by human osteosarcoma cells. *Science (Wash DC)* 226:972–974, 1984.

54. Griffin, C. and Baylin, S. B. Expression of the c-myb oncogene in human small cell lung carcinoma. *Cancer Res.* 45:272–275, 1985.

55. Grosveld, G., Verwoerd, T., Van Agthoven, T., deKlein, A., Ramachandran, K. L., Heisterkamp, N., Stam, K., and Graffen, J. The chronic myelocytic cell line K562 contains a breakpoint in bcr and produces a chimeric bcr/c-abl transcript. *Mol. Cell. Biol.* 6:607–616, 1986.

56. Gu, J., Linnoila, R. I., Seibel, N. L., Gazdar, A. F., Minna, J. D., Brooks, B. J., Hollis, G. F., and Kirsch, I. R. A study of myc-related gene expression in small cell lung cancer by in situ hybridization. *Am. J. Pathol.* 132:13–17, 1988.

57. Hansen, M. F. and Cavenee, W. K. Tumor suppressors: Recessive mutations that lead to cancer. *Cell* 53:172–173, 1988.

58. Hansen, M. F., Koufos, A., Gallie, B. L., Phillips, R. A., Fodstad, O., Brogger, A., Gedde-Dahl, T., and Cavenee, W. K. Osteosarcoma and retinoblastoma: A shared chromosomal mechanism revealing recessive predisposition. *Proc. Natl. Acad. Sci. U.S.A.* 82:6216–6220, 1985.

59. Heighway, J., Thatcher, N., Carny, T., and Hasleton, P. S. Genetic predisposition to human lung cancer. *Br. J. Cancer* 53:453–457, 1986.

60. Hirai, S.-I., Ryseck, R.-P., Mechta, F., Bravo, R., and Yamiu, M. Characterization of *jun*D: A new member of the *jun* proto-oncogene family. *EMBO J.* 8:1433–1439, 1989.

61. Hirayu, H., Dere, W. H., and Rapoport, B. Initiation of normal thyroid cells in primary culture is associated with enhanced c-*myc* messenger ribonucleic acid levels. *Endocrinology* 120:924–928, 1987.

62. Hofler, H., Ruhri, C., Putz, B., Wirnsberger, G., and Hauser, H. Oncogene expression in endocrine pancreatic tumors. *Virchows Arch. (B)* 55:355–361, 1988.

63. Horiguchi, J., Sherman, M. L., Sampson-Johannes, A., Weber, B. L., and Kufe, D. W. CSF-1 and c-fms expression in human carcinoma cell lines. *Biochem. Biophys. Res. Commun.* 157:395–401, 1988.

64. Hunts, J., Ueda, M., Ozawa, S., Abe, O., Pastan, I., and Shimizu, N. Hyperproduction and amplification of the epidermal growth factor receptor in squamous cell carcinomas. *Jpn. J. Cancer Res.* 76:663–666, 1985.

65. Ishitoya, J., Toriyama, M., Oguchi, N., Kitamura, K., Ohshima, M., Asano, K., and Yamamoto, T. Gene amplification and overexpression of EGF receptor in squamous cell carcinoma of the head and neck. *Br. J. Cancer* 59:559–562, 1989.

66. Johnson, B. E., Ihde, D. C., Bunn, P. A., Becker, B., Walsh, T., Weinstein, Z. R., Matthews, M. J., Whang-Peng, J., Makuch, R. W., Johnston-Early, A., Lichter, A. S., Carney, D. N., Cohen, M. H., Glatstein, E., and Minna, J. D. Patients with small cell lung cancer treated with combination chemotherapy with or without irradiation. Data on potential cures, chronic toxicities and late relapses after a five to eleven-year follow-up. *Ann. Intern. Med.* 103:430–438, 1985.

67. Johnson, B. E., Ihde, D. C., Makuch, R. W., Gazdar, A. F., Carney, D. N., Oie, H., Russell, E., Nau, M. M., and Minna, J. D. *myc* family oncogene amplification in tumor cell lines established from small cell lung cancer patients and its relationship to clinical status and course. *J. Clin. Invest.* 79:1629–1634, 1987.

68. Johnson, B. E., Sakaguchi, A. Y., Gazdar, A. F., Minna, J. D., Burch, D., Marshall, A., and Naylor, S. L. Restriction fragment polymorphism studies show consistent loss of chromosome 3p alleles in small cell lung cancer patients' tumors. *J. Clin. Invest.* 82:502–507, 1988.

69. Johnson, T. L., Lloyd, R. V., and Thor, A. Expression of *ras* oncogene p21 antigen in normal and proliferative thyroid tissues. *Am. J. Pathol.* 127:60–65, 1987.

70. Kagimoto, M., Miyoshi, J., Tashiro, K., Naito, Y., Sakaki, Y., Suishi, K., Tanaka, K., and Inamura, T. Isolation and characterization of an activated c-Ha-*ras* gene from a squamous cell lung carcinoma cell line. *Int. J. Cancer* 35:809–812, 1985.

71. Kasid, U., Pfeifer, A., Weichselbaum, R. R., Dritschilo, A., and Mark, G. E. The *raf* oncogene is associated with radiation-resistant human laryngeal cancer. *Science (Wash DC)* 237:1039–1041, 1987.

72. Kawashima, K., Shikama, H., Imoto, K., Izawa, M., *et al.* Close correlation between restriction fragment length polymorphism of the L-myc gene and metastasis of human lung cancer to the lymph nodes and other organs. *Proc. Natl. Acad. Sci. U.S.A.* 85:2353–2356, 1988.

73. Keifer, P. E., Bipler, G., Kubasch, M., and Havemann. Amplification and expression of protooncogenes in human small cell lung cancer cell lines. *Cancer Res.* 77:6236–6241, 1987.

74. Kimura, E. and Armelin, H. A. Role of proto-oncogene c-Ki-*ras* amplification and overexpression in the malignancy of Y-1 adrenocortical tumor cells. *Braz. J. Med. Biol. Res.* 21:189–201, 1988.

75. Klimpfinger, M., Ruhri, C., Putz, B., Pfragner, R., Wirnsberger, G., and Hofler, H. Oncogene expression in a medullary thyroid carcinoma. *Virchows. Arch. (B)* 54:256–259, 1988.

76. Kokai, Y., Cohen, J. A., Drebin, J. A., and Greene, M. I. Stage- and tissue-specific expression of the *neu* oncogene in rat development. *Proc. Natl. Acad. Sci. U.S.A.* 84:8498–8501, 1987.

77. Konieczny, S. F., Drobes, B. L., Menke, S. L., and Taprowsky, E. J. Inhibition of myogenic differentiation by the H-*ras* oncogene is associated with the down regulation of the MyoD1 gene. *Oncogene* 4:473–481, 1989.
78. Koprowski, H., Herlyn, M., Balaban, G., Parmiter, A., Ross, A., and Nowell, P. Expression of the receptor for epidermal growth factor correlates with increased dosage of chromosome 7 in malignant melanoma. *Somatic Cell Mol. Genet.* 11:297–302, 1985.
79. Lacy, J., Sarkar, S. N., and Summers, W. C. Induction of c-*myc* expression in human B lymphocytes by B-cell growth factor and anti-immunoglobulin. *Proc. Natl. Acad. Sci. U.S.A.* 83:1458–1462, 1986.
80. Lavigueur, A., Maltby, V., Mock, D., Rossant, J., Pawson, T., and Bernstein, A. High incidence of lung, bone, and lymphoid tumors in transgenic mice overexpressing mutant alleles of the p53 oncogene. *Mol. Cell. Biol.* 9:3982–3991, 1989.
81. Lee, W.-H., Bookstein, R., and Lee, E. Y.-H. P. Studies on the human retinoblastoma susceptibility gene. *J. Cell. Biochem.* 38:213–227, 1988.
82. Lemoine, N. R., Mayall, E. S., Wyllie, F. W., Farr, C. J., Hughes, D., Padua, R. A., Thurston, V., Williams, E. D., and Wynford-Thomas, D. Activated *ras* oncogenes in human thyroid cancers. *Cancer Res.* 48:4459–4463, 1988.
83. Lemoine, N. R., Mayall, E. S., Wyllie, F. W., Williams, E. D., Goyns, M., Stringer, B., and Wynford-Thomas, D. High frequency of *ras* oncogene activation in all stages of human thyroid tumorigenesis. *Oncogene* 4:159–164, 1989.
84. Leon, J., Kamino, H., Steinberg, J. J., and Pellicer, A. H-*ras* activation in benign and self-regressing skin tumors (keratoacanthomas) in both human and animal model systems. *Mol. Cell. Biol.* 8:786–793, 1988.
85. Little, C. D., Nau, M. M., Carney, D. N., Gazdar, A. F., and Minna, J. D. Amplification and expression of the c-*myc* oncogene in human lung cancer cell lines. *Nature (Lond)* 306:194–196, 1983.
86. Mabry, A., Gross, J. L., and Chen, S. F. Harvey-*ras* induced differentiation of human medullary thyroid carcinoma cells is accompanied by alterations in signal transduction. *Proc. Am. Assoc. Cancer Res.* 30:487, 1989.
87. Mabry, M., Nakagawa, T., Baylin, S., Pettengill, O., Sorenson, G., and Nelkin, B. Insertion of the v-Ha-ras oncogene induces differentiation of calcitonin-producing human small cell lung cancer. *J. Clin. Invest.* 89:196–199, 1989.
88. MacAuley, A. and Pawson, T. Cooperative transforming activities of *ras*, *myc*, and *src* viral oncogenes in nonestablished rat adrenocortical cells. *J. Virol.* 62:4712–4721, 1988.
89. Makinen, T., Pekonen, F., Franssila, K., and Lamberg, B.-A. Receptors for epidermal growth factor and thyrotropin in thyroid cancers. *Acta Endocrinol. (Copenh)* 117:45–50, 1988.
90. Mareel, M. M., Van Roy, F. M., Messiaen, L. M., Boghaert, E. R., and Bruyneel, E. A. Qualitative and quantitative analysis of tumour invasion in vivo and in vitro. *J. Cell Sci. (Suppl)* 8:141–163, 1987.
91. Martin-Zanca, D., Hughes, S. H., and Barbacid, M. A human oncogene formed by the fusion of truncated tropomyosin and protein tyrosine kinase sequences. *Nature (Lond)* 319:743–748, 1986.
92. McCoy, M., Bargmann, C. I., and Weinberg, R. A. Human colon carcinoma Ki-ras2 oncogene and its corresponding proto-oncogene. *Mol. Cell. Biol.* 4:1577–1582, 1984.
93. McCoy, M., Toole, J. J., Cunningham, J. M., Chang, E. H., Lowy, D. R., and Weinberg, R. A. Characterization of human colon/lung carcinoma oncogene. *Nature (Lond)* 302:79–81, 1983.
94. Mesa-Tejada, R., Keydar, I., Ramanarayanan, M., Ohno, T., Fenoglio, C., and Spiegelman, S. Immunohistochemical evidence for RNA virus related components in human breast cancer. *Ann. Clin. Lab. Sci.* 9:202–211, 1979.
95. Minna, J. D., Battey, J. F., Brooks, B. J., Cuttitta, F., Gazdar, A. F., Johnson, B. E., Inde, D.

C., Lebacq-Verheyden, A. M., Mulshine, J., Nau, M. M. Molecular genetic analysis reveals chromosomal deletion, gene amplification, and autocrine growth factor production in the pathogenesis of human lung cancer. *Cold Spring Harbor Symp. Quant. Biol.* *51*:843–853, 1986.

96. Mitani, K., Kurosawa, H., Suzuki, A., Hayashi, Y., Hanada, R., Yamomoto, K., Komatsu, M., Kobayashi, N., and Yamada, M. Amplification of N-myc in rhabdomyosarcoma. *Jpn. J. Cancer Res.* *77*:1062–1065, 1986.

97. Miyaki, M., Sato, C., Matsu, T., Koike, M., Mori, T., Kosaki, G., Takai, S., Tonomura, A., and Tsuchida, N. Amplification and enhanced expression of cellular oncogene c-Ki-ras-2 in a human epidermoid carcinoma of the lung. *Jpn. J. Cancer Res.* *76*:260–265, 1985.

98. Moore, B. E. and Bose, H. R. Expression of the c-*rel* and c-*myc* proto-oncogenes in avian tissue. *Oncogene* *4*:845–852, 1989.

99. Mori, N., Yokota, J., Oshimura, M., Cavenee, W. K., Mizoguchi, H., Noguchi, M., Shimosato, Y., Sugimara, T., and Terada, M. Concordant deletions of chromosome 3p and loss of heterozygosity for chromosomes 13 and 17 in small cell lung carcinoma. *Cancer Res.* *49*:5130–5135, 1989.

100. Mori, S., Akiyama, T., Yamada, Y., Morishita, Y., Sugawara, I., Toyoshima, K., and Yamamoto, T. C-*erb*B-2 gene product, a membrane protein commonly expressed on human fetal epithelial cells. *Lab. Invest.* *61*:93–97, 1989.

101. Muleris, M., Salmon, R. J., Girodet, J., Zafrani, B., and Dutrillaux, B. Recurrent deletions of chromosomes 11q and 3p in anal canal carcinoma. *Int. J. Cancer* *39*:595–598, 1987.

102. Muller, R., Bravo, R., Burckhardt, J., and Curran, T. Induction of c-*fos* gene and protein by growth factors precedes activation of c-*myc*. *Nature (Lond)* *312*:716–720, 1984.

103. Muller, R., Slamon, D. J., Tremblay, J. M., Cline, M. J., and Verma, I. M. Differential expression of cellular oncogenes during pre- and postnatal development of the mouse. *Nature (Lond)* *299*:6640–6644, 1982.

104. Muller, R., Tremblay, J. M., Adamson, E. D., and Verma, I. M. Tissue and cell type-specific expression of two human c-oncogenes. *Nature (Lond)* *304*:484–486, 1983.

105. Nakagawa, T., Mabry, M., deBustros, A., Ihle, J. N., Nelkin, B. D., and Baylin, S. B. Introduction of v-Ha-*ras* oncogene induces differentiation of cultured human medullary thyroid carcinoma cells. *Proc. Natl. Acad. Sci. U.S.A.* *84*:5923–5927, 1987.

106. Nakano, H., Yamamoto, F., Neville, C., Evans, D., Mizuno, T., and Perucho, M. Isolation of transforming sequences of two human lung carcinomas: Structural and functional analysis of the activated c-Ki-*ras* oncogenes. *Proc. Natl. Acad. Sci. U.S.A.* *81*:71–75, 1984.

107. Nelkin, B. D., deBustros, A. C., Mabry, M., and Baylin, S. B. The molecular biology of medullary thyroid carcinoma. A model for cancer development and progression. *JAMA* *261*:3130–3135, 1989.

108. Nelkin, B. D., Nakamura, Y., White, R. W., deBustros, A. C., Herman, J., Wells, S. A., and Baylin, S. B. Low incidence of loss of chromosome 10 in sporadic and hereditary human medullary thyroid carcinoma. *Cancer Res.* *49*:4114–4119, 1989.

109. Noda, M., Ko, M., Ogura, A., Liu, D.-G., Amano, T., Takano, T., and Ikawa, Y. Sarcoma viruses carrying *ras* oncogenes induce differentiation associated properties in a neuronal cell line. *Nature (Lond)* *318*:73–75, 1985.

110. Ogiso, Y., Oikawa, T., Kondo, N., Kuzumaki, N., Sugihara, T., and Ohura, T. Expression of proto-oncogenes in normal and tumor tissues of human skin. *J. Invest. Dermatol.* *90*:841–844, 1988.

111. Olofsson, B., Chardin, P., Touchot, N., Zahraoui, A., and Tavitian, A. Expression of the *ras*-related *ral* A, *rho* 12 and *rab* genes in adult mouse tissues. *Oncogene* *3*:231–234, 1988.

112. Olson, E. N. and Capetanaki, Y. G. Developmental regulation of intermediate filament and actin mRNAs during myogenesis is disrupted by oncogenic *ras* genes. *Oncogene* 4:907–913, 1989.

113. Olson, E. N., Spizz, G., and Tainsky, M. A. The oncogenic forms of N-*ras* or H-*ras* prevent skeletal myoblast differentiation. *Mol. Cell. Biol.* 7:2104–2111, 1987.

114. Parada, L. F., Tabin, C. J., Shih, C., and Weinberg, R. A. Human EJ bladder carcinoma oncogene is homologue of Harvey sarcoma virus *ras* gene. *Nature (Lond)* 297:474–478, 1982.

115. Perlmutter, R. M., Marth, J. M., Ziegler, S. F., Garvin, A. M., Pawar, S., Cooke, M. P., and Abraham, K. M. Specialized protein tyrosine kinase proto-oncogenes in hematopoietic cells. *Biochim. Biophys. Acta* 948:245–262, 1988.

116. Perucho, M., Goldfarb, M., Shimuzu, K., and Lama, C. Human-tumor–derived cell lines contain common and different transforming genes. *Cell* 27:467–476, 1981.

117. Portella, G., Ferulano, G., Santoro, M., Grieco, M., Fusco, A., and Vecchio, G. The Kirsten murine sarcoma virus induces rat thyroid carcinomas in vivo. *Oncogene* 4:181–187, 1989.

118. Pulciani, S., Santos, E., Lauver, A. V., Long, L. K., Aaronson, S. A., and Barbacid, M. Oncogenes in solid human tumours. *Nature (Lond)* 300:539–542, 1982.

119. Pulciani, S., Santos, E., Long, L. K., Sorrentino, V., and Barbacid, M. *ras* gene amplification and malignant transformation. *Mol. Cell. Biol.* 5:2836–2841, 1985.

120. Radice, P., Pierotti, M. A., Borello, M. Q., Illeni, M. T., Rovini, D., and Della Porta, G. HRAS1 Proto-oncogene polymorphisms in human malignant melanoma: Taq 1 defined alleles significantly associated with the disease. *Oncogene* 2:91–95, 1987.

121. Rao, V. N., Huebner, K., Isobe, M., Ar-Rushdi, A., Croce, C., Shyam, E., and Reddy, P. *elk* tissue specific *ets*-related genes on chromosome X and 14 near translocation breakpoints. *Science (Wash DC)* 244:66–70, 1989.

122. Raybaud, F., Noguchi, T., Marics, I., Adelaide, J., Planche, J., Batoz, B., Aubert, C., deLapeyriere, O., and Birnbaum, D. Detection of a low frequency of activated *ras* genes in human melanomas using a tumorigenicity assay. *Cancer Res.* 48:950–953, 1988.

123. Reiss, M., Dibble, C. L., and Narayanan, R. Transcriptional activation of the c-*myc* proto-oncogene in murine keratinocytes enhances the response to epidermal growth factor. *J. Invest. Derm.* 93:136–141, 1989.

124. Reissmann, P. T., Simon, M. A., Lee, W.-H., and Slamon, D. J. Studies of the retinoblastoma genes in human sarcomas. *Oncogene* 4:839–843, 1989.

125. Reynolds, R. K., Hoekzema, G. S., Vogel, J., Hinrichs, S. H., and Jay, G. Multiple endocrine neoplasia induced by the promiscuous expression of a viral oncogene. *Proc. Natl. Acad. Sci. U.S.A.* 85:3135–3139, 1988.

126. Rodenhuis, S., Slebos, R. J., Boot, A. J. M., Evers, S. C., Mooi, W. J., Wagenaar, S. S., van Bodegom, P. C., and Bos, J. L. Incidence and possible significance of K-ras oncogene activation in adenocarcinoma of the human lung. *Cancer Res.* 48:5738–5741, 1988.

127. Rodenhuis, S., van de Wetering, M. L., Moot, M. J., Evers, S. G., van Zendwijk, N., and Bos, J. L. Mutational activation of the K-*ras* oncogene. A possible pathogenetic factor in adenocarcinoma of the lung. *N. Engl. J. Med.* 317:929–935, 1987.

128. Ronai, Z. A., Okin, E., and Weinstein, I. B. Ultraviolet light induces the expression of oncogenes in rat fibroblast and human keratinocyte cells. *Oncogene* 2:201–204, 1988.

129. Ruther, U., Komitowski, D., Schubert, F. R., and Wagner, E. F. c-*fos* induces bone tumors in transgenic mice. *Oncogene* 4:861–866, 1989.

130. Saksella, K., Bergh, J., Lehto, V.-P., Wilsson, K., and Alitalo, K. Amplification of the myc oncogene in a subpopulation of human small cell lung cancers. *Cancer Res.* 45:1823–1827, 1985.

131. Saksella, K., Bergh, J., and Nilsson, K. Amplification of the N-myc oncogene in an adenocarcinoma of the lung. *J. Cell. Biochem.* 31:297–304, 1986.

132. Santos, E., Martin-Zanca, D., and Reddy, E. P. Malignant activation of a Ki-ras oncogene in lung carcinoma but not in normal lung tissue of the same patient. *Science (Wash DC)* 223:661–664, 1984.

133. Schalken, J. A., Roebroek, A. J. M., Oomen, P. P. C. A., Wagenaar, S. S., Debryne, F. M. J., Bloemers, H. P. J., and Van de Ves, W. J. M. *fur* gene expression as a discriminating marker for small cell and non–small cell lung carcinomas. *J. Clin. Invest.* 80:1545–1549, 1987.

134. Schimke, R. T. Gene amplification in cultured animal cells. *Cell* 37:705–713, 1984.

135. Schimke, R. T., Sherwood, S. W., Hill, A. B., and Johnston, R. N. Overreplication and recombination of DNA in higher eukaryotes: Potential consequences and biological implications. *Proc. Natl. Acad. Sci. U.S.A.* 83:2157–2161, 1986.

136. Schmid, P., Schultz, W. A., and Hameister, H. Dynamic expression pattern of the *myc* protooncogene in midgestation mouse embryos. *Science (Wash DC)* 243:226–229, 1989.

137. Schneider, M. D., Perryman, M. B., Payne, P. A., Spizz, G., Roberts, R., and Olson, E. N. Autonomous expression of c-*myc* in BC3H1 cells partially inhibits but does not prevent myogenic differentiation. *Mol. Cell. Biol.* 7:1973–1977, 1987.

138. Schneider, P. M., Hung, M.-C., Chiooca, S. M., Manning, J., Zhao, X., Fang, K., and Roth, J. A. Differential expression of the c-*erb*B-2 gene in human small cell and non–small cell lung cancer. *Cancer Res.* 49:4968–4971, 1989.

139. Schwab, M., Alitalo, K., Varmus, H. E., Bishop, J. M., and George, D. A cellular oncogene (c-Ki-*ras*) is amplified, overexpressed, and located within karyotypic abnormalities in mouse adrenocortical tumour cells. *Nature (Lond)* 303:497–501, 1983.

140. Seifter, E. J., Sansville, E. A., and Battey, J. Comparison of amplified and unamplified c-*myc* gene structure and expression in human small cell lung carcinoma cell lines. *Cancer Res.* 46:2050–2055, 1986.

141. Sekiya, T., Fushimi, H., Hori, H., Hirohashi, S., Nishimura, S., and Sugimura, T. Molecular cloning and total nucleotide sequence of the human c-Ha-*ras*-1 gene activated in a melanoma from a Japanese patient. *Proc. Natl. Acad. Sci. U.S.A.* 81:4771–4775, 1984.

142. Semba, K., Kamata, N., Toyoshima, K., and Yamamoto, T. A v-*erb*B-related protooncogene, c-erbB-2, is distinct from the c-erbB-1/epidermal growth factor-receptor gene and is amplified in a human salivary gland adenocarcinoma. *Proc. Natl. Acad. Sci. U.S.A.* 82:6497–6501, 1985.

143. Semsei, I. M., Ma, S., and Cutler, R. G. Tissue and age specific expression of the *myc* protooncogene family throughout the life span of the C57BL/6J mouse strain. *Oncogene* 4:465–470, 1989.

144. Shih, C., Padhy, L. C., Murray, M., and Weinberg, R. A. Transforming genes of carcinomas and neuroblastomas introduced into mouse fibroblasts. *Nature (Lond)* 290:261–264, 1981.

145. Shih, C. and Weinberg, R. A. Isolation of a transforming sequence from a human bladder carcinoma cell line. *Cell* 29:161–169, 1984.

146. Shih, T. Y., Clanton, D. J., Hattori, S., Ulsh, L. S., and Chen, Z. Structure and function of p21 *ras* proteins: Biochemical, immunochemical, and site directed mutagenesis studies. In: *UCLA Symposium Proceedings: Growth Factors, Tumor Promoters and Cancer Genes*, 1986, p. 124.

147. Shimuzu, K., Birnbaum, D., Ruley, M. A., Fasano, O., Suard, Y., Edlung, L., Taparowsky, E., Goldfarb, M., and Wigler, M. Structure of the Ki-*ras* gene of the human cell lung carcinoma cell line Calu-1. *Nature (Lond)* 304:497–500, 1983.

148. Sinn, E., Muller, W., Pattengale, P., Tepler, I., Wallace, R., and Leder, P. Coexpression of MMTV/v-Ha-ras and MMTV/c-myc genes in transgenic mice: Synergistic action of oncogenes in vivo. *Cell* 49:465–475, 1987.

149. Sithanandan, G., Dean, M., Brennscheidt, U., Beck, T., Gazdar, A., Minna, J. D., Brauch, H., Zbar, B., and Rapp, U. R. Loss of heterozygosity at the c-*raf* locus in small cell lung carcinoma. *Oncogene* 4:451–455, 1989.

150. Slamon, D. J., deKernion, J. B., Verma, I. M., and Cline, M. J. Expression of cellular oncogenes in human malignancies. *Science (Wash DC)* 224:256–262, 1984.

151. Sporn, M. and Todaro, G. Autocrine secretion and malignant transformation of cells. *N. Engl. J. Med.* 303:878–880, 1980.

152. Stenman, G., Mark, J., and Ekedhal, C. Relationships between chromosomal patterns

and proto-oncogenes in human benign mixed salivary gland tumors. *Tumour Biol.* 5:103–117, 1984.

153. Sternberg, E. A., Spizz, G., Perry, M. E., and Olson, E. R. A *ras*-dependent pathway abolishes activity of a muscle-specific enhancer upstream from the muscle creatine kinase gene. *Mol. Cell. Biol.* 9:594–601, 1989.

154. Suarez, T. G., Du Villard, J. A., Caillou, B., Schlumberger, M., Tubiama, M., Parmentier, C., and Monier, R. Detection of activated *ras* oncogenes in human thyroid carcinomas. *Oncogene* 2:403–406, 1988.

155. Tadokoro, K., Ueda, M., Ohshima, T., Fujita, K., Rikimaru, K., Takahashi, N., Enomoto, S., and Tsudhida, N. Activation of oncogenes in human oral cancer cells: A novel codon 13 mutation of c-H-*ras*-1 and concurrent amplification of c-*erb*B-1 and c-*myc*. *Oncogene* 4:499–505, 1989.

156. Taira, M., Yoshida, T., Miyagawa, K., Sakamoto, H., Terada, M., and Sugimura, T. cDNA sequence of human transforming gene *hst* and identification of the coding sequence required for transforming activity. *Proc. Natl. Acad. Sci. U.S.A.* 84:2980–2984, 1987.

157. Tal, M., Wetzler, M., Josefberg, Z., Deutch, A., Gutman, M., Assaf, D., Kris, Y., Shiloh, Y., Givolo, D., and Schlessinger, J. Sporadic amplification of the HER2/*neu* protooncogene in adenocarcinomas of various tissues. *Cancer Res.* 48:1517–1520, 1988.

158. Tanaka, T., Slamon, D. J., Battifora, H., and Cline, M. J. Expression of p21/*ras* oncoproteins in human cancers. *Cancer Res.* 46:1465–1469, 1986.

159. Tatosyan, A. G., Galetski, S. A., Kisseljova, N. P., Asanovca, A., Zborovskaya, I. B., Spitkovsky, D. D., Revasova, E. S., Martin, P., and Kisseljov, P. Oncogene expression in human tumors. *Int. J. Cancer* 35:731–736, 1985.

160. Taya, Y., Hosogai, K., Hirohashi, S., Shimosato, Y., and Tsuchiya, R. A novel combination of K-ras and myc amplification accompanied by point mutational activation of K-ras in a human lung cancer. *EMBO J.* 3:2943–2946, 1984.

161. Theillet, C., Le Roy, X., DeLapyriere, O., Grosgeorges, J., Adnane, J., Raynaud, S. D., Simony-Lafontaine, J., Goldfarb, M., Escot, C., Birnbaum, B., and Gaudry, P. Amplification of *FGF*-related genes in human tumors: Possible involvement of *HST* in breast carcinomas. *Oncogene* 4:915–922, 1989.

162. Thompson, C. B., Challoner, P. B., Neiman, P. E., and Groudine, M. Levels of c-*myc* oncogene mRNA are invariant throughout the cell cycle. *Nature (Lond)* 314:363–366, 1985.

163. Tramontano, D., Chin, W. W., Moses, A. C., and Ingbar, S. H. Thyrotropin and dibutyryl cyclic AMP increase levels of c-*myc* and c-*fos* mRNAs in cultured rat thyroid cells. *J. Biol. Chem.* 261:3919–3922, 1986.

164. Tremblay, P. J., Pothier, F., Hoang, T., Tremblay, G., Brownstein, S., Liszauer, A., and Jolicouer, P. Transgenic mice carrying the mouse mammary tumor virus *ras* fusion gene: Distinct effects on various tissues. *Mol. Cell. Biol.* 9:854–859, 1989.

165. Trent, J. M., Thompson, F. H., and Meyskens, F. L. Identification of a recurring translocation site involving chromosome 6 in human malignant melanoma. *Cancer Res.* 49:420–429, 1989.

166. Tsuda, H., Shimosato, Y., Upton, M. P., Yokata, J., Terada, M., Ohura, M., Sugimura, T., and Hirohashi, S. Retrospective study on amplification of N-myc and c-myc genes in pediatric solid tumors and its association with prognosis and tumor differentiation. *Lab. Invest.* 59:321–327, 1988.

167. Tsukamoto, A. S., Grosschedl, R., Guzman, R. C., Parslow, T., and Varmus, H. E. Expression of the int-1 gene in transgenic mice is associated with mammary gland hyperplasia and adenocarcinomas in male and female mice. *Cell* 55:619–625, 1988.

168. van der Hout, A. H., Kok, K., van der Veen, A. Y., Osinga, J., deLeij, F. F. M. H., and Buys, C. H. C. M. Localization of amplified c-*myc* and N-*myc* in small cell lung cancer cell lines. *Cancer Genet. Cytogenet.* 38:1–8, 1989.

169. Vander Rauwelaert, E., Maisin, J. R., and Merregaert, J. Provirus integration and myc amplification in ⁹⁰SR induced osteosarcomas of CF1 mice. *Oncogene* 2:215–222, 1988.

170. Van't Veer, L. J., Burgering, B. M. Th., Versteeg, R., Boot, A. J., Ruiter, D. J., Osanto, S., Schrier, P. I., and Bos, J. L. N-*ras* mutations in human cutaneous melanoma correlated with sun exposure. *Mol. Cell. Biol.* 9:3114–3116, 1989.

171. Vecchio, G., Cavazzana, A. O., Triche, T. J., Ron, D., Reynolds, C. P., and Eva, A. Expression of the *dbl* proto-oncogene in Ewing's sarcoma. *Oncogene* 4:897–900, 1989.

172. Versnel, M. A., Hagemeijer, A., Bouts, M. J., va der Kwast, T. H., and Hoogsteden, H. C. Expression of c-*sis* (PDGF β chain) and PDGF A chain genes in ten human malignant mesothelioma cell lines derived from primary and metastatic tumors. *Oncogene* 2:601–605, 1988.

173. Weber, S., Zuckerman, J., Bostwick, D. G., and Bensch, K. G. Gastrin releasing peptide is a selective mitogen for small cell lung carcinoma in vitro. *J. Clin. Invest.* 75:306–309, 1985.

174. Westermark, B., Johnsson, A., Paulsson, Y., Betsholtz, C., Heldin, C.-H., Herlyn, M., Rodeck, U., and Koprowski, H. Human melanoma cell lines of primary and metastatic origin express the genes encoding the chains of platelet derived growth factor (PDGF) and produce a PDGF-like growth factor. *Proc. Natl. Acad. Sci. U.S.A.* 83:7197–7200, 1986.

175. Whang-Peng, J., Kao Shen, C. S., Lee, E. C., Bunn, P. A., Carney, N. D., Gazdar, A. F., Portlook, C., and Minna, J. D. Deletion 3p(14-23), double minute chromosomes, and homogeneously staining regions in human small cell lung cancer. In: *Gene Amplification,* edited by R. T. Schimke, Cold Spring Harbor, Cold Spring Harbor Laboratory, 1982, pp. 107–113.

176. Wong, A. J., Ruppert, J. M., Eggleston, J., Hamilton, S. R., Baylin, S. B., and Vogelstein, B. Gene amplification in small cell carcinoma of the lung. *Science (Wash DC)* 233:461–464, 1986.

177. Wong, D. T. W. and Bigwas, D. K. Expression of c-erb β protooncogene during dimethylbenzanthracene-induced tumorigenesis in hamster cheek pouch. *Oncogene* 2:67–72, 1987.

178. Wong, D. T. W., Gertz, R., Chow, P., Chang, A. L. C., McBride, J., Chiang, T., Matossian, K., Gallagher, G., and Shklar, G. Detection of Ki-*ras* messenger RNA in normal and chemically transformed hamster oral keratinocytes. *Cancer Res.* 49:4562–4567, 1989.

179. Yamamoto, T., Kamata, N., Kawano, H. G., Shimizu, S., Kuroki, T., Toyoshima, K., Rikimaru, K., Nomura, N., Ishizaki, R., Pastan, I., Gamou, S., and Shimizu, N. High incidence of amplification of the epidermal growth factor receptor gene in human squamous carcinoma cell lines. *Cancer Res.* 46:414–416, 1986.

180. Yamashita, S., Ong, J., Fagin, J. A., and Melmed, S. Expression of the *myc* cellular proto-oncogene in human thyroid tissue. *J. Clin. Endocrinol. Metab.* 63:1170–1173, 1986.

181. Yasuda, H., Kobayashi, H., Ohkawara, A., and Kuzumaki, N. Differential expression of *ras* oncogene products among types of human melanomas and melanocytic nevi. *J. Invest. Derm.* 93:54–59, 1989.

182. Yoakum, G. H., Lechner, J. F., Gabrielson, E. W., and Korba, B. E. Transformation of human bronchial epithelial cells transfected by Harvey *ras* oncogene. *Science (Lond)* 227:1174–1179, 1985.

183. Yokota, J., Toyoshima, K., Sugimura, T., Yamamoto, T., Terada, M., Battifora, H., and Cline, M. J. Amplification of c-erbB-2 oncogene in human adenocarcinomas *in vivo. Lancet* i:765–767, 1986.

184. Yokata, J., Wada, M., Shimosato, Y., Terada, M., and Sugimura, T. Loss of heterozygosity on chromosomes 3, 13, and 17 in small cell lung carcinoma and on chromosome 3 in adenocarcinoma of the lung. *Proc. Natl. Acad. Sci. U.S.A.* 84:9252–9256, 1987.

185. Yokata, J., Wada, M., Yoshida, T., Noguchi, M., Terasaki, T., Shimosato, Y., Sugimura,

T., and Terada, M. Heterogeneity of lung cancer cells with respect to amplification and rearrangement of *myc* family oncogenes. *Oncogene* 2:607–611, 1988.

186. Yokoyama, T., Tsukahara, T., Nakagawa, C., Kikuchi, T., Minoda, K., and Shimatake, H. The N-*myc* gene product in primary retinoblastomas. *Cancer* 63:2134–2138, 1989.

187. Yoshimoto, K., Hirohashi, S., and Sekiya, T. Increased expression of the c-myc gene without gene amplification in human lung cancer and colon cancer cell lines. *Jpn. J. Cancer Res.* 77:540–545, 1986.

188. Yuasa, Y., Gol, R. A., Chang, A., Chiu, I.-M., Reddy, E. P., Tronick, S. R., and Aaronson, S. A. Mechanism of activation of an N-*ras* oncogene of SW-127, human lung carcinoma cells. *Proc. Natl. Acad. Sci. U.S.A.* 81:3670–3674, 1984.

189. Yuasa, Y., Reddy, E. P., Rhim, J. S., Tronick, S. R., and Aaronson, S. A. Activated N-ras in a human rectal carcinoma cell line associated with clonal homozygosity in myb locus restriction fragment polymorphism. *Jpn. J. Cancer Res.* 77:639–647, 1986.

190. Yuspa, S. H., Vass, W., and Scolnick, E. Altered growth and differentiation of cultured mouse epidermal cells infected with oncogenic viruses and chemicals. *Cancer Res.* 43:6021–6031, 1983.

191. Zhou, D. J., Casey, G., and Cline, M. J. Amplification of human int-2 in breast cancers and squamous carcinomas. *Oncogene* 2:279–282, 1988.

192. Zimmerman, K. A., Yancopoulos, G. D., Collum, R. G., Smith, R. K., Kohl, N. E., Denis, K. A., Nau, M. N., Witte, O. N., Torand-Allerand, D., Gee, C. E., Minna, J. D., and Alt, F. W. Differential expression of *myc* family genes during murine development. *Nature (Lond)* 319:780–783, 1986.

13

Genetic Regulation of Invasion and Metastasis

Lance A. Liotta
Elise C. Kohn

Tumor invasion and metastasis are the major causes of treatment failure for cancer patients. Approximately 70% of patients with newly diagnosed solid tumors (excluding skin cancers other than melanoma) have occult or clinically detectable metastases. This underscores the importance of understanding the biology underlying malignant behavior and the metastatic cascade so as to continually improve our diagnostic methods. The clinical objectives of laboratory research in the field of invasion and metastasis are (*a*) accurate prediction of the metastatic propensity of a patient's tumor; (*b*) localization of clinically silent micrometastases; and (*c*) selective eradication of established metastases during treatment of the primary tumor.

Extracellular Matrix

The mammalian organism is composed of a series of tissue compartments separated from one another by two types of extracellular matrix: basement membrane and interstitial stroma. The matrix determines tissue architecture, performs important biological functions, and exists as a mechanical barrier to invasion.[3,10] During the transition from *in situ* to invasive carcinoma, tumor cells penetrate the epithelial basement membrane and enter the underlying interstitial stroma. Once the tumor cells enter the stroma they gain access to lymphatics and blood vessels for further dissemination. During intravasation and extravasation, tumor cells of any histologic origin must penetrate the subendothelial basement membrane. In the distant organ where metastatic colonies are initiated, extravasated tumor cells must migrate through the perivascular interstitial stroma before tumor colony growth occurs in the organ parenchyma. Therefore, tumor cell interaction with the extracellular matrix occurs at multiple stages in the metastatic cascade.

General and widespread changes occur in the organization, distribution, and quantity of the epithelial basement membrane during the transition from benign to invasive carcinoma. The human breast is a particular example. Benign proliferative disorders of the breast such as fibrocystic disease, sclerosing adenosis, intraductal hyperplasia, fibroadenoma, and intraductal papilloma are all characterized by disorganization of the normal epithelial stromal architecture. Extreme forms can mimic the appearance of invasive carcinoma. However, no matter how extensive the architectural disorganization, these benign disorders are always characterized by a continuous basement membrane separating the epithelium from the stroma. In contrast, invasive ductal carcinoma, invasive lobular carcinoma, and tubular carcinoma consistently possess a defective extracellular basement membrane with zones of basement membrane loss around the invading tumor cells in the stroma. The basement membrane is also markedly defective adjacent to tumor cells in lymph node and organ metastases. In some focal regions of well-differentiated carcinoma, partial basement membrane formation by differentiated structures can be identified. These findings are of direct application to diagnostic problems in surgical pathology, such as the differentiation of tangential sections of *in situ* lesions from true invasion or the differentiation of severe adenosis from invasive carcinoma. Loss of basement membranes in human carcinomas significantly correlates with increased incidence of metastases and poor five-year survival.

Multistep Cascade of Metastasis

The metastatic colony is the end result of a complex series of tumor-host interactions (Table 13.1, Fig. 13.1).[3,10] Primary tumor initiation and progression are followed by the transition from *in situ* lesions to angiogenesis and locally invasive cancer. Newly formed tumor vessels are defective and are easily invaded by tumor cells exiting the primary mass. These advancing tumor cells also have the capacity to invade normal host blood vessels, discharging themselves into the venous circulation singly and in clumps. Millions of tumor cells are shed into the circulation daily. However, fortunately this is a very inefficient process, with less than 0.01% of circulating cells successfully initiating metastatic colonies. Tumor cells can also readily enter the lymphatic circulation and arrest in large lymphatics of the subcapsular sinuses of lymph nodes. Due to extensive lymphatic-hematogenous communications, tumor cells disseminate via the lymphatic and circulatory systems in parallel.

Circulating tumor cells utilize a variety of means to arrest in the vessels of the target organs. Tumor cells, singly or in small clumps, can adhere to exposed subendothelial basement membrane or to the endothelial surface, causing endothelial cell retraction with subsequent exposure of the basement membrane. Once attached to the basement membrane, endothelial cells extend over the tumor cells, reestablishing the endothelial lining of the vessel and isolating the tumor cells from the circulation. Tumor cells then secrete proteolytic enzymes that degrade the basement membrane, opening a rent through which the tumor cell sends pseudopodial processes as the first step in migration into the target tissue parenchyma. Lastly, tumor cell proliferation occurs at the secondary site.

Tumor cells must overcome host defenses at all stages of the metastatic cascade. Although tumor-specific antigens have been identified in animal models, it remains unclear whether similar antigens play a role in human tumors and whether the recognition of these antigens can be boosted by biologic response modifier therapy. Limited effectiveness of immunotherapy may be attributable to tumor antigen heterogeneity, tumor antigen shedding, or absence of tumor cell immunogenicity. Tumor-infiltrating lymphocytes and lymphokine-activated killer cells may be more

Table 13.1.
Potential Mechanisms Underlying Tumor Metastasis

Metastatic Cascade Event	Potential Mechanisms
1. Tumor initiation	Carcinogenic insult, oncogene activation or derepression, chromosome rearrangement
2. Promotion and progression	Karyotypic, genetic, and epigenetic instability, gene amplification; promotion-associated genes and hormones
3. Uncontrolled proliferation	Autocrine growth factors or their receptors, receptors for host hormones such as estrogen
4. Angiogenesis	Multiple angiogenesis factors, including known growth factors
5. Invasion of local tissues, blood, and lymphatic vessels	Serum chemoattractants, autocrine motility factors, attachment receptors, degradative enzymes
6. Circulating tumor cell arrest and extravasation	Tumor cell homotypic or heterotypic aggregation
a. Adherence to endothelium	Tumor cell interaction with fibrin, platelets, and clotting factors; adhesion to RGD type receptors
b. Retraction of endothelium	Platelet factors, tumor cell factors
c. Adhesion to basement membrane	Laminin receptor, thrombospondin receptor
d. Dissolution of basement membrane	Degradative proteases, type IV collagenase, heparanase, cathepsins
e. Locomotion	Autocrine motility factors, chemotaxis factors
7. Colony formation at secondary site	Receptors for local tissue growth factors, angiogenesis factors
8. Evasion of host defenses and resistance to therapy	Resistance to killing by host macrophages, natural killer cells, and activated T cells; failure to express, or blocking of, tumor specific antigens; amplification of drug resistance genes

effective against heterogeneous tumor cell populations. These immunomodulatory cells have been used in clinical trials for their abilities to eliminate circulating tumor cells and destroy existing primary and metastatic masses.

The vascular supply is as important for metastases as for the primary tumor. Angiogenesis is necessary at the beginning and at the end of the metastatic cascade. The vascularization of metastases presents the same metastatic potential as did the tumor vessels in the primary mass—metastases can metastasize. This can result in a geometric progression of disease. In addition, clinical evidence supports the existence of dormant metastases, documented by the fact that lethal metastatic disease can occur more than two decades after extirpation of the primary tumor in breast and colon cancers. Three potential mechanisms of tumor dormancy have been distinguished in animal models: (a) immunologic restraint such that the tumor population death rate equals its growth rate; (b) constitutive dependency of tumor cells on host growth factors; and (c) avascularity, causing the metastasis to be limited in size because of deficient nutrient diffusion.

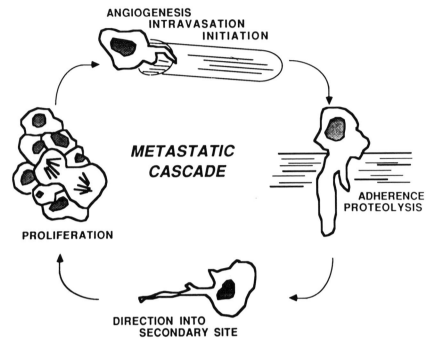

Figure 13.1.   Metastatic cascade. The process of metastasis is complicated and dynamic. Tumor cells must adhere to parenchymal and endothelial basement membrane, locally degrade the basement membrane, exit through those defects, and migrate. Tumor cells must have some sensor functions to direct their migration to favorable secondary sites, such as autocrine motility factors and growth factors, which also play a chemoattractant role. Lastly, tumor cells must be able to proliferate at the metastatic sites.

Three-Step Theory of Invasion

A three-step hypothesis has been proposed describing the sequence of biochemical events during tumor cell invasion of the extracellular matrix.[3,7] Attachment may be mediated by specific glycoproteins, such as laminin and fibronectin, through tumor cell plasma membrane receptors. Following attachment, the tumor cell secretes hydrolytic enzymes (or induces host cells to secrete enzymes) that can locally degrade the matrix as well as the attachment glycoproteins. Matrix lysis most likely occurs in a highly localized region close to the tumor cell surface, where the active enzyme outbalances the natural protease inhibitors present in the serum and in the matrix itself. The third step is tumor cell locomotion into the region of the matrix modified by proteolysis. The direction of the locomotion may be influenced by autocrine motility factors and tumor- and normal cell–secreted cytokines. Autocrine motility factors are a newly described class of proteins that bind to the cell and profoundly stimulate cell migration. In contrast to autocrine growth factors, the motility factors help initiate tumor cell locomotion, whereas other chemoattractant growth factors, serum proteins, or matrix components may influence the organ specificity of metastases. Continued invasion of the matrix takes place by cyclical repetition of these three steps.

Adhesion Protein Receptors

Cell surface receptors for the basement membrane glycoprotein laminin mediate adhesion of tumor cells to the basement membrane prior to invasion. Laminin has a cruciform shape with three short arms and one long arm; all arms have globular end regions that bind to basement membrane type IV collagen. Laminin plays a role in cell attachment and spreading, mitogenesis, neurite outgrowth, morphogenesis, protease secretion, and cell movement. Many types of malignant cells contain high-affinity laminin receptors that bind the short arm of the laminin molecule. Laminin receptors may be altered in number or degree of occupancy in human carcinomas. For example, breast and colon carcinomas contain a higher number of exposed (unoccupied) receptors than do benign breast or colon lesions. In addition, the laminin receptors of normal epithelium are polarized at the basal surface of the cell and are occupied with laminin in the basement membrane, whereas laminin receptors on invading carcinoma cells are amplified and may be distributed over the entire surface of the cell. Laminin receptors also have been shown to coalesce into the leading pseudopodia of migrating cells.[16] The laminin receptor has been shown experimentally to be involved in hematogenous metastases. Treatment of tumor cells with the receptor-binding fragment of laminin at very low concentrations markedly inhibits or abolishes lung metastases from intravenously inoculated tumor cells by blocking the adhesion of circulating tumor cells to the subendothelial basement membrane. Another potential therapeutic application involves the use of adriamycin-containing liposomes bearing monoclonal antibodies to the laminin receptor.[9] These liposomes bind avidly to breast carcinoma cells; in proliferation assays *in vitro*, these targeted liposomes preferentially killed breast carcinoma cells ($>90\%$) compared with normal breast epithelial cells (15%).

Another family of cell surface receptors, the integrins, has been identified. These receptors bind with low affinity to a variety of adhesion proteins including fibronectin, von Willebrand factor, fibrin, type I collagen, and thrombospondin. The function of these heterodimeric receptors can be inhibited by specific peptides related to the Arg-Gly-Asp (RGD) sequence of fibronectin. RGD sequences are present on a wide variety of proteins and may serve as the recognition site for binding of the integrins. Integrin proteins are thought to align adhesion proteins such as fibronectin on the cell surface with cytoskeletal components such as talin and actin, thus altering cell shape. Integrin-type proteins may play an adhesive role in platelet–tumor cell interactions, the binding of lymphoid cells to endothelium, and the interaction of circulating tumor cells with endothelial surfaces and extracellular matrix components. Coinjection of tumor cells with large quantities of RGD peptides will inhibit metastasis formation in animal models by interfering with the adhesion of tumor cells to the endothelial surface.

Tumor Cell Proteinases

Invasion of the extracellular matrix is not due merely to passive growth pressure exerted by the tumor; it also requires active biochemical mechanisms. It has been shown that invasive tumor cells secrete matrix-degrading proteinases. Collagens constitute the structural scaffolding on which the other components of the extracellular matrix are assembled. Different families of tumor-derived collagenases degrade the interstitial collagens (types I–III), and type IV basement membrane collagen. These collagenases are metal ion–dependent enzymes, which also have a specific class of inhibitors, TIMPs, tissue inhibitors of metalloproteinases. Type IV collagenases are augmented in highly metastatic tumor cells and in endothelial cells during angiogenesis. Amplification of type IV collagenase production has been biochemically linked to the genetic induction of metastases in experimental models. These enzymes are secreted as latent enzymes and are activated and function locally.[14] A specific inhibitor of type IV collagenase has recently been described. This novel member of the TIMP family, TIMP-2, secreted by the tumor cell, complexes to the type IV procollagenase and prevents its enzymatic activity.

Antibodies prepared against intact type IV collagenase as well as peptide fragments react specifically upon immunohistochemical staining. Antibody reactivity (Fig. 13.2) is confined to the myoepithelial cells in normal breast tissues, reflecting the participation of these cells in the physiologic turnover of the basement membrane that surrounds normal breast ducts and lobules.[5] Intense epithelial immunoreactivity was demonstrated in carcinoma *in situ*, and strongly in the invading cells seen in the stroma or in lymph node metastases. These results show that human tumor cells produce type IV collagenase, which is necessary for the invasive phenotype.

Two other classes of proteinases have documented importance in tumor progression. Cathepsin B, a cysteine proteinase, is a lysosomal acid hydrolase.[12] This enzyme can cleave a broad array of substrates including myosin, actin, proteoglycans, fibronectin, and laminin. It has been found in the plasma membrane fraction

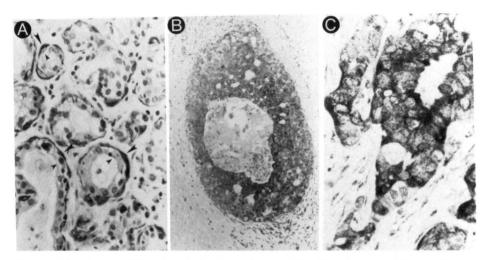

Figure 13.2. Immunohistochemical staining for type IV collagenase. Peptide-specific antibodies against human type IV collagenase were used to stain sections of **A,** normal lactating breast, **B,** comedocarcinoma (carcinoma *in situ*), and **C,** infiltrating ductal carcinoma. In normal lactating breast, staining for type IV collagenase is specific for the myoepithelial cells (*large arrow*) and does not stain the epithelial cells (*small arrow*). In **B** and **C,** only the breast carcinoma cells contain type IV collagenase, and the stroma cells are negative.

of tumor cells and in conditioned media from tumor cell cultures. Urokinase and tissue plasminogen activator, serine proteinases, are increased in cancer cells. Cell transformation has been shown to induce substantial increases in the extracellular release of plasminogen activators. A comparison between human primary lung and breast carcinomas and nonmalignant tissues has demonstrated a several-fold elevation in the expression of urokinase mRNA.

Tumor Cell Motility

The final step in invasion is the process of tumor cell locomotion. The invasive malignant cell must be able to emigrate from the primary site into and out of the circulation, as well as through the interstitial stroma, prior to completing the metastatic cascade by proliferation at the secondary site. To complete this complicated series of events, tumor cells require two types of stimuli: first, an initiating stimulus that can maintain migratory pressures, and second, specific host-derived factors that may act as targeting agents for tumor cells after they have entered the circulation. Autocrine motility factors (AMF) are tumor cell–derived proteins that stimulate random migration, consistent with the process of initiation of motility.[4] The motile response to this cytokine is inhibited by cell treatment with pertussis toxin, indicating that the signaling event associated with this cellular action is guanine nucleotide-binding protein-mediated signal transduction. Stimulated random migration has also been reported with components of the extracellular matrix such as laminin, fibronectin, and type IV collagen. In contrast, autocrine and host-derived growth factors such as the insulin-like growth factors (IGF) have dual physiologic roles. These agents are known mitogens for a broad array of tumor cell types *in vitro*, and have been shown recently to stimulate directed migration of tumor cells in similar concentrations. Unlike AMF, pertussis toxin has no effect on the IGF-induced locomotion, demonstrating that two independent signaling systems are involved in initiation versus direction of tumor cell migration. Studies of urine samples from bladder cancer patients and patients with benign urothelial processes indicate that AMF-like activity is secreted by the bladder cancer cells into the urine and can be shown to stimulate migration of cancer cells *in vitro*.[2] A statistically significant correlation between the levels of AMF activity in the urines and the grade and stage of malignancy was documented. This observation is the basis for the development of an AMF assay as a potential diagnostic test for transitional carcinoma of the bladder.

Organ Tropism for Metastases

The distribution of metastases varies widely, depending upon the histologic type and anatomic location of the primary tumor.[15] As discussed above, both host- and tumor cell–derived chemoattractants play a role in the direction of metastatic spread; however, survival at the secondary site must also be considered. Therefore, both the "homing" of the tumor cells to the secondary site and the "soil" of that location are important determinants in the extent of metastases. The most frequent organ location of distant metastases in many types of cancer appears to be the first capillary bed encountered by the circulating cells. Major pathways of metastases are determined primarily by anatomic, but also by humoral, considerations.

For example:

1. Sarcomas arising in the extremities metastasize primarily to the lungs. Sarcoma cells entering the tumor venous drainage are carried via the vena cava to the pulmonary capillary beds. In addition, the pulmonary parenchyma is an abundant source of chemoattractant growth factors, such as the insulin-like growth factors described above.
2. Lung cancer disseminates widely to multiple organs, including the brain. Lung cancer is the only tumor that has direct access to the general arterial circulation via the pulmonary vein through the left heart.
3. Prostate cancer and breast cancers are notable for their high propensity to metastasize to bone. There is an extensive paravertebral venous plexus, Batson's plexus, which anastomoses to the periprostatic venous plexus and the mammary drainage.
4. Ovarian cancer remains confined for long periods of time in the peritoneal cavity. Local spread occurs to the peritoneal surfaces, the posterior gutters, and to the domes of the diaphragms. With the advent of more successful primary therapies, parenchymal liver and lung metastases are more common.

There are many metastatic sites that cannot be predicted by anatomical considerations alone and can be considered examples of organ tropism. Clear cell carcinoma of the kidney often metastasizes to bone and thyroid, and ocular melanoma frequently metastasizes to the liver. Theoretical mechanisms for organ tropism include (a) tumor cells that disseminate equally in all organs, but preferentially grow only in specific organs; (b) circulating tumor cells that may adhere preferentially to the endothelial lumenal surface only in the target organ; and (c) circulating tumor cells that may respond to soluble attractants as described for the IGFs. Such factors could diffuse locally from the target organ and act as migratory directors. Research with animal models indicates that all of these mechanisms play a role to varying degrees, depending on the tumor model system.

Molecular Genetics of Metastasis

The complicated metastatic cascade must involve multiple gene products, including those for the components already discussed: laminin receptor, autocrine motility factor, attractant growth factors, and interstitial type IV collagenases. A coordinated group of gene products expressed above a certain threshold may be required for a tumor cell to traverse each progressive step in the metastatic process successfully. The crucial gene products may be positive or suppressive modifiers regulating host immune recognition, cell growth, attachment, proteolysis, locomotion, and differentiation. The specific family of gene products necessary for metastases may be different for each histologic type of tumor and is an active area of cancer research.

The evidence linking oncogenes to the induction or maintenance of human malignancies has become increasingly compelling.[6] In the past, oncogenes have been linked to unrestrained tumor growth. Recently, two types of experimental approach have indicated that certain classes or combinations of oncogenes may play a role in the metastatic behavior of tumors. In the first experimental approach, human tumor DNA samples are surveyed for the level of oncogene expression, and this is coordinated with the disease stage. In the second approach, tumor DNA or isolated oncogenes are transfected into recipient cells. The transfected cells are then studied for their metastatic propensity. Notable examples of these two approaches are work with the HER-2/*neu* oncogene in human breast carcinoma and *bcl*-2 in follicular lymphomas, and the transfection of the *ras* oncogene in rodent systems. The HER-2/*neu* (*neu*) oncogene encodes a protein that is a member of the tyrosine kinase family of enzymes and is related to, but distinct from, the epidermal growth-factor receptor gene.[11] At present, the ligand for the *neu* oncogene–encoded receptor protein has not been identified. A significant increase in the incidence of *neu* oncogene amplification is noted in breast cancer patients with more than three axillary lymph nodes containing metastatic disease. Amplification of *neu* is also highly correlated with disease relapse as well as tumor size, and has been shown to be an independent prognostic variable. *bcl*-2 is an oncogene that is involved in the generation of a fusion protein after the t(14;18) that occurs in nearly all follicular lymphomas. Recently, use of the polymerase chain reaction for amplification of *bcl*-2 has helped to document the presence of undetectable residual malignant follicular lymphoma cells. Thus, though the specific functions of these two oncogenes are unknown, the level of expression may provide important prognostic information about the patient's course.

Transfection of members of the *ras* oncogene family into suitable rodent recipient cells, including diploid rat embryo fibroblasts, can induce these cells to progress rapidly to express the complete metastatic phenotype.[8] Other oncogenes, including *myc*, *src*, and *fos* failed to induce metastases in rodent cells. Furthermore, when *ras* was transfected in combination with the adenovirus type 2 E1a oncogene, the recipient cells remained very tumorigenic but lost the metastatic phenotype. Thus, some genes can suppress the action of *ras* to induce metastases. In this same experiment, the production of type IV collagenase paralleled the metastatic phenotype, consistent with the requirements of the three-step hypothesis.

Two novel genes have been described that negatively affect the metastatic process. p53 is a gene on chromosome 17 which, when introduced in a mutant form, is associated with the generation of metastases.[1] In clinical correlation, metastatic colon cancers have been found frequently to have deletions and mutations of the 17p chromosome. A novel gene, nm23, has been described that is expressed in benign and nonmetastatic processes and is down-regulated and/or lost in metastatic tumors.[13] In addition, nm23 has been found to be 78% identical at the amino acid level with the *Drosophila awd* gene (*a*bnormal *w*ing *d*isc). The *awd* gene is a developmental gene, and mutations in *awd* cause abnormal tissue morphology and necrosis with widespread aberrant differentiation in *Drosophila*, analogous to the changes in malignant progression. Thus, the loss of a gene that is highly homologous to a gene necessary for normal development is associated with the metastatic phenotype, underscoring the importance of both positive and negative regulators in normal development and in malignant and metastatic disease. Therefore, the current working hypothesis is that induction of the metastatic phenotype requires a complement of genes. Evolving data demonstrate that these genes may be of both the active and the suppressor forms, such that when genes interact in the correct fashion, a cascade of specific gene products is elaborated which can confer either the malignant or the metastatic phenotype or both.

REFERENCES

1. Baker, S. J., Fearon, E. R., Nigro, J. M., Hamilton, S. R., Preisinger, A. C., Jessup, J. M., vanTuinen, P., Ledbetter, D. H., Barker, D. F., Nakamura, Y., White, R., and Vogelstein, B. Chromosome 17 deletions and p53 gene mutations in colorectal carcinomas. *Science (Wash DC)* 244:217–221, 1989.
2. Guirguis, R., Schiffmann, E., Liu, B., *et al.* Detection of autocrine motility factor(s) in urine as markers of bladder cancer. *J. Natl. Cancer Inst.* 80:1203, 1988.
3. Liotta, L. A. Tumor invasion: Role of the extracellular matrix. *Cancer Res.* 46:1, 1986.
4. Liotta, L. A., Mandler, R., Murano, G., *et al.* Tumor cell autocrine motility factor. *Proc. Natl. Acad. Sci. U.S.A.* 83:3302, 1986.
5. Monteagudo, C., Merino, M. J., San-Juan, J., Liotta, L. A., and Stetler-Stevenson, W. G. Immunohistochemical distribution of type IV collagenase in normal, benign, and malignant breast tissue. *Am. J. Pathol.* in press, 1989.
6. Muschel, R. and Liotta, L. A. Role of oncogenes in metastases. *Carcinogenesis* 9:705, 1988.
7. Nicolson, G. L. Organ specificity of tumor metastasis: Role of preferential adhesion, invasion and growth of malignant cells at specific secondary sites. *Cancer Metastasis Rev.* 7:143, 1988.
8. Pozzatti, R., Muschel, R., Williams, J., *et al.* Primary rat embryo cells transformed by one or two oncogenes show different metastatic potentials. *Science (Wash DC)* 232:223, 1986.
9. Rahman, A., Panneerselvam, M., Guirguis, R., *et al.* Anti-laminin receptor antibody targeting of liposomes with encapsulated doxorubicin to human breast carcinoma cells *in vitro. J. Natl. Cancer Inst.* 81:1794, 1989.
10. Schirrmacher, V. Cancer metastasis: Experimental approaches, theoretical concepts, and impacts for treatment strategies. *Adv. Cancer Res.* 43:1, 1985.
11. Slamon, D. J., Clark, G. M., Wong, S. G., *et al.* Human breast cancer: correlation of relapse and survival with amplification of the HER-2/*neu* oncogene. *Science (Wash DC)* 235:177, 1987.
12. Sloane, B. F., Rozhin, J., Johnson, K., Taylor, H., Crissman, J. D., and Honn, K. V. Cathepsin B: Association with plasma membrane in metastatic tumors. *Proc. Natl. Acad. Sci. U.S.A.* 83:2483, 1986.
13. Steeg, P. S., Bevilacqua, G., Koper, L., *et al.* Evidence for a novel gene associated with low tumor metastatic potential. *J. Natl. Cancer Inst.* 80:200, 1988.
14. Stetler-Stevenson, W. G., Krutsch, H. C., Wacher, M. P., Margulies, I. M. K., and Liotta, L. A. The activation of human type IV collagenase proenzyme: Sequence identification of the major conversion product following organomercurial activation. *J. Biol. Chem.* 264:1353, 1989.
15. Sugarbaker, E. V. Patterns of metastasis in human malignancies. *Cancer Biol. Rev.* 2:2355, 1981.
16. Wewer, U. M., Taraboletti, G., Sobel, M. E., *et al.* Laminin receptor: Role in tumor cell migration. *Cancer Res.* 47:5691, 1987.

14

Analysis of Gene Expression in Endocrine Cells

Ronald A. DeLellis
Hubert J. Wolfe

Introduction

The past decade has witnessed the development and application of a series of stunning technological advances that have formed the basis of studies of gene expression at the cellular level.[19,20,75] For the pathologist, some of the most important and powerful technical approaches have been those that have wed morphology to biochemistry, immunology, and molecular biology. The new technologies not only have provided important investigational approaches for the analysis of normal cells and tissues but also have provided insight into the understanding of a wide variety of disease processes.[75]

Major conceptual advances resulting from these studies are forming the foundations upon which the entire field of endocrine pathophysiology is being reevaluated and rebuilt.[50,52,58,59] In particular, recombinant DNA technology has permitted the study of structural-functional correlations in endocrine cells at the molecular level. One result of such studies, for example, has been the demonstration that the endocrine, nervous, and immune systems share common mechanisms of response and communication.[26]

Gene Expression in Endocrine Cells

Analysis of gene expression in endocrine cells has been accomplished by multiple approaches (Fig. 14.1). Immunohistochemical techniques, for example, have permitted the localization of hormones and their precursors, hormone receptors, structural proteins, oncogenes, and other constituents at the cellular level and have also permitted correlative analyses of morphology with normal and abnormal function. In addition, these methods are now utilized for functional classifications of endocrine neoplasms as well as their precursor lesions, and they have led directly to the formulation of novel clinical and pathological concepts.

Using biochemical and immunohistochemical techniques, it has been possible to decipher the complex series of events involved in the biosynthesis and secretion of biologically active peptides (Fig. 14.2). These studies have shown that preprohormones are synthesized on membrane-bound ribosomes and that the nascent polypeptide chains are inserted into the lumen of the granular endoplasmic

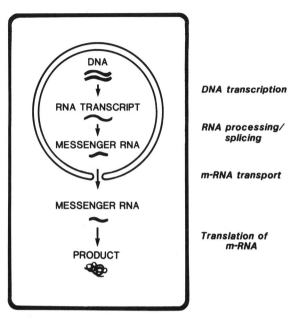

Figure 14.1. Schematic illustration of steps involved in gene expression. DNA is transcribed within the nucleus to primary RNA transcripts, which are then processed and spliced to produce specific messenger RNAs. Messenger RNA leaves the nucleus and enters the cytoplasm where it is translated into the specific gene product. The DNA encoding a particular product, the primary RNA transcript, and the messenger RNA can all be demonstrated by *in situ* hybridization techniques. The translational product of mRNA can be demonstrated by immunohistochemistry.

reticulum via the signal peptide.[49] Additional studies have revealed that the signal peptide of the preprohormone interacts with a signal recognition particle and its receptor in order to initiate the process of vectorial synthesis within the endoplasmic reticulum.[49,54] The attachment of oligosaccharide chains to the hormone precursor occurs either during or very shortly after the synthesis of the polypeptide into the lumen of the endoplasmic reticulum. Other steps including protein folding and disulfide bond formation, which are essential for subsequent steps, occur rapidly.[49]

Hormone precursors are then translocated to the *cis* face of the Golgi region, where many important modifications occur (Fig. 14.2). These include hydroxylation, phosphorylation, glycosylation, oligosaccharide trimming and maturation, sulfation, and endoproteolysis. Although their precise mechanisms are unknown, a series of sorting steps also occurs in the Golgi regions.[49,54] Further biosynthetic events take place within the secretory granules. These include proteolytic cleavage of large precursors and modifications of the NH_2- or COOH-terminal amino acids, including acetylation, amidation, and the formation of pyroglutamic acid residues. Mature secretory granules may be stored for considerable lengths of time prior to exocytosis. In contrast to this so-called regulated pattern of secretion in which hormones are stored in secretory granules prior to their release, other biologically active peptides may bypass this pathway and may be secreted constitutively[49] (Fig. 14.3).

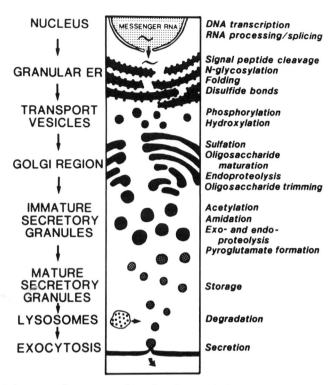

NUCLEUS		DNA transcription
		RNA processing/splicing
↓		
GRANULAR ER		Signal peptide cleavage
		N-glycosylation
		Folding
↓		Disulfide bonds
TRANSPORT		Phosphorylation
VESICLES		Hydroxylation
↓		
		Sulfation
		Oligosaccharide
GOLGI REGION		maturation
		Endoproteolysis
↓		Oligosaccharide trimming
IMMATURE		Acetylation
SECRETORY		Amidation
GRANULES		Exo- and endo-
		proteolysis
↓		Pyroglutamate formation
MATURE		
SECRETORY		
GRANULES		Storage
↓		
LYSOSOMES		Degradation
↓		
EXOCYTOSIS		Secretion

Figure 14.2.   Schematic illustration of regulated peptide hormone biosynthesis. The synthesis of peptide hormones involves a complex series of steps that begins with the formation of amino acid chains on ribosomes and eventuates in the secretion of the mature hormone. Hormones that are secreted constitutively bypass this complex series of steps and are not stored within the cytoplasm to any extent.

Analysis of gene expression at the levels of genomic DNA and messenger RNA provides a pivotal approach to the understanding of the pathobiology of the endocrine system. The development of recombinant DNA technologies and methods for detecting nucleic acid hybrids has offered the unique opportunity to accomplish these analyses at multiple levels of organization. *In situ* hybridization techniques, in particular, have been effective for the correlation of morphological features with specific patterns of gene expression at the cellular level (Fig. 14.3). The methods of *in situ* hybridization are derived from Southern and Northern blotting procedures and are complementary to these techniques. The *in situ* hybridization technology, however, offers the unique opportunity of localizing specific genomic DNA and mRNA sequences at the microscopic and submicroscopic levels in tissues, single cells, or chromosomal preparations.

In contrast to immunohistochemistry, which is dependent on the peptide content of cells, hybridization analyses offer the possibility of identifying cells on the basis of their contents of specific messenger RNAs[11] (Fig. 14.3). There are many advantages to this approach. For example, an acutely stimulated endocrine cell that is in the process of active secretion will often produce a negative immunohistochemical reaction for a particular peptide; however, correlative *in situ* hybridization

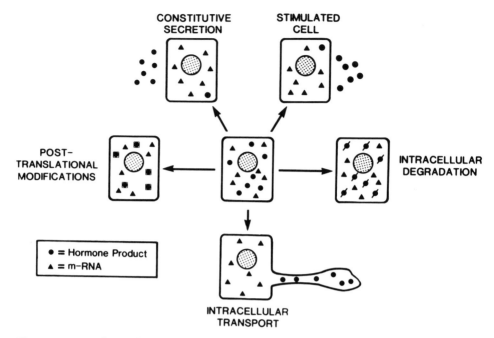

Figure 14.3. Relationships between levels of hormonal product (*closed circles*) and messenger RNA (*closed triangles*) in endocrine cells. In a *resting endocrine cell (center),* there is a balance between the relative amounts of product and corresponding mRNA. In a *simulated cell,* most of the product is released into the circulation. In the process of *constitutive* secretion, the hormonal product is released as soon as it is synthesized. In both instances, immunohistochemistry for the product may give negative results, while hybridization analyses give positive results. Hormones may be *degraded* by the action of lysosomal enzymes. Extensive *posttranslational modifications* may result in the formation of a product (*closed square*) that may change its reactivity with antibodies that recognize the native hormone. In neuronal cells, the hormone is synthesized within cell bodies and is rapidly transported to cell processes that may be quite distant from their corresponding cell bodies.

studies using probes for the corresponding peptide mRNAs will often give an intensely positive signal in the same cells. Similarly, cells that are secreting their products constitutively may give only equivocally positive or negative immunohistochemical results with strongly positive hybridization signals.

Extensive posttranslational processing may result in situations in which immunohistochemical results may be negative or equivocal while hybridization analyses give strong positive signals (Fig. 14.3). In such instances, antibodies raised against natural or synthetic hormones may fail to react with epitopes on hormone precursors, which may have been modified extensively within the endoplasmic reticulum, Golgi regions, or secretory granules. This phenomenon may occur in certain actively synthesizing endocrine neoplasms such as insulinomas, which may secrete considerably more proinsulin than insulin.[13] Extensive intracellular degradation of peptides may also lead to instances in which there are negative immunohistochemical results but positive hybridization signals. Finally, *in situ* hybridization is the method of choice to differentiate *de novo* synthesis from uptake of hormonal peptides.

An emerging theme for many studies employing hybridization techniques for analyses of endocrine tissues is that variations in biosynthetic activities are reflected in changes in mRNA levels for a given gene product.[32] High levels of a particular hormonal product, as determined by radioimmunoassays of tissue extracts, or extensive staining, as determined by immunohistochemistry, could represent increased synthesis or decreased synthesis and transport with accumulation of the particular product within the cell.[32] Changes in mRNA levels as determined by Northern blot analysis and quantitative *in situ* hybridization assays can, therefore, more closely correlate with functional activity than those determined by immunohistochemical techniques. It should be noted, moreover, that the combination of immunohistochemistry and *in situ* hybridization in the same section has the potential for providing the maximal amount of information on the highly dynamic processes of gene transcription and translation.[31]

Although this chapter highlights the contributions of *in situ* hybridization analyses for studies of endocrine cell systems, other molecular approaches, which are also of considerable value. The polymerase chain reaction (PCR), for example, has presented a particularly important approach for studies of neurotensin (NT). Messenger RNA encoding neurotensin is difficult to detect in the gut by *in situ* hybridization, probably because of the abundance of ribonuclease in this tissue. In the rat gut and hypothalamus, NT messenger RNA can be detected by the PCR technique without using radioisotopes, but not by Northern blotting.[73] Studies of normal rat adrenal medulla and the rat pheochromocytoma line PC12 have shown that it is possible to detect NT mRNA after 30 cycles in the PCR technique but not after 25 cycles. It should be apparent, therefore, that the choice of a specific molecular technology should be tailored to the needs of the particular problem that is to be resolved.

In Situ Hybridization

TISSUE FIXATION

The efficacy of *in situ* hybridization for the analysis of gene expression depends on numerous factors including the numbers of copies of mRNA expressed in the cells of interest and the extent to which the tissue preservation preserves the mRNA and permits its interaction with the probe.[28,63,75] Many fixatives have been evaluated for *in situ* hybridization analyses; however, most workers have concluded that mRNA is optimally preserved by cross-linking aldehyde fixatives. Non-cross-linking fixatives, including Bouin's, Carnoy's, osmium tetroxide, and Zenker's, provide considerably less retention of messenger RNA than does paraformaldehyde. Buffered 4% paraformaldehyde has been the fixative of choice in most studies.[63] Tissues should be fixed immediately after removal for two to three hours. Following fixation, the tissues can be placed in a 30% sucrose solution for 18 hours to remove residual fixative and to decrease freeze artifacts upon subsequent quenching of the tissues in liquid nitrogen. The tissues can then be stored in liquid nitrogen prior to sectioning. Alternatively, fresh tissues can be frozen in liquid nitrogen, cut in a cryostat, and post-fixed in 4% paraformaldehyde. Cytocentrifuge preparations of cell cultures may also be used for these analyses. In some instances, formalin-fixed and paraffin-embedded tissue samples have been used for the demonstration of some high copy number mRNAs. The latter approach has, for example, been particularly effective for the demonstration of parathyroid hormone mRNA in samples of human and rat parathyroid tissue.[69]

Pretreatment of tissue sections with detergents such as Triton X-100 and with proteolytic enzymes (proteinase K or pronase) just before hybridization permits the interaction of probes with cell-bound mRNA and DNA.[63] In some hybridization protocols, however, these pretreatment procedures have been omitted. Pretreatment of sections with acetic anhydride is effective in decreasing nonspecific binding of the probe to tissue sections. Treatment of sections with salmon sperm DNA or with RNA has also been used to decrease nonspecific background staining.

PROBES AND PROBE LABELING

Many types of probes have been used for *in situ* hybridization studies. These include nick-translated double-stranded DNA, randomly primed DNA, synthetic single-stranded oligonucleotides, single-stranded cDNA prepared in the M13 vector, and single-stranded antisense or complementary RNA (cRNA) probes.[10,12,56,63,74] The advantages of cRNA probes over nick-translated DNA probes include higher specific activities, higher hybridization efficiency, constant probe size, lack of back hybridization, and the ability to reduce background by treatment of sections with RNAse for the removal of nonhybridized single-stranded RNA.[31] Transcription vectors for the synthesis of both sense and antisense RNA probes are now available from a variety of commercial sources.[31]

The most commonly used probe labeling techniques employ radioisotopes combined with autoradiography.[19] The highest degree of resolution can be obtained with the use of tritium; however, exposure times are quite prolonged with this isotope. ^{32}P-labeled probes can be used with very short exposure times; however, the resolution is low. ^{35}S has been particularly valuable for *in situ* hybridization analyses, since the resolution is high and exposure times are relatively short. In each case, the type of radioactive probe employed depends on the goals of the study and the numbers of target nucleic acids within the cells of interest.

Biotin and photobiotin labeling have been employed with nick-translated DNA probes, oligonucleotide probes, and to a lesser extent with antisense RNA probes.[63] Probes directly labeled with fluorochromes or enzymes have also been used for some applications of *in situ* hybridization technology.[18,28] In general, the sensitivity of fluorochrome labels is low, but these methods are of particular value for flow cytofluorometric applications of *in situ* hybridization. Enzymes, including alkaline phosphatase and horseradish peroxidase, have been used for 3' or 5' labeling of synthetic oligonucleotide probes. Haptenated probes have been used together with antibodies directed against the specific haptens, followed by immunofluorescence or immunoperoxidase procedures. Bromodeoxyuridine, sulfonation, and digoxigenin provide alternative methods for probe labeling. Finally, antibodies directed against DNA-RNA or RNA-RNA complexes have been utilized for some *in situ* hybridization protocols.[18]

CONTROLS

The results of *in situ* hybridization are subject to the influence of a large number of variables that must be controlled if reliable and reproducible results are to be obtained. Generally, detection of gene expression by *in situ* hybridization should include the use of alternative molecular methods including Northern and Southern blotting and the polymerase chain reaction. Of particular value as a positive control is the use of normal or neoplastic tissues known to highly express a specific gene, or the use of cell lines genetically engineered to express the gene of interest. Normal or neoplastic tissues or genetically engineered cell lines that do not express the specific gene of interest represent important negative controls. Cell lines genetically engineered to overexpress a related but distinct gene provide important controls of specificity. Other controls include the use of probes that have been prehybridized with their complementary nucleic acids, hybridization with nonspecific vector sequences, and in the case of antisense RNA probes, the use of the corresponding sense probe, which should fail to hybridize with the nucleic acid.[75]

An additional control is the combined use of immunohistochemistry for the localization of the gene protein product and the corresponding messenger RNA with *in situ* hybridization.[31] These techniques may be performed in serial sections or in the same section using combined immunohistochemistry and *in situ* hybridization. In some protocols, *in situ* hybridization is performed before immunohistochemical staining, while in other protocols immunohistochemistry is performed before hybridization. When immunohistochemistry is performed first, the RNAse activity of the antisera should be first inhibited by the addition of heparin or vanadyl ribonucleoside complex. On the other hand, if *in situ* hybridization is performed first, the hybridization buffer must be free of dextran sulfate, since this reagent may lead to high background staining. When radioactive probes are being employed, the absence of dextran sulfate considerably increases the exposure time. Moreover, the proteinase K, which is used in the hybridization procedure, may result in diminished immunoreactivity of the antigen under study. In the procedure developed by Hofler and co-workers, immunohistochemistry is performed immediately after hybridization, but before the slides are dipped in autoradiographic emulsion.[31]

Applications of Hybridization Technologies to Endocrine Pathology

HYPOTHALAMUS

In situ hybridization methodologies have been used extensively for the localization of hypothalamic mRNAs encoding regulatory peptides.[7,21,39,65,77] The high density of neurons containing somatostatin mRNA in the periventricular nucleus and parvicellular portions of the paraventricular nucleus corresponds closely to the distribution of somatostatin peptide as determined by immunohistochemistry (Fig. 14.4). The demonstration of somatostatin mRNA in neurons of the arcuate nucleus

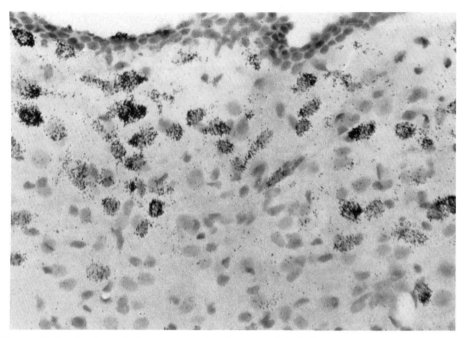

Figure 14.4.   Rat hypothalamus reacted with tritium-labeled antisense RNA somatostatin probe. Autoradiographic grains are present over the cell bodies of hypothalamic neurons. Cell processes are negative. (250X)

and portions of the lateral and anterior hypothalamus illustrates the value of *in situ* hybridization studies, since these areas can be demonstrated to contain somatostatin peptide only after colchicine treatment, which is known to block the release of peptides from cells.[30] Somatostatin mRNA can also be localized to cell bodies of the ventral premammillary nucleus. This area contains high concentrations of somatostatin in cell processes, but neuronal cell bodies have been consistently negative for the peptide, even after colchicine treatment.

In addition to somatostatin, mRNAs encoding vasopressin, oxytocin, corticotropin-releasing hormone (CRH), gonadotropin-releasing hormone, and thyrotropin-releasing hormone (TRH) have also been localized within the hypothalamus.[7,21,39,65,77] Probes for the mRNAs encoding these peptides have been used not only for precise anatomic localization of the different neuronal groups, but also for studies of regulation of their biosynthesis and secretion. For example, it is well known that the biosynthesis of oxytocin and vasopressin are increased in response to increases in plasma osmolarity. Such increases in oxytocin and vasopressin cellular mRNA content in response to salt loading can be visualized by quantitative *in situ* hybridization analyses. Similarly, increases in CRH mRNA have been seen in the parvicellular region of the periventricular nucleus following adrenalectomy, while increases in TRH mRNA have been noted in the paraventricular nucleus following thyroidectomy.[39] These studies clearly indicate that *in situ* hybridization is a technique with which both the presence and the relative quantities of specific mRNAs can be determined with a high degree of resolution and specificity.

In addition, they show that changes in the levels of mRNA encoding neuroendocrine peptides reflect changes in biosynthesis and storage.

PITUITARY

In situ hybridization analyses have been effective in demonstrating virtually all of the major anterior pituitary hormone mRNAs.[11] Studies in experimental animals have revealed that this approach is useful not only for identifying cells of specific types, but also for estimating the amounts of mRNA encoding particular hormones in single cells.[66,76] Long-term estrogen treatment in the rat, for example, is associated with a marked increase in prolactin cells and with a four to five fold increase in the hybridization signal in individual prolactin cells.[48] *In situ* hybridization analyses with a probe recognizing only full-length pro-opiomelanocortin (POMC) heterogeneous nuclear RNA have also proven to be effective in studying the regulation of this gene in individual cell nuclei within the rat pituitary.[22,56] Adrenalectomy increases both the average hybridization intensity over individual corticotroph cell nuclei and the numbers of corticotroph cells. Dexamethasone administration, on the other hand, results in a rapid disappearance of hybridization signal over the corticotrophs of the anterior lobe, suggesting that POMC transcription had been inhibited.

Hybridization analyses also promise to be useful for studies of human pituitary tumors.[52] Approximately 25% of all pituitary adenomas have been categorized as nonfunctioning or null cell types. Using oligonucleotide probes and Northern blot analyses, expression of one or more of the anterior pituitary hormone genes was found in 86% of apparent null cell tumors.[35,36] Expression of one or more of the glycoprotein hormone genes (α, LH β, FSH β, TSH β) was identified in 79% of cases, with expression of multiple β subunit genes in many cases. Expression of α-subunit mRNA was found in each of the adenomas from patients expressing one of the β-subunit mRNAs and in three patients without detectable β-subunit mRNA. These findings clearly indicate that the majority of clinically nonfunctional pituitary adenomas are actually glycoprotein hormone cell adenomas.

Studies of so-called silent somatotroph adenomas have revealed that these tumors contain little or no immunoreactive growth hormone, but apparently correspond at the ultrastructural level to sparsely granulated growth hormone cells.[45,68] *In situ* hybridization analyses have clearly revealed, however, that these neoplasms contain growth hormone mRNA. Studies of silent corticotroph adenomas have revealed positive signals for POMC mRNA in about 70% of cases, while virtually all functioning corticotroph adenomas revealed positive hybridization signals.[47]

Using a combination of immunohistochemistry and *in situ* hybridization, Lloyd and co-workers have demonstrated that most acidophils in the pituitary express either growth hormone or prolactin and their respective mRNAs.[45] Occasional acidophils, however, contain growth hormone messenger RNA and prolactin. Tumors from patients with prolactinomas only had cells with prolactin mRNA, while tumors from most patients with acromegaly or gigantism had both growth hormone and prolactin mRNAs. In most cases, prolactin and growth hormone were present in different cells within the tumor, but five cases demonstrated that both hormones originated from the same cell. The latter tumors, therefore, may be classified as true mammosomatotropic adenomas. In one case reported by Lloyd and co-workers, an adenoma from a patient with acromegaly had immunoreactive prolactin, but was negative for growth hormone by immunohistochemistry.[45] *In situ* hybridization in this case revealed messenger RNAs both for growth hormone and for prolactin. Approximately one-third of patients with acromegaly have evidence of hyperprolactinemia due to a pituitary stalk section effect secondary to the neoplasm. The results of hybridization analysis, however, suggest that hyperprolactinemia in these patients may result in part from secretion of prolactin directly from the tumor.[45]

Messenger RNAs encoding the chromogranin proteins have also been analyzed in the pituitary using both Northern blot and *in situ* hybridization techniques.[16,46] Some cells of the anterior pituitary have shown both positive hybridization signals for chromogranin A messenger RNA and positive staining for the chromogranin A protein. In adjacent sections stained for anterior pituitary hormones, the chromogranin A–positive cells corresponded to cells containing FSH or LH. Growth hormone adenomas had variable expression of chromogranin A protein and also expressed chromogranin mRNA, while null cell adenomas expressed both chromogranin A protein and the corresponding mRNA. Chromogranin B mRNA was found in most cells and tumors that expressed chromogranin A mRNA, including certain normal cells of the anterior pituitary and both null cell and growth hormone–producing adenomas. Prolactinomas also expressed chromogranin B mRNA but were negative for chromogranin A mRNA and its corresponding protein. This observation suggests that the various chromogranin genes are differentially expressed in endocrine cells.

Recent studies suggest that the chromogranin proteins may serve as precursor molecules for biologically active peptides. Two peptides (GAWK and CCB) that have been isolated from pituitary extracts, represent portions of the chromogranin B molecule.[5,46,60] The GAWK and CCB peptides have been localized to GH and TSH cells that also express both chromogranin A and B mRNAs.

THYROID

In situ hybridization analyses in the thyroid have been performed for studies both of the thyroglobulin-producing follicular cells and the calcitonin-producing C cells. Using a tritium-labeled cDNA thyroglobulin probe, Bergé-Le Franc *et al.* demonstrated strong hybridization signals in the columnar cells of hyperplastic follicles found in Graves' disease and toxic adenomas, while the flattened follicular cells encountered in diffuse nontoxic goiter and colloid adenoma showed significantly less hybridization signal.[3] These findings indicate that *in situ* hybridization permits the correlation of the morphofunctional state of the follicles with their content of thyroglobulin mRNA.

Cell lines derived from normal rat thyroid have also been studied with respect to thyroglobulin synthesis and their contents of thyroglobulin mRNA.[2] One cell line, designated FRTL-5, synthesizes sevenfold less thyroglobulin than FRTL-424, even though the levels of thyroglobulin mRNA are comparable in the two lines. It has been suggested that a posttranscriptional step may be impaired in the translation of thyroglobulin mRNA in FRTL-5 cells. Studies of cell lines derived from

experimentally induced thyroid tumors have revealed an absence or markedly reduced expression of thyroglobulin mRNA, suggesting that neoplastic transformation results in a substantial reduction in thyroglobulin gene expression. Preliminary
results indicate that quantitation of mRNAs encoding thyroglobulin may be useful
in assessing differentiation in human thyroid tumors. Bergé-Le Franc et al. have
studied a series of thyroid follicular cell neoplasms, and have shown that moderately differentiated thyroid carcinomas contained 2–3 times less thyroglobulin
mRNA than well-differentiated tumors.[4] Additional studies will be required, however, to confirm these observations.

Calcitonin peptide and calcitonin mRNA have been demonstrated in both normal and neoplastic C cells in humans and in a variety of laboratory animals.[34,78]
Previous studies employing immunohistochemical techniques have revealed considerable heterogeneity in staining for calcitonin in medullary thyroid carcinomas,
and one study has suggested that tumors showing areas of absent staining have a
more ominous prognosis than those tumors that are uniformly positive for this
peptide.[42] Hybridization analyses using probes for calcitonin mRNA, however,
have revealed positive signals in virtually all of the tumor cells of these cases.

The calcitonin gene generates two different mRNAs, which encode calcitonin
and the calcitonin gene–related peptide (CGRP). These distinct mRNAs are generated by differential RNA processing from a single genomic locus.[59] The patterns of
expression of these distinct mRNAs are tissue specific, so that calcitonin is produced primarily in thyroid C cells, while CGRP is produced primarily in the brain,
spinal cord, and cranial nerve ganglia. Both calcitonin and CGRP mRNA transcripts
and their respective peptides are coproduced in normal thyroid C cells in a ratio of
about 95:1. CGRP (but not calcitonin) has also been identified within intrathyroidal
nerve fibers. Hybridization analyses have revealed that CGRP mRNA is present in
the same cells as calcitonin mRNA in cases of medullary thyroid carcinoma.

Blot hybridization techniques have revealed that metastases of some medullary
thyroid carcinomas also express pro-opiomelanocortin (POMC) sequences that differ to some extent from POMC sequences found in normal pituitary gland.[67]

Gastrin-releasing peptide (GRP), the mammalian analogue of bombesin, has
been analyzed in C cells both by immunohistochemistry and by in situ hybridization techniques.[70,71] Studies of thyroid glands at various developmental stages have
shown GRP mRNAs and peptide in up to 80% of C cells in fetuses and neonates,
but in less than 10% of C cells in normal adult glands. Prominent expression of the
GRP gene has also been found in cases of medullary thyroid carcinoma. Sunday
and co-workers have demonstrated that over 90% of the cells in three medullary
carcinomas contained GRP mRNA as demonstrated by in situ hybridization. Moreover, hyperplastic C cells adjacent to follicular adenomas and papillary carcinomas
show an increased frequency of GRP immunostaining and an increase in GRP
mRNA levels. These observations suggest a potential role for GRP gene activation
in thyroid development and possibly in C cell growth, which may be analogous to
the role of GRP in pulmonary development.[70,71] Messenger RNAs encoding both
chromogranin A and B have also been localized to normal and neoplastic C cells.
Somatostatin mRNA has been localized to a subset of the C cells (Fig. 14.5).

Messenger RNAs encoding oncogene proteins have also been detected in medullary thyroid carcinomas.[37] The studies of Hofler and co-workers, for example,

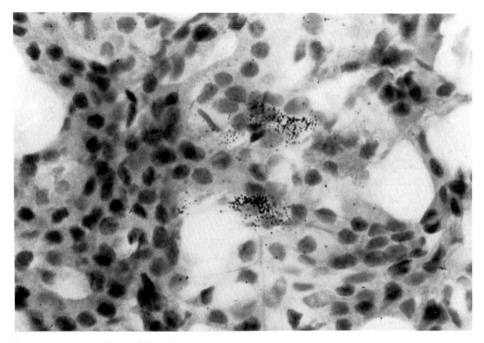

Figure 14.5.   Rat thyroid gland reacted with tritium-labeled antisense RNA somatostatin probe. Autoradiographic grains are concentrated over individual C cells. Follicular cells are negative. (250X). (From DeLellis, R. A., and Wolfe, H. J. The application of *in situ* hybridization techniques to endocrine pathology: An overview. In: *Endocrine Pathology Update*, edited by J. Lechago and J. Kameya, Philadelphia, Field and Wood Publishers, 1990, pp. 293–310.)

have shown that some of these tumors contain mRNAs encoding the C-*myc*, Ha-*ras* and C-*fms* oncogene proteins. In addition, the studies of Boultwood and co-workers have shown that N-*myc* mRNA is readily demonstrable by *in situ* hybridization in about 30% of medullary thyroid carcinomas, but is undetectable in normal C cell populatons.[8] Previous studies have implicated the N-*myc* oncogene in the development and progression of neuroblastoma, small (oat) cell bronchogenic carcinoma, retinoblastoma, and Wilm's tumor.[9]

PARATHYROID

Chromogranin A, which is similar to parathyroid secretory protein I, has been detected within the parathyroid gland both by radioimmunoassays of tissue extracts and by immunohistochemistry. In general, however, the intensity of immunohistochemical staining is low. *In situ* hybridization studies using a chromogranin A probe have revealed intense hybridization signals within the gland.[61] An mRNA encoding a novel parathyroid hormone–like peptide has been detected in a variety of normal and neoplastic tissues.[33] The human parathyroid hormone–like peptide was initially isolated and cloned from tumors associated with humoral hypercalcemia of malignancy. Northern analysis of mRNAs from abnormal human parathyroid tissue has revealed an overexpression of transcripts for the PTH-like peptide, which appears specific for adenomatous or autonomous glands. Additional studies will be required to confirm these observations.

The immunohistochemical demonstration of parathyroid hormone and its precursors has been hampered because of the rapid secretion of hormone from the gland and because of the lack of availability of reliable antibodies for use in cellular localization studies. The studies of Stork and co-workers, however, have demonstrated the possibility of demonstrating parathyroid hormone messenger RNA with ^{35}S-labeled antisense probes both in fresh-frozen and formalin-fixed and paraffin-embedded sections.[69] In normal glands, chief cells and transitional clear cells showed high steady-state expression of parathyroid hormone messenger RNA. Oxyphil cells, on the other hand, showed a low level of hybridization signal, while stromal fibroblasts and fat cells were negative. Parathyroid cells in cases of primary chief cell hyperplasia, similar to normal chief cells, also showed high levels of expression. All cases of secondary hyperplasia revealed varying degrees of hybridization in different nodules within the same gland. Cells comprising adenomas showed lower levels of hybridization than the adjacent rims of compressed normal or suppressed parathyroid tissue. The increased secretion of parathyroid hormone by adenomas may be explained by the increased numbers of cells comprising the adenomas rather than by increased secretion of individual cells. These observations indicate that chronic hyperparathyroidism with chronic hypercalcemia is insufficient to lower the steady state of parathyroid hormone within the portions of parathyroid glands uninvolved by adenoma.[70] From a purely pragmatic point of view, the demonstration of increased hybridization over the parathyroid tissue uninvolved by adenoma may be a useful criterion to establish a diagnosis of adenoma in the appropriate clinical setting.

ADRENAL GLAND

Hybridization analyses have been used to a somewhat limited extent for analysis of regulatory peptides and other products within the adrenal medulla. *In situ* hybridization studies with probes for chromogranin A mRNA have revealed a distinctive distribution of signal within the bovine medulla.[61] These studies have revealed a 3-fold greater hybridization intensity in the outer or peripheral medulla as compared to more central regions. Correlative studies using probes for enkephalin and phenylethanolamine-N-methyltransferase mRNAs also revealed strong hybridization over the peripheral medullary region, while the central medulla showed very little signal. Neuropeptide Y and POMC mRNAs are present within normal adrenal medullary cells and pheochromocytomas. Although calcitonin has been reported to be present in extracts of the adrenal, hybridization analyses have not confirmed the presence of either calcitonin or CGRP mRNAs at this site.

OTHER ENDOCRINE CELLS AND ENDOCRINE TUMORS

In situ hybridization techniques have been used for studies of the distribution of peptide-encoding mRNAs in a variety of tissues.[11] For example, somatostatin mRNA has been localized to some thyroid C cells, endocrine cells of the pancreas and gastroenteric mucosa, and neurons of the submucous and myenteric plexi of the gastrointestinal tract.[30] Comparative immunohistochemical and *in situ* hybridization analyses have generally shown good correlations between the localization of peptide and the corresponding mRNA. However, positive hybridization results have been obtained for certain submucous and myenteric neurons that were negative for somatostatin peptide, even after colchicine blockade. The distribution of vasoactive intestinal peptide mRNA has also been studied by *in situ* hybridization within the gastrointestinal tract (Fig. 14.6). These studies have revealed the presence of VIP mRNA primarily within neuronal cells.[72] A similar localization has been observed for both GRP and VIP mRNAs.

In the pancreas, the mRNA encoding pancreatic polypeptide has also been demonstrated by *in situ* hybridization techniques[40] (Fig. 14.7). Comparative immunohistochemical and hybridization analyses have revealed a higher sensitivity for the *in situ* approach, particularly in studies of early developmental stages.

The value of *in situ* hybridization analyses for the demonstration of peptide–encoding messenger RNAs is especially well demonstrated in the case of peptide YY (PYY).[41] Peptide YY, a 36–amino acid peptide, is a member of the pancreatic polypeptide (PP) family. The functions of PYY include inhibition of gastric and intestinal motility, suppression of pancreatic exocrine secretion, and suppression of gastric secretion. Amino acid homologies between PYY and other members of the PP family have made studies of their localization and biosynthesis extremely difficult. All of the precursors have signal peptides of 27–31 amino acids in length at

Figure 14.6. Human colon reacted with tritium-labeled antisense RNA probe for VIP. Autoradiographic grains are concentrated over neurons in the submucous plexus. (250X). (From DeLellis, R. A., and Wolfe, H. J. The application of *in situ* hybridization techniques to endocrine pathology: An overview. In: *Endocrine Pathology Update*, edited by J. Lechago and J. Kameya, Philadelphia, Field and Wood Publishers, 1990, pp. 293–310.)

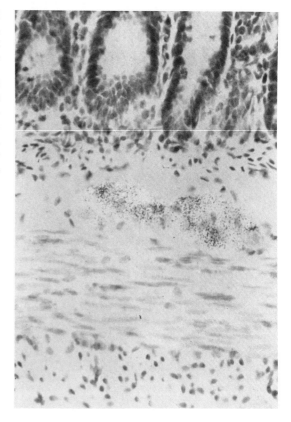

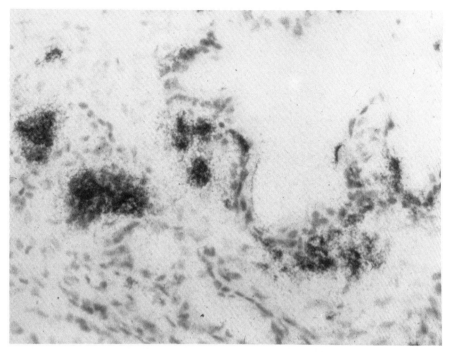

Figure 14.7. Human pancreas reacted with 32P-labeled probe for human pancreatic polypeptide. Both single cells and cell clusters (some budding from ducts) can be seen with this technique. Serial sections stained for PP by immunohistochemistry showed fewer positive cells and cell clusters. (250X). (From DeLellis, R. A., and Wolfe, H. J. The application of *in situ* hybridization techniques to endocrine pathology: An overview. In: *Endocrine Pathology Update*, edited by J. Lechago and J. Kameya, Philadelphia, Field and Wood Publishers, 1990, pp. 293–310.)

their amino–terminal ends, followed by a 36–amino acid sequence representing the mature hormone. This is followed by an identical 3–amino acid sequence that terminates in a carboxy-terminus extension peptide of 27–30 amino acids. With the cloning of the PYY cDNA, it became possible to generate molecular probes that recognized the PYY messenger RNA specifically. Northern blot analyses revealed high levels of PYY mRNA in the ileum, cecum, colon, and pancreas and lower levels in the stomach and rectum. *In situ* hybridization studies revealed that PYY was present not only in the islets but also in two to three cell clusters throughout the parenchyma. In the gastrointestinal tract, single positive cells were dispersed within the mucosa.

Immunohistochemistry and *in situ* hybridization analyses have been used for developmental studies of neuroendocrine cell populations. Sunday and co-workers have shown, for example, that levels of GRP mRNA are markedly elevated in fetal lung from about 16 to 24 weeks of gestation.[70] Thereafter, the levels in fetal lung are quite low, approaching those seen in the adult. In contrast, GRP peptide levels remain elevated until several months after birth. By *in situ* hybridization studies, hybridizing cells appeared earlier than peptide-positive cells. At about 16–24 weeks, the numbers of hybridizing cells were equivalent to peptide-positive cells and were

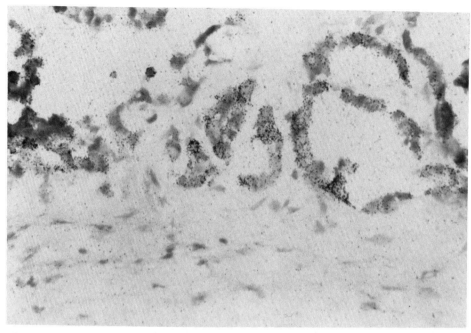

Figure 14.8. Pancreatic endocrine tumor associated with the Verner-Morrison syndrome reacted with tritium-labeled antisense probe for pancreatic polypeptide. This tumor showed a relatively weak reaction for pancreatic polypeptide by immunohistochemistry. By *in situ* hybridization, some cells contain abundant message for PP, as determined by the density of autoradiographic grains, while other cells contain little or no message. The stroma is completely unreactive. (250X). (From DeLellis, R. A., and Wolfe, H. J. The application of *in situ* hybridization techniques to endocrine pathology: An overview. In: *Endocrine Pathology Update*, edited by J. Lechago and J. Kameya, Philadelphia, Field and Wood Publishers, 1990, pp. 293–310.)

present primarily in proximal airways. Ultimately, positive peptide staining was observed, but hybridizing foci were markedly decreased. The transient expression of high levels of GRP mRNA suggests that secretion of GRP may play an important regulatory role in the development of the lung.

The studies of Cuttitta and co-workers and other groups have demonstrated that GRP-associated peptides are commonly expressed in small cell carcinomas of the lung. Sunday and co-workers and Hamid *et al.*[25] have studied neuroendocrine lung tumors with respect to the expression of GRP mRNA and GRP. Positive stains for GRP peptide and positive signals for GRP mRNA were observed in 3 of 4 typical carcinoid tumors, 2 of 3 atypical carcinoids, and 4 of 8 small cell carcinomas as reported by Sunday *et al.*[70] In one of the two atypical carcinoids and 3 of 4 small cell carcinomas, GRP peptide was only weakly stained, but GRP mRNA was strongly expressed. The sensitivity of detection of GRP immunoreactivity varies somewhat, depending on the antisera employed. This is likely due to the fact that some tumors contain a predominance of pro-GRP, as determined by immunoreactivity for the GRP C-terminal extension peptides. Some groups have reported concordant immunostaining for GRP and the C-terminal extension peptides in about 50% of

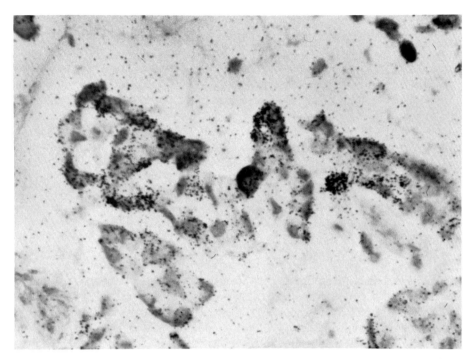

Figure 14.9. Pancreatic endocrine tumor associated with the Verner-Morrison syndrome. Combined *in situ* hybridization for VIP with a tritium-labeled antisense VIP probe and immunohistochemistry for VIP. Some cells (*center*) show strong staining for the peptide with a relatively weak hybridization signal, while others show a strong hybridization signal but weak staining for the peptide. (250X). (From DeLellis, R. A., and Wolfe, H. J. The application of *in situ* hybridization techniques to endocrine pathology: An overview. In: *Endocrine Pathology Update*, edited by J. Lechago and J. Kameya, Philadelphia, Field and Wood Publishers, 1990, pp. 293–310.)

small cell lung carcinomas. Other groups have reported that almost all small cell carcinomas contain GRP mRNA by *in situ* hybridization.[25]

In situ hybridization analyses are of value in understanding mechanisms of ectopic hormone production.[15] This technique enables the pathologist to determine whether tumor cells are concentrating hormone products from the circulation and surrounding tissues, or whether they are actively transcribing the specific mRNA for peptide synthesis. Moreover, in instances where hormone levels are elevated in association with a particular tumor, it is possible to determine if this is actual peptide production by the tumor or an indirect result. As discussed earlier, certain pituitary adenomas may be associated with hyperprolactinemia because of impingement of the tumor on the pituitary stalk. The absence of prolactin mRNA within the tumor would provide evidence that the source of the increased prolactin was the nontumorous portion of the gland, which was stimulated to produce prolactin because of a lack of prolactin inhibitory factor (dopamine).

Hybridization analyses are also of importance in analyzing tumors producing multiple peptides.[57] We have studied in detail a pancreatic neoplasm associated with the production of both VIP and pancreatic polypeptide (PP)[29,31] (Figs. 14.8 and 14.9). By immunohistochemistry, VIP and PP were localized in varying proportions of the tumor cells.[57] Combined immunohistochemistry and *in situ* hybridization revealed some cells with strong hybridization signals but no staining for PP or VIP, while other cells showed strong immunohistochemical staining but negative hybridization results. Most cells, however, showed some immunohistochemical staining and some hybridization signal. The combination of *in situ* hybridization and immunohistochemistry was of value in establishing whether the same cell had the capacity to synthesize and store both peptides. For these studies, tissue sections processed for the localization of PP mRNA were stained by immunohistochemistry for VIP. Some cells that were positive for VIP peptide contained PP mRNA. Additional studies established that cells that contained PP peptide also contained VIP mRNA.

Summary and Conclusions

Although this chapter has focused on the applications of *in situ* hybridization technology to studies of normal and neoplastic endocrine cells, there are other major applications of recombinant DNA technology in endocrinology. A number of genes responsible for known hereditary disorders of the endocrine system have now been cloned, and molecular diagnostic tests are now available for their specific identification[1,64] (Table 14.1).

In those instances where the precise molecular defect remains unknown, it is possible, using linkage studies, to indirectly detect the abnormal genes. The multiple endocrine neoplasia syndromes are among those in which such linkage studies have made a substantial impact. Multiple endocrine neoplasia type 2, a disorder associated with medullary thyroid carcinoma, parathyroid hyperplasia, and pheochromocytoma is inherited as an autosomal dominant trait. The American branch of a large Swedish family with this disorder has been under study at the New England Medical Center for nearly twenty years.[23] Provocative serum calcitonin testing not only has demonstrated the utility of calcitonin as a marker for early medullary thyroid carcinoma but also has allowed the detection of a precursor lesion of medullary thyroid carcinoma (C cell hyperplasia). This kindred has been traced through five generations representing over 100 family members, and a total of 34 patients have been successfully treated for a combination of either occult medullary thyroid carcinoma and/or its precursor lesion, C cell hyperplasia, as well as adrenal medullary proliferative disorders.

Table 14.1.
Cloned Genes Responsible for Some Hereditary Endocrine Disorders

Gene	Chromosome	Disorder
Steroid 21-hydroxylase	6p	Congenital adrenal hyperplasia
Thyroglobulin	8q	Hereditary congenital hypothyroidism
Steroid 17-hydroxylase/17, 20-lyase	10	Congenital adrenal hyperplasia
Insulin	11p	Diabetes mellitus due to abnormal insulins
Parathyroid hormone	11p	Familial hypoparathyroidism
Growth hormone	17q	Isolated familial growth hormone deficiency

Considerable progress has been made by several groups toward the cloning and mapping of the gene responsible for this inherited disorder. The best information, to date, suggests that the defect is localized to chromosome 10, and multilinkage data from the work of two groups indicated that multiple endocrine neoplasia type 2 may lie somewhere within a four centimorgan distance around the centromere.[51, 62] The ultimate detection of the gene responsible for this condition may offer us new insights into the fundamental molecular mechanisms of endocrine cell regulation as well as create invaluable molecular probes for the detection and classification of heritable disorders.

Similar progress is being made in the understanding of multiple endocrine neoplasia type I. This disorder affects the parathyroid glands, anterior pituitary, and pancreatic islets. Recent studies indicate that a single inherited locus on chromosome 11, and q13, is responsible for the development of MEN I and that the monoclonal development of pancreatic and parathyroid tumors in affected patients involves similar allelic deletions on chromosome 11.[38]

REFERENCES

1. Antonarakis, S. A. Diagnosis of genetic disorders at the DNA level. *N. Engl. J. Med.* 320:153–163, 1989.
2. Avvidimento, V. E., Monticelli, A., Tramontano, D., Polistina, C., Nitsch, D., and DiLauro, R. Differential expression of thyroglobulin gene in normal and transformed thyroid cells. *J. Biochem.* 149:467–472, 1985.
3. Bergé-LeFranc, J.-L., Cartouzou, G., Bignon, C., and Lissitzky, S. Quantitative *in situ* hybridization of ³H-labeled cDNA to the messenger ribonucleic acid of thyroglobulin in human thyroid tissues. *J. Clin. Endocrinol. Metab.* 57:470–476, 1985.
4. Bergé-LeFranc, J.-L., Cartouzou, G., DeMicco, C., Fragu, P., and Lissitzky, S. Quantitation of thyroglobulin mRNA by *in situ* hybridization in differentiated thyroid cancers: Difference between well differentiated and poorly differentiated histologic types. *Cancer* 56:345–350. 1985.
5. Bishop, A. E., Sekiya, K., and Salahuddin, M. J. The distribution of GAWK-like immunoreactivity in neuroendocrine cells of the human gut, pancreas, adrenal and pituitary glands and it co-localization with chromogranin B. *Histochemistry* 90:475–483, 1989.
6. Block, B., LeGuellec, D., and DeKeyzer, Y. Detection of the messenger RNAs coding for the opioid peptide precursors in pituitary and adrenal by *in situ* hybridization. *Neurosci. Lett.* 53:141–148, 1985.
7. Block, B., Popovici, T., LeQuellec, D., Normand, E., Chouham, S., Guitteny, A. F., and Bohlen, P. *In situ* hybridization histochemistry for the analysis of gene expression in the endocrine and central nervous system tissues: A 3 year experience. *J. Neurosci. Res.* 16:183–200, 1986.
8. Boultwood, J., Wyllie, F. S., Williams, E. D., and Wynford-Thomas, D. N-myc expression in neoplasia of human thyroid C-cells. *Cancer Res.* 48:4073–4077, 1988.
9. Brodeur, G. M., Seeger, R. C., Sather, H., *et al.* Clinical implications of oncogene activation in human neuroblastomas. *Cancer* 58:541–545, 1986.
10. Caruthers, M. H. Gene synthesis machines: DNA chemistry and its uses. *Science (Wash DC)* 230:281–285, 1985.
11. Coghlan, J. P., Aldred, P., Haralambidis, J., *et al.* Hybridization histochemistry. *Anal. Biochem.* 149:1–28, 1985.
12. Cox, K. H., DeLeon, D. V., Angerer, L. M., and Angerer, R. C. Detection of mRNAs in sea

urchin embryos by *in situ* hybridization using asymmetric RNA probes. *Develop. Biol.* 101:485–502, 1984.

13. Creutzfeldt, W., Arnold, R., Creutzfeldt, C., *et al.* Biochemical and morphological investigations of 30 insulinomas. *Diabetologia* 9:217–231, 1973.

14. Cuttitta, F., Fedorko, J., Gu, J., Lebacq-Verheyden, A.-M., Linnoila, R. I., and Battery, J. F. Gastrin releasing peptide gene associated peptides are expressed in normal fetal lung and small cell lung cancer: A novel peptide family found in man. *J. Clin. Endocrinol. Metab.* 67:576–583, 1988.

15. DeBold, C. R., Menefee, J. K., Nicholson, W. E., and Orth, D. N. Pro-opiomelanocortin gene is expressed in many normal human tumors not associated with ectopic adenocorticotropin syndrome. *Mol. Endocrinol.* 2:862–870, 1988.

16. Deftos, L. J., Murray, S. S., Burton, D. W., Parmer, R. J., and O'Connor, D. T. A cloned chromogranin A cDNA detects a 2.3 Kb mRNA in diverse neuroendocrine tissues. *Biophys. Res. Comm.* 137:418–423, 1986.

17. DeLellis, R. A., and Wolfe, H. J. Contributions of immunohistochemical and molecular biological techniques to endocrine pathology. *J. Histochem. Cytochem.* 35:1347–1351, 1987.

18. DeLellis, R. A., and Wolfe, H. J. The application of *in situ* hybridization techniques to endocrine pathology: An overview. In: *Endocrine Pathology Update*, edited by J. Lechago and J. Kameya, Philadelphia, Field and Wood Publishers, 1990, pp. 293–310.

19. DeLellis, R. A., and Wolfe, H. J. New techniques in gene product analysis. *Arch. Pathol. Lab. Med.* 111:620–627, 1987.

20. Fenoglio-Preiser, C., and Willman, C. Molecular biology and the pathologist: General principles and applications. *Arch. Pathol. Lab. Med.* 111:601–619, 1987.

21. Fink, J. S., Montminy, M. R., Tsukada, T., *et al. In situ* hybridization of somatostatin and vasoactive intestinal peptide mRNA in the rat nervous system. Contrasting patterns of ontogeny. In: *In Situ Hybridization in Brain*, edited by G. R. Uhl., N.Y., Plenum Press, 1986, p. 181.

22. Fremeau, R. T., Lundblad, J. R., Pritchett, D. B., Wilcox, J. N., and Roberts, J. L. Regulation of pro-opiomelanocortin gene transcription in individual cell nuclei. *Science (Wash DC)* 234:1265–1269, 1986.

23. Gagel, R. F., Tashjian, A. H., Cummings, T., *et al.* The clinical outcome of prospective screening for multiple endocrine neoplasia type 2a. An 18 year experience. *N. Engl. J. Med.* 318:478–484, 1988.

24. Hagn, C., Schmid, K. W., Fischer-Colbrie, R., and Winkler, H. Chromogranin A, B and C in human adrenal medullary and endocrine tissue. *Lab. Invest.* 55:405–441, 1986.

25. Hamid, Q. A., Bishop, A. E., Springall, D. R., *et al.* Detection of human probombesin mRNA in neuroendocrine (small cell) carcinoma of the lung. *In situ* hybridization with a cRNA probe. *Cancer* 63:266–271, 1989.

26. Harnson, L. C., and Campbell, I. L. Cytokines: An expanding network of immunoinflammatory hormones. *Mol. Endocrinol.* 1151–1156, 1988.

27. Helman, L. J., Gazdar, A. F., Park, J. G., Cohen, P. S., Cotelingam, J. D., and Israel, M. A. Chromogranin A expression in normal and malignant human tissues. *J. Clin. Invest.* 82:686–690, 1988.

28. Hofler, H. What's new in *in situ* hybridization? *Pathol. Res. Pract.* 182:421–430, 1987.

29. Hofler, H., Childers, H., Dayal, Y., Leiter, A., Goodman, R., DeLellis, R., Tischler, A., and Wolfe, H. Detection of neuroendocrine gene expression of tumor cells by combined *in situ* hybridization and immunocytochemistry. *Verh. Dtsch. Ges. Pathol.* (Stuttgart) 70:211–216, 1986.

30. Hofler, H., Childers, H., Montminy, M. R., Lechan, R. M., Goodman, R. H., and Wolfe, H. J. *In situ* hybridization methods for the detection of somatostatin mRNA in tissue sections using antisense mRNA probes. *Histochem. J.* 18:597–604, 1986.

31. Hofler, H., DeLellis, R. A., and Wolfe, H. J. *In situ* hybridization and immunohistochem-

istry. In: *Advances in Immunohistochemistry,* edited by R. A. DeLellis, New York, Raven Press, 1988, pp. 47–66.

32. Hudson, P., Penschow, J., Shine, J., Ryan, G., Niall, H., and Coghlan, J. Hybridization histochemistry: Use of recombinant DNA as a homing probe for tissue localization of specific messenger ribonucleic acid populations. *Endocrinology* 108:353–356, 1981.

33. Ikeda, K., Weir, E. C., Mangin, M., Dannies, P. S., Kinder, B., Deftos, L. J., Brown, E. M., and Broadus, A. E. Expression of messenger ribonucleic acids encoding a parathyroid hormone like peptide in normal human and animal tissues with abnormal expression in human parathyroid adenomas. *Mol. Endocrinol.* 2:1230–1236, 1988.

34. Jacobs, J. W., Simpson, E., Penschow, J., Hudson, P., Coghlan, J., and Niall, H. Characterization and localization of calcitonin mRNA in rat thyroid. *Endocrinology* 113:1616–1622, 1983.

35. Jameson, J. L., Klibanski, A., Black, P., Zervas, N. T., Lindell, C. M., Hsu, D. W., Ridgeway, E. C., and Habener, J. F. Glycoprotein hormone genes are expressed in clinically non-functional pituitary adenomas. *J. Clin. Invest.* 80:1472–1478, 1980.

36. Jameson, J. L., Lindell, C. M., and Habener, J. F. Gonadotropin and thyrotropin alpha beta subunit gene expression in normal and neoplastic tissues characterized using specific messenger ribonucleic acid hybridization probes. *J. Clin. Endocrinol. Metab.* 64:319–327, 1987.

37. Klimpfinger, M., Rutori, C., Putz, B., Pfranger, R., Wirnsberger, G., and Hofler, H. Oncogene expression in a medullary thyroid carcinoma. *Virchows Arch.(B)* 54:256–259, 1988.

38. Larsson, C., Skogseid, B., Oberg, K. Multiple endocrine neoplasia type 1 gene maps to chromosome 11 and is lost in insulinoma. *Nature (Lond)* 332:85–87, 1988.

39. Lechan, R. M., Wu, P., Jackson, I., et al. Thyrotropin releasing hormone precursor: Characterization in rat brain. *Science (Wash DC)* 231:159–161, 1986.

40. Leiter, A. B., Keutmann, H. T., and Goodman, R. H. Structure of a precursor to human pancreatic polypeptide. *J. Biol. Chem.* 259:14702–14705, 1984.

41. Leiter, A., Toder, A., Wolfe, H. J., et al. Peptide YY. Structure of the precursor and expression in exocrine pancreas. *J. Biol. Chem.* 262:12984–12988, 1987.

42. Lippman, S. M., Mendelsohn, G. Trump, D. L., et al. The prognostic and biological significance of cellular heterogeneity in medullary thyroid carcinoma. *J. Clin. Endocrinol. Metab.* 54:233–241, 1982.

43. Lloyd, R. V. Molecular probes and endocrine disease. *Am. J. Surg. Pathol.* 14 (Supplement 1):34–44, 1990.

44. Lloyd, R. V. Use of molecular probes in the study of endocrine studies. *Hum. Pathol.* 18:1199–1211, 1987.

45. Lloyd, R. V., Cano, M., Chandler, W. F., Barkan, A. L., Horvath, E., and Kovacs, K. Human growth hormone and prolactin secreting pituitary adenomas analyzed by *in situ* hybridization. *Am. J. Pathol.* 134:605–613, 1989.

46. Lloyd, R. V., Iancangelo, A., and Eiden, L. E. Chromogranin A and B messenger ribonucleic acids in pituitary and other normal and neoplastic human endocrine tissues. *Lab. Invest.* 60:548–556, 1989.

47. Lloyd, R. V., Jin, L., Horvath, E., and Kovacs, K. *In situ* hybridization analysis of functioning and silent corticotroph adenomas of the human pituitary. *Lab. Invest.* 60:54A(abstract), 1989.

48. Lloyd, R. V., and Landefeld, T. D. Detection of prolactin messenger RNA in rat anterior pituitary by *in situ* hybridization. *Am. J. Pathol.* 125:35–44, 1986.

49. Mains, R. E., Cullen, E. I., May, V., and Eipper, B. A. The role of secretory granules in peptide biosynthesis. *Ann. N.Y. Acad. Sci.* 493:278–291, 1987.

50. Martin, J. B. Molecular genetics: Application to the clinical neurosciences. *Science (Wash DC)* 238:765–771, 1987.

51. Matthew, C. G. P., Chin, K. S., Easton, D. F., *et al.* A linked genetic marker for multiple endocrine neoplasia type 2A on chromosome 10. *Nature (Lond)* 328:527–528, 1987.
52. Melmed, S., Braunstein, G. D., Chavez, R. J., and Becker, D. P. Pituitary tumors secreting growth hormone and prolactin. *Ann. Int. Med.* 105:238–253, 1986.
53. Moller, D. E., and Flier, J. S. Detection of an alteration in the insulin receptor gene in a patient with insulin resistance, acanthosis nigricans and the polycystic ovary syndrome. *N. Engl. J. Med.* 319:1526–1529, 1988.
54. Moore, H. P. H. Factors controlling packaging of peptide hormones into secretory granules. *Ann. N.Y. Acad. Sci.* 493:50–61, 1987.
55. Pardue, M. L., and Gall, J. G. Molecular hybridization of radioactive DNA to the DNA of cytological preparations. *Proc. Natl. Acad. Sci. U.S.A.* 64:600–604, 1969.
56. Roberts, J. L., Lundblad, J. R., Eberwine, J. H., *et al.* Hormonal regulation of POMC gene expression in pituitary. *Ann. N.Y. Acad. Sci.* 512:275–285, 1985.
57. Rood, R. P., DeLellis, R. A., Dayal, Y., and Donowitz, M. Pancreatic cholera syndrome due to a vasoactive intestinal peptide producing tumor. Further insights into the pathophysiology. *Gastroenterology* 94:813–818, 1988.
58. Rosenfeld, M. G., Leff, S. E., Russo, A. F., and Evans, R. M. Neuron specific alternative RNA processing in neuroendocrine gene expression. *Biochem. Soc. Trans.* 15:128–131, 1987.
59. Sabate, M. I., Stolarsky, L. S., Polak, J. M., Bloom, S. R., Varndell, I. M., Ghatei, M. A., Evans, R. M., and Rosenfeld, M. G. Regulation of neuroendocrine gene expression by alternative RNA processing. *J. Biol. Chem.* 260:2589–2592, 1985.
60. Schmidt, W. E., Seigel, E. G., Lamberts, R., Gallwitz, B., and Creutzfeldt, W. Pancreastatin: Molecular and immunocytochemical characterization of a novel peptide in porcine and human tissues. *Endocrinology* 123:1395–1404, 1988.
61. Siegel, R. E., Iacangelo, A., Park, J., and Eiden, L. E. Chromogranin A biosynthetic cell populations in bovine endocrine and neuronal tissues. Detection by *in situ* hybridization histochemistry. *Mol. Endocrinol.* 2:368–374, 1988.
62. Simpson, N. E., Kidd, K. K., Goodfellow, P. J., *et al.* Assignment of multiple endocrine neoplasia type 2a to chromosome 10 by linkage. *Nature (Lond)* 328:528–529, 1987.
63. Singer, R. H., Lawrence, J. B., and Villname, C. Optimization of *in situ* hybridization using isotopic and non-isotopic detection methods. *Biotechnology* 4:230–244, 1986.
64. Sobol, H., Narod, S. A., Nakamura, Y., *et al.* Screening for multiple endocrine neoplasia type 2a with DNA polymorphism analysis. *N. Engl. J. Med.* 321:996–1001, 1989.
65. Standish, L. J., Adams, L. A., Vician, L., Clifton, D. K., and Steiner, R. A. Neuroanatomical localization of cells containing gonadotropin releasing hormone messenger RNA in the primate brain by *in situ* hybridization histochemistry. *Mol. Endocrinol.* 1:371–376, 1987.
66. Steel, J. H., Hamid, Q., Van Noorden, S., Jones, P., Denny, P., Burrin, J., Legon, S., Bloom, S. R., and Polak, J. M. Combined use of *in situ* hybridization and immunohistochemistry for the investigation of prolactin gene expression in immature, pubertal, pregnant, lactating and ovariectomised rats. *Histochemistry* 89:75–80, 1988.
67. Steenbergh, P. H., Hoppener, J. W. M., Zandberg, J., Roos, B. A., Jansz, H. S., and Lips, C. J. M. Expression of the proopio-melanocortin gene in human medullary thyroid carcinoma. *J. Clin. Endocrinol. Metab.* 58:904–908, 1984.
68. Stefaneanu, L., Lloyd, R., Horvath, E., Kovacs, K., Asa, S. L., Killinger, D. W., and Smyth, H. S. A morphologic study of silent somatotroph adenomas of the human pituitary. *Lab. Invest.* 60:92A (abstract), 1989.
69. Stork, P. J., Herteaux, C., Frazier, R., Kronenburg, H., and Wolfe, H. J. Expression and distribution of parathyroid hormone and parathyroid hormone messenger RNA in pathological conditions of the parathyroid. *Lab. Invest.* 60:92A (Abstract), 1989.
70. Sunday, M. E., Kaplan, M. E., Motoyama, E., Chin, W., and Spindel, E. R. Gastrin releas-

ing peptide (mammalian bombesin) gene expression in health and disease. *Lab. Invest.* *59*:5–24, 1988.

71. Sunday, M. E., Wolfe, H. J., Roos, B. A., and Spindel, E. R. Gastrin releasing peptide gene expression in developing, hyperplastic and neoplastic human human thyroid C-cells. *Endocrinology 122*:1551–1558, 1988.

72. Tsukada, T., Horovitch, S. J., Montiminy, M. R., and Goodman, R. H. Structure of the human vasoactive intestinal peptide gene. DNA *4*:293–300, 1985.

73. Tsutsumi, Y., Tischler, A., Stork, P. J., and Wolfe, H. J. Demonstration of neurotensin mRNA in pheochromocytoma cells using a synthetic oligonucleoide probe. *Lab. Invest.* *60*:98A (abstract), 1989.

74. Varndell, I. M., Polak, J. M., Sikri, K. L., *et al.* Visualization of messenger RNA directing peptide synthesis by *in situ* hybridization using a novel single standard cDNA probe: Potential for the investigation of gene expression and endocrine cell activity. *Histochemistry 81*:597–601, 1984.

75. Wolfe, H. J. DNA probes in diagnostic pathology. *Am. J. Clin. Pathol. 90*:340–344, 1988.

76. Yamamoto, N., Seo, H., Suganuma, N., Matsui, N., Nakane, T., Kuwayama, A., and Kageyama, N. Effect of estrogen on prolactin mRNA in the rat pituitary. *Neuroendocrinology 42*:494–497, 1986.

77. Young, W. S., and Zoeller, R. T. Neuroendocrine gene expression in the hypothalamus: *In situ* hybridization histochemical studies. *Cell. Mol. Neurobiol. 7*:353–356, 1987.

78. Zajac, J. D., Penschow, J., Mason, T., Tregear, G., Coghlan, J., and Martin, T. J. Identification of calcitonin and calcitonin gene related peptide messenger RNA in medullary thyroid carcinomas by hybridization histochemistry. *J. Clin. Endocrinol. Metab. 62*: 1037–1043. 1986.

15

Application of Molecular Techniques to the Diagnosis of Viral Disease

James K. McDougall

Introduction

This chapter describes the application of molecular biological procedures to the diagnosis of some infectious viral diseases. Molecular hybridization technology has already made a highly significant impact on the practice of diagnostic pathology, for example the progress made in developing an understanding of the human immunodeficiency viruses (HIV) and in the introduction of sensitive and specific diagnostic tests. An explosion of interest in human papillomaviruses (HPV) has followed the identification of viral nucleic acid sequences in squamous carcinoma cells from a number of sites, resulting in the availability of diagnostic tests based upon molecular technology. While some currently available techniques are too complex and costly for routine diagnostic use, other methods already provide valuable and accurate data. The use of recombinant DNA–based technology in the detection and study of HPV, a situation in which virus isolation or serological testing are not currently available, demonstrates both the power and utility of these procedures.

Examples and Applications in Virus Diagnosis

HERPESVIRUSES

Light or electron microscopy, immunofluorescence, or immunoassay are all valuable techniques that can provide relatively rapid diagnosis. All of these methods have attendant caveats, *e.g.*, sufficient cells for cytology; inability to distinguish between different herpes viruses at the electron microscope level; nonspecific reactions in immunoassays. The use of restriction endonucleases to cleave viral DNA into specific fragments that provide distinctive patterns after electrophoresis allows herpes simplex virus type 1 (HSV-1) isolates to be distinguished from HSV-2.[4,41] This technique can also be of major value in epidemiological studies and can be further enhanced by the hybridization of specific cloned sequences to restriction enzyme–cleaved viral DNA transferred to nitrocellulose to identify polymorphisms.

CYTOMEGALOVIRUS

The transmission and variability of human cytomegalovirus (HCMV) has been studied in this way.[5,6,28] The extent to which the diversity found among HCMV strains is reflected in biological and functional differences is not known. The potential for HCMV strain differences to influence disease, perhaps by variation in the efficiency of replication, or differences in cell tropism or the ability to establish persistent or latent infection, has not been explored. As an initial step in studying the significance of HCMV strain variation we mapped and compared the EcoRI and HindIII restriction sites in the long and short unique regions of the genome among a series of 20 low-passage HCMV isolates and four widely used high-passage laboratory strains. Mapping was done by hybridizing HCMV restriction fragments with a series of subgenomic cloned fragments of HCMV strain AD169 (Fig. 15.1). In this way we documented where variation has occurred in the HCMV genome and compared patterns of variation among strains in specific regions. We identified restriction sites conserved among all strains studied, and sites and genomic regions that vary among few or many strains.[6]

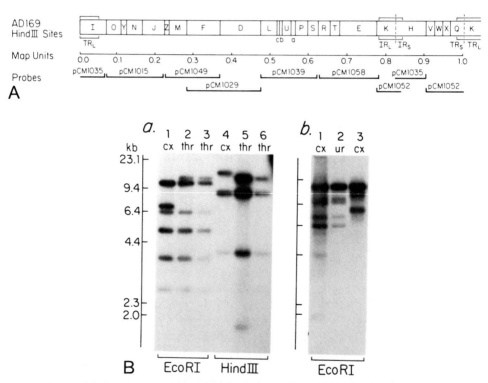

Figure 15.1. **A,** Restriction endonuclease *Hind*III cleavage map of HCMV strain AD169 and subgenomic fragments[68] used as hybridization probes. **B,** (*a*) Viral DNA from cervix (*cx*) and throat (*thr*) isolates cleaved with *Eco*RI or *Hind*III and hybridized with pCM 1015. (*b*) Cervical and urine (*ur*) isolates cleaved with *Eco*RI and hybridized with pCM 1052. Southern blots show variation in isolates by restriction enzyme "fingerprints."

In the study by Chandler *et al.*[5] we examined serial isolates from the cervix, urine, and throat of eight women attending a clinic for sexually transmitted diseases (STD) and seven women receiving routine prenatal care. Isolates were compared by restriction enzyme "fingerprinting" of viral DNA. Four of the eight women from the STD clinic were infected with more than one strain of HCMV. Two women shed different strains in serial isolates, and two women shed different strains simultaneously from different body sites. These results suggest that exogenous reinfection is not uncommon in women with a high probability of exposure to HCMV.

The cervical isolate from patient 1 differed from both the throat isolate obtained on the same date and the subsequent throat isolate. Differences were observed in the regions of the genome hybridizing with the probes pCM1035, pCM1015 (Fig. 15.1*B*, *a*), pCM1039, pCM1058, and pCM1052.

A second patient's cervical and urine isolates were identical, but the cervical isolate obtained five months later was different. Differences were observed when each of the probes shown in Figure 15.1 was used.

The initial urine isolate from patient 3 differed from the cervical isolate obtained on the same date and from all subsequent isolates. She reported sexual contact with three partners in the preceding six months. Urine and throat isolates obtained from her newborn baby were identical to her later isolates. Differences were observed with all probes used.

Patient 4 had two identical cervical isolates one month apart. Four months after her initial visit, a strain of HCMV was recovered from her urine; this isolate differed from the cervical isolates. HCMV was isolated from the urine of her sex partner on two occasions; both isolates were identical to the cervical isolates.

The polymerase chain reaction (PCR) has been used to detect HCMV in tissues or fluids from various sources. For example, using primers synthesized from both immediate-early and late gene regions of the virus, amplified products have been detected in urine from newborn infants[17] and from peripheral blood.[56] Sensitivity and specificity in these diagnostic tests were excellent, and the procedure provides results rapidly. This may be of particular value in the diagnosis of infections in the immunocompromised patient.

VARICELLA-ZOSTER VIRUS

As with HSV and HCMV, varicella-zoster virus (VZV) can reactivate from a latent state to impose a further burden on the immunosuppressed individual.[11] For example, the brain of an AIDS patient who had been admitted to a hospital for evaluation of left hemiparesis and altered mental status 1 month prior to death was studied using *in situ* hybridization.[9] One year before admission the patient had a herpes zoster infection in the dermatome of the ophthalmic branch of the left trigeminal nerve. There was subsequent spread of the zoster infection to the left eye, with resultant keratitis, iritis, and eventual retinitis. This progressed to complete loss of vision in the left eye. In addition there was a loss of visual acuity in the nasal portion of the right eye, thought to represent a postchiasmal lesion. Left hemiparesis (upper and lower extremities and face) developed with total blindness. A CT scan revealed a low-density region in the right posterior internal capsule.

During the next several weeks mental and respiratory status deteriorated, with continued fevers and hypotension until death. The brain revealed extensive encephalitis, most likely due to VZV infection limited to the visual system. Electron microscopy showed evidence of herpesvirus virions. The lesions were centered on the visual pathways extending from the eyes to the calcarine cortex, suggesting anterograde progression of infection (Fig. 15.2). There was an absence of cytomegaly (for HCMV), lack of astrocytic atypia and characteristic oligodendroglial inclusions (for papovavirus, seen in progressive multifocal leukoencephalopathy), and the absence of neural inclusions (seen in subacute sclerosing panencephalitis caused by measles virus). Antibodies to herpes simplex types I and II were not reactive with the intranuclear inclusions.

In situ hybridization experiments were conducted using probes for VZV, HCMV, HSV, and HIV. The results demonstrated VZV nucleic acid sequences in cells within the lesions (Fig. 15.3). The other DNA virus probes did not provide positive results in these lesions.

Figure 15.2. Areas of necrosis (*arrowed*) in left calcarine cortex.

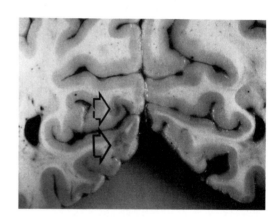

Figure 15.3. *In situ* hybridization with ³H-labeled varicella zoster virus DNA probe in area of necrosis.

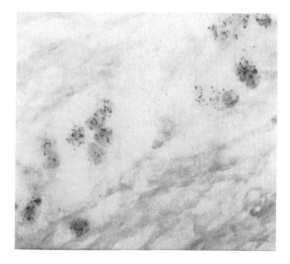

HUMAN IMMUNODEFICIENCY VIRUSES (HIV)

Diagnosis of HIV-1 and HIV-2 currently depends mainly upon serological assays of antibody response to p24, gp41, and/or gp120, which are the *gag* protein, the *env* transmembrane glycoprotein, and the major external glycoprotein of the virion respectively.[1,26,64] The enzyme-linked immunosorbent assay (ELISA), usually confirmed by Western blot, is the most suitable procedure.[63] The antigens for such tests can be prepared by recombinant DNA methods and can be further refined by developing synthetic peptides that react with defined viral epitopes.[25] These methods should eventually provide the increased specificity and sensitivity needed for early, accurate diagnosis. Multiple techniques, including viral transmission studies, viral cultures of cerebrospinal fluid (CSF) and brain, *in situ* hybridization, Southern blot analysis, electron microscopy, analysis of CSF HIV antibody production, and immunohistochemistry have revealed evidence of direct CNS involvement by HIV.[55,65]

We[48] studied autopsied brains from 20 adult patients who expired from AIDS to determine the relationship of human immunodeficiency virus (HIV) infection to white matter lesions and to clinical findings. In four patients with dementia/encephalopathy and abnormalities of the white matter, there was evidence of HIV infection as shown by *in situ* hybridization. In contrast, the remaining 16 patients who had no evidence of white matter degeneration revealed no hybridization to the HIV probe.

Cases 1-4.   In 4 patients the brains revealed abnormalities of the white matter, which could not be attributed to the other conditions discussed above. Two distinctive types of leukoencephalopathy were seen. The first type was present in three (cases 1, 3, and 4) and consisted of patchy white matter pallor, perivascular monocytes and macrophages, multinucleated cells, and gliosis. These lesions were found diffusely in the cerebral and/or cerebellar white matter. The second type of white matter abnormality was vacuolar degeneration of the white matter and was present in four patients (cases 1, 2, 3, and 4). The lesions were characterized by a vacuolar or spongiform change in the white matter, with loss of axons and their myelin sheaths, and a few scattered macrophages and hypertrophic astrocytes. Frequently, small foci of mineralization in the form of both iron and calcium deposits were found within these lesions. Occasional multinucleated giant cells and astrocytes with atypical features were present. These vacuolar changes were found in the following locations: internal capsule (case 1), superficial lateral columns of the cervicothoracic cord and diffusely throughout the cerebral, cerebellar, and brainstem white matter (case 2), frontal white matter (case 3), medullary pyramids and frontal white matter (case 4). No organism was identified, and the lesions were not in patterns or regions suggestive of traumatic or ischemic injury, or wallerian degeneration. One of these brains (case 3) had a simultaneous cryptococcal infection that had no relationship to the white matter lesions.

In situ hybridization demonstrated HIV infection in these 4 brains. The positive areas were within or adjacent to white matter abnormalities, particularly areas of vacuolar degeneration. Nearly all of the HIV-infected cells could be identified with certainty as capillary endothelial cells, perivascular macrophages/monocytes, or rare multinucleated cells. Macrophages and monocytes were easily recognized on HE sections by their plump, granular PAS-positive cytoplasm, oval to round cleaved nuclei, as well as their size and location. Nearly all the macrophages/monocytes that stained positive with the HIV probe were perivascular in location. Rare multinucleated cells also demonstrated HIV infection (see Fig. 15.4). Endothelial cells were likewise easily

Figure 15.4. HIV-infected endothelial cells, macrophages, and multinucleated cells shown by *in situ*
hybridization with ³H-labeled HIV
probe.

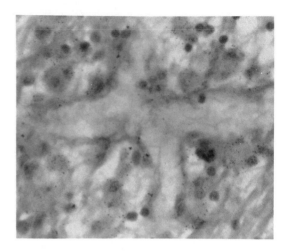

discerned by their "reactive" appearance (*i.e.*, enlarged vesicular nuclei and thinned cytoplasm lining the capillary lumen). Nissl and LFB-PAS-H stains were routinely used
to differentiate between macrophages and neuroglial cells. Neurons, glia, ependyma,
meningeal and endothelial cells of vessels larger in caliber than capillaries failed to
show HIV infection.

VIRAL HEPATITIS

Diagnosis of viral hepatitis is routinely achieved by using a panel of serological
tests for IgM-specific anti–hepatitis A virus (HAV), hepatitis B surface antigen
(HBsAg), and total IgG/IgM hepatitis B core antigen (HBc). The combination of results from these tests allows interpretation of viral status. Recombinant DNA
probes have proven useful adjuncts in a number of situations, *e.g.*, in the detection
of integrated HBV DNA in hepatoma tissue[52,53] and in the quantitative analysis of
HBV DNA in serum, saliva, and semen.[31] *In situ* hybridization has been used to detect viral DNA in formalin-fixed liver specimens, using a biotin-labeled probe, and
the results compared with serological and histological features.[47] As a result of this
comparison it was suggested that the recombinant DNA approach has sufficient
sensitivity and specificity to be of clinical value.

HUMAN PAPILLOMAVIRUS

Recombinant DNA methodology has enabled the cloning and sequencing of
HPV deoxyribonucleic acids (DNA)[39] so that we can now recognize multiple types
of HPV, which exhibit a range of tissue tropisms and clinical manifestations. More
than 50 distinct types of human papillomaviruses have been described, each of
which shares less than 50% overall DNA homology with other types, using liquid
hybridization assays.[10] Subtypes are defined as isolates that are closely defined by
hybridization.

Human papillomaviruses infect epithelial cells of skin and mucosal surfaces.
Virus is rarely, if ever, demonstrated in basal cells, but viral DNA can be detected in

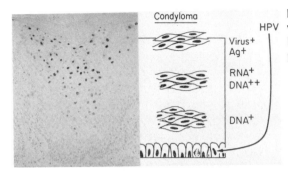

Condyloma

HPV

Virus+
Ag+

RNA+
DNA++

DNA+

Figure 15.5. *In situ* hybridization with biotin-labeled HPV-6 DNA. Detection of viral DNA increases in parallel with cell differentiation.

suprabasal cells by *in situ* hybridization[24] and in the successive cell layers (Fig. 15.5). Viral RNA expression is generally detectable in the strata spinosum and granulosum,[59] and viral protein is detected in the upper layers, *i.e.*, strata granulosum and corneum.[38,54] Thus virus replication can be seen to depend on the host gene expression program of the differentiating epithelial cells. Mature virions are only found in the superficial layers of cells. This sequence of viral and cellular events can lead to a spectrum of pathomorphology from the common wart to invasive carcinoma. Much current research is designed to determine the extent of the role of HPV in the conversion of cells to malignancy and to assess the contribution of other factors in HPV oncogenesis.[66]

HPV GENOME

The genome of papillomaviruses is a double-stranded circular DNA molecule of approximately 7900 bp. The overall genetic organization is very similar among all the types, both human and animal. Among the types whose complete DNA sequence has been determined, a series of conserved open reading frames (ORFs) have been found on one strand of the DNA, and the other strand apparently is noncoding.[7,15,51] Consistent with this observation is the finding that all viral RNAs detected to date are complementary to only one strand of DNA. By analogy to SV40 and polyoma, those ORFs that are expressed in transformed cells are called "early" (E), whereas those ORFs only expressed in productive infections are called "late" (L). The former are presumed to be involved in viral replication and/or transformation, and the latter in virus maturation and assembly. As many as eight ORFs have been described in the early region (E1-E8), and two ORFs in the late region (L1 and L2). Between the 3' end of the late region ORFs and the 5' end of the early region ORFs is located a noncoding region (NCR), alternatively called the upstream regulatory region (URR) or the long-control region (LCR). The NCR contains a number of transcriptional and replicative regulatory elements.

DNA sequence comparisons have shown a greater degree of homology between some HPV types than was measured by reassociation kinetics, *e.g.*, HPV6b and HPV11 share 82% homology overall.[16] The extent of DNA homology is not spread evenly throughout the genome. The E1, L2, and L1 ORFs are the most highly conserved; the NCR and the region between E2 and L2, are the most divergent.

HPV DIAGNOSIS

Although HPV capsid antigen (L1 ORF) can be detected in cells and tissue sections (Fig. 15.6) using antiserum against the genus-specific antigen of papillomaviruses,[34] this does not distinguish virus types, and the antigen may not be expressed in all lesions or tumors associated with HPV. Definitive diagnostic results can be obtained by various molecular hybridization methods, *i.e.*, Southern blotting,[58] *in situ* hybridization,[24,35] dot blot,[8] and Northern blotting.[62] By these means it can be shown that HPV types have limited site-specificity and differ in their association with benign or malignant disease.[12,33,67]

Although nitrocellulose or similar substrate blotting procedures have provided the most accurate and sensitive method for detecting and characterizing viral nucleic acid sequences in tumors and experimentally transformed cells[36] (Fig. 15.7), recent improvements in cytological hybridization methods allow rapid detection of virus and analysis of HPV type directly in biopsied tissue (Fig. 15.8) and in cervical smears.[14,27,42,59] In particular, these *in situ* hybridization procedures facilitate retrospective studies of stored specimens.[42]

Dot blot and Southern hybridization have been employed in conjunction with cytology to determine the prevalence of HPV, for example in groups having different risks of exposure to genital HPV infection.[37]

The polymerase chain reaction (PCR) technique[50] is proving to be of great value in epidemiologic studies of the prevalence of HPV infection and will have a significant impact on the diagnosis of many infectious diseases.

This is a technique in which small amounts of DNA can be amplified and then identified by gel analysis and/or by Southern transfer hybridization. Typically, the double–stranded DNA is heat-denatured, and primers are annealed at low temperature and extended at an intermediate temperature using thermostable DNA polymerase. PCR is based on the repetition of such cycles of denaturing, annealing, and extension of DNA, and can yield a theoretical amplification efficiency of 2^n, which for 40 cycles is 10^{12}, but which in practice usually results in amplification between 10^6 and 10^8.

We have compared Southern filter hybridizations and the polymerase chain

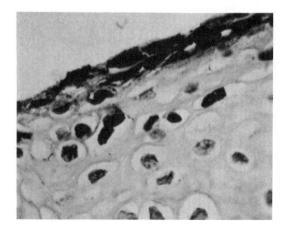

Figure 15.6. Immunoperoxidase staining showing papillomavirus genus-specific antigen in koilocytotic nuclei.

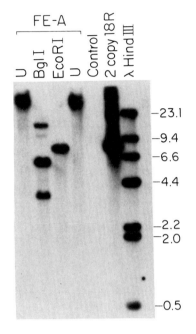

Figure 15.7. Southern blot of DNA from HPV-18 immortalized human keratinocytes. DNA extracted from FE-A cell line either undigested (*U*) or cleaved with *Bgl*I or *Eco*RI. Control is normal human keratinocyte DNA.

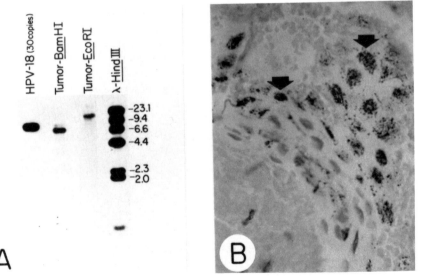

Figure 15.8. **A,** Southern blot of cervical carcinoma DNA cleaved with *Bam*HI or *Eco*RI and hybridized with HPV-18 ^{32}P-DNA. **B,** Biotin-labeled HPV-6 DNA hybridized *in situ* to adenosquamous carcinoma.

reaction (PCR) in detection of HPV-6 or HPV-16 DNA in cervical specimens. Southern filter hybridizations were performed on total DNAs extracted from exfoliated cells. They were carried out under stringent hybridization conditions with either ^{32}P-labeled HPV-6 or HPV-16 DNA as the probe. PCRs were done with 25-mer primer pairs specific for the HPV-6 or HPV-16 E6/E7 region with amplification for 35 cycles, followed by Southern filter hybridizations of amplified DNA using an internal oligonucleotide (40-mer) endlabeled with γ ^{32}P-ATP as the probe. Pap smears were prepared for PCR by extraction of DNA from cells removed from the slides. Biopsies were deparaffinized and boiled, and the PCR was performed directly on that material. Preliminary analyses of specimens from 48 women showed a sensitivity of 66% (comparing Southerns on exfoliated cells to PCRs on Pap smears). When comparing Southerns to PCRs on either Pap smears or biopsies, the sensitivity was 39.5%. In each instance, the specificity of the PCR was 100%.

IMMUNE RESPONSE TO HPV

In most viral infections antibody is detectable in serum between 3 and 12 days after the onset of infection. Indeed antibody to a wart virus in human sera was first reported by Almeida and Goffe in 1965.[2] Several studies have examined the prevalence of HPV-reactive antibodies among human subjects with plantar warts (usually associated with HPV-1) or common skin warts (usually associated with HPV-2). Undisrupted HPV virions prepared from pooled skin wart biopsies were used as targets. Using type-specified HPV-1 virions as targets in a radioimmunoassay, Pfister and Zur Hausen[46] found that 52 of 110 patients with plantar warts and/or common skin warts had detectable HPV-1-reactive antibodies. Similar antibody prevalences were observed in age-matched populations selected without regard for history of plantar warts or skin warts. The authors concluded that HPV-1 infections induced low levels of antibodies, which were detectable in approximately 50% of wart patient sera, and that the similar antibody prevalences in nonselected subjects reflected a high rate of asymptomatic infection in those populations.

To determine which viral antigens of HPV-6 and HPV-16 elicited immune response, all of the open reading frames were expressed as bacterial fusion proteins using a trpE vector[22] and were used as targets in Western blot assays to detect serum antibodies.[30] Serum specimens were obtained from patients with genital warts (condylomata acuminata) and their sexual partners. Reactivities were detected against the HPV-6 L1 fusion protein in 37 of 75 sera, against HPV-6 L2 in 14 of 75 sera, and against HPV-6 E2 in 3 of 75 sera; no reactivities were seen with other HPV-6 constructs. The antibodies directed against the HPV-6b fusion protein showed no cross-reactivity with comparable regions of the HPV-16 ORFs.

The location of the epitope on the HPV-6 capsid antigen (L1 ORF) that is recognized by human sera has been mapped to a 21 amino acid segment using a nested set of unidirectional 3' to 5' deleted expression plasmids.[32] The type-common epitope on the HPV-6 capsid antigen recognized by antisera raised against SDS-disrupted BPV-1 virions (Dako, Inc) has been mapped to a distinct site on the HPV-6 L1 ORF. These results are summarized in Jenison et al.[32]

As an initial assessment of which of the HPV-16-encoded fusion proteins are immunoreactive, 53 sera were obtained from women attending a sexually transmitted diseases clinic. When tested against the fusion proteins in Western blot assays, 1 of 53 was positive for L1, 23 of 53 for L2, 8 of 53 for E7 and 5 of 53 for E4. No reactivities to the E1, E2, or E6 fusion proteins were seen (Jenison, Yu, Valentine, and Galloway, unpublished results).

There is reason to be optimistic that these strategies will lead to the development of serologic assays for specific HPV infections. At this time it is unclear what the significance of humoral antibody response to various viral antigens might be, and whether humoral antibodies play a role in the regression of HPV infection.

HPV AND DISEASE

Genital warts are the most commonly diagnosed viral sexually transmitted disease in the U.S.A.[3] Data from the National Disease and Therapeutic Index suggest an increase in cases of condyloma accuminata of 4- to 5-fold between 1966 and 1984, with most of that increase occurring before 1976. In a study from Rochester, Minn., Chuang et al.[9] reported an 8-fold increase in condylomata between 1950 and 1978. An increase in the incidence of genital warts has been recorded in Britain,[18] with a 2.5-fold increase reported. Of the four most frequently detected HPV types in genital lesions, types 6 and 11 have been associated mainly with benign lesions and types 16 and 18 with high-grade neoplasia and carcinoma. Munoz et al.[44] analyzed data from 20 studies of the prevalence of infection with HPV types 6, 11, 16, or 18, and while pointing out the deficiencies of the studies in epidemiological terms, showed an increase in type 16/18 positivity in invasive squamous carcinoma over that found in CIN lesions or the normal cervix. Types 6 and 11 decreased from an overall prevalence of 18% in CIN to <1% in invasive cancer. The data accumulated by Munoz et al.[44] also show an overall prevalence of 15% for types 16 and 18 in the normal cervix. In a large study with over 9000 patients, de Villiers et al.[19] found about 10% of women with normal cytological smears to be HPV positive, the majority hybridizing with types 16/18. In a combined molecular hybridization and histological analysis of cervical biopsies,[13] it was shown that HPV-16 infection correlated with morphological aberrations found in lesions with a high risk of progression and in carcinomas. In the same study, HPV 6/11 were identified only in condylomas, and infection with these types was characterized by koilocytotic atypia. In contrast, a study by Kiviat et al.[37] showed an equal prevalence of koilocytes among women with HPV 6/11 and those with HPV 16/18 infections.

Current evidence suggests a long latency between acquisition of genital papillomavirus infection and development of malignancy, with infection commonly occurring during the second decade, and cervical cancer from the fifth decade.[19] Data from one study indicate that invasive cancer is no more aggressive in younger than in older women,[49] but it is not yet known whether a younger age of acquisition of HPV is associated with an increased risk for developing anogenital tumors. An important observation is that cancers of the vulva, vagina, penis, and anus occur less frequently than cervical cancer. Despite the fact that HPV 16 or 18

DNA is detected in a similar proportion of these tumors, the incidence of cervical cancer is significantly greater than that of the other tumors. Similar disparities in rates of cervical versus penile cancer have been reported in all countries studied to date.[60] Since cervical cancer arises in areas of columnar epithelium that naturally undergo metaplasia,[57] it is possible that these cells are more likely to be transformed than are cells of the stratified squamous epithelium.

The ability of papillomaviruses to induce benign cellular proliferations both in animals and humans has been demonstrated and accepted for many years. Conversion of cells in virus-induced papillomas to a malignant phenotype has been observed and tested experimentally, and it has been shown that papillomavirus nucleic acid sequences and virus-specified proteins usually persist in malignant cells. In the case of human disease, extensive studies on anogenital cancer and epidermodysplasia verruciformis have identified subsets of HPV types which have tropism for particular tissues and which can be further subdivided based upon the frequency with which individual types are associated with progression to malignant disease.

That genital cancer may result from a synergistic interaction between virus infection and initiating events (which could also be mediated by a second virus infection) has been hypothesized by Zur Hausen[66] and, from a different bias, by Fenoglio *et al.*[21] This is consistent with the data from animal tumors[29,61] and other human carcinomas.[20,40,45] Further development of the hypothesis led Zur Hausen[68] to propose that the factors interacting with HPV may modify cellular genes that would otherwise exert control over viral expression and thereby tumor progression. This could explain why carcinomas develop in only a small number of individuals, even though there is a high prevalence of HPV infection.

The recognition of multiple types of human papillomaviruses has resulted in remarkable progress in the detection of persisting viral nucleic acid sequences in carcinomas.

A preferential association of some types with benign lesions, while others may be frequently found in malignant tumors, has been observed. HPV types 5 and 8 in epidermodysplasia verruciformis patients and types 16, 18, 31, 33, etc. in genital lesions are most frequently associated with progression to malignancy, whereas other types (*e.g.*, HPV-6, -10, -11, -20) are regularly identified in benign warts. Such distinctions are not absolute, but they provide the initial steps toward establishing a causal role for some human papillomaviruses in carcinomas. The data from human and animal studies indicate that papillomaviruses contribute significantly to the development of many, if not all, carcinomas, but we do not yet have a clear understanding of the importance of other interacting viral, chemical, or cellular factors. The application of gene cloning and nonstringent hybridization[39] has provided us with an apparently ever-increasing catalogue of human papillomaviruses. More effort is now required to establish their prevalence, the natural history of infection, and the mechanism of neoplastic transformation.

Conclusions

The development of new technology based upon recombinant DNA methodology has opened up new vistas in disease diagnosis and experimental pathology.

Infectious diseases, particularly of viral etiology, provide excellent examples of the power of these techniques, when compared to many of the routine diagnostic methods which can be time-consuming and expensive.

There are many commercially available probes and test systems which utilize nonisotopic label for detection and can be used without the need for highly sophisticated laboratory facilities. The development of the polymerase chain reaction to amplify target sequences and provide increased sensitivity and rapid results opens up new possibilities in a rapidly developing area that will impact all aspects of pathology.

Acknowledgments

The studies reported from this laboratory were done in collaboration with Drs. D. A. Galloway, A. M. Beckmann, S. H. Chandler, and S. W. Rostad. I thank Drs. H. zur Hausen, B. Fleckenstein, and D. Dina for viral probes, also M. Wright for manuscript preparation. Supported by PHS grants CA 29350 and CA 42792 awarded by NCI.

REFERENCES

1. Allan, I. S., Coligan, J. E., Barin, F., *et al.* Major glycoprotein antigens that induce antibodies in AIDS patients are encoded by HTLV III. *Science (Wash DC) 228*:1091–1094, 1985.
2. Almeida, J. D. and Goffe, A. P. Antibody to wart virus in human sera demonstrated by electron microscopy and precipitin tests. *Lancet 2*:1205–1207, 1965.
3. Becker, T. M., Stone, K. M., and Alexander, E. R. Genital human papillomavirus infection: A growing concern. *Obstet. Gynecol. Clin. N. Am. 14*:389–396, 1987.
4. Buchman, T. G., Roizman, B., Adams, G., and Stover, B. H. Restriction endonuclease fingerprinting of herpes simplex virus DNA: A novel epidemiological tool applied to a nosocomial outbreak. *J. Infect. Dis. 138*:488–498, 1978.
5. Chandler, S. H., Handsfield, H. H., and McDougall, J. K. Isolation of multiple strains of cytomegalovirus from women attending a clinic for sexually transmitted diseases. *J. Inf. Dis. 155*:655–660, 1987.
6. Chandler, S. H. and McDougall, J. K. Comparisons of restriction site polymorphisms among clinical isolates and laboratory strains of human cytomegalovirus. *J. Gen. Virol. 67*:2179–2192, 1986.
7. Chen, E. Y., Howley, P. M., Levinson, A. D., and Seeburg, P. H. The primary structure and genetic organization of the bovine papillomavirus type 1 genome. *Nature (Lond) 299*:529–534, 1982.
8. Chomczynski, P. and Qasba, P. Alkaline transfer of DNA to plastic membrane. *Biochem. Biophys. Res. Commun. 122*:340–344, 1984.
9. Chuang, T. Y., Perry, H. O., Kurland, L. T., and Ilstrup, D. M. Condyloma acuminatum in Rochester, Minn, 1950–1978. Parts I and II. *Arch. Dermatol. 120*:469–483, 1984.
10. Coggin, J. R. and Zur Hausen, H. Workshop on papillomaviruses and cancer. *Cancer Res. 39*:545–546, 1979.
11. Cole, E. L., Meisler, D. M., and Calabrese, L. H. Herpes zoster ophthalmicus and acquired immune deficiency syndrome. *Arch. Ophthalmol. 102*:1027–1029, 1984.
12. Crum, C. P. and Levine, R. U. Human papillomavirus infection and cervical neoplasia: New perspectives. *Int. J. Gynecol. Pathol. 3*:376–388, 1984.
13. Crum, C. P., Mitao, M., Levine, R. U., and Silverstein, S. Cervical papillomaviruses segregate within morphologically distinct precancerous lesions. *J. Virol. 54*:675–681, 1985.
14. Crum, C. P., Nagai, N., Levine, R. U., and Silverstein, S. In situ hybridization analysis of HPV-16 DNA sequences in early cervical neoplasia. *Am. J. Pathol. 123*:174–182, 1986.
15. Danos, O., Georges, E., Orth, G., and Yaniv, M. Fine structure of the Cottontail Rabbit

papillomavirus mRNAs expressed in the transplantable VX2 carcinoma. *J. Virol.* 53:735–741, 1985.

16. Dartmann, K., Schwarz, E., Gissmann, L., and Zur Hausen, H. The nucleotide sequence and genome organization of human papillomavirus type II. *Virology* 151:124–130, 1986.
17. Demmler, G. J., Buffone, G. J., Schimbor, C. M., and May, R. A. Detection of cytomegalovirus in urine from newborns by using polymerase chain reaction DNA amplification. *J. Infect. Dis.* 158:1177–1184, 1988.
18. Department of Health and Social Security: Annual Reports of the Chief Medical Office. *Genitourinary Med.* 1985.
19. de Villiers, E. M., Wagner, D., Schneider, A., *et al.* Human papillomavirus infections in women with and without abnormal cervical cytology. *Lancet* 2:703–706, 1987.
20. Duff, T. B. Laryngeal papillomatosis. *J. Laryngol. Otol.* 85:947–956, 1971.
21. Fenoglio, C. M., Galloway, D. A., Crum, C. P., Levine, R. U., Richart, R. M., and McDougall, J. K. Herpes simplex virus and cervical neoplasia. *Prog. Surg. Pathol.* 4:45–82, 1982.
22. Firzlaff, J. M., Hsia, C. N. L., Halbert, C. P. H. L., Jenison, S. A., and Galloway, D. A. Polyclonal antibodies to human papillomavirus type 6b and type 16 bacterially derived fusion proteins. In: Papillomaviruses, edited by B. M. Steinberg, J. L. Brandsma, and L. B. Taichman. Cancer Cells 5. New York, Cold Spring Harbor Laboratory, 1987:105–113.
23. Fleckenstein, B., Muller, I., and Collins, J. Cloning of the complete human cytomegalovirus genome in cosmids. *Gene* 18:39–46, 1982.
24. Gall, J. G. and Pardue, M. L. Formation and detection of RNA-DNA hybrid molecules in cytological preparations. *Proc. Natl. Acad. Sci. U.S.A.* 63:378–382, 1969.
25. Gnann, J. W., Schwimmbeck, P. L., Nelson, J. A., Truax, A. B., and Oldstone, M. B. A. Diagnosis of AIDS using a 12-amino acid peptide representing an immunodominant epitope of the human immunodeficiency virus. *J. Infect. Dis.* 156:261–267, 1987.
26. Groopman, J. E., Chen, F. W., Hope, J. A. *et al.* Serological characterization of HTLV III infection in AIDS and related disorders. *J. Infect. Dis.* 153:736–742, 1986.
27. Gupta, J., Gendelman, H. E., Naghashfar, Z., *et al.* Specific identification of human papillomavirus type in cervical smears and paraffin sections by in situ hybridization with radioactive probes. *Int. J. Gynecol. Pathol.* 4:211–218, 1985.
28. Handsfield, H. H., Chandler, S. H., Caine, V. A. *et al.* Cytomegalovirus infection in sex partners: Evidence for sexual transmission. *J. Infect. Dis.* 151:344–348, 1985.
29. Jarrett, W. F. H., McNeill, P. E., Laird, H. M., *et al.* Papilloma viruses in benign and malignant tumors of cattle. *Lancet* 2:703–706, 1987.
30. Jenison, S. A., Firzlaff, J. M., Langenberg, A., and Galloway, D. A. Identification of immunoreactive antigens of human papillomavirus type 6b by using *Eschericia coli*-expressed fusion proteins. *J. Virol.* 62:2115–2123, 1988.
31. Jenison, S. A., Lemon, S. M., Baker, L. N., and Newbold, J. E. Quantitative analysis of hepatitis B virus DNA in saliva and semen of chronically infected homosexual men. *J. Infect. Dis.* 156:299–307, 1987.
32. Jenison, S. A., Yu X.-P., Valentine, J., and Galloway, D. A. Human antibodies react with an epitope of the human papillomavirus type 6b L1 open reading frame which is distinct from the type-common epitope. *J. Virol.* 63:809–818, 1989.
33. Jenson, A. B., Lim, L. Y., and Lancaster, W. D. Role of papillomavirus in proliferative squamous lesions. *Surv. Synth. Pathol. Res.* 4:8–13, 1985.
34. Jenson, A. B., Rosenthal, J. R., Olson, C., Pass, F., Lancaster, W. D., and Shah, K. Immunologic relatedness of papillomaviruses from different species. *J. Natl. Cancer Inst.* 64:495–500, 1980.
35. John, H. L., Birnsteil, M. L., and Jones, K. W. RNA-DNA hybrids at the cytological level. *Nature (Lond)* 223:582–587, 1969.
36. Kaur, P. and McDougall, J. K. Characterization of primary human keratinocytes transformed by human papillomavirus type 18. *J. Virol.* 62:1917–1924, 1988.
37. Kiviat, N. B., Koutsky, L. A., Paavonen, J. A., *et al.* Prevalence of genital papillomavirus

infection among women attending a college student health clinic or a sexually transmitted disease clinic. *J. Infect. Dis.* 159:293–301, 1989.

38. Kurman, R. J., Sanz, L. E., Jenson, A. B., Perry, S., and Lancaster, W. D. Papillomavirus infection of the cervix 1. Correlation of histology with viral structural antigens and DNA sequences. *Int. J. Gynecol. Pathol.* 1:17–28, 1982.

39. Law, M. F., Lancaster, W. D., and Howley, P. M. Conserved polynucleotide sequences among the genomes of papillomaviruses. *J. Virol.* 32:199–211, 1979.

40. Le Jeune, F. E. The story of Warren Bell. *Ann. Otol. Rhinol. Laryngol.* 50:905–908, 1941.

41. Lonsdale, D. M. A rapid technique for distinguishing herpes simplex virus type 1 from type 2 by restriction-enzyme technology. *Lancet* 1:849–852, 1979.

42. McDougall, J. K., Beckmann, A. M., and Kiviat, N. B. Methods for diagnosing papillomavirus infection. In: *Papillomaviruses*, edited by D. Evered and S. Clark, London, CIBA Found. Symp. 120:86–103, 1986.

43. McDougall, J. K., Rostad, S. W., Shaw, C. M., Olson, K., Alvord, E. C., and Buonaguro, F. M. HIV and herpesviruses in immunodeficiency and cancer. *Aids and Associated Cancers in Africa*, edited by G. Giraldo, Basel, Karger, 1988, pp. 260–271.

44. Munoz, N., Bosch, X., and Kaldor, J. M. Does human papillomavirus cause cervical cancer? The state of the epidemiological evidence. *Br. J. Cancer* 57:1–5, 1988.

45. Orth, G., Favre, M., Breitburd, F., *et al.* Epidermodysplasia verruciformis: A model for the role of papilloma viruses in human cancer. Cold Spring Harbor Conf Cell Prolif 7:259–282, 1980.

46. Pfister, H. and Zur Hausen, H. Characterization of proteins of human papilloma viruses (HPV) and antibody response to HPV-1. *Med. Microbiol. Immunol.* 166:13–19, 1978.

47. Rijntjes, P. J. M., Van Ditzhuijsen, T. J. M., Van Loon, A. M., Van Haelst, U. J. G. M., Bronkhorst, F. B., and Yap, S. H. Hepatitis B virus DNA detected in formalin-fixed liver specimens and its relation to serologic markers and histopathologic features in chronic liver disease. *Am. J. Pathol.* 120:411–418, 1985.

48. Rostad, S. W., Sumi, S. M., Shaw, C. M., Olson, K., and McDougall, J. K. Human immunodeficiency virus (HIV) infection in brains with AIDS-related leukoencephalopathy. *AIDS Res. Human Retrov.* 3:363–373, 1987.

49. Russell, J. M., Blair, V., and Hunter, R. D. Cervical carcinoma: Prognosis in younger patients. *Br. Med. J.* 295:300–303, 1987.

50. Saiki, R. K., Bugawan, T. L., Horn, G. T., Mullis, K. B., and Erlich, H. A. Analysis of enzymatically amplified β-globin and HLA-DQα DNA with allele-specific oligonucleotide probes. *Nature (Lond)* 324:163–166, 1986.

51. Schwarz, E., Durst, M., Demankowski, C., *et al.* DNA sequence and genome organization of genital human papillomavirus type 6B. *EMBO J.* 2:2341–2348, 1983.

52. Shafritz, D. A. and Kew, M. C. Identification of integrated hepatitis B virus DNA sequences in human hepatocellular carcinomas. *Hepatology* 1:1–8, 1981.

53. Shafritz, D. A., Shouval, D., and Sherman, H. J. Integration of hepatitis B virus DNA into the genome of liver cells in chronic liver disease and hepatocellular carcinoma. *N. Engl. J. Med.* 305:1067–1073, 1981.

54. Shah, K. H., Lewis, M. G., Jenson, A. B., Kurman, R. J., and Lancaster, W. D. Papillomavirus and cervical dysplasia. *Lancet* 2:1190–1192, 1980.

55. Shaw, G. M., Harper, M. E., Hahn, B. H., *et al.* HTLV III infection in brains of children and adults with AIDS encephalopathy. *Science (Wash DC)* 227:177–182, 1985.

56. Shibata, D., Martin, W. J., Appleman, M. D., Causey, D. M., Leedom, J. M., and Armheim, N. Detection of cytomegalovirus DNA in peripheral blood of patients infected with human immunodeficiency virus. *J. Infect. Dis.* 158:1185–1192, 1988.

57. Singer, A. The uterine cervix from adolescence to the menopause. *Br. J. Obstet. Gynecol.* 82:81–89, 1975.

58. Southern, E. M. Detection of specific sequences among DNA fragments separated by gel electrophoresis. *J. Mol. Biol.* 98:503–517, 1975.

59. Stoler, M. H. and Broker, T. R. In situ hybridization detection of human papillomavirus

DNAs and messenger RNAs in genital condylomas and a cervical carcinoma. *Hum. Pathol.* 17:1250–1258, 1986.

60. Syrjanen, K. J. Biology of human papillomavirus (HPV) infections and their role in squamous cell carcinogenesis. *Med. Biol.* 65:21–29, 1987.

61. Syverton, J. T. The pathogenesis of the rabbit papilloma-to-carcinoma sequence. *Ann. N.Y. Acad. Sci.* 54:1126–1137, 1952.

62. Thomas, P. Hybridization of denatured RNA and small DNA fragments transferred to nitrocellulose. *Proc. Natl. Acad. Sci. U.S.A.* 77:5201–5205, 1980.

63. Veronese, F. D. M., Copeland, T. D., De Vico, A. L., *et al.* Characterization of highly immunogenic p66/p51 as the reverse transcriptase of HTLV III/LAV. *Science (Wash DC)* 231:1289–1291, 1986.

64. Veronese, F. D. M., De Vico, A. L., Copeland, T. D., Oroszlan, S., Gallo, R. C., and Sarngadharan, M. G. Characterization of gp 41 as the transmembrane protein coded by the HTLV III/LAV envelope gene. *Science (Wash DC)* 229:1402–1405, 1985.

65. Wiley, C. A., Schrier, R. D., Nelson, J. A., Lampert, P. W., and Oldstone, M. B. A. Cellular localization of human immunodeficiency virus infection within the brains of acquired immune deficiency syndrome patients. *Proc. Natl. Acad. Sci. U.S.A.* 83:7089–7093, 1986.

66. Zur Hausen, H. Human genital cancer: Synergism between two virus infections or synergism between a virus infection and initiating events. *Lancet* 2:1370–1373, 1982.

67. Zur Hausen, H. Genital papillomavirus infections. *Prog. Med. Virol.* 32:15–21, 1985.

68. Zur Hausen, H. Intracellular surveillance of persisting viral infections. *Lancet* 2:489–491, 1986.

16

Application of Molecular Techniques to Forensic Pathology

Ross E. Zumwalt

Introduction

The tremendous potential for the use of molecular biological techniques in solving crimes was first recognized and used by Jeffreys.[18] In 1985 he had developed DNA probes that recognized hypervariable regions of human DNA. Using the probes and standard molecular biology techniques he was able to make DNA "fingerprints" unique to individuals. Recognizing the potential to conclusively match a biological stain to an individual he agreed to make a DNA fingerprint from semen DNA recovered from a rape murder victim and try to match it to DNA fingerprints from a large population of men in the community where the murder had occurred. During the process he was able to exonerate a mentally retarded suspect who had confessed to the crime[13] and eventually to match the fingerprint to the DNA of the perpetrator.[28]

The ability to make a DNA fingerprint for forensic use was the outgrowth of intensive research in molecular biology that was directed not toward forensic issues, but toward identification of genetic markers for specific genes.[10] These molecular techniques were actively and successfully being used in the study of inherited and infectious diseases and malignancies,[8,39] as discussed in the other chapters in this book.

DNA Fingerprinting Background and Techniques

The genetic information of all organisms is coded in the sequence of the nucleic acid building blocks of DNA (or RNA). The order in which the four purine or pyrimidine bases, adenosine, thymine, cytosine, and guanine (ATCG) occur in the DNA provides the information required to construct the necessary proteins that form and regulate the organism. The amount and order of DNA is different for each species and for individuals within most species, such as man.

The human genome is complex and enormous, arranged in 46 separate packages or chromosomes collectively containing approximately three billion bases. Many segments of the genome are responsible for coding for specific proteins and are called genes. The base pair coding sequences of genes (the exons) are usually constant among individuals of a species. Indeed, alterations in the sequence of bases in a gene may produce disease. In humans, disease such as sickle cell anemia and Duchenne's muscular dystrophy are the result of alterations in base sequence.

In their search for the genetic basis for diseases, scientists have developed refined techniques to isolate, purify, and sequence specific areas of the human and nonhuman genomes.[5,35]

Through this intensive research, scientists have found that genes make up only a small proportion of the human genome; much of the genome does not specifically code for proteins. The function of this "junk" or noncoding DNA is not well understood. It is in these noncoding areas that most of the variations in sequence of DNA bases occurs.

The overall heterozygosity of the human genome is quite low with polymorphisms or variations in the base sequence averaging approximately one per 500 bases.[6] Of course the variable sites are not uniformly distributed along the chromosomes. Some areas are hypervariable with many differences between individuals.

Polymorphisms may be the result of a single base substitution, deletion, or addition. Other polymorphisms are more complex, such as segmental deletions, additions, inversions, or duplications (Fig. 16.1). In many instances polymorphisms are the result of variable numbers of tandem repeats (VNTRs) (Fig. 16.2).[17,30,37,40,41,43]

C−A−T−A−**G**−C−G   C−A−T−A−**G**−C−G

C−A−T−A− **T**−C−G   C−A−T−A−C−G

a. substitution   b. deletion

C−A−T−A−G−C−G   C−A−**T**−A−**G**−C−G

C−A−T−A−G−**G**−C−G   C−A−**G**−A−**T**−C−G

c. addition   d. inversion

Figure 16.1. Types of base sequence variations in DNA. *a. Substitution:* A guanine base (G) of the DNA sequence of the individual represented on the top line is replaced by a thymine base (T) in the sequence of the individual on the bottom line. *b. Deletion:* A guanine base (G) of an individual is not present in the sequence of another. *c. Addition:* There is an additional guanine base (G) present in a sequence in an individual which is not present in the standard sequence. *d. Inversion:* A sequence of 3 DNA base pairs is inverted and inserted into an area of DNA in reverse order. (Inversions may be many base pairs long.)

C−C− T−A−G−A−C−A−T− T−A−G−A−C−A−T− T−A−G−A−C−A−T− C−C

C−C− T−A−G−A−C−A−T− T−A−G−A−C−A−T− T−A−G−A−C−A−T−

T−A−G−A−C−A−T− T−A−G−A−C−A−T− C−C

Figure 16.2. Variable number of tandem repeats (VNTRs). Many areas of the genome have repeated sequences following one after another. In the individual represented by the *top line,* the 7-base sequence T-A-G-A-C-A-T is repeated three times. In the individual represented by the *bottom line* the same 7-base sequence is repeated 5 times.

For example, an individual might have 15 repeats of a DNA segment in a specific VNTR area on one chromosome of a pair and five repeats on the other. Another individual might have ten repeats and twenty repeats for the same VNTR area. Many such polymorphic areas have been identified, and the number of different repeats documented in the human population exceeds 100 for many of them.[3] The identification and documentation of VNTRs polymorphisms were used by forensic scientists to create DNA fingerprints unique to individuals, and they are being used in practical forensic situations.[12,20]

Polymorphisms are identified by restriction endonucleases, enzymes that cut DNA into segments at specific base sequences. If a variation in DNA sequence between individuals results in an addition or loss of a restriction endonuclease site, the result will be DNA segments of different lengths from different individuals for certain areas of the genome. The different segment lengths are known as restriction fragment length polymorphisms (RFLPs). Furthermore a restriction endonuclease that cuts DNA at the ends of a VNTR, but not within the VNTR will produce DNA segments of different lengths from different individuals. The RFLPs are detected by the standard molecular biology techniques of gel electrophoresis, Southern blot transfer, and hybridization with oligonucleotide probes specific for the sequences or repeats (Fig. 16.3).

The first DNA fingerprints were produced by Jeffreys using multilocus VNTR probes that identified homologous VNTRs on many chromosomes,[18] which he called minisatellites. Jeffreys' probes contained core sequences of base pairs that were similar, but not identical, to the base sequences of tandem repeats found at many different areas of the genome. The probes were similar enough to the differ-

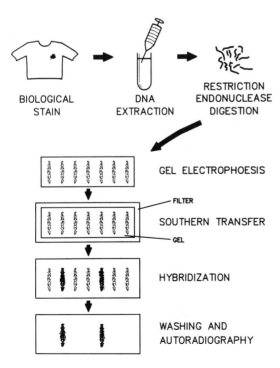

Figure 16.3. DNA profile, Southern blot technique. A biological stain or tissue of potential forensic importance is identified and isolated. High molecular weight DNA is extracted and digested with restriction endonucleases, resulting in DNA fragments of varying lengths. The DNA fragments are subjected to gel electrophoresis, which separates them according to size. A buffer is passed through the gel by capillary action to transfer the DNA to a nitrocellulose filter (Southern transfer). Hybridization is accomplished by exposing a radioactive probe to the filter. The filter is washed to remove nonspecifically bound probe and exposed to x-ray film. The resulting banded autoradiograph becomes the DNA fingerprint or profile.

ent tandem sequences to hybridize with them under low-stringency hybridization conditions. These multilocus VNTR probes produced highly sensitive, but complex, multibanded DNA fingerprints (Fig. 16.4).[20] The chance of two individuals having the same banding pattern, or fingerprint, was less than the world's population.[12] The only exception is identical twins, who do have identical DNA and therefore identical DNA fingerprints.

Subsequently, single locus VNTR probes were developed that identified polymorphic tandem repeat sequences found in only one location in the genome. Using these probes on an individual's DNA (processed as before), resulted in the identification of two bands per individual, based on the two alleles at that chromosome location.[17] The two bands were highly specific for the individual, but clearly not as sensitive in eliminating other individuals as the multibanded fingerprint from a multilocus probe. However, the DNA could be probed with several different single-locus VNTR probes, yielding two additional bands per probe.[4,41] Then, a mathematical calculation using the known numbers of alleles and their frequency at each VNTRs locus could determine the probability of another individual matching all of the bands from the several probes.[3]

For practical purposes, the multilocus VNTR probes that produce a highly sensitive, densely banded pattern or fingerprint began being used for paternity testing, immigration disputes, and other questions of relatedness.[19] In contrast, the single-locus probes with their high specificity were used in criminal cases to profile a biological stain or tissue for excluding or identifying a suspect.[28]

Polymorphic areas in the DNA other than VNTR areas can be utilized for DNA profiling in forensic situations. Single-locus probes are available to detect polymorphic areas such as deletions, additions, and inversions.[37] Using appropriate restriction endonucleases, profiling these polymorphisms can demonstrate differences

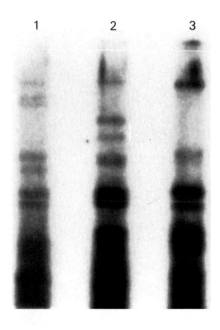

Figure 16.4. DNA fingerprints from 3 individuals. DNA from each was digested with endonuclease Hae III and hybridized with M13 bacteriophage minisatellite probe as described in reference 7. Each individual has a distinct banding pattern.

among individuals. Since these polymorphisms are not as complex as the VNTRs, they are not as hypervariable, and there may be only a small number of variable patterns in the population; some may have only two possible sequences and therefore yield only two possible patterns on gel electrophoresis. Probes to these polymorphic areas are specific, but not nearly as sensitive as VNTR probes, and a larger number of these probes are required for positive identification.

Other methods of DNA profiling do not use the Southern Blot technique, but identify specific sequences of DNA from a polymorphic area as either present or not present. A biological sample can be screened with a series of single-locus probes, each specific for a DNA sequence of known frequency in the population. Using a dot blot technique, an individual's DNA can be profiled, depending on how many and which of the standard sequences were present in his genome (Fig. 16.5).

Finally, although not currently utilized in practical forensic situations, certain hypervariable polymorphic segments in the genome can be identified and isolated from individuals and biological samples, and that segment sequenced for the specific DNA base order.[42] Depending on the variability and length of that segment in the population, the finding of identical sequences in two samples may suggest that they came from the same source.

The choice of the molecular biological technique for DNA profiling depends on the source and quality of the biological sample to be tested and the forensic question to be answered (Table 16.1). Since the sequence of DNA is the same in all

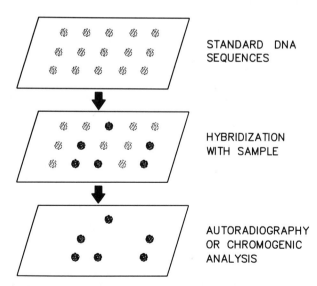

STANDARD DNA
SEQUENCES

HYBRIDIZATION
WITH SAMPLE

AUTORADIOGRAPHY
OR CHROMOGENIC
ANALYSIS

Figure 16.5.   Dot blot technique. A filter spotted with DNA fragments containing sequences representing one of two or more possible sequences from polymorphic areas in the genome are hybridized with labeled bulk single-stranded DNA extracted from a biological specimen. Hybridization on the filter occurs only where labeled DNA specifically matches the spotted sequences. After washing, hybridized dots are demonstrated either by autoradiography or by chromogenic analysis for nonisotopic probes to produce a dot blot pattern specific to each individual.

Table 16.1.
DNA Analysis Methods

Targeted DNA	Molecular Biology Techniques	Result	Advantages	Limitations
Multiple VNTRS: (variable number of tandem repeats (VNTRs) on many chromosomes)	Southern blot technique; multilocus probes	DNA fingerprint: multibanded autoradiogram	Highly sensitive; good for paternity and relatedness testing	Need large amounts of DNA; DNA must be high molecular weight, nondegraded; not applicable for putrefied, degraded, or formalin-fixed material
Single locus VNTRs: one area of the genome having a variable number of tandem repeats (VNTRs)	Southern blot technique; single-locus probes	DNA profile: double-banded autoradiogram	Highly specific; can exclude a wrong suspect; good for criminal cases	DNA must be high molecular weight, non-degraded. Not applicable for putrefied, degraded, formalin-fixed material, or for augmentation techniques
Single locus polymorphism: single variable segment of the genome due to deletions, additions, or inversions, but not VNTRs	Southern blot technique; single-locus probe	DNA profile single or double banded autoradiogram	Highly specific; can be performed on small amounts of DNA or degraded DNA; suitable for DNA augmentation techniques	Not very sensitive; requires many probes to different polymorphisms for identification
Single locus polymorphism	Dot blot technique; single-locus probe	Positive or negative dot blot (a single variant of a polymorphic area is either present or not)	Technique is much simpler and cost effective than Southern blot technique; good for very small amounts of DNA or degraded DNA; suitable for augmentation techniques	Not sensitive; requires probes to many different polymorphic areas for identification

Table 16.1. *(continued)*
DNA Analysis Methods

Targeted DNA	Molecular Biology Techniques	Result	Advantages	Limitations
Single locus polymorphism	DNA sequencing	Direct readout of DNA base sequence	Highly specific; potentially highly sensitive; can be used on mitochondrial DNA	Suitable area of nuclear DNA or mitochondrial DNA that is variable for all individuals; has not been standardized

cells, including sperm (some tumors excepted), DNA profiling can be done on any fluid or tissue where DNA can be extracted.[14] The amount and quality of the DNA, however, may determine which techniques can be used for profiling (Table 16.1). Although sufficient DNA can be recovered from as little as 1 µl of blood to screen with single-locus probes,[41] larger amounts are needed for DNA fingerprinting with VNTRs. Because VNTRs represent relatively long segments of DNA, some of the segments may be broken by putrefaction or during recovery. If the VNTR segment is broken by putrefaction, its individualizing characteristic is lost. Fortunately, drying of cells or DNA does not significantly alter the DNA or the ability to extract high molecular weight DNA for fingerprint analysis.[12] High humidity and temperature are, however, particularly favorable for DNA degradation. Therefore, DNA in blood or tissue of a decomposing body will tend to degrade or break up, while a fast-drying blood spatter on a wall may retain undegraded DNA for long periods of time. High molecular weight nondegraded DNA has been extracted from four-year-old blood stains.[12] Although DNA has been recovered from Egyptian mummies, it has been degraded.[32]

Preliminary studies have suggested that once blood stains are dried, exposure to humidity, heat, and ultraviolet light does not alter the RFLP patterns obtained.[26] However, soil contamination affects the DNA integrity of blood, and suitable DNA for profiling cannot be obtained.[27] Generally, DNA can be isolated from biological stains on synthetic and natural fibers, plastic, and wood or paper products.[27] Additional studies are needed to clarify the effects of environment, bacterial contamination, and substrata on biological stains.

Even in the face of moderate putrefaction with DNA degradation, some DNA profiling may be possible using single-locus probes to non-VNTR polymorphic areas. Since many of these polymorphic areas represent short segments of DNA, degradation with some DNA fragmentation would not affect the integrity of the polymorphic segment.

A breakthrough in utilizing very small amounts of recovered DNA from forensic cases has been the development of the polymerase chain reaction (PCR), an amplification technique to enzymatically multiply a specific segment of DNA as many as 10,000 times.[36] The multiplied segments can then be analyzed by single-locus probes. To amplify a segment of DNA, small primer sequences of DNA specific to the 5' and 3' ends of a segment of DNA containing a polymorphic sequence, are annealed to as little as 1 ng of denatured genomic DNA. Theoretically only one copy of the segment to be studied need be present in the DNA sample. DNA polymerase I is then added to the sample to synthesize copies of the DNA from the primers. Repeated cycles of DNA denaturation, primer annealing, and DNA synthesis results in amplification (Fig. 16.6). The amplified segment can then be tested with the appropriate single-locus probes.[24] Since the PCR technique is currently reliable only for short DNA segments, long segments representing VNTRs have not been amplified.

DNA profiling using restriction fragment length polymorphisms results in a banding pattern on x-ray film. A determination of identity of a sample requires comparison of the banding patterns of known and suspect samples. Unless the known and suspect samples are run side by side, it may be difficult to determine differences in fragment lengths, even with good control makers. Therefore, DNA profiling using Southern blot technique requires not only good laboratory technique, but also expert interpretation.

In contrast, the dot blot technique is simpler than Southern transfer and can be highly sensitive, particularly when coupled with PCR. Amplified DNA segments may also be detected by less sensitive nonradioisotopic hybridization techniques, eliminating the need for a laboratory with radio-labeling capabilities.[36] Amplification and hybridization using nonradioisotopic labeling not only allows hybridization to be carried out in a laboratory without radioactive-labeling capabilities, but has been performed directly on DNA amplified from crude cell lysates, eliminating the need for DNA purification.

Since blood contains mostly red blood cells, which do not contain DNA, blood stains must be large enough to contain sufficient white blood cells for recovery of adequate DNA. Dried blood stains of 50 μl of blood are adequate for profiling with multilocus VNTR probes. Seminal stains need not be nearly as large, because of the large number of sperm in semen. Sperm recovered from a vaginal swab of a sexual assault victim must be separated from the victim's vaginal and cervical cells to avoid contamination of the suspect's DNA with the victim's. Fortunately, this can easily be done by first lysing the female cells with proteinase K, which does not affect the sperm.[12] Centrifugation then separates the sperm from the lysate.

DNA can be extracted from hair roots, as well as from stains and tissues. A freshly plucked hair root may contain as much as 200 ng of DNA, 10 to 20 times that of a shed hair.[16] A single hair therefore does not contain enough DNA for profiling of VNTRs, and augmentation techniques are usually required.[16] However, if a victim has pulled many hairs from a suspect at the time of an assault, the combined amount of DNA may be adequate for fingerprinting.

Although hair shafts contain no cells and therefore no nuclear DNA, they do contain mitochondrial DNA. Polymorphisms have been demonstrated from mitochondrial DNA by RFLP techniques and by direct sequencing. Mitochondrial DNA has been amplified from single hair shafts.[16,42] Since there are many more copies of mitochondrial DNA per cell than nuclear DNA, polymorphisms in mitochondrial DNA may be more available for profiling when the forensic specimen is very small. The potential to distinguish individuals by their mitochondrial DNA now exists and awaits further refinement.

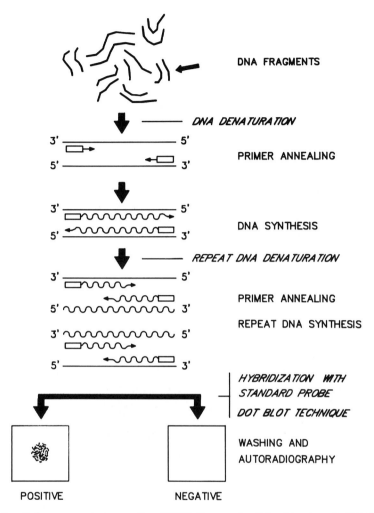

Figure 16.6. Polymerase chain reaction (PCR). Segments of double-stranded DNA from a small biological sample or from a degraded or putrefied biological sample are denatured into single strands. Short primer sequences of DNA, corresponding exactly to the 3' and 5' ends of a targeted segment, are annealed to the DNA. DNA polymerase I initiates the formation of new double-stranded DNA segments. As many as 25 cycles of DNA denaturation, primer annealing, and DNA synthesis may be performed to amplify the DNA segment. Hybridization of the amplified DNA segment with a labeled probe directed toward a possible sequence in a polymorphic area within the segment is performed by dot blot technique. If the amplified segment contains the specific sequence, hybridization occurs, and the signal is positive. If the amplified segment contains a different sequence in the polymorphic portion of the segment, no hybridization occurs, and the signal is negative.

Practical Applications of Molecular Biology to Forensic Pathology

HOMICIDE

The prosecution of a person suspected of murder frequently revolves around proving a temporal or physical association between victim and suspect. Any biological evidence that could prove that the suspect had physical contact with the victim might implicate the suspect in the crime. When two individuals or objects come into contact with each other there is a potential that trace materials will interchange. If there is an interchange of body fluids or tissue, then there is the potential of profiling the DNA from the biological sample. If, for example, the suspect leaves biological material such as blood, semen, hair, or tissue on the victim, on the victim's clothing, or at the scene of the crime, there is a possibility of recovering the DNA, profiling the polymorphic areas, and matching it with a known sample of DNA taken from the suspect. Similarly, a biological specimen from the victim, such as blood stains, hair, or tissue samples might link the victim to a suspect's clothing, apartment, or automobile, providing incriminating evidence. Therefore, identification and recovery of any biological stains or tissues associated with a victim, a crime scene, or a potential suspect may have tremendous importance.

SEXUAL ASSAULT

In no other crime should molecular techniques be as crucial as in the evaluation of crimes of sexual assault. DNA fingerprinting of sperm recovered in sexual assault victims can conclusively identify the source of the sperm. This information will be evidence of sexual intercourse between two individuals. Where sperm has been recovered from a victim of sexual assault and a suspect has been identified, there should be no need to argue mistaken identity. DNA fingerprinting should answer the question of whose sperm was recovered, and in stranger-to-stranger sexual assaults this should lead to more convictions and more guilty pleas. Another powerful use of the technology will be the speedy exoneration of the innocent suspect.

DNA profiling in sexual assault can also detect more than one assailant. Using a single locus VNTR probe on DNA recovered from sperm from a vaginal swab, the number of persons having recent sexual relations would be the number of bands divided by two. Using several different single locus VNTR probes and mathematical probability calculations, positive identifications of multiple assailants in sexual assault is possible. Similarly, such DNA profiling can separate the DNA of a husband or boyfriend from that of a suspect, if an assault has occurred temporally related to consensual sexual relations.

PARENTAGE

DNA analysis should replace traditional serology in paternity disputes, by providing positive confirmation or elimination of a putative father. In situations where the mother, child, and putative father are available for testing, multilocus VNTR probes resulting in multibanded DNA fingerprints are many times more sensitive than conventional serologic tests. Even when one parent is unavailable for testing, parentage can be established between a father and child or mother and child by this procedure.[31] Therefore, DNA analysis will be helpful in cases of abandoned newborns and infanticide where a suspect mother has been identified. DNA fingerprinting can be done on placental as well as fetal tissue.

DNA fingerprinting is being used routinely and successfully in establishing relationships in immigration disputes. In establishing relationships more distant than parents and children, DNA fingerprinting can be particularly helpful if several members of the known family can be tested and compared to the pattern of the putative relative.[33,34] For example, direct sequencing of mitochondrial DNA has helped establish grandparentage in cases of children stolen during Argentina's "dirty war."

HIT AND RUN

Frequently in a hit-and-run motor vehicle accident, regardless of whether the victim is killed or merely injured, there is biological evidence retained on the vehicle in the form of blood stains or tissue. A positive match can be made between the stains or tissue and the victim if the vehicle is found.

DRIVER VERSUS PASSENGER

In multiple fatalities in motor vehicle accidents it is occasionally difficult, but important, to determine who was driving at the time of the impact. The question is raised if occupants are unbelted, are ejected during the crash, or if passersby remove the occupants to provide first aid and then leave the scene before their activities can be recorded. Comparison of DNA from blood stains found in the drivers compartment or from the passenger's side may point compellingly to the correct seating arrangement.

MASS DISASTERS

The identification of mutilated remains from disasters such as airplane crashes has generally been based on the traditional forensic techniques of dental and x-ray comparisons, forensic anthropological evaluation, and circumstantial evidence. DNA profiling can be useful if these techniques fail. A DNA fingerprint from tissue from a mutilated body could be compared to that of putative relatives. Theoretically, DNA comparisons could be made with such specimen as hair from the suspected person's hair brush or from a retained surgical specimen maintained in a hospital pathology archive. DNA has been recovered from formalin-fixed surgical specimens and even from paraffin blocks of surgical specimens.[9,15] Although the DNA is usually fragmented, precluding analysis of VNTRs, single-locus probe analysis would be possible.

UNIDENTIFIED PUTREFIED REMAINS

Decomposition often results in visibly unidentifiable bodies. When traditional forensic odontology and anthropology fail in positive identification, it may be possible to utilize DNA analysis. Putrefaction causes DNA degradation and precludes DNA fingerprints, but adequate DNA for augmentation techniques and single locus probes can often be retrieved. Mitochondrial DNA, for example, has been retrieved from relatively fresh bone fragments.[34]

ANALYSIS OF UNKNOWN BIOLOGICAL MATERIAL

It may be alleged that blood stains at a crime scene or on a suspect's clothing are of animal origin. However, probes are available to distinguish human from nonhuman DNA.[38] Other probes to the human Y chromosome can determine the sex of a human stain or tissue.[11,22] Some polymorphic sequences are much more prevalent in certain ethnic groups, and probes to these DNA segments can suggest the population group of a human stain.[2] As DNA analysis is done on more and more individuals, further documentation of population groups and their DNA characteristics will be available.

NONHUMAN FORENSIC CONSIDERATIONS

All organisms contain DNA (or RNA). Polymorphisms are present in all higher species, and DNA analysis can be used to identify lineage in show dogs or race horses or to identify species in cases of illegal poaching, where suspicious stains are present on a suspect or his vehicle.[21]

COLLECTING AND PRESERVING BIOLOGICAL SAMPLES FOR MOLECULAR TECHNIQUES

The successful extraction of usable DNA is related to the size and age of the specimen, the environmental conditions that the specimen was subjected to prior to collection, and the conditions under which the sample is stored.[26,27] Since moisture and high temperature are most damaging to DNA, specimens in general should be stored dried and frozen.

Biological stains and vaginal swabs should be thoroughly air dried at room temperature and stored in a freezer. For long storage, freezing at -70°C. is preferred. Liquid blood, unlike tissue and stains, should be stored with preservative, EDTA (Na+) or EDTA (K+), and refrigerated not frozen.

From autopsied bodies, the success of extraction of nondegraded DNA depends on the postmortem interval. For fresh bodies, in addition to refrigerated preserved blood, spleen and muscle have been successfully utilized. A 1 cm sample frozen at -70°C is suggested. In decomposed bodies, spongy bone such as rib may be the best sample. It is, however, always wise to ask the laboratory doing the analysis for their tissue preference.

USE OF MOLECULAR TECHNIQUES IN CRIMINAL PROCEEDINGS

Although just recently introduced, DNA fingerprints and DNA profiles have been used in criminal prosecutions in a number of states. In most situations the results of the DNA analysis have been accepted by the courts with little challenge. For the results of a new scientific test to be accepted by a court it must pass judicial scrutiny to determine its suitability for a jury. In general, the criteria for acceptance of new scientific evidence were set up following a 1923 case that has since been known as the Frye hearing.[23] The decision in that case essentially said that the scientific technology by which the evidence was obtained must be sufficiently established so as to have gained general acceptance in the particular field in which it belongs.

In the subsequent years, new scientific tests have been subjected to the three-part Frye test.[23] (1) The validity of the underlying theory of the scientific test must be proved. (2) The scientific technique applying the theory must be valid. (3) There must be general acceptance of the proper way to perform the technique for particular situations.

Although DNA analyses have generally been accepted by courts and have passed the Frye test in many states, there has been some disagreement as to whether DNA profiling meets the third criterion of the Frye test.[1,25] Since different laboratories use different probes (protected by trademark) and since there are differences of opinion on what constitutes adequate quality control, some scientists have argued that there is not general acceptance in the scientific community of the proper way to perform the technique for particular situations. Since different laboratories use different probes, specific DNA fingerprints or profiles developed in one laboratory cannot be duplicated in another.

In response to this challenge, attempts are being made to initiate standardization of DNA analysis for forensic situations. A committee to study the problem is being considered by the National Academy of Sciences,[1] and the Federal Bureau of Investigation (FBI) has published guidelines for a quality assurance program for DNA RFLP analysis.[29]

Challenges to the results of specific DNA analyses will eventually result in standardization of testing. In order to avoid legal challenges, laboratories will probably adhere to strict guidelines for quality control of the procedures involved in DNA analysis and will probably agree to use only certain readily available probes, so that results from one laboratory can be confirmed by others.

LIMITATIONS OF DNA ANALYSES IN FORENSIC SITUATIONS

Limitations on the use of molecular techniques in forensic pathology include the time and expense involved in the procedures, the limited number of laboratories offering the technology, the need to solidify legal acceptance of the science, and the need to train forensic pathologists in the sampling techniques.

Conclusion

Molecular techniques were first used to help solve a crime in 1987. With the potential to conclusively identify the source of a biological stain, molecular techniques were quickly promoted as a revolutionary advance in forensic medicine. In short order molecular techniques were applied to many aspects of forensic medicine including paternity testing, immigration disputes, sexual assault, homicide, and identification of unknown remains. The technology is still evolving; there is as yet no standardization among laboratories offering forensic molecular testing. Improvements in automation of the techniques and the ability to use smaller and less pure specimen continues.

With the fast-paced progress of recombinant DNA techniques applied to forensic pathology have come problems. Newly developed techniques become outdated quickly and are replaced by more efficient, more specific techniques. Different laboratories offering molecular techniques to solve forensic cases utilize different protocols, different probes, and different standards. Results generated in one laboratory may not be reproducible in another using different reagents, probes, and techniques. With competition among commercial laboratories to provide state-of-the-art molecular techniques, quality assurance and quality control procedures may not meet the standard of a strict scientific research community. The potential of the practical application of molecular techniques to forensic pathology may thereby suffer because of the rapid advancement of the science. Those who do not understand the science may try to discredit it by focusing on discrepancies in results, brought about by over-interpretation of data and lack of ideal quality control procedures.

Nonetheless, the scientific potential of molecular techniques to solve practical forensic problems is becoming established and publicized. Efforts are already being made in some states to establish DNA data banks from violent offenders. Those practicing forensic pathology should become acquainted with the potential and limitations of molecular techniques for solving practical forensic problems and should be familiar with the proper collection, handling, and storage of biological samples.

REFERENCES

1. Anderson, A. New techniques on trial. *Nature (Lond)* 339:408, 1989.
2. Baird, M., Balazs, I., Giusti, A., *et al.* Allele frequency distribution of two highly polymorphic DNA sequences in three ethnic groups and its application to the determination of paternity. *Am. J. Hum. Genet.* 39:489–510, 1986.
3. Baird, M., Giusti, A., Meade, E., *et al. The Application of DNA-PRINT (tm) for Identification from Forensic Biological Material, Advances in Forensic Haemogenetics 2.* New York, Springer-Verlag, 1988.
4. Baird, M., Wexler, K., Clyne, M., *et al. The Application of DNA-PRINT (tm) for the Estimation of Paternity, Advanced in Forensic Haemogenetics 2.* New York, Springer-Verlag, 1988.
5. Caskey, C. T. Disease diagnosis by recombinant DNA methods. *Science (Wash DC)* 236:1223–1229, 1987.
6. Cooper, D. N. and Schmidtke, J. DNA restriction fragment length polymorphisms and heterozygosity in the human genome. *Hum. Genet.* 66:1–16, 1984.
7. Devor, E. J., Ivanovich, J. M., Hickok, J. M., and Todd, R. D. A rapid method for confirming cell line identity: DNA "fingerprinting" with a minisatellite probe from M13 bacteriophage. *BioTechniques* 6:200, 1988.
8. Donis-Keller, H., Green, P., Helms, C., *et al.* A genetic linkage map of the human genome. *Cell* 51:319–337, 1987.
9. Dubeau, L., Chandler, L. A., Gralow, J. R., Nichols, P. W., and Jones, P. A. Southern blot analysis of DNA extracted from formalin fixed pathology specimens. *Cancer Res.* 46:2964–2969, 1986.
10. Fenoglio-Preiser, C. M. and Willman, C. L. Molecular biology and the pathologist. *Arch. Pathol. Lab. Med.* III:601–619, 1987.
11. Fukushima, H., Hasekura, H., and Nagai, K. Identification of male bloodstains by dot hybridization of human Y chromosome-specific deoxribonucleic acid (DNA) probe. *J. Forensic Sci.* 33:621–627, 1988.

12. Gill, P., Jeffreys, A. J., and Werrett, D. J. Forensic application of DNA "fingerprints". *Nature (Lond) 318*:577–579, 1985.

13. Gill, P. and Werrett, D. J. Exclusion of a man charged with murder by DNA fingerprinting. *Forensic Sci. Int. 35*:145–148, 1987.

14. Giusti, A., Baird, M., Pasquale, S., Balazs, I., and Glassberg, J. Application of deoxyribonucleic acid (DNA) polymorphisms to the analysis of DNA recovered from sperm. *J. Forensic Sci. 31*:409–417, 1986.

15. Goelz, S. E., Hamilton, S. R., and Vogelstein, V. Parification of DNA from formaldehyde fixed and paraffin embedded human tissue. *Biochem. Biophys. Res. Commun. 130*:118–126, 1985.

16. Higuchi, R., VonBeroldingen, C. H., Sensabaugh, G. F., and Erlich, H. A. DNA typing from single hairs. *Nature (Lond) 332*:543–546, 1988.

17. Jeffreys, A. J., Wilson, V., and Thein, S. L. Hypervariable "minisatellite" regions in human DNA. *Nature (Lond) 314*:67–73, 1985.

18. Jeffreys, A. J., Wilson, V., and Thein, S. L. Individual-specific "fingerprints" of human DNA. *Nature (Lond) 316*:76–79, 1985.

19. Jeffreys, A. J., Brookfield, J. F. Y., and Semeonoff, R. Positive identification of an immigration test-case using human DNA fingerprints. *Nature (Lond) 317*:818–819, 1985.

20. Jeffreys, A. J., Wilson, V., Thein, S. L., Weatherall, D. J., and Ponder, B. A. J. DNA fingerprints and segregation analysis of multiple markers in human pedigrees. *Am. J. Hum. Genet. 39*:11–24, 1986.

21. Jeffreys, A. J. and Morton, D. B. DNA fingerprints of dogs and cats. *Anim. Genet. 18*:1–15, 1986.

22. Kobayashi, R., Nakauchi, H., Nakahori, Y., Nakagomek, Y., and Matsuzawa, S. Sex identification in fresh blood and dried bloodstains by a non-isotopic deoxyribonucleic acid (DNA) analyzing technique. *J. Forensic Sci. 33*:613–620, 1988.

23. Levin, R. DNA typing on the witness stand. *Science (Wash DC) 244*:1033–1035, 1989.

24. Mary, J. L. Multiplying genes by leaps and bounds, *Science (Wash DC) 240*:1408–1410, 1988.

25. McGourty, C. Genetic tests made official by UK courts. *Nature (Lond) 339*:408, 1989.

26. McNally, L., Shaler, R. C., Baird, M., Balasz, I., DeForest, P., and Kobilinsky, L. Evaluation of deoxyribonucleic acid (DNA) isolated from human bloodstains exposed to ultraviolet light, heat, humidity and soil contamination. *J. Forensia Sci. 34*:1059–1069, 1989.

27. McNally, L., Shaler, R. C., Baird, M., Balasz, I., Kobilinsky, L., and DeForest, P. The effect of environment and substrata on deoxyribonucleic acid (DNA): The use of casework samples from New York City. *J. Forensic Sci. 34*:1070—1077, 1989.

28. Merz, B. DNA fingerprints come to court. *JAMA 259*:2193–2194, 1988.

29. Mudd, J., Hartmann, J., Kuo, M., *et al.* Guidelines for a quality assurance program for DNA restriction fragment length polymorphism analysis. *Crime Lab Dig. 16*:40–59, 1989.

30. Nakamura, Y., *et al.* Variable number of tandem repeat (VNTR) markers for human gene mapping. *Science (Wash DC) 235*:1616–1622, 1987.

31. Odelberg, S. J., Demers, D. B., Westin, E. H., and Hossaini, A. A. Establishing paternity using minisatellite DNA probes when the putative father is unavailable for testing. *J. Forensic Sci. 33*:921–928, 1988.

32. Paabo, S. Molecular cloning of ancient egyptian mummy DNA. *Nature (Lond) 314*:644–645, 1985.

33. Raymond, C. Forensic experts tackle task of identifying thousands of "disappeared" victims. *JAMA 261*:1388–1389, 1989.

34. Raymond, C. Where are the children of the "dirty war"? Science seeks answers for their relatives. *JAMA 261*:1389–1393, 1989.

35. Reeders, S. T., Breuning, M. N., Davies, K. E., Nicholls, R. D., Jarman, A. P., Higgs, D. R., Pearson, P. L., and Weatherall, D. J. A highly polymorphic DNA marker linked to adult polycystic kidney disease on chromosome 16. *Nature (Lond) 317*:542–544, 1985.

36. Saiki, R. K., Bugawan, T. L., Horn, G. T., Mullis, K. B., and Erlich, H. A. Analysis of enzymatically amplified B-globin and HLA-DQX DNA with allele-specific oligonucleotide probes. *Nature (Lond) 324*:163–166, 1986.
37. Schumm, J. W., Knowlton, R., Braman, J., *et al.* Identification of more than 500 RFLPs by screening random genomic clones. *Am. J. Hum. Genet. 42*:143–159, 1988.
38. Tyler, M. G., Kirby, L. T., Wood, S., Vernon, S., and Ferris, J. A. J. Human blood stain identification and sex determination in dried blood stains using recombinant DNA techniques. *Forensic Sci. Int. 31*:267–272, 1986.
39. White, R. and Lalouel, J. M. Chromosome mapping with DNA markers. *Sci. Am. 258*:40–48, 1988.
40. Wong, Z., Wilson, V., Jeffreys, A. J., and Thein, S. L. Cloning a selected fragment from a human DNA "fingerprint", isolation of an extremely polymorphic minisatellite. *Nuc. Acids Res. 14*:4605–4616, 1986.
41. Wong, Z., Wilson, V., Patel, I., Povey, S., and Jeffreys, A. J. Characterization of a panel of highly variable minisatellites cloned from human DNA. *Ann. Hum. Genet. 51*:269–288, 1987.
42. Wrischnik, L. A., Higuchi, R. G., Stoneking, M., Erlich, H. A., Arnheim, N., and Wilson, A. C. Length mutations in human mitochondrial DNA: Direct sequencing of enzymatically amplified DNA. *Nucleic Acids Res. 15*:529–542, 1987.
43. Wyman, A. R. and White, R. A highly polymorphic locus in human DNA. *Proc. Natl. Acad. Sci. U.S.A. 77*:6754–6758, 1980.

APPENDIX A

ABBREVIATIONS

A	Adenine
AML	Acute myeloid leukemia
ANLL	Acute nonlymphocytic leukemia; same as acute myeloid leukemia
APL	Acute promyelocytic leukemia
A:T bonding	Adenosine and thymidine bonding
β-globin	Beta-globin
bp	Base pair
C	Cysteine
C	Complement (the protein)
cDNA	Complementary DNA
CF	Cystic fibrosis
CGRP	Calcitonin gene-related peptide
CML	Chronic myelogenous leukemia
COOH	Carboxyl group, carboxy terminus of a protein
CRH	Corticotropin-releasing hormone
CSF	Colony stimulating factor
CVS	Chorionic villus sampling
DBA	Dimethylbenzanthracene
DGGE	Denaturing gradient gel electrophoresis
DMs	Double minutes
DNA	Deoxyribonucleic acid
DNP	Dinitrophenol
dNTP	Deoxynucleotide triphosphates
EBV	Epstein-Barr virus
EGF	Epidermal growth factor
EGFR	Epidermal growth factor receptor
ER	Estrogen receptor
FAP	Familial adenomatous polyposis
FSH	Follicle-stimulating hormone
G	Guanosine
G:C bonding	Guanosine and cytosine bonding
GCG	Genetic Computer Group
GNBLs	Ganglioneuroblastomas
GRP	Gastrin-releasing peptide
HD	Hodgkin's disease

HBV	Hepatitis B virus
HIV	Human immunodeficiency virus
HNP	Hereditary nonpolyposis
HPLC	High-pressure liquid chromatography
HPV	Human papilloma virus
HSRs	Homogeneously staining regions
HTLV-I	Human T cell lymphotropic virus I
HTLV-II	Human T cell lymphotropic virus II
IGF-1	Insulin growth factor-1
ISH	*In situ* hybridization
J_H	"Joining" region exons of the immunoglobulin heavy chain
kb	Kilobase
KCl	Potassium chloride
kD	Kilodaltons
LH	Luteinizing hormone
LTR	Long terminal repeats
M	Mitosis
MEN 1	Multiple endocrine adenomatosis
$MgCl_2$	Magnesium chloride
MMTV	Mouse mammary tumor virus
mRNA	Messenger RNA
NaOH	Sodium hydroxide
NBL	Neuroblastoma
NH_2	Amino group; amino terminus of a protein
NSCLC	Non-small cell lung cancer
NT	Neurotensin
21-OH	21-Hydroxyl
^{32}P	Radioactive phosphorous
PCR	Polymerase chain reaction
PDGF	Platelet-derived growth factor
pg	Picogram
POMC	Pro-opiomelanocortin
PP	Pancreatic polypeptide
PR	Progesterone receptor
PYY	Peptide YY
RB	Retinoblastoma
RFLP	Restriction fragment length polymorphism
RIA	Radioimmunoassay
RNA	Ribonucleic acid
^{35}S	Radioactive sulfur
SCLC	Small cell lung cancer
T	Thymidine
Taq	Thermus aquaticus (DNA polymerase I)
TCR	T cell receptor
TGF	Transforming growth factor
TGF-β	Transforming growth factor-beta
TPA	Tetradecanoylphorbol acetate

TRH	Thyrotropin-releasing hormone
μg	Microgram
μM	Micromolar
UV	Ultraviolet radiation
VIP	Vasoactive intestinal polypeptide
VNTRs	Variable number tandem repeats
VTR	Variable tandem repeats

GLOSSARY

ALLELE - One of several alternative forms of a DNA sequence occupying a given locus on the chromosome.

ALLELE-SPECIFIC OLIGONUCLEO-TIDE PROBES (ASO Probes) - DNA probes ranging in size from 15 to 25 nucleotide base pairs exactly complementary to a normal gene sequence or a sequence with a point mutation.

AMPLIFICATION - Production of additional copies of a chromosomal sequence, usually in a tandem array.

ANEUPLOIDY - A karyotypic abnormality in which the amount of chromosomal DNA is altered; often seen in malignant tumors.

ANNEAL - Association of two complementary single-stranded nucleic acid strands through hydrogen bond formation.

AUTOSOME - All chromosomes excluding sex (X or Y) chromosomes.

BACTERIOPHAGE (Phage) - Viruses that infect bacteria.

BASE PAIR (bp) - A hydrogen-bonded complex of A and T (or U in RNA) or of C and G, between two complementary nucleic acid strands.

***cis* ACTIVATION (*cis*-acting locus)** - Affecting the activity of DNA sequences on its own molecule of DNA.

cDNA - A single-stranded DNA complementary to a mRNA molecule, synthesized from it by reverse transcription.

cDNA CLONE - A double-stranded DNA sequence derived from an mRNA, carried in a cloning vector; cDNA clones contain only exon sequences.

cDNA LIBRARIES - A collection of cDNA clones derived from a particular cell or tissue, propagated in cloning vectors.

CENTROMERIC DNA - DNA in the constricted region of a chromosome including the attachment site to the mitotic spindle.

CLONE - A large number of identical cells or molecules, derived from a single ancestor.

CLONING - The mechanism by which one attempts to isolate a unique nucleic acid sequence.

CLONING VECTOR - Any plasmid or phage into which a foreign DNA may be inserted and propagated in a heterologous host (such as bacterial, viral, or eucaryotic cells).

CODON - A triplet of nucleotides representing an amino acid or termination signal.

CONSTITUTIVE EXPRESSION - Genes that are expressed as a function of the interaction of RNA polymerase with a promoter, without additional regulation; also referred to as "housekeeping genes" for genes expressed in all cells.

CHROMATIN - Complex of DNA and proteins in interphase nuclei.

CHROMOSOMAL BANDS - Variation in the staining of DNA due to differential binding of dyes along the length of each DNA molecule.

COSMIDS - Plasmids into which phage λ *cos* sites have been inserted; as a result, the plasmid DNA can be packaged *in vitro* in the phage coat.

DENATURATION - Conversion from the double-stranded state.

DENATURING GELS - Gels containing a denaturant (often formaldehyde) which allow differential migration of DNA duplexes (in DNA gels) or maintain the linearity of RNA (in RNA gels).

DOUBLE MINUTE CHROMOSOME (DMs) - A tiny portion of extrachromosomal DNA (often found in tumor cells) that lacks a centromere but is still able to replicate and segregate. DMs may develop due to selection pressure placed on the cell, such as for drug resistance.

DOWNSTREAM - DNA sequences proceeding farther down from the direction of initiation of expression (*i.e.*, from the "start" site of transcription).

DUPLEX - Double-stranded nucleic acids (DNA-DNA, DNA-RNA, RNA-RNA) formed through hydrogen bonding between complementary nucleotides.

ENDONUCLEASES - Nucleases that cleave bonds within a nucleic acid; they may be specific for RNA or DNA.

ENHANCER - A *cis*-acting sequence that increases the utilization of eucaryotic promoters and may function in either orientation or location (downstream or upstream from the promoter).

EPISOMES - Acentric, circular, self-replicating DNA molecules.

EUCARYOTIC - Organisms whose cells have limiting membranes around nuclear material.

EXON - The actual coding sequences in DNA, which are interrupted by noncoding sequences (introns).

EXONUCLEASES - Nucleases that cleave nucleotides one at a time from the end of a polynucleotide chain; they may be specific for either the 5' or 3' end of DNA or RNA.

EXPRESSION VECTOR - A vector in which the inserted DNA molecule is placed near a promoter, allowing mRNA and protein expression of the cloned sequence in a heterologous host.

FILTER HYBRIDIZATION - Hybridization performed by incubating a denatured DNA preparation immobilized on a nitrocellulose filter with a solution of radioactively labeled RNA or DNA.

"FINGERPRINTING" - The use of polymorphic probes in Southern analysis to analyze multiple polymorphic loci, thus creating a "genetic fingerprint," unique to an individual.

G_2 - Second growth phase in the cell cycle.

GENE (CISTRON) - A segment of DNA that codes for protein; it includes regions preceding and following the coding region (leader and trailer) as well as intervening sequences (introns) between individual coding segments (exons).

GENE DELETIONS - Excision and loss of DNA sequences.

GENETIC LINKAGE ANALYSIS - The analysis of cosegregation of polymorphic DNA sequences in the population with disease phenotypes, thus establishing "linkage" between the genetic polymorphism (and its chromosomal location) and the disease-associated gene.

GENETIC MARKER - A trait whose inheritance can be followed within a family.

GENOMIC DNA CLONES (CHROMOSOMAL) - Genomic DNA sequences (containing introns and exons) inserted in a cloning vector.

GENOTYPE - The genetic constitution of an organism.

GUANOSINE TRIPHOSPHATE (GTP) - The phosphate donor for the proteins encoded by the *ras* gene, which bind and then remove one phosphate moiety from this phosphate carrier, thus converting GTP to GDP (guanine disphosphate).

HAPLOID GENOME ("n") - A set of chromosomes containing one of each of the autosomes and one sex chromosome.

HETEROGENOUS NUCLEAR RNA - Initially transcribed mRNA containing all exons and introns in a coding sequence; located only in the nucleus.

HOMOGENEOUSLY STAINING REGION - An extended region of a chromosome that has a homogeneous banding pattern and is thought to be a site of gene amplification.

HOMOLOGY - Degree of complementarity between two nucleic acid strands or two coding sequences.

HYBRIDIZATION - The pairing of complementary RNA and DNA strands to give an RNA-DNA hybrid, through the formation of hydrogen bonds between complementary nucleotides.

INTERSPERSED REPEATS - Families of DNA sequences that are present in a single copy in multiple locations throughout the genome.

INTRON - The noncoding or intervening sequences between exons in a gene. These are removed during the processing of messenger RNA, so that mature messenger RNA contains only coding exon sequences.

KARYOTYPING - Visualization of the entire chromosome complement of a cell or species during mitosis.

KILOBASE (kb) - 1000 base pairs of DNA or 1000 bases of RNA.

LEADING STRAND - Strand of DNA synthesized continuously in the 5'-3' direction.

LIBRARY - A set of cloned DNA fragments all derived from the same cell or tissue or organism.

LIGASES - Enzymes that promote formation of phosphodiester bonds between the adjacent nucleotides.

LINKAGE - The tendency of genes to be inherited together as a result of their location on the same chromosome; measured by percentage of recombination between loci.

LINKAGE GROUP - All loci that can be connected (directly or indirectly) by linkage relationships; equivalent to a chromosome.

LIQUID (SOLUTION) HYBRIDIZATION - A reaction between complementary nucleic acid strands performed in solution.

LOCUS - The position on a chromosome at which the gene for a particular trait resides.

LOD SCORE - The statistical assessment of linkage by calculating the ratio between the likelihood of linkage over the likelihood of random assortment. This calculation is the logarithm of the odds ratio. LOD scores of 3.0 or greater (odds of 1000:1 in favor of linkage over random assortment) are considered as being statistically significant, while an LOD score of -2.0 (odds of 100:1 against linkage) statistically excludes two markers as being linked. LOD scores between -2.0 and +3.0 are considered inconclusive.

MAJOR HISTOCOMPATIBILITY LOCUS (MHC) - A large chromosome region containing a cluster of genes that code for transplantation antigens and proteins involved in immune recognition.

MEIOTIC RECOMBINATION - Recombination of the DNA sequences between two chromosomes at meiosis.

MELTING - Denaturation (of DNA).

MELTING TEMPERATURE (T_m) - The midpoint of the temperature range over which DNA is denatured.

MESSENGER RNA (mRNA) - The RNA molecule copied from a DNA template containing a gene that will encode a protein.

METAPHASE CHROMOSOMES - Condensed chromosomes at metaphase, attached to the mitotic spindle through centromeres.

MUTATION - Any change in the sequence of a nucleic acid strand.

NICK - The absence in duplex DNA and RNA of a phosphodiester bond between two adjacent nucleotides on one strand.

NICK TRANSLATION - The ability of DNA polymerase I to use a nick as a starting point from which one strand of a duplex DNA can be degraded and replaced by resynthesis of new material; it is used to introduce radioactively labeled nucleotides into DNA *in vitro*.

NORTHERN BLOTTING - A technique for transferring RNA from an agarose gel to a nitrocellulose filter on which it can be hybridized to a complementary probe.

NUCLEOSOME - The basic structural unit of chromatin, consisting of approximately 160 bp of duplex DNA associated with an octamer of histone proteins.

ONCOGENE - A genetic sequence involved in the control of growth and differentiation; disruption of these sequences may be associated with malignant transformation. Cellular oncogenes (transforming genes) are those identified in tumors or tumor cell lines, while viral oncogenes are those that have been transduced from eucaryotic cells by a retrovirus.

OLIGONUCLEOTIDES - A short series of nucleotide base pairs used as primers or probes.

pBR322 - One of the standard plasmid cloning vectors.

PHAGE (BACTERIOPHAGE) - A bacterial virus.

PHENOTYPE - The appearance or other characteristics of an organism, resulting from the interaction of its genetic constitution with the environment.

PLASMID - An autonomous self-replicating extrachromosomal circular DNA molecule.

POLYMERASE - An enzyme catalyzing the addition of nucleotides (DNA or RNA) to a free 3'OH residue, thus extending a nucleic acid strand.

POLYMERASE CHAIN REACTION (PCR) - The technique by which target DNA sequences may be selectively replicated (amplified) *in vitro*. DNA primers flanking a site of interest (which hybridize to opposite strands of the duplex) are added to the denatured target; followed by the addition of DNA polymerase which uses the primer to extend the DNA sequence of the target. After multiple cycles of denaturation, primer binding, and DNA synthesis, a target sequence may be replicated 10^6-fold *in vitro*.

POLYMORPHISM - The simultaneous occurrence in the population of genomes showing allelic variations, *i.e.*, variation in the amount or sequence of DNA between two individuals.

PRIMARY TRANSCRIPT - The original unmodified RNA product corresponding to a transcription unit containing exons and introns.

PRIMER - A short sequence (often of RNA) that is paired with one strand of DNA and provides a free 3'-OH end at which a DNA polymerase starts synthesis of a deoxyribonucleotide chain, using the full length strand as a template.

PROMOTER - A region of DNA immediately upstream of the start site of transcription of a gene, which is involved in binding RNA polymerase to initiate transcription.

PROTO-ONCOGENE - A normal cellular gene capable of being activated to become an oncogene by retrovirus transduction or by alterations in its structure or expression.

PULSED-FIELD GEL ELECTROPHORESIS (PULSED-FIELD) - An electrophoresis technique in which the electrical field in a gel system constantly alternates direction (in contrast to traditional horizontal unidirectional electrophoresis), allowing for the separation of very large DNA fragments and resolution of individual chromosomes.

RARE BASE CUTTERS - Restriction enzymes that have an 8 bp recognition sequence that will occur only with rare frequency (every 65,000 bp) in the genome. These enzymes cleave DNA to create very long fragments.

REANNEAL - (See Anneal, Reassociation, Renaturation).

REASSOCIATION OF DNA - The pairing of complementary single strands to form a double helix, through hydrogen bond formation between complementary nucleotides.

RENATURATION - The reassociation of denatured complementary single-stranded nucleic acid strands.

RESTRICTION ENDONUCLEASES - Enzymes that recognize short DNA sequences and cleave the DNA at these target sites.

RESTRICTION ENZYMES - See Restriction endonuclease.

RESTRICTION FRAGMENT LENGTH POLYMORPHISM (RFLP) - A type of variation in DNA sequence, usually a point mutation in the DNA sequence which creates or abolishes a restriction enzyme recognition site.

RESTRICTION MAP - A linear array of sites on DNA cleaved by various restriction enzymes.

REVERSE GENETICS - The process (usually using DNA polymorphisms and genetic linkage analysis) in which a disease-associated gene is isolated for a disease phenotype even though the protein product and its biochemical function is unknown.

REVERSE TRANSCRIPTION - Synthesis of DNA on a template of mRNA; accomplished by reverse transcriptase.

RNA-DRIVEN HYBRIDIZATION REACTIONS - Reactions that use an excess of RNA to react with all complementary sequences in a single-stranded preparation of DNA.

SOLUTION HYBRIDIZATION - Liquid hybridization.

SOMATIC CELL - All cells of an organism with the exception of germ cells.

SOUTHERN BLOTTING (SOUTHERN ANALYSIS) - The procedure for transferring DNA from an agarose gel to a filter where it can be hybridized with a complementary nucleic acid probe.

SPLICING - The process whereby intron sequences are removed from an mRNA molecule.

SPLICEOSOME - The complex of small nuclear ribonucleoproteins (SNRPs) that mediate the excision of group II introns (introns in most mRNAs).

STICKY ENDS - Complementary single strands of DNA that protrude from opposite ends of a duplex or from ends of different duplex molecules; generated by staggered cutting of restriction endonuclease target sequences.

STRINGENCY - The conditions of the hybridization reaction that determine the stability of the duplex; 75% stringency implies that 25% of the bp of the two strands may be mismatched, but a duplex will still form; 100% stringency implies that two strands must be perfectly complementary for a duplex to form.

TEMPLATE - A nucleic acid sequence (usually DNA) that is used as a guide to direct the synthesis of a complementary strand of nucleotides linked by phosphodiester bonds (as in DNA replication or transcription).

trans **ACTIVATION** - The activation of a genetic locus by a protein encoded by a different genetic locus on a different DNA molecule (chromosome).

TRANSCRIPTION - The copying of a DNA sequence into mRNA, performed by an enzyme known as RNA polymerase II. The mRNA is then processed and spliced to remove introns, then moved to the cytoplasm, where it is translated into a polypeptide chain.

TRANSCRIPTIONAL PROMOTER - See Promoter.

TRANSFECTION - The transfer of foreign DNA into heterologous cells.

TRANSLATION - Synthesis of protein on the ribosomes using mRNA as the template.

TRANSLOCATION - A rearrangement in which part of a chromosome is detached by breakage and becomes reattached to another chromosome.

TRIPLET CODE - The sequence of each of three successive bp in nucleotides in mRNA which associate with the anticodon of a transfer RNA (tRNA) molecule carrying a specific amino acid.

UPSTREAM - Sequences proceeding in the opposite direction from the "start site" of transcription or gene expression.

VARIABLE NUMBER TANDEM REPEATS (VNTRs) - A type of genetic polymorphism resulting from a variable number of tandem (side-by-side) repeats of a core sequence of nucleotides at a particular genetic locus.

VECTOR - See Cloning Vector.

Index

Page numbers in *italics* denote figures; those followed by "t" denote tables.

365